# MILITARY PYROTECHNICS

# MILITARY PYROTECHNICS
## Principles and Practices

Ajoy K. Bose

**CRC Press**
Taylor & Francis Group
Boca Raton London New York

CRC Press is an imprint of the
Taylor & Francis Group, an **informa** business

First edition published 2022
by CRC Press
6000 Broken Sound Parkway NW, Suite 300, Boca Raton, FL 33487-2742

and by CRC Press
2 Park Square, Milton Park, Abingdon, Oxon, OX14 4RN

© 2022 Taylor & Francis Group, LLC

CRC Press is an imprint of Taylor & Francis Group, LLC

*Library of Congress Cataloging-in-Publication Data*
Names: Bose, Ajoy K., 1949- author.
Title: Military pyrotechnics : principles and practices / Ajoy K Bose.
Description: First edition. I Boca Raton : CRC Press, 2022. I Includes bibliographical
references and index.
Identifiers: LCCN 2021022134 (print) I LCCN 2021022135 (ebook) I ISBN
9780367554118 (hardback) I ISBN 9780367554125 (paperback) I ISBN 9781003093404
(ebook)
Subjects: LCSH: Military fireworks. I Explosives, Military.
Classification: LCC UF860 .B67 2022 (print) I LCC UF860 (ebook) I DDC 623.4/52--dc23
LC record available at https://lccn.loc.gov/2021022134
LC ebook record available at https://lccn.loc.gov/2021022135


ISBN: 978-0-367-55411-8 (hbk)
ISBN: 978-0-367-55412-5 (pbk)
ISBN: 978-1-003-09340-4 (ebk)


Access the Support Material: https://www.routledge.com/9780367554118

DOI: 10.1201/9781003093404


Typeset in Times
by MPS Limited, Dehradun

# Dedication

---

*Dedicated to*
*My mother Late Smt. Bonolata Bose*
*and*
*My father Late Shri Kedar Nath Bose,*
*Allahabad (UP) India.*

*Cover design by Amit Bose*
*Figures by Ajoy K. Bose (excluding reprinted)*

# Contents

## Section 1   Military Pyrotechnic Compositions

# Section 2   Pyrotechnic Ammunitions and Devices

# Preface

This book is intended to serve as a guide to military pyrotechnics for professionals of ammunition industries, inspection & quality assurance, ammunition training centres and armed forces establishments.

Military pyrotechnics is a topic that does not have an accepted standard syllabus for teaching in universities barring a few institutes in the world and the number of published books are very few. Although there has been a lot of research work done over several decades, there is not a single book covering full range of military pyrotechnics and providing answers to *"know-what"*, *"know-why"* and *"know-how"* of various principles and practices of military pyrotechnics in a lucid language.

This book covers the entire spectrum of military pyrotechnics with relevant and updated topics for an easy comprehension at elementary level with contents presented in a format that trainers find easier to teach in a resourceful way. All figures (excluding reprinted) have been made by me for clarity of the intricacies. Complex mathematical derivations have been deliberately avoided while preparing this book to facilitate easy understanding. Historical developments in evolution of pyrotechnics have been consciously avoided so as to provide more relevant technical content.

The book is focussed on various aspects of military pyrotechnics like sensitivity, combustion, factors affecting performance, determination of performance parameters, ingredients, various composition manufacturing methods, various composition details, ignition devices, painting and marking of components, ammunition and packages; basic requirements of pyrotechnic ammunitions, ammunition design and development, nomenclature, proof requirements, storage, shelf life, compatibility, various ammunitions and devices.

Advancements in ammunitions like infrared have taken shape only a few decades ago. Recent developments in instrumentation techniques have unfolded many nuances associated with performance of compositions. Latest trend is that of the 'green pyrotechnics' for environment-friendly military pyrotechnics. The book covers these latest advancements also.

I believe this book will definitely meet the expectations of all professionals associated with production, inspection, training and deployment in the field of military pyrotechnics.

*Dr Ajoy K. Bose*
*Pune, India*

# Acknowledgements

My profound thanks to all scientists in the field of military pyrotechnics whose relentless dedication has made military pyrotechnics a special branch of high energy materials with a large variety of compositions, ammunitions and devices. Without their enormous contribution in the field of pyrotechnics through research papers, technical reports and books, probably this book would not have taken shape.

I am extremely obliged to copyright owners for their permission to reproduce some of their content as text, tables and figures.

My sincere thanks to all my colleagues from various ordnance factories, proof establishments, quality assurance establishments and defence research establishments for sharing their knowledge during my long stint of more than 34 years at Ordnance Factories.

I am grateful to Director General Ordnance Factories (DGOF) and Chairman, Ordnance Factory Board (OFB) for approving inclusion of Ordnance Factory products. I thank Director, High Energy Materials Research Laboratory (HEMRL) for approving inclusion of their research papers and teaching materials. I also thank General Manager, Border Security Force (BSF) for permitting me to include their products. My thanks to M/s Commercial Explosives (India) Private Limited for approving inclusion of their products in the book.

I would like to thank my wife Mrs. Deepashree Bose (MA Economics) and sons Amit Bose (B. Tech) and Ankit Bose (B. Tech) for their motivational support throughout while preparing the manuscript. I am also thankful to my son Amit Bose for preparing the cover design of this book.

I acknowledge the valuable support extended by Dr. Gagandeep Singh, Senior Publisher (Engineering) CRC Press, Taylor & Francis Books India Pvt Ltd.

# About the author

**Dr. Ajoy K. Bose, M.Sc., D.Phil., I.O.F.S (Retd.)**

**Dr. Ajoy K. Bose** was born on 25<sup>th</sup> June 1949 in Allahabad (now Prayagraj) Uttar Pradesh (UP), India. He completed Masters of Science in Physical Chemistry in 1970 and obtained a Doctorate in Chemistry for his thesis *'Kinetics and Mechanism of Oxidation by Chloramine-T'* in 1974 from Allahabad University, Allahabad, UP (India) and published 13 scientific papers in international journals. He got associated with explosives and ammunition industry in 1975 under Ministry of Defence Production, Government of India by joining as Assistant Manager (Probation) under Directorate General, Indian Ordnance Factories.

The author had the privilege to work in ammunition and explosive industry for over 34 years (including 8 years exclusively in pyrotechnic ammunition industry) during which he was involved with production, quality control and research and development of various ammunitions, propellants and pyrotechnics. He was deeply involved in solving a large number of teething technical problems in ammunition manufacture and product improvement through innovations in processes and materials over the years. During his career, the author worked in various ammunition, propellant and pyrotechnic factories in India. Over these years, the author had close interactions with various ammunition and explosive factories, inspectorates and controllerates of quality assurance (for army, navy, air force and para military force) and controllerate of military explosives as well as various proof establishments and Defence Research and Development Organisation (DRDO).

The author is recipient of prestigious awards like *'Ayudh Bhushan'* for his outstanding services in solving technical problems in production and product improvements. The author was involved with design of igniter for which Ordnance Factory Dehuroad received citation *'Raksha Mantri Award for Excellence for Design Efforts'* under Ministry of Defence (Production) and cash award by India's Defence Minister. The author received citation and cash award from Director General Ordnance Factories (DGOF) and Chairman Ordnance Factory Board (OFB) for outstanding design of the igniter.

The factory during his tenure as General Manager (Higher Administrative Grade) received the *'Raksha Mantri Award for Excellence 2008–09 for Best Performing Factory of OFB'* from the Ministry of Defence production. He contributed significantly in achieving again the *'Raksha Mantri Award for Excellence 2009–10 for Best Performing Factory of OFB'* during his part tenure prior to retirement.

He retired from service as General Manager (HAG) in June 2009 after more than three decades of illustrious career from one of the Asia's biggest pyrotechnic factories at Pune (Maharashtra) in India.

# Section I

## Military Pyrotechnic Compositions

# 1 Introduction

## 1.1 GENERAL

Pyrotechnics are used in the manufacture of firecrackers to a variety of military pyrotechnic ammunitions and devices for various purposes, including the most advanced aerospace vehicles being developed worldwide.

The word pyrotechnic is derived from Greek word *pyros* (fire/heat) and *techne* (technique or art or skill) and, therefore, pyrotechnic manufacture can be referred to as the art of making fire. This can be partially relevant only to the firework industry but not for the military pyrotechnics industry where military pyrotechnics is science and not art. The manufacture of pyrotechnic ammunitions and devices requires a thorough theoretical and practical knowledge of ammunition technology. Therefore, one must remove the perception of art and approach the subject through the knowledge of modern science and technology.

The differences between fireworks and military pyrotechnics are given in Table 1.1.

The word "pyrotechnics" used in this book refers mainly to military pyrotechnic compositions, ammunitions and devices, which include all pyrotechnics used by armed forces as well civil administration and merchant navy.

Ammunition technology involves the usage of large number of materials like metals, non-metals, explosives, textiles, plastics, rubber, leather, paper, wood, adhesives, sealants, lubricants and many chemicals. Advances in pyrotechnic research and advanced manufacturing processes of materials and their analysis and advanced testing and proof equipment for ammunition have led to the development of new, improved pyrotechnic ammunitions over the years.

Safety is an important parameter in the manufacture of pyrotechnic ammunitions/devices, for safety of personnel, plant and equipment. Therefore emphasis on the safety aspect is required during production.

Pyrotechnic compositions are used in almost all the ammunitions. There are several definitions of pyrotechnics in terms of energetic systems or components, science of materials, technology, art and science, mixture of substance, mixture of chemicals and science, which are given below.

    a. "Pyrotechnics are *energetic systems or components* that provide visible light, infrared light, sound, smoke, gas, heat, or other pyrotechnic effects by the release of their chemical energy during combustion."

    b. "Pyrotechnics is the *science of materials* capable of self-contained and self-sustained exothermic chemical reactions for pyrotechnic effects like heat, light, gas, smoke and sound."

DOI: 10.1201/9781003093404-1

**TABLE 1.1**

**Differences Between Fireworks and Military Pyrotechnics**

| Parameters | Fireworks | Military pyrotechnics |
|---|---|---|
| Purpose | Celebration/ entertainment | Military operations for variety of purposes like signalling, distress signalling, infrared illumination, screening smoke, colour smoke, infrared smoke, infrared decoys, incendiary, photoflash for photography, riot control, etc. and a very few for civil purpose like merchant navy and civil administration for riots |
| Special effects | Visual radiation, sound and flash | In addition to visual radiation, sound and flash, military pyrotechnics also have infrared illumination, photoflash illumination, infrared smoke, screening smoke, colour smoke, etc. |
| Place of firing | Ground | Ground, aircraft, ship and submarine |
| Place of special effects | On ground or in air | On ground, in air and over the sea |
| Mode of firing | Mostly hand | Hand, hand-held projectors, mortars, howitzers, guns and dispensers in aircraft |
| Range | Short range on ground or in air | Short to long range on ground, air and over the sea |

c. "Pyrotechnics is the *technology* of utilising exothermic chemical reactions that generally speaking, are non-explosive, relatively slow, self-sustaining and self-contained [1]."

d. "Pyrotechnics is an *energetic mixture* that produces large amounts of light, heat, smoke, and/or sound when ignited by means of a deflagration event rather than a detonation" [2].

e. "Pyrotechnics is the *art and science* of creating and utilising the heat effects and products from exothermally reacting, predominantly solid mixtures or compounds when the reaction is, with some exceptions, non-explosive and relatively slow, self-sustaining, and self-contained" [3].

f. "Pyrotechnic compositions are a *mixture of substances* that when ignited, undergo an energetic chemical reaction at a controlled rate intended to produce on demand and in various combinations, specific time delays or quantities of heat, noise, smoke, light or infrared radiation."

g. "A pyrotechnic composition is a *substance or mixture of substances* designed to produce an effect by heat, light, sound, gas/smoke or a combination of these, as a result of non-detonative self-sustaining exothermic chemical reactions."

h. "Pyrotechnic *compositions (a mixture of various ingredients)* are capable of producing certain controlled special effects like heat, light, smoke, gas or delay or a mix of these special effects, by self-sustaining and non-detonating exothermic chemical combustion between the ingredients of the composition. This controlled special effect is not possible by use of other high energy materials."

i. "A *mixture of chemicals* which, when ignited, is capable of reacting exothermically to produce light, heat, smoke, sound or gas (US DoD)" [4].

j. "The *science* of controlled exothermic chemical reactions which are used to create timing devices, sound effects, aerosol dispersions, high pressure gas, intense heat, electromagnetic radiation, or combinations of these, and produce the maximum effect from the least volume. High explosives are excluded, but initiators are included" [5].

Military pyrotechnic compositions may be defined as follows:

*Military pyrotechnic compositions are mostly solid mixtures of two necessary ingredients (fuel and oxidiser) with other ingredients, which on suitable ignition stimulus, undergoes a controlled, mostly non-detonative and self-sustained exothermic chemical reaction, producing combustion products with special effects (visible radiation, infrared radiation, heat, smoke, sound and gas); the presence, type, quality and magnitude of special effects depends upon characteristics of composition, design features of ammunition and type of environment.* (Ajoy K. Bose).

It must be noted that all the special effects, like flame, light, sound, smoke, etc., are not desired in the same pyrotechnic combustion and hence suitable compositions are developed as per required special effects. For example, in smoke composition intended to emit smoke, there is practically no light, flame or sound while an illuminating composition provides visible radiation, but there is no sound or appreciable gas or smoke.

Pyrotechnics belong to the family of high energetic materials. The energetic material may be defined as under.

*"The high energetic materials are metastable substances having chemical energy stored in them, which, on subjecting to an external stimulus undergo self-sustained exothermic reactions yielding gaseous and or condensed products, heat, smoke, sound, radiation and pressure/detonation; the presence, type, quality and magnitude of each of these special effects depends upon the characteristics of energetic material, design features of ammunition and type of environment"* (Ajoy K. Bose).

Design features of ammunition plays important role in deciding the special effect of high energy materials. For example, a high explosive like TNT in powder form shall burn in open, a pressed TNT pellet shall exhibit blasting effect, RDX/TNT filled HEAT ammunition with copper cone shall exhibit hollow charge effect by penetrating the armour plate or RDX/WAX in HESH ammunition shall exhibit scab effect on armour plate.

High energy materials are a group of pure substances and compositions, classified as explosives, propellants and pyrotechnics. High energy materials have been further subclassified as under.

   a. Primary explosives (e.g., lead azide, lead styphnate, tetrazene and mercury fulminate).
   b. Secondary explosives (e.g., TNT, RDX, PETN and CE).
   c. Composite explosives (e.g., pentolite – a mixture of PETN and TNT, cyclotol – a mixture of RDX and TNT and amatol – a mixture of ammonium nitrate and TNT)
   d. Propellants (e.g., single base, double space, triple base and composite).
   e. Pyrotechnic compositions (e.g., illuminating, screening smoke, infrared smoke, signalling colour smoke, signalling flare, riot, delay, photoflash, tracer, incendiary, simulating, flash and smoke, infrared decoy flare, infrared illuminating, priming, booster, squib, friction, stab, percussion, igniter, gunpowder, electrical sensitive and miscellaneous pyrotechnic compositions).

The classification of energetic materials is shown in Figure 1.1 with a few examples.

High energy materials can also be classified as military explosives and civil explosives.

It is interesting to understand the difference between pyrotechnic composition and other high energy materials like primary explosives (initiators), secondary explosives, composite explosives and propellants, which is given below.

   a. *Primary explosives (Initiators):* These are high energy materials, having low energy of activation, highly sensitive to external stimuli like spark, friction, impact and heat and can initiate the detonation of a relatively insensitive explosive. Examples are lead azide, mercury fulminate, lead styphnate, etc.
   b. *Secondary and composite explosives (High explosives):* These are high energy materials, having high enegy of activation, least sensitive to external stimuli, detonate on suitable initiation to create ground burst or airburst,

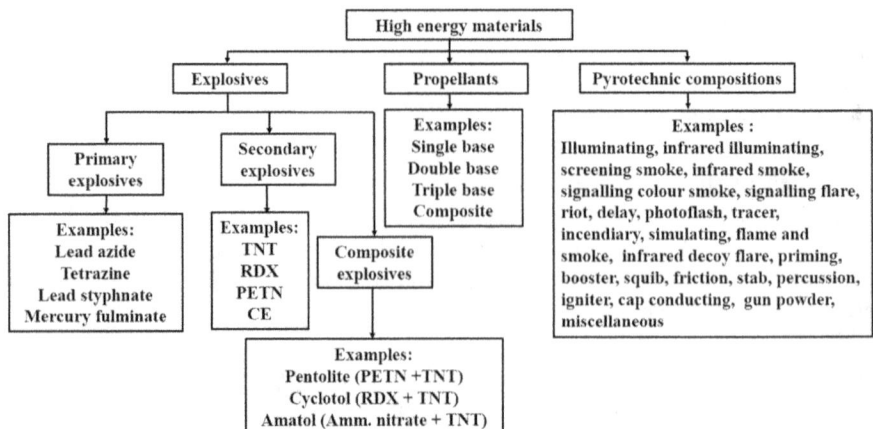

**FIGURE 1.1**   Classification of Energetic Materials.

create shockwaves, shatter and penetrate. Examples are TNT, RDX, PETN, CE(Tetryl), pentolite, cyclotol etc.

c. *Propellants:* These are high energy materials, sensitive to external stimuli, and on initiation burn to propel projectiles (bombs, shells, etc.) and rockets and pressurise piston devices or rotate turbines and gyroscopes. Examples are single base propellants (contains nitrocellulose (NC) as main ingredient), double base propellants (contains NC and nitroglycerine (NG) as main ingredients), triple base propellants (contains NC, NG and Picrite (nitroguanidine) as main ingredients), high energy propellants (contains RDX with nitrocellulose), composite propellants (contains NC, NG, AP, Al and HTPB), etc.

d. *Pyrotechnics:* These are high energy materials, generally less sensitive and produce special effects like visible radiation, infrared radiation, heat (fire/flame/flash), gas, smoke and sound, etc. Such a wide variety of special effects is not exhibited by any other high energy material.

A comparison of high energy materials is shown in Table 1.2

Another comparison of high energy material is shown in Table 1.3.

Deflagration and detonation depend upon the rate of combustion with respect to speed of sound. Detonation occurs when the rate of combustion is higher than the speed of sound, while deflagration occurs when the rate of combustion is lower than speed of sound. Deflagration moves layer by layer while detonation involves shockwaves in the explosive.

A comparison of magnesium based illuminating composition (48% magnesium, 48% sodium nitrate and 4% organic binder) in a 10 mm diameter candle with similar column of TNT is given in Table 1.4.

---

**TABLE 1.2**

**A Comparison of Some of the Characteristics of Pyrotechnics, Propellants and Explosives [6]**

| Parameters | Pyrotechnics | Propellants | High explosives |
|---|---|---|---|
| Type of reaction | Progressive burning | Propagative burning | Adiabatic compression |
| Type of ingredients | Solid | Solid and/or liquid | Solid and/or liquid |
| Reaction by products | Some gas, solid residue | Some gas, some residue | Gas |
| Ease of initiation | Minimum to moderate | Moderate | Maximum |
| Requires oxygen | No | Yes | Yes |
| Output | Flame, glow, gas, pressure, sound, smoke and flash | Gas pressure | Extreme heat, pressure |
| Rate of reaction | Slow | Rapid | Extremely rapid |
| Brisance | Minimum | Moderate | Maximum |

## TABLE 1.3

### Comparative Data on Classification of Energetic Materials [7]

| Parameters | Pyrolants (Pyrotechnics) | Propellants | High explosives |
|---|---|---|---|
| Process | Burning | Deflagration | Detonation |
| Speed of reaction | Subsonic* $< 1$m s$^{-1}$ | Subsonic*1–1,000 m s$^{-1}$ | Supersonic* $\gg 1,000$ m s$^{-1}$ |
| Reaction products | Mainly condensed | Mainly gaseous | Mainly gaseous |
| Oxygen balance | Fuel rich | Balanced | Balanced-fuel rich |
| Combustion enthalpy | 1–30 kJg$^{-1}$ 5–50 kJcm$^{-3}$ | 5–10 kJg$^{-1}$ 10–20 kJcm$^{-3}$ | 5–15 kJg$^{-1}$ 15–25 kJcm$^{-3}$ |
| Density range | 2–10 gcm$^{-3}$ | 1.5–2.5 gcm$^{-3}$ | $< 2$ gcm$^{-3\#}$ |

*Notes*

* With respect to speed of sound of energetic material
# Not taking into account metallised formulations and heavy metal-based primary explosives

## TABLE 1.4

### Comparison of Energy Output of a Pyrotechnic with that of an Explosive [8]

| | Pyrotechnic | TNT |
|---|---|---|
| Propagation rate (m.s$^{-1}$) | $2.3 \times 10^{-3}$ | 6,950 |
| Energy (J.kg$^{-1}$) | 8,600 | 4,800 |
| Density (kg.m$^{-3}$) | 2.4 | 1.64 |
| Power (w) | $3.7 \times 10^3$ | $4.3 \times 10^9$ |

The combustion of a pressed pyrotechnic composition with burning layer (with or without flame) differs from that of a high explosive or a high explosive mixture in the absence of pressure differential in the combustion zone. The energy content in pyrotechnic composition may be comparable with high explosives but the rate of release of energy is much slower and hence energy is released for a considerable duration of time. The primary explosives (initiators) and high explosives are mostly single compounds (though high explosive mixtures may be made like cyclotol (RDX/TNT)), whereas pyrotechnics are mostly a mixture of finely divided combustible solid substances (and sometimes incendiary gel).

High explosives, being single compounds, comprise fuel and oxidisers in the same compound and hence the combustion is rapid, exceeding the speed of sound, which results into detonation. Pyrotechnics on the other hand are intimate mixtures of fuel and oxidisers, which are required to come together for combustion and hence the rate of combustion is comparatively slow.

The pyrotechnic combustion does not depend entirely on ambient air as the oxidiser provides the necessary oxygen for combustion. Pyrotechnic compositions are mixtures and thereby differ from compounds. The difference between a mixture and a compound can be understood with the help of the following two examples.

a. Antimony and sulphur can be mixed to form a mixture in any proportion. The properties of antimony and sulphur are retained in the mixture. They can also be segregated by simple means. On the other hand, the compound antimony trisulphide $Sb_2S_3$ can be formed by the chemical reaction of two moles of antimony to three moles of sulphur forming one mole of antimony trisulphide. The compound antimony trisulphide does not possess any properties of antimony or sulphur. It is having altogether different physical and chemical properties than antimony and sulphur, in melting point, boiling point and density, as shown in Table 1.5.

b. Iron (II) sulphide is made of iron and sulphur. Whereas iron is attracted by a magnet and sulphur can be dissolved in carbon disulphide in a mixture, iron (II) sulphide is neither attracted by a magnet nor can be dissolved in carbon disulphide.

## 1.2 PYROTECHNIC COMBUSTION VIS A VIS CHEMICAL REACTION IN LIQUID PHASE

Let us understand how the combustion of a pressed pyrotechnic composition differs from that of a chemical reaction in liquid phase. The differences are as under.

a. The rate of reaction in a liquid chemical reaction is uniform throughout the mass of the liquid; while in a pressed pyrotechnic combustion the reaction zone moves layer by layer.
b. The whole mass of liquid is having same temperature (being uniform reaction) in a liquid chemical reaction, while in a pressed pyrotechnic combustion; only the reaction zone is having high temperature.

**TABLE 1.5**
**Variation in Properties of Ingredient and Compound**

| Material | Atomic weight/ molecular weight | Melting point $^0C$ | Boiling point $^0C$ | Density g.cm$^{-3}$ |
|---|---|---|---|---|
| Antimony | 121.77 | 630.63 | 1,587 | 6.697 |
| Sulphur | 32.065 | 115.21 | 444.6 | 2.079 |
| Antimony trisulphide | 339.735 | 550 | 1,080 | 4.64 |

c. The rate of reaction in a liquid chemical reaction cannot be modified at a certain stage while in a pressed pyrotechnic combustion; it is possible to modify the rate of reaction by having a layer of desired composition (say delay composition) in the pressed mass.
d. The particles of reactants and the products in a liquid chemical reaction have free space to move in the whole liquid mass while the particles of reactant and products in a pressed pyrotechnic combustion have restricted space (porosity) to move.

However, the above difference does not apply in case of loose pyrotechnic composition (i.e. non-compacted composition), which is filled in some ammunition like those of photoflash composition or loose pyrotechnic composition like gunpowder used as a propellant to eject star or candle or canister where the entire composition burns in a few milliseconds.

## 1.3   ADVANTAGES AND DISADVANTAGES OF PYROTECHNIC COMPOSITIONS

The advantages of pyrotechnic compositions are:

a. Ingredients used are of low cost
b. Ease of manufacture due to simple plant and machineries and processes
c. Low energy requirement for composition ignition
d. Can be initiated by simple mechanical/electrical/frictional/spark/laser methods
e. High energy content per unit mass
f. Variety of sizes possible with desired output energy/special effects
g. Possibility of maneuvering of ingredients and the composition for desired effect, resulting in a large range of compositions for:
    i. Energy output in variety of forms like visible radiation, infrared radiation, heat, sound, gas, etc.
    ii. Low heat output to very high heat output
    iii. Gasless to very high-pressure gas output
    iv. Burn time from milliseconds to several minutes
h. Capability to deliver more energy in short time
i. Long-term storable energy in the form of composition (under proper storage conditions)
j. Capable of withstanding reasonable extreme temperatures ($-40^0$C to $+60^0$C)
k. Wide variety of applications as almost 90% of ammunitions contain pyrotechnic compositions in some form or other

The disadvantages of pyrotechnic compositions are:

a. Contain high energy materials requiring special precautions
b. Safety hazard like inadvertent functioning, static charges

    c. Individual filled component performance cannot be checked (since one time use)

    d. Occasional failures in performance

    e. Various factors affect performance (see Chapter 4)

    f. Toxicity issues of burning of pyrotechnics [9]

However, despite some disadvantages, the advantages are predominant and hence pyrotechnic compositions are used in large numbers of ammunitions to produce desired special effects.

## 1.4 PYROTECHNIC COMPOSITION SPECIAL EFFECTS

The pyrotechnic compositions may be subdivided into three types as under.

    a. The two-component composition (basic)

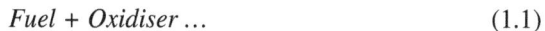

$$Fuel + Oxidiser\ldots \tag{1.1}$$

    b. The three-component composition (simple)

$$Fuel + Oxidiser + Binder\ldots \tag{1.2}$$

    c. The multi-component composition (complex)

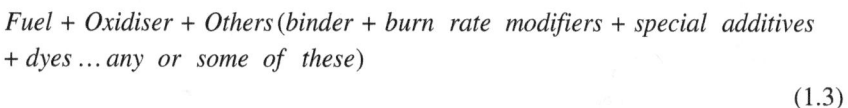

$$Fuel + Oxidiser + Others\,(binder + burn\ rate\ modifiers + special\ additives$$
$$+ dyes\ldots any\ or\ some\ of\ these)$$
$$\tag{1.3}$$

Pyrotechnic combustion is an oxidation–reduction reaction (redox reaction) where the fuel is oxidised while the oxidiser is reduced. The complex pyrotechnic composition combustion is as given below.

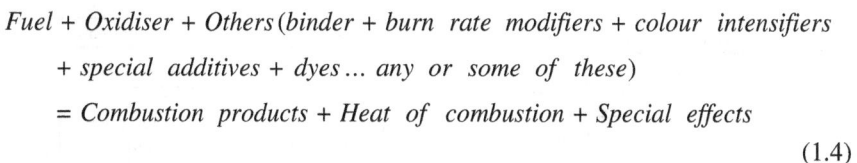

$$Fuel + Oxidiser + Others\,(binder + burn\ rate\ modifiers + colour\ intensifiers$$
$$+ special\ additives + dyes\ldots any\ or\ some\ of\ these)$$
$$= Combustion\ products + Heat\ of\ combustion + Special\ effects$$
$$\tag{1.4}$$

High explosives, propellants and initiators have four crucial elements C, H, O and N, but pyrotechnic ingredients not only contain some of the above elements but also contain a large number of other elements, making it possible to manufacture large number of compositions. The addition of any one ingredient like burn rate modifier or colour intensifier or dyes or additives or variation of composition percentage further increases the number of possible pyrotechnic compositions. Though all such compositions may not have significant use in ammunitions, the figure is too large compared to primary explosives (initiators), secondary explosives, composite explosives and propellants.

Pyrotechnic compositions produce one or more of the special effects like radiation, smoke, heat (fire/flame/flash), sound and gas. (Figure 1.2).

Figure 1.3 gives the types of combustion outputs that form the approximate basis of classification of pyrotechnic compositions as shown below dotted lines.

Pyrotechnic combustion outputs like gas/aerosol, hot slag, high heat and low heat and electromagnetic radiation form the basis of classification of pyrotechnic compositions and these combustion products are utilised as follows.

a. Gas/aerosols are used as sound-producing compositions, or smoke-producing compositions like screening smoke, signalling colour smoke, infrared smoke and also used for expelling stars, candles and canisters.

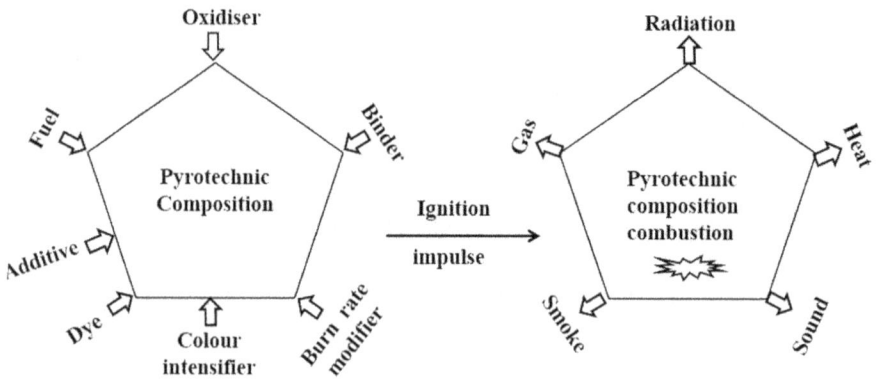

**FIGURE 1.2**  Pyrotechnic Composition Special Effects.

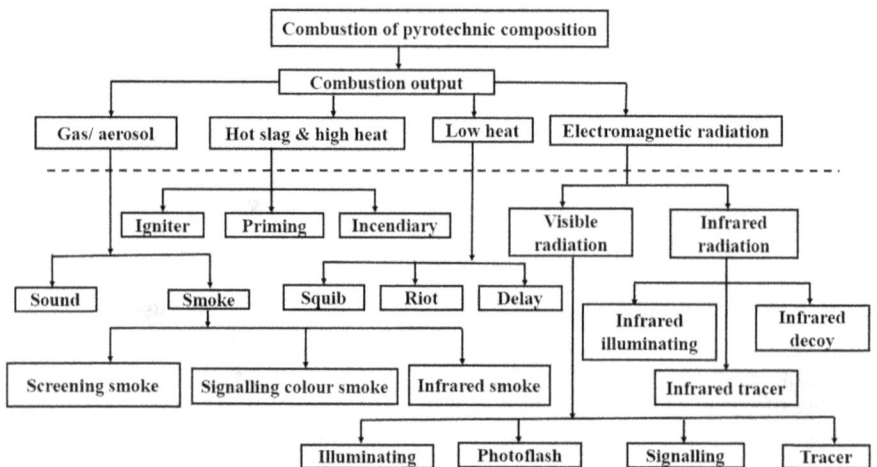

**FIGURE 1.3**  Combustion Outputs and Pyrotechnic Classification.

b. Hot slag and high heat are used as igniter, priming and incendiary compositions. The heat is used up for vapourising the metallic fuel, and heating further layers of composition. The solid hot slag also assists in heating further layers of the composition.
c. Low heat is used in squibs, riot and delay compositions.
d. Electromagnetic radiation as visible radiation is used in illuminating, signalling, photoflash and tracer compositions. Electromagnetic radiation in infrared region is used as infrared illuminating, infrared decoy and infrared tracer compositions.

### 1.4.1 Special Effects – Radiation

These composition on combustion emit radiation, either visible or infrared, for a variety of purposes.

a. *Visible Radiation (Visible light)*:
    i. *Illuminating Compositions:* These contain pyrotechnic composition pressed in the form of a bare pellet (star) or as a candle (used with parachute). These compositions on burning give visible bright white light and illuminates the battle field of specified area.
    ii. *Photoflash Compositions:* These compositions are used as loose composition (uncompacted) for faster burning of composition. These pyrotechnic compositions produce brilliant flash light for a few milliseconds for night time aerial photography.
    iii. *Signalling Flare Compositions:* These compositions are pressed in the form of pellets (stars) or candles and assembled in the signalling ammunitions. These compositions on burning in bare pellet (star) form or candle with parachute give desired visible flame colours (red, green, yellow, white, etc.) for signalling purpose.
    iv. *Tracer Compositions:* These compositions are pressed at higher load (to counter excess pressure of propellants) and burn for few seconds to trace the path of projectiles.
    v. *Spotting Compositions:* These compositions are used in small arms ammunition for spotting the location of hit of the ammunition round.

b. Infrared Radiation:

These compositions are of following types.

i. *Infrared Decoy Composition*s: infrared radiation output as decoys for heat seeking missiles so as to save the aircraft, helicopter, ship etc.
ii. *Infrared Illuminant Compositions*: infrared radiation output for illuminating the target for covert illumination and observed through night vision devices
iii. *Infrared Tracer Compositions*: Infrared output for tracing the tracer path through night vision devices
iv. *Infrared Flare Tracking Compositions*: For tracking the path of wire guided missile

### 1.4.2  SPECIAL EFFECTS – SMOKE

These pyrotechnic compositions on initiation produce smoke. Smoke is produced due to dispersion of *aerosols* in the air during combustion of composition or condensation of colour smoke particles.

There are three types of smokes.

a. *Screening Smoke Compositions:* These compositions on combustion emit screening smoke. The combustion produces gases that expel the combustion products in the air. Dense white-grey smoke hampers visibility of troops, tanks, guns, military vehicles or movement of troops due to its screening effect. Black smoke is generally not used due to its low density, which is unable to produce screening effect.

b. *Signalling Colour Smoke Compositions:* The signalling colour smoke is used for communication/signalling like location of the area for dropping foods or ammunition, equipment, etc. in snow-bound areas or in sea for rescue operations or for target practice like training ammunitions for indicating the point of impact by the projectile. These compositions on combustion give signalling colour smoke (red, orange, blue or green) due to presence of colour dyes as ingredients in the composition.

c. *Infrared Smoke Compositions:* These compositions produce infrared smoke as thermal attenuation screen in infrared region for guns, tanks, military vehicles, etc.

### 1.4.3  SPECIAL EFFECTS – HEAT

These pyrotechnic compositions on combustion produce hot combustion products with intense heat or low heat.

a. *Intense Heat*:
   i. *Incendiary Compositions:* These compositions burn vigorously producing high thermal energy, i.e., incendiary effect and are used against targets on the ground or in the air as well as for destruction of war equipment and documents.
   ii. *Igniter Compositions:* These compositions are used for initiation of ignition of propellants basically used in missiles. The igniter composition is also used for burning of base bleed propellant so as to obtain higher range of the projectile.
   iii. *Priming and Booster Compositions:* The priming (or first fire charges) and booster compositions produce heat for initiating the burning of layer of main composition.

b. *Low Heat*:

These compositions produce low heat as energy released is much smaller in magnitude and thus leads to slow ignition of the composition. Examples are

  i. *Riot Compositions:* These are used for riot control purposes.

  ii. *Delay Compositions:* These are used to provide delay between two events in the ignition train.

  iii. *Squib Compositions:* These are used for initiating ignition through squibs.

### 1.4.4 SPECIAL EFFECTS – SOUND AND FLASH

*Simulating compositions* or *flash and sound compositions* are used to obtain cracking and bursting effects, thus simulating the flash and sound of a battle field. The production of whistling sound is due to fast evolution of gases through a small opening during combustion making waves in the air at the speed of sound.

### 1.4.5 SPECIAL EFFECTS – GAS

All pyrotechnics compositions produce varying quantity of gases. Amongst pyrotechnic compositions, smoke and sound and flash compositions produce fairly good amount of gases. Some initiating pyrotechnic compositions in combination with propellant produce a fairly large quantity of gases from power cartridges, which are used for mechanical works such as pushing a piston or driving a turbine. Such devices are also used for ejectors, cutters, removers, and thrusters. These compositions are also used for activating primary battery for missiles and torpedoes as well as for car airbags.

Table 1.6 shows special effects of pyrotechnic compositions and their purpose.

The pyrotechnic special effects have been successfully used to create a variety of ammunitions for specific requirements like aiming the targets, camouflaging the vital equipment (tanks, military vehicles), communications, signalling, destruction of equipment and documents, decoys, training and lighting the area for aerial photography, etc. A large quantity of these pyrotechnic ammunitions and devices are being used by the army, air force, navy, paramilitary forces, merchant navy, police and space agencies around the world. It is also used for safety purposes in seat ejection for para jumping, railway tracks safety indicators, distress signalling for rescue operations and riot control measures, etc.

These pyrotechnics continue to be vital components in all aerospace personnel escape and armament deployment systems. These have been used extensively in space exploration vehicles, like Mercury, Gemini, Saturn, Apollo, etc., for initiating various devices (Table 1.7).

Military pyrotechnics are intended for use in non-lethal military combat operations. It provides a wide range of special effects. Therefore, a majority of pyrotechnic ammunition and devices are generally non-destructive or non-lethal. However, there are a few pyrotechnic ammunitions-like incendiaries and flares (flares made to function on ground) that are lethal.

## 1.5 MAIN CHARACTERISTICS OF PYROTECHNIC COMPOSITIONS

Following are the main characteristics of pyrotechnic compositions.

  a. Pyrotechnic compositions are high energy materials.

**TABLE 1.6**

**Special Effects of Pyrotechnic Compositions and their Purpose**

| Special effects | Composition classification | Broad purpose |
|---|---|---|
| Radiation | Illuminating | Visible radiation |
| | Infrared illumination | Covert illumination |
| | Photoflash | Night photography |
| | Signalling | Signalling (including distress) |
| | Tracer | Tracing with visible radiation |
| | Infrared tracer | Tracing with infrared radiation |
| | Infrared flare decoys | Infrared radiation for decoy |
| Smoke | Screening smoke | Screening of battle field |
| | Signalling colour smoke | Signalling (including distress) |
| | Infrared smoke | Thermal attenuation screen |
| Intense heat | Incendiary | Destruction of documents, depots |
| | Igniter | Ignition of propellants |
| Low heat | Riot | Mob control |
| | Delay | Time interval |
| | Initiatory | Initiation of ignition |
| | Squib | Initiation of ignition |
| | Priming and booster | Initiation of ignition in composition |
| Flash and sound | Simulating | Battlefield sound and flash |
| Gas | Gas (combination of pyrotechnic and propellant) | Mechanical work, reduce base drag |

**TABLE 1.7**

**Number of Spacecraft-Installed Pyrotechnic Devices [10]**

| Programme | Number |
|---|---|
| Mercury | 46 |
| Gemini | 139 |
| Saturn | Approx. 150 |
| Apollo (CSM/SLA/LM) | 314 |
| Apollo (CSM/LA) FOR SKYLAB | 249 |

b. Pyrotechnic compositions are mechanical/heterogeneous mixtures of two or more than two ingredients. The two main reactants are fuel and oxidiser while the other ingredients are binders, burn rate modifiers, colour intensifiers, special additives, dyes, etc.

c. Pyrotechnic compositions are sensitive to external stimuli. Combustion in pyrotechnic compositions takes place only when an external stimulus of appropriate magnitude is provided. The sensitivity of some of the pyrotechnic compositions is much higher than secondary explosives and hence proper care is to be taken in handling the composition. The sensitivity depends upon various factors (see Chapter 2).

d. The pyrotechnic compositions can be used to produce special effects like visible radiation, infrared radiation, smoke, heat, gas, sound and flash.

e. Pyrotechnic compositions have energy content comparable with high explosives but its rate of release of energy is slower.

f. Pyrotechnic compositions generally contain their own oxygen for burning when suitably initiated and can burn in the absence of air. Unlike wood or paper, which burn by taking oxygen from the air, pyrotechnic compositions do not need oxygen from the air for burning as oxygen is made available through oxidisers present in the compositions (few exceptions like gel incendiary compositions or phosphorous use atmospheric oxygen for burning). Thus, the burning of a pyrotechnic composition cannot be quenched by suffocating through a blanket of dry fire extinguishing powder or fire extinguishing foam. The composition would continue to burn till it gets oxygen from the oxidiser.

g. Pyrotechnic compositions burn fast when in lose form. However, the linear burn rate is reduced when consolidated as a pellet.

h. It is possible to obtain a pressed pyrotechnic composition to burn layer by layer (similar to burning of a cigarette) to ensure consistent burning so as to obtain desired burn rate.

i. They can be filled by pouring, pressing, manual ramming, cast filling, screw feeding, or extruding (see Chapter 24).

j. Pyrotechnic compositions are mostly hygroscopic and hence need special care against humidity/moisture during manufacture, packing and storage (see Section 4.2.15).

k. Pyrotechnic compositions have a shelf life in ammunition, which vary from composition to composition (see Section 27.3).

## 1.6 GENERAL ESSENTIAL REQUIREMENTS OF MILITARY PYROTECHNIC COMPOSITIONS

Following are the general essential requirements of a military pyrotechnic composition for its proper performance in ammunitions. Before selection of any composition, these factors must be considered.

a. The ingredients of the composition should be stable, both physically and chemically, at normal temperature.

b. The ingredients of the composition should be easily available at reasonable cost.

c. The composition should be safe, stable and should not degrade in all environmental and mechanical handling effects during storage and have a reasonable shelf life within the ammunition.

d. The composition should be compatible with other ingredients as well as components of the ammunitions (see Chapter 27).

e. The composition preferably should not generate static electricity during handling.

f. The composition manufacture process should not be complex for mass manufacture.

g. The composition should be adequately sensitive to initiation by external stimuli but not too sensitive so as to be unsafe for processing, handling, transporting and using in ammunition. The composition should be insensitive to shock, jolt, vibration during manufacture, transportation and storage. Ideally, it preferably should be insensitive to moisture.

h. The composition must be homogenous and give reproducible performance. Its rate of heat generation should be more than the rate of heat loss to the surrounding and heat loss due to special effects i.e., the composition should have required exothermicity so that the transmission of heat should take place to the subsequent layer for ignition and provide reproducible performance.

i. The composition combustion must be exothermic, self-sustained and self-contained and should proceed at a specified rate except a few compositions that require burning in milliseconds like those of photoflash compositions.

j. The composition should preferably be free flowing to allow it to be homogeneously blended and filled.

k. The compositions filled by pouring, manual tamping, cast filling, screw feeding, pressing or extruding as star or candle should be able to withstand the firing stresses like setback force, set forward force, acceleration, deceleration, rotation/spin, etc. in ammunitions.

l. The composition rate of burning should be controllable to suit different applications and burn uniformly/consistently throughout its specified duration of burning

m. It should give the desired special effect with optimum amount of composition.

n. It should not be affected by extremes of temperature variations and have stability during long storage. The composition should withstand extreme climatic temperatures like very low ($-40\ °C$) and very high temperature ($+60\ °C$) to ensure that ammunition is functional at extreme cold conditions like snow-bound areas and at very high altitudes as well as hot places like desert and may also be stored without any adverse effect.

o. The composition as well as its combustion products should preferably be non-corrosive to weapon as well as less toxic and should not be detrimental to the environment and human beings.

p. It should not detonate.

Special requirements of individual military pyrotechnic compositions are given in individual composition chapters (Chapters 8–22).

## 1.7 IMPORTANT PARAMETERS OF MILITARY PYROTECHNIC COMPOSITIONS

Important parameters of military pyrotechnic compositions (depending upon the type of composition) are listed below.

a. Sensitivity to external stimuli (ignitability)
b. Moisture content
c. Apparent bulk density and loading density
d. Ignition temperature
e. Heat of combustion
f. Flame temperature
g. Luminosity and burn rate
h. Infrared flare intensity and emission time
i. Flame length of composition
j. Ignition delay
k. Burn rate
l. Ignition energy
m. Compaction strength of pressed composition
n. Infrared attenuation by smoke
o. Smoke obscuration power
p. Sound intensity
q. Electrical conductivity
r. Thermal conductivity
s. Volume of gas production
t. Efficiency of composition
u. Colour quality

The measurement of these parameters has been given in brief in Chapter 5.

## 1.8 PYROTECHNIC AMMUNITIONS AND DEVICES WITH THEIR APPLICATIONS

Military pyrotechnic ammunitions have been defined [11] by Hart (1955) as a *"Category of ammunition employed primarily for the production of light, heat, smoke and sound for such typical non-destructive purposes as battlefield illumination, signalling, marking, tracking, tracing, spotting, ignition, simulation and aerial night photography... produced as a result of chemical reactions caused by the application of proper stimuli to chemical elements or compounds alone or in intimate mixtures. Military pyrotechnics are major aids and accessories in tactical operations for communications, warning, reconnaissance and the effective application of destructive firepower, in strategic operations for intelligence, in supporting activities such as rescue operations and troop training, in research and development of rocket powders and propellants, and in exploration of the upper atmosphere"*

The desired performance of pyrotechnic ammunition depends upon the combustion of the main composition. Thus, a signalling flare composition would show a trace of bright-coloured light while a screening smoke composition would exhibit screening smoke.

Pyrotechnics have been found to be extremely useful in making a large variety of ammunitions and devices with special effects. These pyrotechnic compositions are filled in grenades, bombs, canisters, pots, containers, cartridges, shells, etc. The special effects exhibited by pyrotechnic composition in ammunitions and devices are energies in various forms produced during combustion.

The duration of special effects depends upon the type of ammunition. For example, photoflash ammunitions give high intense flash for milliseconds duration, tracers give few seconds duration and illuminating ammunition gives illumination for few minutes duration.

There are several components in ammunitions, devices and missiles that contain pyrotechnic compositions like caps, primers, tracers, delays, igniters, squibs, impulse cartridge, power cartridge etc. Though these find wide applications in various ammunitions, all these ammunitions may or may not be pyrotechnic ammunitions. For example, high explosive anti-tank (HEAT) ammunition contains cap, primer and tracer, but is not a pyrotechnic ammunition since its end performance/special effects are different than pyrotechnic ammunition.

Some pyrotechnic ammunitions and devices with their special effects and applications are summarised in Table 1.8.

---

**TABLE 1.8**

**Pyrotechnic Ammunitions and Devices with their Applications**

| Ammunitions, devices and filled components | Special effects/function | Applications |
|---|---|---|
| Illuminating flares (illuminating shells, bombs, cartridges, ground indicating flares) | Production of visible radiation | Battle field illumination for target identification, bombardment, landing of war planes and materials at night, emergency airstrip location & identification, miscellaneous devices like flare trip wire |
| Infrared flares (cartridges, shells, decoys) | Production of infrared radiation | Infrared illumination, infrared decoys for heat seeking missile, covert photography, tracking the path of wire guided missile |
| Infrared smoke (containers/pots, grenades, bombs and shells) | Production of infrared smoke | Thermal attenuating screen for infrared radiation of tanks, guns etc. |
| Signalling devices (cartridges, canisters grenades, shells) | Production of coloured light and signalling colour smoke | Communication signals (for day and night) in military operations, distress signal during search and rescue operation from ships and rafts, |

**TABLE 1.8** *(Continued)*
**Pyrotechnic Ammunitions and Devices with their Applications**

| Ammunitions, devices and filled components | Special effects/function | Applications |
|---|---|---|
| | | submarines movement indication |
| Photoflash (bombs and cartridges) | Production of high intensity light | Aerial night photography |
| Incendiaries (Small arms bullets, bombs, grenades and shells) | Production of intense heat | Destruction of targets (aircraft, ammunition depots, oil depots, equipment and documents) |
| Tracers | Projectile flight tracker, self-destruction, | Used in small arms, medium calibre and large calibre ammunitions |
| Screening smoke (bombs, pots, grenades and shells) | Production of screening smoke | Screening/masking of troop movement, tank and vehicles, tracking and acquisition by tracing path of space vehicle |
| Training/practice ammunitions (bombs, pots, grenades and shells) | Training | Training of personnel in handling, loading and firing of ammunition |
| Simulators (air burst, ground burst, whistling and blank cartridges) | Simulate the ammunition-weapon firing | Deceiving the enemy with flash and sound of battle field |
| Riot control (cartridges, bombs, grenades and shells) | Production of irritating aerosol, production of flash and bang, spraying of indelible ink, ejection of plastic/rubber balls | Riot control for civil administration, marking of riot personnel, rescue of hostage from captors |
| Delays | Time interval between two successive events in ignition train | Various ammunitions and devices like delay in ignition train and delay in fuze pyrotechnic train |
| Initiation of ignition (miscellaneous caps, primers, igniters and squibs,) | Production of flame and hot solid or hot fluid slags for ignition | Burning of pyrotechnics and propellants in cartridges, bombs, shells and missiles |
| Power cartridges (piston and bellow devices, explosive bolts and explosive nuts, actuators, thrusters, pin pullers, cable and hose cutters, line cutters, valves, switches etc. | Production of gas | Mechanical works |
| (a) Aircraft pyrotechnic devices | Maneuvering aircraft | Seat ejection system for pilot, release systems for tanks and materials, fire suppress systems |

*(Continued)*

**TABLE 1.8** *(Continued)*
**Pyrotechnic Ammunitions and Devices with their Applications**

| Ammunitions, devices and filled components | Special effects/function | Applications |
| --- | --- | --- |
| (b) Spacecraft pyrotechnic devices | Space exploration and recovery | Launch and control systems, time delays, emergency systems, stage separation systems, fairing release systems, recovery and landing systems |
| (c) Missile pyrotechnic devices | Missile launching, missile trajectory, missile function | Ignition system, time delays, control system, safety and arming system, destruction systems |

## REFERENCES

[1]. "*Military Explosives*", TM9-1300-214, Headquarter, Department of the Army, Washington DC, 25 September 1990.

[2]. Dr. Jesse J. Sabatini, "*A Review of Illuminating Pyrotechnics*", Propellants, Explosives, Pyrotechnics, January 2018, Volume 43, Issue 1, Page 28–37, Copyright Wiley-VCH GmbH. Reproduced with permission.

[3]. Dr. Herbert Ellern, "*Military and Civilian Pyrotechnics*", © Chemical Publishing Co., New York, 1968.

[4]. https://www.militaryfactory.com/dictionary/military-terms-defined.asp?term_id=4327

[5]. Robert M. Blunt, "*Proceedings of the Seventh International Pyrotechnics Seminar*", 14–18 July 1980, Volume I, IIT Research Institute, Chicago, Illinois 60616.

[6]. F.L. McIntyre," *A Compilation of Hazard and Test Data for Pyrotechnic Compositions*", Contractor Report ARLCD-CR-80047, October 1980. U.S. Army Research and Development Command, Large Caliber Weapon System Laboratory, Dover, New Jersey.

[7]. Ernst-Christian Koch, "*Metal Fluorocarbon Based Energetic Materials*", Page 2, 2012, Wiley-VCH Verlag GmbH & Co KGaA. Reproduced with permission.

[8]. Dr N. Davies, "*Pyrotechnics Handbook*", Department of Environmental and Ordnance Systems, © Cranfield University, Royal Military College of Science, January 2008.

[9]. Guy Marlair, Richard Turcotte, Queenie Kwok, and Ruddy Branka. "*Toxicity Issues Pertaining to Burning Pyrotechnics*". 33. International Pyrotechnics Seminar (IPS 2006), Jul 2006, pp. 467–483, Fort Collins, United States.

[10]. Laurence J. Bement and Morry L. Schimmel, "*A Manual for Pyrotechnic Design, Development and Qualification*", NASA Technical Memorandum 110172 by National Aeronautics and Space Administration, Langley Research Centre, Hampton, Virginia 23681-0001.

[11]. "*Flare Effectiveness Factors: A Guide to Improved Utilization for Visual Target Acquisition*", Target Acquisition Working Group Report 61 JTCG/ME-74-10, November 1973.

# 2 Pyrotechnic Composition Sensitivity

## 2.1 GENERAL

The sensitivity of a pyrotechnic composition is defined as *"the measure of achieving the ignition temperature of the composition (or achieving the energy of activation of the composition) so as to stimulate ignition"*. The pyrotechnic compositions are sensitive to the following external stimuli that can be leveraged for igniting compositions:

a. Mechanical energy (includes impact, percussion and stab energy)
b. Frictional energy (includes safety matches, strike anywhere matches)
c. Electrical energy (includes electric current, electrostatic discharge, i.e., spark energy)
d. Thermal energy (includes primer/igniter, port fire friction, laser beam, fuze)

The use of above-mentioned external stimuli for ignition of pyrotechnic composition layers may lead to the emission of *flame, hot gaseous products, hot solid products, hot fluid products*, etc., which are responsible for further combustion of composition layers.

The pyrotechnic composition, ideally, should only be sensitive to the above energies, namely mechanical, friction, electrical and thermal, which are used for initiation of ignition. On the other hand, composition should be sufficiently insensitive to static electricity, shock, jolt and vibration. It is essential to carry out sensitivity tests of any new composition and endorse them in the *"Safety Certificate"* of the composition. These are required to be understood and proper precautions are required to be taken during manufacture of compositions, filling/pressing of the compositions and assembly of ammunition.

Each pyrotechnic composition has its own sensitivity to external stimulus. The sensitivity of pyrotechnic compositions varies from each other due to difference in ingredients and their characteristics, their proportion, sieve size, moisture content and physical state like loose or pressed, etc.

The sensitivity determination is important for following aspects:

a. It provides energy required to ignite the compositions
b. It provides risks associated in manufacture of the composition
c. It provides data on sensitivity, which is necessary to keep margin of safety during manufacture, handling and use. It provides to consider prevention of premature ignition and reduction in the associated hazards.
d. It helps determine alternatives for composition material, if sensitivity values are very high

DOI: 10.1201/9781003093404-2

## 2.2 FACTORS AFFECTING SENSITIVITY OF PYROTECHNIC COMPOSITIONS

Several factors that affect the sensitivity of the pyrotechnic composition are discussed below.

### 2.2.1 ENERGY OF ACTIVATION

The lower the energy required for ignition of a composition, the higher is the sensitivity of the composition. This energy requirement is known as *Energy of Activation* and is explained in detail in Section 3.4.

Some compositions need low energy to attain their respective ignition temperature for combustion. These compositions are therefore highly sensitive to external stimuli like flash, flame, heat, impact, friction, etc. Examples are priming compositions.

On the other hand, some compositions require more energy and do not ignite at lower energy and its combustion takes place with difficulty. These are less sensitive to flash, flame impact and friction, etc. Examples are incendiary compositions requiring high temperature flash (mostly from detonators).

### 2.2.2 MELTING AND DECOMPOSITION TEMPERATURE OF OXIDISER

The sensitivity of compositions depends upon the melting and decomposition temperature of the oxidiser used; the less the melting point and decomposition temperature, the less the energy required for combustion to take place and hence the higher the sensitivity of composition containing such oxidisers as lower melting point of oxidiser allows it to melt rapidly causing intimate contact with the fuel.

Contrary to this, oxidisers with higher melting point and higher decomposition temperature require more energy for combustion to take place and hence compositions containing such oxidisers are less sensitive.

It can be seen in Table 2.1 that lead dioxide, sodium nitrate, potassium nitrate, potassium chlorate and barium peroxide have low melting and decomposition temperatures than iron (III) oxide and iron (II, III) oxide, and hence compositions containing the former are more sensitive than the latter.

### 2.2.3 MELTING POINT, BOILING POINT AND THERMAL CONDUCTIVITY OF FUELS

A higher sensitivity is expected from low melting and low boiling point and high thermal conductivity of fuel. The reason being that low melting and low boiling point fuels lead to better contact of reacting ingredients as the fuel is molten and/or in vapour phase while high thermal conductivity leads to better transmission of heat energy to next layer. Higher thermal conductivity of metals increases the spark sensitivity. Hence compositions containing metal fuels with oxidisers and other ingredients in compacted form give better conduction. Table 2.2 provides the approximate melting point, boiling point and thermal conductivity of fuels.

## TABLE 2.1
## Melting and Decomposition Temperatures of Some Oxidisers

| Nomenclature | Melting temperature °C | Decomposition temperature °C |
|---|---|---|
| Lead dioxide | – | 290 |
| Sodium nitrate | 307 | 520 |
| Potassium nitrate | 334 | 630 |
| Potassium chlorate | 356 | 400 |
| Barium peroxide | 450 | 795 |
| Lead tetroxide (red lead) | – | 500 |
| Strontium nitrate | 570 | 670 |
| Barium nitrate | 592 | 600 |
| Potassium perchlorate | 610 | 620 |
| Lead chromate | 844 | 600 (starts) |
| Lead oxide | 886 | – |
| Iron (iii) oxide or iron sesquioxide | 1,565 | – |
| Iron (ii, iii) oxide or ferroso ferric oxide | 1,594 | – |

## TABLE 2.2
## Melting Point, Boiling Point and Thermal Conductivity of Some Fuels

| Nomenclature | Melting point °C | Boiling point °C | Thermal conductivity (watts. $m^{-1}.°K^{-1}$) |
|---|---|---|---|
| Phosphorous yellow | 44 | – | 0.236 |
| Sulphur | 119 | 445 | 0.205 |
| Magnesium | 649 | 1,107 | 160 |
| Aluminium | 660 | 2,467 | 235 |
| Copper | 1,084.62 | 2,927 | 400 |
| Iron | 1,535 | 2,750 | 80 |
| Titanium | 1,660 | 3,287 | 22 |
| Zirconium | 1,852 | 4,377 | 23 |
| Boron | 2,300 | 2,550 | 27 |
| Tungsten | 3,410 | 5,660 | 170 |
| Molybdenum | 2,623 | 4,639 | 139 |

### 2.2.4  HEAT OF COMBUSTION OF FUELS AND HEAT OF DECOMPOSITION OF OXIDISERS

The sensitivity increases with increase in heat of combustion of fuel (Table 6.1) and increase in heat of decomposition of oxidiser (Table 6.4). Thus, a composition containing boron as a fuel (high heat of combustion 14.0 $kcal.g^{-1}$) and potassium chlorate as an oxidiser (exothermic heat of decomposition $(-)10.6$ $kcal.mole^{-1}$) will be highly sensitive as compared to highly insensitive composition containing sulphur as fuel (heat of combustion 2.2 $kcal.g^{-1}$) and sodium nitrate as oxidiser (endothermic heat of decomposition $(+)$ 60.5 $kcal.mole^{-1}$).

### 2.2.5  STOICHIOMETRIC RATIO OF FUEL AND OXIDISER IN COMPOSITIONS

Pyrotechnic compositions have *neutral/zero, positive or negative oxygen balance* (see Section 4.2.8), depending upon the proportion of oxidiser and fuel in the pyrotechnic composition. The stoichiometric composition with just sufficient oxygen to oxidise all fuel (known as neutral/zero oxygen balance) or slightly less fuel is more sensitive to external stimuli.

### 2.2.6  DEGREE OF FINENESS OF INGREDIENTS

The degree of fineness of the ingredients affects sensitivity. Extremely fine particles of ingredients increase the sensitivity of the composition due to more surface area having intimate contact between the particles of the ingredients for combustion to take place.

### 2.2.7  HARDNESS AND SHAPE OF INGREDIENTS

Some ingredients like sand, ground glass powder and silica are hard materials with angular projections. These materials enhance the friction sensitivity of the composition including mechanical percussion sensitivity and stab sensitivity of the composition. This could be ascribed to the hardness of these materials where the external stimuli (mechanical energy or friction energy) get concentrated leading to formation of "*hotspots*" (see Section 22.3.2), thus making the composition sensitive.

### 2.2.8  DENSITY OF THE COMPOSITIONS

A higher density of composition has lower heat sensitivity than low density composition since same amount of heat energy is expended over a large mass of ingredients and thus the effective energy gain on an individual grain becomes less.

However, sensitivity to percussion or stab is high with higher density since higher density offers more resistance to the striker restricted over a small mass of composition for making "*hotspots*" (see Section 22.3.2).

**TABLE 2.3**

**Effect of Temperature on Impact Sensitivity of Ground, Chemically Pure Ammonium Nitrate [1]**

| Temperature °C | Picatinny Arsenal impact test with 2 kg weight (in.) |
|:---:|:---:|
| 25 | 31 |
| 75 | 28 |
| 100 | 27 |
| 150 | 27 |
| 175 | 12 |

## 2.2.9 Ambient Temperature

A higher ambient temperature enhances the sensitivity of the composition as the difference between the ambient temperature and the ignition temperature of the composition is reduced. This necessitates special precautions during the drying of composition due to increased sensitivity of the composition at higher temperatures inside the dryer (see Section 7.7). Table 2.3 provides the effect of temperature on impact sensitivity of ground, chemically pure ammonium nitrate.

## 2.2.10 Atmospheric Pressure

A higher atmospheric pressure would lead to higher sensitivity of composition due to the large concentration of combustion products near the surface, which impart heat energy to the layer of the composition for ignition (see Section 4.4.1).

## 2.2.11 Binding/Coating Materials

Binder reduces the luminosity of illuminating compositions and affects sensitivity to impact, friction and spark. Addition of some ingredients like paraffin, stearin (glyceryl tristearate), oils, etc. reduces the friction sensitivity of the composition. This could be ascribed to the fact that friction is distributed to a larger mass of composition and hence does not get concentrated at one point. It also reduces the flame sensitivity as the coating allows the heat energy, with difficulty, to reach the ingredients as it absorbs some energy in transforming its state from solid to fluid phase.

However, some binders also increase the sensitivity of pyrotechnic compositions.

## 2.2.12 Inert Materials

Addition of inert materials in pyrotechnic composition reduces the sensitivity of the composition as these inert materials do not take part in the combustion process but occupy the space in the pressed mass and take away heat energy. They are interspersed between the reacting ingredients, thereby reducing their intimate contact to a certain extent, and hence reducing the combustion rate.

### 2.2.13  MOISTURE/SOLVENT

Moisture/solvent reduces the sensitivity of the composition since a fraction of the ignition energy is used in evaporating the moisture/solvent content in the composition (see Section 4.2.15).

### 2.2.14  HOMOGENEITY OF COMPOSITION

Higher homogeneity of composition increases the sensitivity of the composition.

### 2.2.15  SPECIFIC HEAT OF COMPOSITION

A higher specific heat of composition lowers the sensitivity of the composition because more energy is required to raise the temperature of the composition.

The above effects have been summarised in Table 2.4.

## 2.3  SENSITIVITY AND HEAT OF COMBUSTION

A higher sensitivity of composition does not necessarily mean that the composition has a very higher heat of combustion. Thus, a composition with high sensitivity may exhibit low heat of combustion and a composition with low sensitivity may exhibit high heat of combustion. However, there are some compositions that exhibit high

---

**TABLE 2.4**
**Parameters Affecting the Sensitivity of Pyrotechnic Compositions**

| Parameters | Sensitivity |
|---|---|
| Lower melting point and decomposition temperature of oxidiser | Higher sensitivity |
| Lower melting point, boiling point and higher thermal conductivity of fuel | Higher sensitivity |
| Stoichiometric proportion of fuel and oxidiser | Higher sensitivity |
| Higher degree of fineness of ingredients | Higher sensitivity |
| Higher hardness of the ingredients (for stab or percussion cap) | Higher sensitivity |
| Higher density of composition (for stab or percussion cap) | Higher sensitivity |
| Higher ambient temperature | Higher sensitivity |
| Higher ambient pressure | Higher sensitivity |
| Higher homogeneity of composition | Higher sensitivity |
| Higher specific heat of composition | Lower sensitivity |
| Inert material | Lower sensitivity |
| Higher moisture/solvent | Lower sensitivity |

sensitivity as well as high heat of combustion and need special precautions during production like "sound and flash" compositions.

## 2.4 MEASUREMENT OF SENSITIVITY OF PYROTECHNIC COMPOSITIONS

Following types of measurement of sensitivity of pyrotechnic compositions is done:

a. Impact sensitivity
b. Percussion sensitivity
c. Friction sensitivity
d. Thermal (flame) sensitivity
e. Electrostatic discharge (Spark) sensitivity

These parameters are measured and recorded for implementing safety measures in handling the compositions.

There are several types of equipment used to measure sensitivity, differing from each other and also yielding different values, but provide the same order of sensitivity.

### 2.4.1 IMPACT SENSITIVITY

It is the measure of tendency of the composition to ignite under impact stimulus. A small sample of composition is kept on a die set with an anvil just placed over it. A standard weight is dropped from a height to cause ignition. The composition gets compacted due to impact and when the impact energy is sufficient, it causes ignition. It is measured as the height from which the specified weight is dropped leading to ignition.

There are two extreme values as given below [2].

a. The minimum drop height at which 95 ±5% ignition takes place, in 25 parallel tests, termed as upper sensitivity limit
b. The maximum drop height at which no ignition (misfire) takes place, termed as the lower sensitivity limit

A composition is considered comparatively more sensitive when the height of drop is less since a low kinetic energy is sufficient to cause ignition. Contrary to this, a composition is comparatively considered less sensitive when the height of drop is more since a high kinetic energy is required to cause ignition.

A 10 kg load is used over 0.5 cm$^2$ area of 0.05 g of composition. A number of trials are carried out with different heights and the performance is recorded. A graph between drop height and probability of functioning is plotted and the height for 50% functioning is noted for calculation for sensitivity. Impact sensitivity can be expressed as work done per unit area (kg m.cm$^{-2}$) as

$$\text{Sensitivity} = \frac{\text{Weight of the load (in kg)} \times \text{Height of fall (metre)}}{\text{Surface area (cm}^2)} \quad \dots \quad (2.1)$$

Impact sensitivity data on some oxidiser–fuel in stoichiometric proportions are shown in Tables 2.5 and 2.6.

There are various other devices that can be used to measure impact sensitivity:

a. Picatinny Arsenal Impact Apparatus (PA)
b. Bureau of Explosive Apparatus (BoE)
c. Bureau of Mines Apparatus (BoM)
d. Explosive Research Laboratory Apparatus (ERL)

The general arrangement for Picatinny Arsenal Impact Sensitivity Apparatus is given in Figure 2.1.

A 20 mg of sample is kept on a die set between two 0.06 mm silver foils with striker placed over it. A standard 2 kg weight (Hammer) is dropped over the striker from a fixed height. The functioning or otherwise of the composition is noted for 20 samples. The probability of functioning (number of times functioned divided by number of samples tested, multiplied by 100) is noted against height.

**TABLE 2.5**
**Impact Work in kg m.cm$^{-2}$ for Binary Pyrotechnic Mixtures [2]**

| Oxidiser | Sulphur | Lactose | Charcoal | Magnesium powder | Aluminium dust |
|---|---|---|---|---|---|
| Potassium chlorate | 1.1 | 1.8 | 3.2 | 4.5 | 4.5 |
| Potassium perchlorate | 1.2 | 2.9 | 4.2 | 4.4 | 5 |
| Potassium nitrate | 3.6 | 5 | 5 | 4.6 | 5 |

**TABLE 2.6**
**Impact Sensitivity of Binary Mixtures with Potassium Chlorate [2]**

| Fuel | Impact work (kg m.cm$^{-2}$) | Fuel | Impact work (kg m.cm$^{-2}$) |
|---|---|---|---|
| Potassium thiocyanate | 0.5 | Naphthalene | 1.3 |
| Realgar | 0.6 | Potassium ferrocyanide | 2.2 |
| Paraffin | 1.1 | Antimony sulphide | 3.5 |
| Sulphur | 1.1 | Graphite | 1.0 |

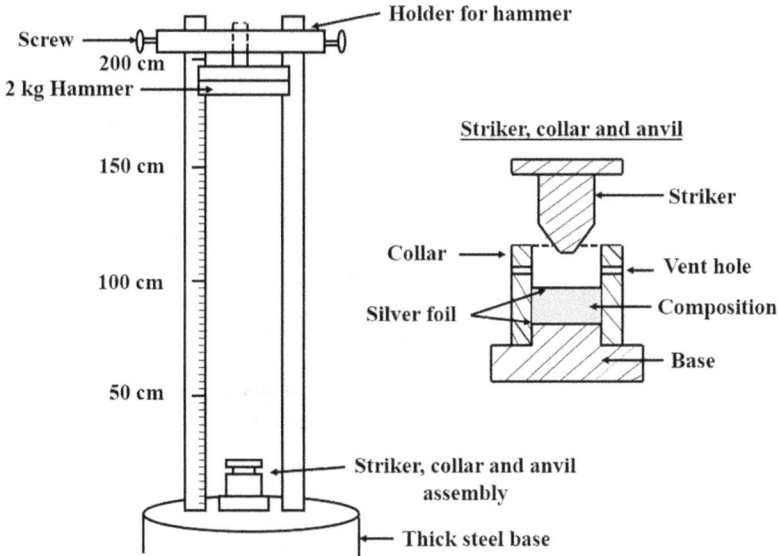

**FIGURE 2.1**  Impact Sensitivity Apparatus.

The height is varied and similar tests and calculations, as above, are carried out. A graph is plotted for height (in cm) on x-axis versus probability of functioning on y-axis which is almost a straight line. The height of fall in centimetres for 50% probability of functioning is noted as given in Figure 2.2.

The drop height, at which the ignition obtained is 50%, is used for calculating the sensitivity. The impact energy by the drop weight is given by

$$E = m \times g \times h \ldots \tag{2.2}$$

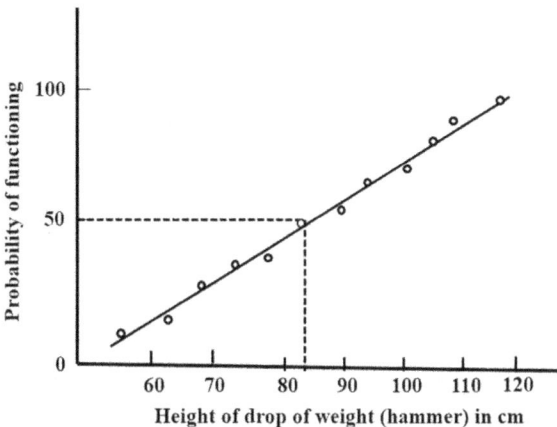

**FIGURE 2.2**  Probability of Functioning with Weight Drop Height.

where $E$ = impact energy, $m$ = mass of the drop weight, $h$ = drop height, $g$ = acceleration due to gravity.

In this method, figure of insensitiveness (FOI) is determined by comparing the results as obtained against tetryl (Methyl(2,4,6-trinitrophenyl)nitramide) with assigned FOI as 70. The FOI is given as

$$\text{FOI} = \frac{\text{Median height for 50\% functioning of composition}}{\text{Median height for 50\% functioning of tetryl}} \times 70 \dots (2.3)$$

The test values of FOI values of few high energy materials is shown in Table 2.7.

All compositions having FOI values less than 70 may be considered sensitive. Approximate "*Figure of Insensitivity*" of some pyrotechnic compositions are at Table 2.8.

Thus, compositions SR 252, SR 232, SR 563 are more sensitive than CE (tetryl) since their FOI is less than 70.

The mechanism of functioning of composition under the impact test is as follows. Pyrotechnic compositions during impact test are subjected to sudden high pressure, causing re-arrangement of ingredients to higher compaction density, thereby causing adiabatic compression of voids in the composition, leading to "*hotspots*" (see Section 22.3.2).

### TABLE 2.7
### Figure of Insensitivity (FOI) of High Energy Materials

| High energy material | Approximate figure of insensitivity (FOI) |
|---|---|
| Initiators | 10 To 25 |
| High explosives | 40 To 200 |
| Propellants | 20 To 50 |
| Pyrotechnics | 20 TO 120 |

### TABLE 2.8
### Figure of Insensitivity of Some Pyrotechnic Compositions

| Composition | Approximate figure of insensitivity (FOI) |
|---|---|
| SR 252 priming granulated | 40 |
| SR 232 signalling red composition | 43 |
| SR563 signalling white composition | 45 |
| SR 252 priming | 56 |
| SR 524 flash composition | 90 |
| SR 214 priming composition (smoke) | 91 |
| SR 700 signalling green composition | 105 |
| SR 264A(M) smoke composition | Over 120 |

## 2.4.2 PERCUSSION SENSITIVITY

This is carried out for arriving at a percussion sensitivity of initiators. A steel collar and striker (tip 0.1 sq. inches) assembly is kept with the 3 mg sample. Steel ball is allowed to fall from a height over the striker and the functioning or otherwise of the composition is noted as shown in Figure 2.3. The sensitivity to percussion should be more than 0.74 Joules.cm$^{-2}$, thus ensuring safety in processing the initiator for initiatory compositions.

To have consistency in results, specified steel balls are used and the height (say × inches) at which no functioning on 3 mg sample takes place for five consecutive tests is recorded.

Now, height is increased inch by inch and the height at which all is functioning is recorded for five consecutive tests (say y inch). The mean height (Z) in inches is calculated as (X + Y)/2. The figure of sensitivity to percussion is calculated as

$$\text{Figure of Sensitivity to Percussion} = (F \times Z)/12 \text{ (ft. lb. (sq. inch)}^{-1}) \dots \text{ (2.4)}$$

Where $F$ = factor of steel ball, $Z$ = mean height in inches. This value may also be calculated in joules as 1 ft.lb. (sq. inch)$^{-1}$ = 0.2102 J cm$^{-2}$

Thus, Figure of Sensitivity to Percussion = F × Z × 0.2102/12 (Joules.cm$^{-2}$) (2.5)

Therefore, for example, the diameter of steel balls ¾" (28.1 gms.) for mercury fulminate, 7/8" (44.64 gms.) for lead azide and lead styphnate and 1" (66.68 gms.) for tetrazene should not function below 21", 13.5", 13.5" and 9.1" respectively, thus ensuring that the sensitivity to percussion is more than 0.74 Joules. cm$^{-2}$.

It has not been established as to which factor like the velocity, momentum or kinetic energy of the steel balls (used for imparting the blow) is important in determining sensitiveness to percussion.Initiatory compositions are not subjected to

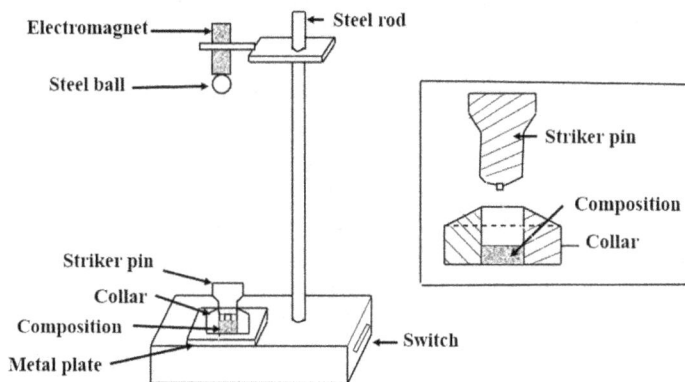

**FIGURE 2.3** Percussion Sensitivity Apparatus.

**TABLE 2.9**

**Steel Balls Dimension and its Range of Utility with Factor for the Ball**

| Diameter of balls (inches) | Range of figure of sensitiveness to percussion which is used (ft.lb. (sq. inch)$^{-1}$) | F factor of the ball (lbs. (sq. inch)$^{-1}$) |
|---|---|---|
| 3/8 | Up to 3 | 1.15 |
| ¾ | 3–5 | 2.00 |
| 7/8 | 5–8 | 3.15 |
| 1 | 8–12 | 4.65 |
| 1–1/8 | 12–17 | 6.70 |
| 1–1/4 | 17–23 | 9.05 |
| 2–3/8 | 23–30 | 12.05 |

*Note: F = mass of balls in lbs./area of impact (0.1 sq. inch)*

percussion sensitivity test but caps filled with initiatory compositions are subjected to sensitivity and insensitivity tests in separate test equipment.

The details of steel balls used in the percussion sensitivity apparatus are given in Table 2.9.

### 2.4.3 FRICTION SENSITIVITY

Friction restricts the motion of material in contact with another material. The amount of friction depends upon the roughness of the two surfaces, area of contact, adhesion between materials and extent of deformation of material.

Friction sensitivity is a measure of tendency of the composition to ignite under friction stimulus. It is measured as the force that must be applied between a pair of frictional surfaces. To determine the friction sensitivity created due to pyrotechnic composition being nipped and crushed between two surfaces, certain friction sensitivity tests have been devised. The old method known as the Mallet Test for determining friction sensitivity is shown in Figure 2.4.

This test [3] classifies explosives into "*very sensitive,*" "*sensitive*" and "*comparatively insensitive*" by use of three materials, namely softwood, hardwood and York stone. A 0.15 g sample of composition (spread over 12.7 mm long and 6.4 mm wide) is kept over each three different material, namely softwood, hardwood and York stone. The composition is given a *swinging/glancing blow* by the boxwood mallet and crack, spark or flash or non-ignition is noted. The observations are recorded as the number of functions per 10 samples as under.

| | |
|---|---|
| 0 ignition out of 10 tests | 0% |
| Up to 6 ignitions out of 10 tests | 50% |
| More than 6 ignitions out of 10 tests | 100% |

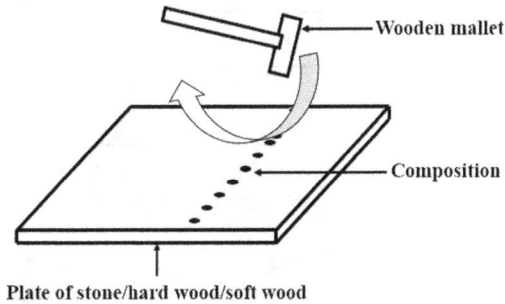

**FIGURE 2.4**   Mallet Test Apparatus.

Explosives that yield some ignition with a boxwood mallet and a softwood anvil and which have FOI of 30 or less are classified as *"very sensitive."* Those which do not give any ignition with a boxwood mallet and have a FOI more than 90 are classified as *"comparatively insensitive."* The test is mainly used for safety certificate of compositions.

Tests are also done in some laboratories with a mild steel mallet on anvils of mild steel.

The drawback of Mallet Test is that the impact energy varies with the height, strength and experience of the person carrying out the test.

There are other methods [4] like *Friction Pendulum Apparatus* where steel or fibre shoes are used or *Rotary Friction Device* where composition is subjected to friction generated between stationary wheel and the sliding anvil surface. Another Mallet Impact Test method using electronic instrument has been reported [5] as under.

A small quantity of explosive material, typically 10 mg, is placed on a steel anvil and is hit with a steel mallet. The mass of the steel mallet and anvil were approximately 145 g and 650 g, respectively; the impact energy was measured through a data acquisition controller as shown in Figure 2.5.

The mass of the anvil had no influence on the peak force values recorded as the force transducer was zeroed even before the testing started. The Kistler 9031A transducer was attached to an aluminium plate [5] where the anvil was positioned above. The base plate was present to give stability to the setup. The transducer was connected to a Kistler charge attenuator 10:1 type 5361A that, in turn, was connected to the Imatek C3008 data acquisition controller. The C3008 controller had a Kistler charge amplifier type 5041, which could be used for selecting 6 kN, 60 kN or 600 N output. The C3008 was connected to a laptop, via Ethernet cable, to enable visualisation and recording of the data.

However, there is no consistency in test results due to tests conducted by different persons, difference in their age, height, strength and experience.

Friction sensitivity of composition is also determined by a machine developed by BAM and manufactured by Julius Peters Company (Germany), as shown in Figure 2.6.

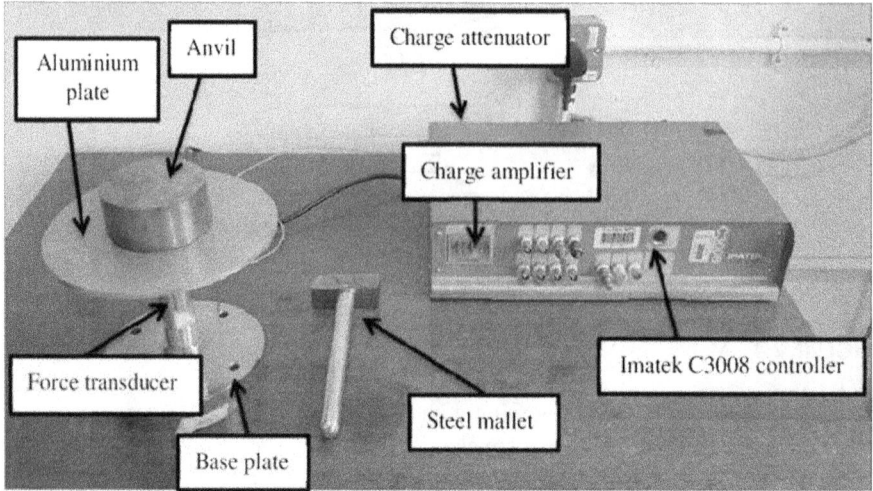

**FIGURE 2.5**   Mallet Test Arrangement for Impact Sensitivity Test (Reprinted from 5).

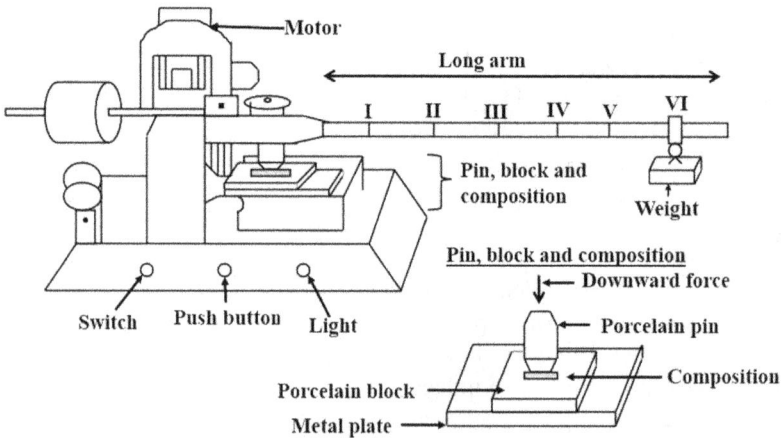

**FIGURE 2.6**   Friction Sensitivity by Julius Peter Apparatus.

A 10 mg sample composition is kept over an abrasive porcelain plate (of size 25.6 × 25.6 × 5 mm with standard curvature and roughness) which is kept over a metallic plate attached to a motor. An abrasive porcelain pin (15 mm long, 10 mm diametre of standard roughness) is attached below a metallic plate and kept over the composition. The arm has six notches for hanging weights. Tests are done by using weights (one out of nine numbers) suspended on the loading arm notches marked at equidistance. This allows the load on the porcelain pin to be varied from 0.5 kg up to 36 kg. As the weight acts over composition through the porcelain pin, the metal plate is moved horizontally with the help of motor. This creates friction in the composition. If the

**TABLE 2.10**

**Load (in kilograms) on Loading Arm of Friction Sensitivity Apparatus [6]**

| Weight | Notch number | | | | | |
|---|---|---|---|---|---|---|
| | *I* | *II* | *III* | *IV* | *V* | *VI* |
| 1 | 0.5 | 0.6 | 0.7 | 0.8 | 0.9 | 1.0 |
| 2 | 1 | 1.2 | 1.4 | 1.6 | 1.8 | 2.0 |
| 3 | 2 | 2.4 | 2.8 | 3.2 | 3.6 | 4.0 |
| 4 | 3 | 3.6 | 4.2 | 4.8 | 5.4 | 6.0 |
| 5 | 4 | 4.8 | 5.6 | 6.4 | 7.2 | 8.0 |
| 6 | 6 | 7.2 | 8.4 | 9.6 | 10.8 | 12.0 |
| 7 | 8 | 9.6 | 11.2 | 12.8 | 14.4 | 16.0 |
| 8 | 12 | 14.4 | 16.8 | 19.2 | 21.6 | 24.0 |
| 9 | 18 | 21.6 | 25.2 | 28.8 | 32.4 | 36.0 |

load is sufficient, the composition emits spark, smoke, etc. Compositions functioning below 36 kg are considered friction-sensitive.

The manufacturer provides an index to choose a weight with corresponding resultant loads (kg) on the sample taken. Thus, load may be varied with different weights by hanging the same at different positions of notches. Table 2.10 gives the load in kilograms kept on marked distances I, II, III, IV, V and VI on the loading arm of the apparatus.

The surfaces of the pin are damaged during use and each pin can be used two times, one from each side while the plate of bigger size can carry 5–10 tests on each side.

An electric push button switch causes one horizontal forward and reverse movement of 10 mm of the porcelain plate containing pyrotechnic composition, thus creating friction between the porcelain pin, composition and porcelain plate. When the frictional energy is high, it leads to ignition of the composition and is exhibited as smoking, burning, spark, flash, explosion, etc.

The principle of working is that when the frictional force is applied, hotspots are formed, which, in turn, depends upon:

a. Weight suspended over loading arm notch
b. Rate of movement between the porcelain pin and composition with porcelain block
c. Heat loss to surrounding
d. Friction coefficient
e. Moisture content of composition
f. Ambient temperature, humidity and pressure

**TABLE 2.11**

**Friction Sensitivity of High Energy Materials**

| High energy material | Approximate friction sensitivity (Kg) |
|---|---|
| High explosives | 14.4 TO 25.2 |
| Propellants | 10.8 TO 24.0 |
| Pyrotechnics | 8.0 TO 36.0 |

The final relative measure of sensitivity is recorded as the smallest load (kg) at which reaction occurred for an 1-in-I0 series of attempts. The lower the load values, the higher the friction sensitivity.

The test values of friction sensitivity of few high energy materials is given in Table 2.11.

To avoid any hazard due to friction, care is taken not to drag any explosive or ammunition as well as ensuring a perfect alignment of punch and mould in pressing equipment. Not only punch and mould surface finish are ensured but lubricants like zinc stearate in composition or graphite on punch and mould is used to achieve smooth movement of punch in mould.

### 2.4.4 THERMAL SENSITIVITY

a. *The Flame Sensitivity* [2] is measured as distance of the surface of the composition from the flash (through Bickford fuse), which leads to 100% ignition or non-ignition of the surface of composition. A small sample is taken in a hard glass test tube and a Bickford fuze is clamped over the surface of the composition. The other end of safety fuze is ignited by a match. There are two sensitivity values obtained as given below.

  I. Upper limit is measured as maximum distance of the surface of the composition from the flash (through Bickford fuze) which leads to 100% ignition of the composition.

  II. Lower limit is measured as a minimum distance from the surface of the composition which results in 100% failure in ignition.

  The reliability factor, i.e., the slope of the straight line connecting the points obtained experimentally on a graph where abscissa (x-axis) represents the distance between the section of the Bickford fuse and the surface of the composition, and the ordinate (y-axis) represents the percent ignition. The values for some compositions are given in Table 2.12.

b. The *Ease of ignition by flash* [3] is recorded by keeping the Bickford fuse in touch with over 3 g of composition in a 14–16 mm diameter 229 mm length standard glass tube in an approved safety cupboard and recording the effects

**TABLE 2.12**
**Sensitivity of Composition to Flame Impulse [2]**

| Composition | Upper limit (cm) | Lower limit (cm) | Reliability factor |
|---|---|---|---|
| Black powder Gr1 | 2 | 15 | 0.7 |
| Igniter | 3 | 13 | 0.8 |
| Red light | 0 | 2 | 3.0 |
| Illuminating | 0 | 3 | 1.0 |

     like fails to ignite, ignites and burns quietly, ignites and burns vigorously, explodes, etc.

c. The *Train test or behaviour on inflammation* [3] is recorded by filling the composition loosely (flush with top edge) on a long trough of semi cylindrical mild steel tube of 12.7 mm diameter and 30.5 cm long and igniting one end with luminous gas flame or a propellant stick. The effect is recorded as fails to ignite, just ignites but fails to support the train beyond "x" distance, or ignites and supports the train steadily, ignites and supports the train vigorously throughout, explodes, etc.

A good measure of thermal sensitivity is determination of ignition temperature of the composition (see Section 5.1.4).

## 2.4.5 ELECTROSTATIC CHARGE (SPARK) SENSITIVITY

A sample of pyrotechnic composition kept in a groove is subjected to specified value of discharge from a capacitor (0.045J, 0.45J and 4.5J). The arrangement [7] is shown in Figure 2.7.

    The pyrotechnic composition is taken loosely in a groove of approximately 10 mm in a non-conducting material like teflon. The top electrode is formed by placing a copper foil of 15 mm square with 0.2 mm thickness over the Teflon block

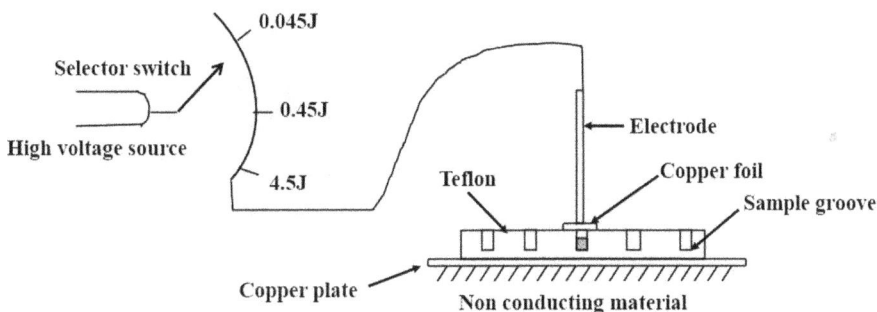

**FIGURE 2.7** Electrostatic Charge (Spark) Sensitivity Apparatus (Reprinted from [7]).

containing composition and connected to the electrode. A copper plate is placed at the bottom of the Teflon block and suitably earthed.

The electrostatic charge is allowed to pass through the composition and its functioning or otherwise is noted from its effect on copper foil. Functioning in 1 out of 10 discharges is considered as sensitive to the corresponding charge. The results are recorded as ignition at 0.045 J, ignition at 0.45 J but not at 0.045 J, ignition at 4.5 J and not at 0.45 J and no ignition at 4.5 J.

The mechanism of test is that a charged capacitor is discharged through the sample material placed in the spark gap between two electrodes and the energy of the spark is calculated using the equation

$$E = CV^2/2 \qquad\qquad (2.6)$$

Where $E$ = Spark Energy, $C$ = Capacitance and $V$ = Voltage

The value of the ignition energy obtained is largely influenced by factors like shape of the electrode and space between the two electrodes.

McLain and Praham [8] used platinum-rhodium filament to generate short level spark of reproducible temperature. The power of filament was varied till ignition occurred. The lower the value of filament power, the higher is the value of ease of ignition. The results are shown in the Table 2.13.

It could be seen that there is no relationship between ease of ignition and ignition temperature. Since the actual ignition system rarely uses spark for initiation of ignition, the values in Table 2.13 are of academic interest.

**TABLE 2.13**

**Ignition Sensitivity to Hot Filament [8]**

| Mixture | Ingredient | Percentage | Filament (watts) | Ignition temp. $^0$C | Ignition ease 1 (easiest) −6 |
|---|---|---|---|---|---|
| Sulphurless mealed gunpowder | $KNO_3$ Charcoal | 9010 | 2.46 | 400 | 1 |
| Red lead starter | $Pb_3O_4$ Si | 90 10 | 2.86 | 555 | 2 |
| Red lead starter | $Pb_3O_4$MnSi | 54.2 34.2 11.6 | 2.86 | 540 | 2 |
| Litharge silicon | PbO Si Fuller's Earth | 78.4 19.6 2.0 | 3.25 | 621 | 3 |
| British starter mix | $KNO_3$ Si Charcoal | 54 40 6 | 3.30 | 560 | 3 |
| Black powder A5 | – | – | 3.95 | 457 | 4 |
| British thermite mix | British Starter mix $Fe_2O_3Al$ (grained) | 65 22 13 | 5.49 | 545 | 5 |
| Red lead starter | $Pb_3O_4$ Mn Si | 48.1 48.1 3.8 | 6.33 | 625 | 6 |
| Red lead starter | $Pb_3O_4$ Mn Si | 78.1 20.8 1.1 | 6 (approx.) | – | No Ignition |

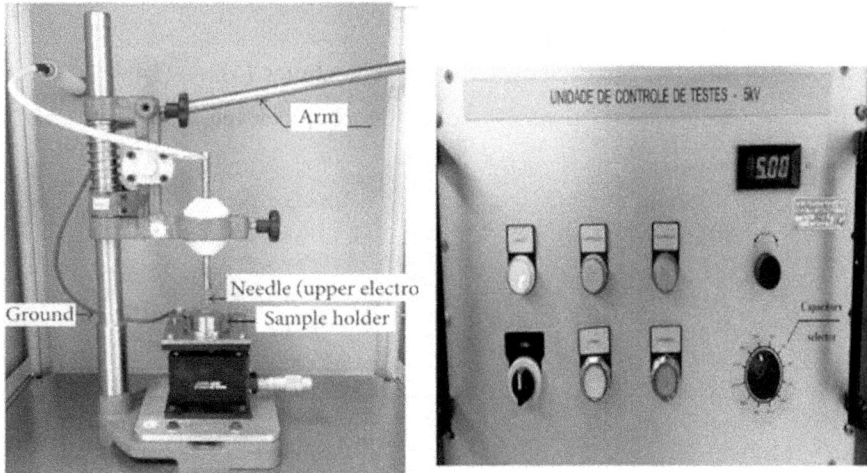

**FIGURE 2.8**   Energetic Material Test Unit and Test Unit of 5 kV (Reprinted from 9).

Another method of electrostatic discharge sensitivity is given in [9]. This test determines the energy limit needed to ignite an explosive by electrostatic stimulus of various intensities. The test is run in an Energetic Materials Test Unit, integrated to a Test Unit of 5 kV of the Electrostatic Discharge Sensitivity (ESD) Tester ESD-100 (Equatorial Sistemas) as shown in Figure 2.8. The initial energy is 0.25 J, charged by a capacitor of 0.02 μF connected to a discharge circuit charged with 5 kV. The gap between the upper electrode (needle) and the metal sample holder is fixed at 0.018 mm. A sample of 30 mg is put inside a Teflon washer placed on the top of the grounded sample holder, and a Mylar sheet (insulating material) is placed on the top of the washer to confine the powder. The needle is charged and manually moved down to the adjusted gap, piercing the Mylar sheet and discharging into the material. A positive reaction is defined as a flash, spark, burn or noise.

McIntyre [4] has provided the electric spark sensitivity of pyrotechnics given in Table 2.14.

McIntyre [4] found that electrical spark sensitivity varied between groups and within group.

The following factors affect electric spark sensitivity.

a. Conductivity of materials (higher conducting materials like zirconium, boron, aluminium and magnesium metals increase spark sensitivity)
b. Presence of non-conducting materials like diatomaceous earth reduces spark sensitivity
c. Loose compositions have higher spark sensitivity than pressed compositions
d. Oxide coating on metallic fuels lead to reduced spark sensitivity
e. Moisture content of composition reduces spark sensitivity

**TABLE 2.14**
**Electric Spark Sensitivity of Pyrotechnics [4]**

| Pyrotechnics | Electric spark sensitivity (J) |
|---|---|
| Initiatory | $0.038 \pm 0.02$ |
| Illuminants | $33 \pm 23$ |
| Smoke | $10.5 \pm 19.8$ |
| Gas | $13 \pm 25$ |
| Sound | $0.6 \pm 0.4$ |
| Heat | $1.72 \pm 2.55$ |
| Time (delay) | $0.80 \pm 1.04$ |

As dust and inflammable liquid vapour are at higher risk of electric spark ignition, precautions must be placed to reduce dust and remove inflammable vapours from the process buildings. As human body can have static energy and processing may generate static electricity, it is necessary to dissipate the same before entering the process building

## 2.5  SAFETY INDICES OF PYROTECHNIC COMPOSITIONS

Data of safety indices like *Figure of Insensitivity, Temperature of Ignition, Ignition by Spark* and *Friction Sensitivity* for a few compositions are given in Table 2.15.

The data of ignition by flame, behaviour on inflammation and mallet test of compositions shown in Table 2.16.

The safety indices on Infrared MTV pyrotechnic compositions (containing magnesium, PTFE and Viton) are given in Table 2.17 [10].

Table 2.18 illustrates friction and impact sensitivity class of energetic materials [11].

Sensitivity of few pyrotechnic compositions has been found (compiled [4]) as under.

(a) HC smoke is sensitive to electrical spark, moderately sensitive to impact and insensitive to friction
(b) Photoflash compositions are sensitive to electrical spark, friction and impact
(c) Initiatory compositions are sensitive to all stimuli, most to electrical spark
(d) Tracer compositions are insensitive to friction, impact and electrical spark
(e) Gas producing compositions are insensitive to friction but highly sensitive to impact

However, the above should be taken as a general guidelines and sensitivity data must be obtained for concerned composition.Inconsistency in sensitivity test results

**TABLE 2.15**
**Safety Indices of Pyrotechnic Compositions**

| Composition type | Figure of insensitivity (impact) | Temperature of ignition | Ignition by electric spark | Friction sensitivity (JP method) | |
|---|---|---|---|---|---|
| | | | | Function at kg | Not function at kg |
| SR 592 signal | 38 | Above 350 | No ignition at 4.5 J | 36 | 32.4 |
| SR 170A signal | 66 | Above 350 | No ignition at 4.5 J | 11.2 | 10.8 |
| SR 269(M) smoke | 59 | Above 350 | No ignition at 4.5 J | ----- | 36 |
| SR 560(M) signal | 70.5 | Above 300 | ---------- | 32.4 | 28.8 |
| SR 252 priming granulated | 40 | Above 350 | ---------- | ---- | 36 |
| SR 252 priming | 56 | Above 350 | --------- | ------ | 36 |
| SR 214 priming | 91 | Above 330 | No ignition at 4.5 J | ------- | 36 |

amongst various laboratories for same composition (prepared and tested at their own laboratories) are expected due to variation in following parameters.

a. Moisture content of the sample (extent of drying of composition)
b. Sample preparations like sample's physical state (ingredients' size, shape, surface area, purity, proportion, moisture, homogeneity of composition etc
c. Ageing of the composition
d. Laboratory room temperature, ambient pressure and humidity
e. Different equipment and different mode of testing (different types of equipment, variation in test procedure like variation in mass of drop weight and sample mass)
f. Non-standardisation/calibration of equipment
g. Personnel doing the sensitivity test especially for Mallet Test.
h. The impact sensitivity is reported in BOM (Bureau of Mines Apparatus) as functioning at minimum drop height while in BOE (Bureau of Explosives apparatus), it is reported at two drop heights. Also, the impact sensitivity is expressed in various units like work done per unit area (kg.m.cm$^{-2}$), FOI, drop heights (cm) or energy (joules).

It could be seen that there are many types of equipment to measure sensitivity, each varying from other and also give different values, but provide the same order of sensitivity.

**TABLE 2.16**

**Safety Indices of Pyrotechnic Compositions**

| Composition type | Ignition by flame | Behaviour on inflammation | Ignition by electric spark | Mallet test (stone/ hardwood/ softwood) |
|---|---|---|---|---|
| SR 592 signal | Ignites and burns vigorously | Ignites and supports the train vigorously throughout | No ignition at 4.5 J | 0/0/0 |
| SR 170A signal | Ignites and burns vigorously | Ignites and supports the train vigorously throughout | No ignition at 4.5 J | 0/0/0 |
| SR 269(M) smoke | Fails to ignite | Just ignites but fails to support the train | No ignition at 4.5 J | 0/0/0 |
| SR 560(M) signal | Explodes | Ignites and supports the train vigorously throughout | ……........... | 0/0/0 |
| SR 252 priming granulated | Ignites and burns vigorously | Ignites and supports the train vigorously throughout | …………… | 0/0/0 |
| SR 252 priming | Ignites and burns quietly | Ignites and supports the train vigorously throughout | …………… | 0/0/0 |
| SR 214 priming | Ignites and burns vigorously | Ignites and supports the train vigorously throughout | No ignition at 4.5 J | 0/0/0 |

**TABLE 2.17**

**Safety Data on MTV Payload [10]**

| Parameter | Values |
|---|---|
| Friction sensitivity | > 360 N |
| Impact sensitivity | 10 Nm |
| Thermal decomposition | 263°C |
| Ignition temperature | > 360°C |
| electrostatic discharge | 0.05 J |

**TABLE 2.18**
**Sensitivity Class for Energetic Materials to Friction and Impact [11]**

| Stimuli | Experimental sensitivity | Sensitivity class |
|---------|--------------------------|-------------------|
| Friction | >360 N | Insensitive |
|  | 80–360 N | Moderately sensitive |
|  | 10–80 N | Sensitive |
|  | <10 N | Very sensitive |
| Impact | > 40 J | Insensitive |
|  | 35–40 J | Moderately sensitive |
|  | 4–35 J | Sensitive |
|  | < 4 J | Very sensitive |

## 2.6  SAFETY CERTIFICATE OF PYROTECHNIC COMPOSITIONS

Each explosive/explosive mixture has one important safety document, namely *Safety Certificate*, which provides safety data as under.

a. Composition specification and its nomenclature
b. Sensitiveness to mechanical shock, i.e., figure of insensitivity
c. Sensitiveness to friction by mallet in percent and Julius Peter in kg
d. Temperature of ignition
e. Inflammability, i.e., ignition by flash
f. Behaviour on inflammation by ignition by spark
g. Incompatibility with other materials
h. Chemical stability
i. Poisonous ingredients
j. General precautions during manufacture, storage and use
k. Method of preparation, mixing and filling
l. UN classification
m. Disposal

This certificate must be available for all explosive compositions and must be read and understood by the personnel for ensuring safety. Readers may see reference [12] on ignitability of pyrotechnic compositions.

## REFERENCES

[1]. *"Properties of Materials used in Pyrotechnic Compositions"*, Engineering Design Handbook-Military Pyrotechnic Series, Part Three, Headquarters, US Army Material Command, Washington D.C., October 1963.
[2]. A.A. Shidlovskiy, *"Principles of Pyrotechnics"*, 3rd Edition, Moscow, 1964. (Translated by Foreign Technology Division, Wright-Patterson Air Force Base, Ohio, 1974). American Fireworks News, 1 July 1997. © 1997, Rex E. & S.P., Inc.

[3]. *"Service Textbook of Explosives"* March 1972, JSP 333, Ministry of Defence, UK, Reprinted in India in 1976.

[4]. F. L. McIntyre," *A Compilation of Hazard and Test Data for Pyrotechnic Compositions"*, Contractor Report ARLCD-CR-80047, October 1980. U.S. Army Research and Development Command, Large Caliber Weapon System Laboratory, Dover, New Jersey.

[5]. Matthew Weaver, Lisa H. Blair, Nathan Flood, and Christopher Stennett, "A *Review of the Mallet Impact Test for Small Scale Explosive Formulations."* Presented at 19th Seminar on New Trends of Energetic Materials (NTREM 2016); 20-22/04/2016, Pardubice Czech Republic.

[6]. L. Richard Simpson and M. Frances Foltz, *"Lawrence Livermore National Laboratory Small Scale Friction Sensitivity (BAM) Test"*, National Technical Information Service, U.S. Department of Commerce, June 1996. Report Number UCRL-ID—124563.

[7]. H. C. Dwivedi, *"Safety Hazard and Sensitivity Evaluation,"* Continued Education Programme Course on Pyrotechnics 4-8, August 2003, High Energy Materials Research Laboratory, Pune, India.

[8]. J.H. McLain and A.L. Praham, *"A New Method of Comparing Ignition Sensitivity of Pyrotechnic Mixtures"*, TDMR 882 (1944), Edgewood Arsenal, Maryland.

[9]. Luciana de Barros, Afonso Paulo Monteiro Pinheiro, Josemar da Encarnação Câmara, and Koshun Iha, *"Qualification of Magnesium/Teflon/Viton Pyrotechnic Composition Used in Rocket Motors Ignition System"*, Journal of Aerospace Technology and Management, 2016, Volume 8, Issue 2, Pages 130–136. 10.5028/jatm.v8i2.596

[10]. Ernst-Christian Koch, *"On the Sensitivity of Pyrotechnic Countermeasure Ammunitions"*. 35th International Pyrotechnics Seminar at Fort Colling, Colorado, USA, July 2008, Page 532, Copyright Wiley-VCH GmbH. Reproduced with permission.

[11]. Davin G. Piercey and Thomas M. Klapötke, *"Nanoscale Aluminum – Metal Oxide (Thermite) Reactions for Application in Energetic Materials"*, Central European Journal of Energetic Materials, 2010, Volume 7, Issue 2, 115–129.

[12]. L.V. De Yong, *"A Review of Methods to determine Ignitability of Pyrotechnic Compositions"*, Report MRL-R-989, Department of Defence, Materials Research Laboratories, Melbourne, Victoria.

# 3 Combustion of Pyrotechnic Compositions

## 3.1 GENERAL

The process of initiation of combustion of pressed pyrotechnic composition in ammunition is done by an external energy stimulus (see Chapters 2 and 22). Combustion of pyrotechnic composition is an intricate process wherein ingredients and products including atmospheric gases react in solid, liquid, gas and vapor forms over the surface of the composition. Some ingredients on receiving external stimuli become in liquid, gas and vapour phase while combustion leads to products in solid, liquid, gas or vapour phase. Pyrotechnic ammunitions and devices work based on sequence of designed quantum of energy transfer from layer to layer of pressed pyrotechnic composition for combustion. Any disruption in this energy transfer shall cause failure of the ammunition system.

## 3.2 PROCESS OF IGNITION

Combustion is the oxidation and reduction reaction where the fuels are oxidised and the oxidisers are reduced. The process of initiation, localised ignition, inflammation and propagation is shown in Figure 3.1 (with exaggerated thickness of composition layer for ease of understanding) where a pyrotechnic composition is pressed inside a container and is subjected to an external stimulus.

The process of ignition (thermo-chemical) of a pressed pyrotechnic composition involves the following sequence

a. *Initiation by external stimuli:*
   It is the ability of the pyrotechnic composition's thermal property to ignite under external stimulus. The external stimulus transfers energy to a limited area on the surface of the composition at certain temperature for a certain time. This is followed by
b. *Heat absorption leading to pre-ignition (Localised ignition):*
   The surface of the composition absorbs heat causing "*hotspot*" (see Section 22.3.2), leading to some events like melting or boiling or vaporisation, dehydration, crystalline transition and thermal decomposition of few ingredients and their interactions leading to initiation of *pre-ignition* of the composition in a localised point. This pre-ignition depends upon the strength of the external stimuli, heat loss through conduction, convection and radiation, duration of the external stimuli and the sensitivity of the composition

DOI: 10.1201/9781003093404-3

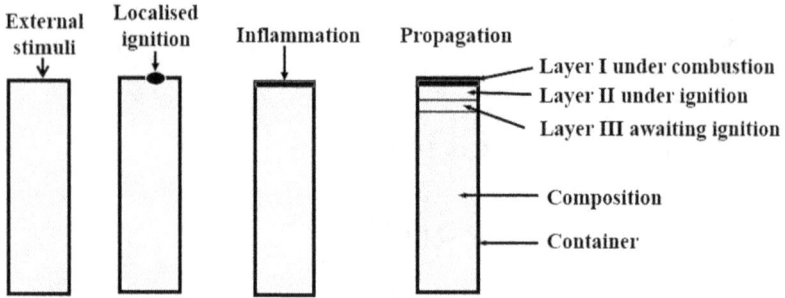

**FIGURE 3.1**   Combustion of Pyrotechnic Composition.

towards the stimuli to achieve *energy of activation* for ignition. This sensitivity of composition depends upon several factors like type and proportion of ingredients in the composition, degree of fineness of ingredients, their densities, density of the composition, moisture content, specific heat, temperature of the layer of composition and external pressure over the composition layer. This is followed by

c. *Inflammation:*

The inflammation of the top layer depends upon heat produced during pre-ignition, heat loss during inflammation and hot products of combustion and porosity of the composition. This is followed by

d. *Propagation:* This is the ability of the once ignited composition, to sustain (or propagate) the burning process through the remainder of the composition. This transmission of burning to the next layer of composition is done by heat transfer through conduction, convection and radiation and is assisted by the porosity of the composition.

For combustion to take place it is essential that

a. The atoms/molecules of the composition should be on the surface of the grain to gain energy from external stimulus

b. The atoms/molecules of reacting ingredients of the composition should be in contact with each other for combustion.

c. The atoms/molecules of the composition should gain very high internal energy sufficient enough for combustion to take place.

Ingredients are made to come together for combustion by proper mixing the composition and ignition stimulus makes melting of one of the ingredients like oxidiser which allows molten material and or gaseous hot combustion products to move inside the porous composition for combustion.

If the external stimulus is removed, a true pyrotechnic composition should be *self-sustained in burning*. The flame observed in combustion is due to combustion of gaseous products just above the reacting layer. The flame changes its shape and appears as moving due to movement of burning gaseous products.

The performance of a true pyrotechnic composition therefore depends upon

a. Pyrotechnic composition's ability to get ignited by an external stimulus with ease
b. Pyrotechnic composition's ability, once ignited, to sustain propagation in the remainder of composition

### 3.2.1  Factors Affecting Quantum of Heat Transfer

The quantum of heat transfer between two adjacent layers of the pyrotechnic composition depends upon several parameters as under.

a. Basic Ingredient Factors:
   i. Fuel and oxidiser properties and percentage
   ii. Presence of binder and percentage
   iii. Presence of burn rate modifiers and percentage
   iv. Presence of dyes and percentage
   v. Presence of special additives and percentage

b. *Physical Factors*:
   i. Ingredients' size, shape and surface area
   ii. Consolidation density
   iii. Porosity

c. *Thermal Properties Factors*:
   i. Ignition temperature of composition
   ii. Specific heat of composition
   iii. Heat of combustion of composition
   iv. Mode of transfer by combustion products like hot solid, molten slag as fluid or gaseous products and their temperatures
   v. Temperature difference between two adjacent layers of composition and surroundings
   vi. Quantum of heat and rate of heat production by composition

### 3.2.2  Burning and Burn Rates

Burning of pyrotechnic compositions consumes the composition and the mass of the composition reduces as burning progresses. Burning varies as per the state of the composition and is of the following types:

a. *En Masse Burning*: The composition in loose form burns *en masse* as hot gases and flames travel easily in interspace between the granules of the composition due to its high porosity. This takes extremely less time to burn the composition and may even cause an explosion, if confined. Examples are photoflash compositions, incendiary compositions and gunpowder compositions used in ejection of signal flares. The burn time is usually in milliseconds.

b. *Radial Burning*: The exposed burning surface in pressed form as pellet increases with burning time since more area is exposed during radial burning (see Section 4.2.16). Example is screening smoke canisters and colour smoke canisters. The burn rate is usually few seconds.

c. *Shaped Burning/All Side Burning*: The exposed burning surface shape in pressed form as pellet remains constant though it decreases in its size as it burns from all directions. Examples are signal flares and infrared flares. The burn time is usually few seconds.

d. *Convex/Oval Burning*: This is observed in tracer burning (see Section 4.3.5). The burn time is few seconds.

e. *Parallel Burning/Uni-Directional Burning*: The composition in pressed form burns layer by layer where one layer burns, while the next layer is under initiation. This takes more time to burn the composition. The exposed burning surface area remains constant throughout burning period as burning is from one end only. Examples are illuminating candles and delay columns. The burn time is usually few seconds or minutes.

In parallel burning, the direction of combustion, direction of flame, direction of heat transfer and direction of gaseous combustion products take place as explained below.

a. *Inward Direction*:
   i. Linear burning moving inward to next layer of composition
   ii. Flame front moving inward to next layer of composition
   iii. Heat transfer moving inward to next layer of composition by conduction, convection and radiation

b. *Outward Direction*:
   i. Partial heat transfer to surrounding atmosphere from burning surface
   ii. Partial heat transfer to container material (if metallic) and through metallic container to surrounding atmosphere
   iii. Main heat transfer for special effects
   iv. Gaseous products leaving the burning surface

Burn rate determines the rate of release of energy by the composition. Rate of burning of a pressed pyrotechnic composition in a cylindrical column can be expressed in several ways.

a. *Linear Burn Rate* (cm. $sec^{-1}$): It is the length of the composition burnt per unit time and is expressed as length.$time^{-1}$

Linear burn rate = length of composition burnt(cm)/burn time(sec)   (3.1)

b. *Inverse Burn Rate* (sec. $cm^{-1}$): It is the inverse of linear burn rate and is the time of burning of the composition per unit length, expressed as time.$length^{-1}$

Inverse burn rate = time(sec)/length of composition burnt(cm)    (3.2)

c. *Mass Burn Rate*: This may be expressed in two ways.

i. *Mass Burnt Per Unit Time* (g. sec$^{-1}$)

Mass of composition burnt per unit time

$$= \text{Mass/time} = (\text{volume} \times \text{density})/\text{time}$$
$$= (\text{area} \times \text{length} \times \text{density})/\text{time}$$
$$= \text{area} \times \text{density} \times \text{linear burn rate}$$
$$= \text{area}(\text{cm}^2) \times \text{density}(\text{g. cm}^{-3}) \times \text{linear burn rate}(\text{cm. sec}^{-1})    (3.3)$$

Thus, it is calculated as product of area, density and linear burn rate.
ii. Mass Burnt Per Unit Time Per Unit Area (g cm$^{-2}$sec$^{-1}$)

$$= \text{density}(\text{g cm}^{-3}) \times \text{linear burn rate}(\text{cm sec}^{-1})    (3.4)$$

Thus, it is calculated as product of linear burn rate and density of the composition.

d. *Volume Burnt Per Unit Time(cm$^3$.sec$^{-1}$)*:

Volume burnt per unit time

$$= \text{area} \times \text{length/time}$$
$$= \text{area}(\text{cm}^2) \times \text{linear burn rate}(\text{cm. sec}^{-1})    (3.5)$$

Thus, it is calculated as product of area and linear burn rate.
Consider a pressed pyrotechnic composition with density "$\rho$," height "h" and diameter "d" inside a cylindrical container, burning layer by layer for "t" seconds as in Figure 3.2. The various burn rates would be:

i. Linear burn rate = $h/t$
ii. Inverse burn rate = $t/h$
iii. Mass burnt per unit time = $\pi d^2 \rho h/(4t)$
iv. Mass burnt per unit time per unit area = $\rho h/t$
v. Volume burnt per unit time = $\pi d^2 h/(4t)$

Burn rates provides the following:

a. Provides an idea of the length of the composition column required in the ammunition
b. Provides the duration of burning and hence the duration of special effects
c. Helps in overall design and specification of the pyrotechnic ammunition

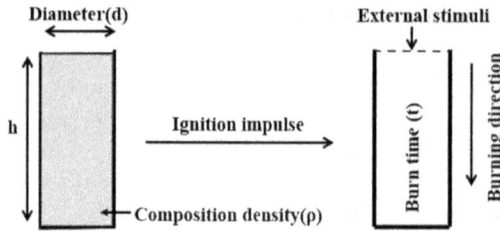

**FIGURE 3.2** Burn Rates.

Linear burn rate and inverse burn rates are used most often. However, as these are inverse of each other, a mere statement that burn rate increases with addition of (say) certain ingredient is not sufficient. It is necessary to state whether linear burn rate or inverse burn rate increases or to state burn rate $(cm.s^{-1})$ or burn rate $(s.cm^{-1})$ increases.

## 3.3 PROBABILITY OF IGNITION

Whether the ignition of a pyrotechnic composition can take place, depends upon:

a. Strength of ignition stimuli
b. Duration of ignition stimuli
c. Energy of activation of the composition

Ignition would take place if:

a. Ignition stimuli is strong and is provided even for a short time
b. Ignition stimuli is weak (not too weak) but is provided for a long time

However, it may be noted that ignition cannot take place under the following extreme conditions:

a. If a too strong ignition stimulus is provided for short duration, then only a few grains of composition may react violently (causing composition surface blown off) but the heat generated would not be sufficient to propagate the burning to the next layer.
b. If the ignition stimulus is too weak, then the heat influx would dissipate in the surrounding and composition layer would not achieve ignition temperature.

In general, when the ignition stimulus is weak, the probability of ignition is low while at higher stronger ignition stimulus values, the probability of ignition increases. Probability of ignition as a function of ignition stimulus, (also known as sigmoid curve or S-curve or ignition sensitivity curve) is shown in Figure 3.3.

This type of curves enables the ammunition designer to arrive at a certain ignition stimuli at which the ignition would not occur as a safety measure while at

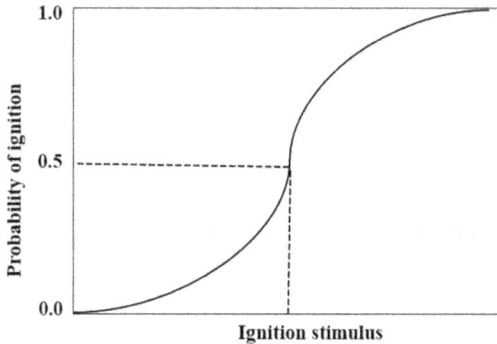

**FIGURE 3.3** Sigmoid Curve or S-Curve.

higher value of ignition stimuli, the ignition definitely would be initiated. It is simple to understand the functioning of stab caps or percussion caps on dropping weight from different heights for arriving at non-functioning and functioning of caps in sensitivity tests. At smaller heights, the stimulus energy of the falling weight is less and hence no cap functioning would occur while at much higher heights, the stimulus energy of the falling weight is much higher ensuring cap to function. This helps in designing the strength of striker force required for ignition. Similarly, another example is the electrical energy stimulus for maximum *"No fire current"(NFC)* and minimum *"All fire current"(AFC)* in squibs and impulse cartridges where varying electrical impulse energy is used to assess functioning or otherwise of the squibs and impulse cartridges and the required impulse is arrived . In view of the fact that there is a probability of functioning even at low stimulus,all pyrotechnic compositions need strong safety measures as ignition is possible in activities like transferring, filling or pressing of compositions.

Ignition of pyrotechnic composition may be discussed, based on concept of *Energy of Activation and Heat of Combustion*, as follows.

## 3.4 ENERGY OF ACTIVATION

The ingredients of pyrotechnic compositions are metastable. They all possess some energy but not enough to cause stretching, bending, distorting and finally breaking the bonds to form newer bonds. Thus, they do not react on their own. They need an external stimulus to increase their energy to break bonds and form newer bonds, i.e., initiation of ignition.

Energy of activation may be defined as *"the minimum additional amount of energy (from sources such as thermal energy, mechanical force induced energy, friction force induced energy or electrical induced energy including spark energy) absorbed by the composition enabling them to achieve energy equal to a threshold value, at which the combustion would take place "*

Activation energy can also be considered as the amount of thermal energy absorbed by the composition to make an exothermic reaction within a small portion of the composition, known as *"hotspot."*

It may, therefore, be concluded that the *energy of activation* is the difference between the threshold energy at which combustion takes place and the average energy of the reactant atoms/molecules of the ingredients of the composition and may be written as:

*Energy of activation*

= [*Threshold energy at which combustion takes place in the composition*]
−[*Average energy of the reactant atoms/molecules of the composition*]

(3.6)

At ambient temperatures, most of the reactant atoms/molecules of the composition have energy less than the threshold value, and hence no combustion takes place. Let us consider A and B as reactants and C and D as combustion products. As the input energy is provided, the energy content of the system increases till it attains its energy of activation when the *Transition state (or activated complex)* is formed following which combustion takes place forming the combustion products as shown in Figure 3.4.

The difference between the peak and the initial base energy level is *Energy* of *Activation* represented as $Ea$ while the difference in energy content from the initial base energy level and the final energy level is *heat of combustion*, represented as $(-) \Delta H$.

Let us now understand through Figure 3.5 as to how the energy of activation is achieved by the atoms/molecules of the composition. Each grain of oxidiser and fuel consists of billions of atoms and/or molecules. These individual atoms/molecules have certain internal energy. These internal energies are continuously being transferred from one atom/molecule to another atom/molecule of the same type.

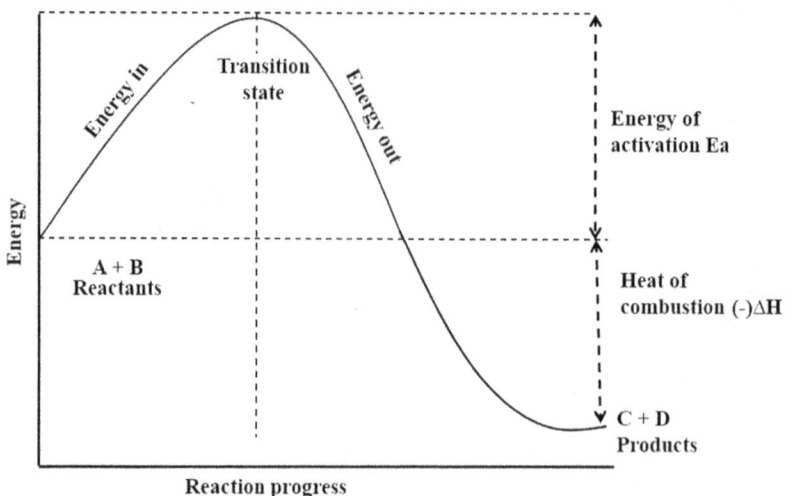

**FIGURE 3.4**  Energy of Activation and Heat of Combustion.

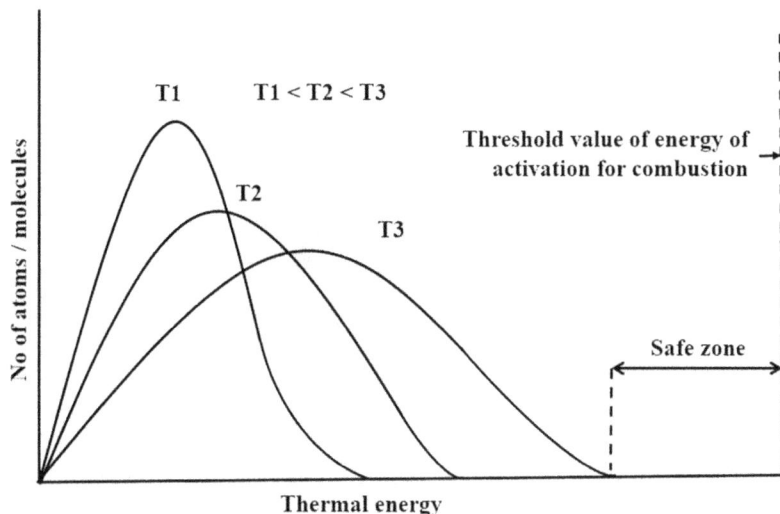

**FIGURE 3.5**    Energy Distribution in Atoms/Molecules.

Thus, an atom/molecule having high energy at a particular instance may have low energy due to transfer of its energy to another atom/molecule at another instance. The internal energy of the atoms/molecules in a small portion of a grain at extreme cold ambient temperature (like snow bound areas) is shown as curve T1. It could be seen that the internal energy distribution of atoms/molecules follow a normal distribution curve. The curve shows that the number of atoms/molecules with very low and as well as very high internal energies is very less in number and majority of the atoms/molecules are near mean energy levels.

As the ambient temperature is increased to T2 (medium ambient temperature) and then to T3 (extreme high ambient temperature like in deserts), the internal energy of all the atoms/molecules increases, as shown by curve T2 and T3. However, the increase in the ambient temperature is not sufficient to raise the internal energy level of the atoms/molecules to threshold value of energy of activation for combustion, as shown in the graph as a vertical dotted line away from the curve T1, T2 and T3. Hence, the atoms/molecules of the grain remain inactive (i.e., no combustion takes place) even at extreme ambient high temperatures.

This is a relief in safety during manufacture, storage and use of the pyrotechnic composition since pyrotechnic compositions are safe even at reasonably extreme high ambient temperatures. The higher the gap between the energy at extreme high ambient temperature and the threshold value of energy of activation, the more the composition is safe for drying. However, if this gap is less (i.e., the energy of activation of the composition is low), then the composition is considered very sensitive to heat stimulus and would need special safety precautions during drying, may be drying at lower temperature for more duration.

When external ignition stimuli having sufficiently high energy (in the form of flash from a fuse, safety match or squib or friction or percussion or stab) are

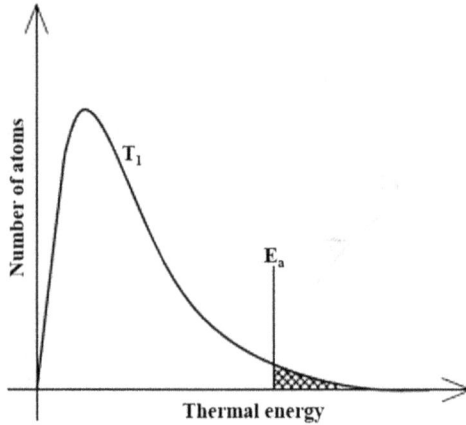

**FIGURE 3.6** Thermal Energy Distribution of Atoms in a Pyrotechnic Composition (Reprinted from [1]).

provided to the composition grain, the internal energies of atoms/molecules on the top surface of the composition increases substantially and exceeds the energy of activation as shown [1] in Figure 3.6.

The atoms/molecules in the unshaded portion of the curve have internal energies less than the energy of activation while those in the shaded portion have internal energies more than the energy of activation. These atoms/molecules in the shaded portion react by breaking their bonds and forming new and stronger bonds with formation of new combustion products and release of energy, known as heat of combustion. If only a few atoms/molecules react, then the heat of combustion would be extremely low and the combustion may not be self-sustained. For the combustion to proceed on self-sustained basis, it is necessary to have required heat of combustion. Military pyrotechnic compositions on ignition always have overall sufficient exothermic heat output to provide heat energy for further ignition in the next layer of the composition.

## 3.5   IGNITION TEMPERATURE

All pyrotechnic compositions need steady heat to increase internal energy of atoms/molecule to cause combustion. *The energy of activation is the minimum energy required for the atoms/molecules in the composition to react while the corresponding temperature at which a pyrotechnic composition (of specified shape, size and boundary constraints) starts igniting is termed as ignition temperature.*

The process of ignition can also be defined as *"the 'initiation of an exothermic reaction by increasing the temperature of a portion of the composition to a point where the reaction becomes self-sustained"*. The temperature at which this occurs is termed as ignition temperature of the composition.

Another definition of ignition temperature is that *"it is the lowest temperature at which the heat output is just sufficient to burn the next layer of composition"*.

Duane M. Johnson [2] has defined ignition temperature as *"that temperature to which a given solid composition must be heated to ignite and continue to combust without further heating by an external source."*

Schidlovskiy [3] has defined "self-ignition temperature" as *"the lowest temperature to which a composition must be heated so as to cause its spontaneous inflammation, associated with a clearly visible luminous, sound or smoke effect."*

Thus, no pyrotechnic composition ignites until the ignition temperature of the composition is achieved. Therefore, compositions which are difficult to ignite have higher ignition temperature while those compositions that are ignited easily have a low ignition temperature. For ignition to occur, the following are essential:

a. The composition must have its own achievable ignition temperature
b. The external stimulus (heat, flame, spark, friction and impact) should be able to initiate heating of composition
c. The composition must be heated to its ignition temperature

Ignition temperature is an important parameter for designing any new composition as well as for safety in processing and handling and storage. It also indicates that a composition with high ignition temperature requires a priming composition for ignition while a composition with low ignition temperature shall pose safety hazards..

### 3.5.1 FACTORS AFFECTING IGNITION TEMPERATURE AND ENERGY OF ACTIVATION

The ignition temperature and energy of activation depends upon the energy required for combustion. A higher energy of activation means lower sensitivity of composition.

The ignition temperature and activation energy of various pyrotechnic compositions differ from each other due to different ingredients as well as their proportion, particle size, moisture content, compaction density, etc. The compositions with organic materials have low ignition temperature while compositions with metal powders have a high ignition temperature.

Size and duration of hotspot (see Section 22.3.2) also affect the ignition temperature.

The three major factors that affect ignition temperature and energy of activation are specific surface of ingredients, decomposition temperature of oxidiser and melting point of fuel and additives. These are discussed below with some examples.

#### 3.5.1.1   The Higher the Specific Surface (Small Particles), the Lower shall be the Ignition Temperature

This may be extremely clear from the following report.

The relationship between the specific surface of magnesium powder and the spontaneous reaction temperature of the pyrotechnic formulation was studied by DTA-TG using five magnesium powders with different granularities and three typical oxidants [4]. The Mg/oxidiser ratios were kept at 63/37, 65/35 and 57/43 for oxidisers $KNO_3$, $Sr(NO_3)_2$ and $KClO_4$, making the composition zero oxygen

**TABLE 3.1**

**Average Spontaneous Reaction Temperature of Different Pyrotechnic Formulations [4]**

| Magnesium specific surface (m²g⁻¹) | Magnesium mesh | $KNO_3$ ($^0C$) | $Sr(NO_3)_2$ ($^0C$) | $KClO_4$ ($^0C$) |
|---|---|---|---|---|
| 25.7 | <100 | 589.1 | 583.8 | 566.5 |
| 44.38 | 200–325 | 579.0 | 577.8 | 563.5 |
| 54.35 | <200 | 577.5 | 577.2 | 558.5 |
| 78.88 | <325 | 579.8 | 553.6 | 549.4 |
| 97.99 | <400 | 573.1 | 566.1 | 543.0 |

balance. It is shown that for all three tested oxidants, the spontaneous reaction temperature of the pyrotechnic formulation steadily decreases as the specific surface area of the magnesium powder increases.

That is, with smaller average particle size of magnesium powder, the ignition temperature of the mixture becomes lower and the combustion reaction proceeds more easily (Table 3.1).

### 3.5.1.2 Decomposition Temperature of Oxidiser and Melting Point of Fuel

The Figure 3.7 shows the pressed composition of oxidiser and fuel where the oxidiser and fuel are partially in contact with each other. On providing suitable external stimuli, the internal energy of the oxidiser and fuel increases. However, when the oxidiser melts, it causes improved contact between the fuel and oxidiser

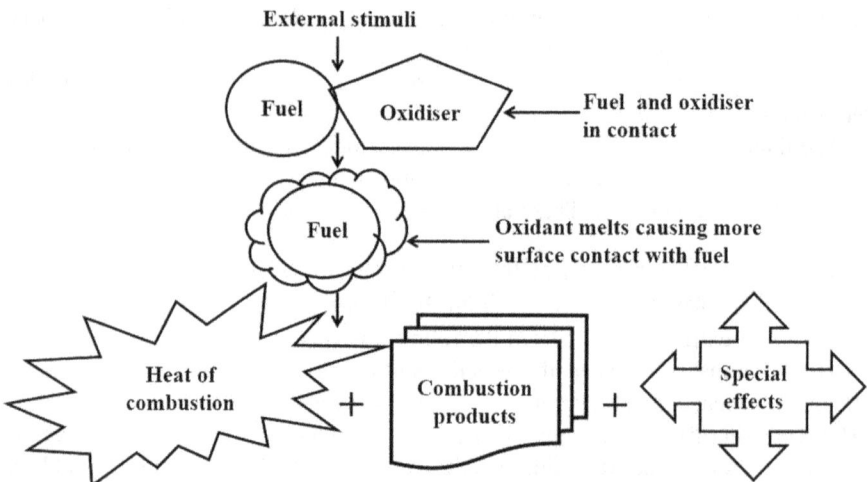

**FIGURE 3.7**   Ignition of Pyrotechnic Compositions.

and the sensitivity increases substantially due to more intimate contact. For the ignition to take place, it is necessary that in most of the cases, the oxidiser must first decompose followed by the combustible fuel. For example, in the case of magnesium and sodium nitrate combustion by laser ignition, the oxidiser melts followed by its decomposition, thereby releasing oxygen, which diffuses through the oxide layer on magnesium and reacts with magnesium. The combustion products move both inward and outward of the fuel surface. Some part of the heat of combustion is transferred to the surface of the metal causing more vaporisation of fuel while a part is lost to the atmosphere.

The decomposition temperature and the heat evolved (exothermic) or consumed (endothermic) in the process by the oxidiser and the melting point and boiling point of the combustible fuel determine the ignition temperature of the composition. Table 3.2 provides ignition temperature of some stoichiometric compositions of oxidisers potassium chlorate, potassium perchlorate and potassium nitrate and fuels sulphur, lactose, magnesium and aluminium.

The cause of varying ignition temperature may be understood from the heat of decomposition of these three oxidisers and the melting and boiling points of these four fuels (Table 3.3).

$KNO_3$ has endothermic decomposition (+)75.5 kcal. $mole^{-1}$ and hence takes away the heat from the system, thereby increasing the ignition temperature. However, $KClO_3$ has exothermic decomposition (–)10. 6 kcal. $mole^{-1}$ and hence releases heat and leads to early ignition, i.e., lower ignition temperature.

---

**TABLE 3.2**
**Self-Ignition Temperature of Binary Mixture in °C [3]**

| Oxidiser | Fuel | | | |
|---|---|---|---|---|
| | Sulphur | Lactose | Magnesium powder | Aluminium dust |
| Potassium chlorate | 220 | 195 | 540 | 785 |
| Potassium perchlorate | 560 | 315 | 460 | 765 |
| Potassium nitrate | 440 | 390 | 565 | 890 |

---

**TABLE 3.3**
**Heat of Decomposition of Oxidisers and Melting and Boiling Point of Fuels**

| Heat of decomposition (kcal/mole) | Sulphur | Lactose | Magnesium powder | Aluminium dust |
|---|---|---|---|---|
| Potassium chlorate (–) 10.6 | m.p. 119 °C | m.p. 202 °C | m.p. 649 °C | m.p. 660 °C |
| Potassium perchlorate (–) 0.68 | b.p. 445 °C | b.p. 669c °C | b.p. 1091 °C | b.p. 2467 °C |
| Potassium nitrate (+) 75.5 | | | | |

Similarly, a low melting point fuel gives lower ignition temperature (say sulphur with melting point $119^0C$ and lactose with melting point $202^0C$) while a higher melting fuel (say magnesium with melting point $649^0C$ and aluminium with melting point $660^0C$) gives higher ignition temperature.

Another example is ignition temperatures for sucrose with $KClO_3$ and $KClO_4$. When $KClO_3$ is replaced by $KClO_4$ as oxidiser of the fuel sucrose, the sensitivity of the mixture decreases and elevates the ignition temperature of the mixture from 472 $^0C$ to 592 $^0C$.

It may be inferred that higher sensitivity and lower ignition temperature are exhibited by compositions containing potassium chlorate than compositions containing potassium nitrate while least sensitivity and high ignition temperature are exhibited by compositions containing aluminium and magnesium due to their higher melting and boiling points. The organic fuels and sulphur lower the ignition temperature of the composition.

### 3.5.1.3   Some Additives Reduce Ignition Temperature

Dr Herbert Ellern [5] has given ignition temperatures for some pyrotechnic compositions and also where the ignition temperature is lowered considerably by addition of a small percentage of sulphur, rosin or asphalt (Table 3.4).

**TABLE 3.4**
**Ignition Temperatures of Some Pyrotechnic Compositions [5]**

| Pyrotechnic composition | Ignition temperature $^0C$ |
|---|---|
| $Zr/Fe_2O_3/SiO_2$ (A1A) | 300 |
| $Si/PbO_2/CuO$ | 540 |
| $Mg/BaO_2$ 12/88* | 570 |
| $Mg/PbO_2$20/80 | 600 |
| $Mg/PbCrO_4$ 5/95 | 620 |
| Delay Powders | |
| $Se/BaO_2$ | >265 |
| $Cr/BaCrO_4/KClO_4$ | 340 |
| $W/BaCrO_4/KClO_4$ | 430 |
| Zr-Ni Alloys/$BaCrO_4/KClO_4$ | 495 |
| $Mn/BaCrO_4/PbCrO_4$ (D-16) | 382–522 |
| $B/BaCrO_4$, 10/90% | 685 |
| $B/BaCrO_4$, 5/95% | 700 |

*Notes*

* Ignition point is depressed to 320°C, 360°C or 395°C, respectively, after addition of 2% of sulphur, rosin or asphalt.

**TABLE 3.5**

**Heat of Combustion, Autoignition and Decomposition Temperature of Pyrotechnic Compositions [6]**

| Pyrotechnics | Heat of combustion (cal. $g^{-1}$) | Autoignition temperature ($^0$C) | Decomposition temperature ($^0$C) |
|---|---|---|---|
| Initiatory | 2,619 ± 623 | 255 ± 96 | 277 ± 102 |
| Illuminants | 2,728 ± 1,514 | 497 ± 123 | 561 ± 135 |
| Smoke | 2,794 ± 887 | 180 ± 66 | 205 ± 75 |
| Gas | 2,261 ± 1,104 | 162 ± 16 | 182 ± 24 |
| Sound | 2,666 ± 789 | 506 ± 169 | 550 ± 168 |
| Heat | 1,746 ± 1,198 | 447± 199 | 505 ± 224 |
| Time (Delay) | 682 ± 222 | 448 ± 159 | 517 ± 153 |

Fred. L. McIntyre [6] has investigated 180 different types of pyrotechnic compositions and reported heat of combustion, autoignition temperature and decomposition temperature (Table 3.5).

The autoignition temperatures of some friction priming pyrotechnic compositions are given in Table 22.17.

### 3.5.1.4 Factors Affecting Ignition Temperature During Determination of Ignition Temperature

Determination of ignition temperature for the same composition can vary depending upon the quantity of composition, rate of heating the composition and degree of confinement of composition, ambient pressure, temperature and variation in characteristics of the composition.

A higher mass of sample would lead to low ignition temperature since a higher mass of sample will have less exposed surface area compared to its volume and hence less heat loss to the surrounding, leading to lower ignition temperature.

Ignition temperature depends upon the rate of heating and the type of composition. For example, a higher rate of heating would lead to higher ignition temperature for coloured smoke composition [6] (as per Figure 3.8 of DTA data) while a higher rate of heating reduces the ignition temperature for tracer priming composition containing magnesium, barium peroxide and acaroid resin. The standard heating rate for differential thermal analysis (DTA) is 5 °C.min$^{-1.}$

Ignition temperature also depends upon the degree of confinement of composition during testing. The ignition temperature of pyrotechnic compositions varies with type of ingredients. The compositions with metallic fuels give very high ignition temperature compared to organic fuels.

Thermal, kinetic and ignition behaviours of three pyrotechnic mixtures comprising Al + Ba $(NO_3)_2$, Mg + $NH_4ClO_4$ and Mg + $KMnO_4$ [7] on the effect of heating rate on the decomposition peak temperature are shown in Table 3.6.

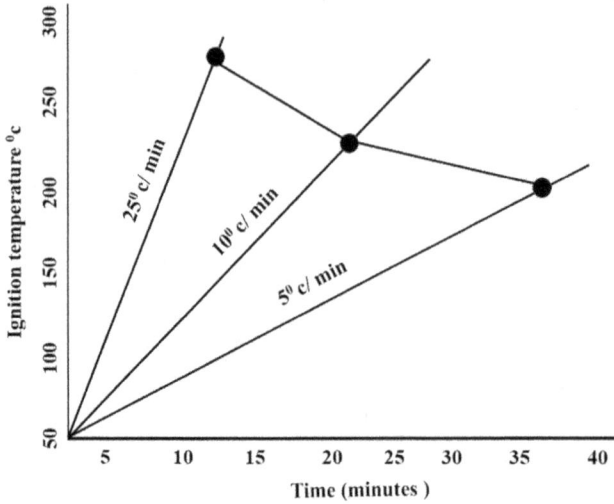

**FIGURE 3.8**   Effect of Rate of Heating on Ignition Temperature (Adapted from [6]).

**TABLE 3.6**
**Effect of Heating Rate on the Decomposition Peak Temperature of the Pyrotechnic Mixtures [7]**

| Composition | Heating rate ($^\circ$C.min$^{-1}$) | Peak temperature ($^\circ$C) |
|---|---|---|
| Al +Ba (NO$_3$)$_2$ | 10 | 600.6 |
| | 20 | 624.9 |
| | 30 | 641.6 |
| | 40 | 655.7 |
| Mg +NH$_4$ClO$_4$ | 10 | 338.1 |
| | 20 | 346.4 |
| | 30 | 352.5 |
| | 40 | 358.1 |
| Mg +KMnlO$_4$ | 10 | 292.5 |
| | 20 | 302.9 |
| | 30 | 310.8 |
| | 40 | 319.5 |

The above table shows that the decomposition peak temperature increases with increase in rate of heating.

The reaction proceeds as under [7].

$$2\ Al + Ba(NO_3)_2 = BaO + 2NO + Al_2O_3 \qquad (3.7)$$

$$5 \ Mg + 2 \ NH_4ClO_4 \rightarrow 5 \ MgO + 3 \ H_2O + 2 \ HCl + N_2 \qquad (3.8)$$

$$Mg + 2 \ KMnO_4 \rightarrow MgO + K_2O + 2 \ MnO_2 + O_2 \qquad (3.9)$$

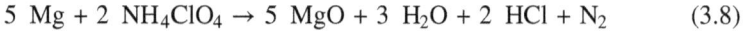

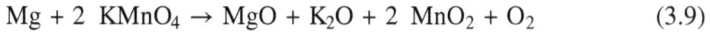

## 3.6 FACTORS AFFECTING TIME TO IGNITION (OR TIME LAG)

There is always a time lag for the composition to pick up ignition from external stimuli and burn independently without any external stimuli. This may be defined [2] as "*the time required to raise the temperature of the surface of the composition from some ambient temperature to the ignition temperature.*"

Consider a pyrotechnic composition kept in an oven with a higher temperature. The temperature of the composition shall begin to rise with time till it achieves the oven temperature. If the oven temperature T1 is below the thermal run-away temperature $T_r$ (see Section 3.11), the composition temperature would reach up to the oven temperature but ignition of the composition would not take place [1].

If the oven temperature is above the thermal run-away temperature $T_r$, the composition temperature would rise, leading to ignition of the composition. This rise in temperature is rapid at higher oven temperature. The same is shown in Figure 3.9 where the time to ignition is lowered as the oven temperature increases.

This time to ignition (or time lag) depends upon the various factors as given below.

a. *Ignition temperature*: The higher the ignition temperature, more time is required to achieve the ignition temperature.
b. *Strength of external stimulus*: The more the strength of external stimulus, the less the time to ignition.
c. *Thermal conductivity of the composition*: Higher the thermal conductivity, more thermal energy is conducted and distributed to the whole mass of

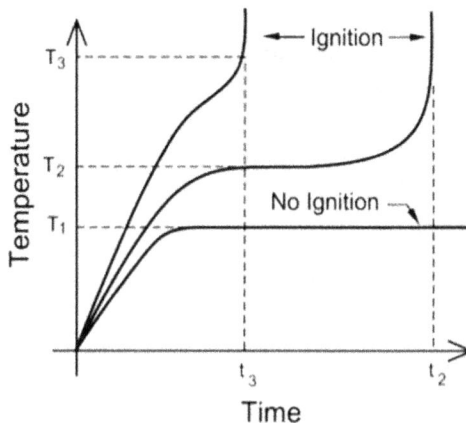

FIGURE 3.9 Time to Ignition (Reprinted from [1]).

composition, thereby less increase in temperature at the localised point of the composition and hence higher the time to ignition.

d. *Density of the composition*: Higher the density, more heat is dissipated to more mass of the composition, thereby less increase in the temperature of the composition at localised point and hence higher the time to ignition.

e. *Specific heat capacity of composition*: Higher the specific heat capacity, higher is the thermal energy required to increase the temperature of the composition and hence higher the time to ignition.

f. *Exposed cross-section area of composition*: Higher the cross-section area, more heat loss to the surrounding and hence higher the time to ignition.

g. *Energy loss to surrounding vis a vis energy feedback for sustaining burning*: More heat loss to surrounding would make the composition to take more time to achieve the ignition temperature.

h. *Moisture*: Higher the moisture content, higher the time to ignition since initial energy would be first utilised for expelling moisture content.

i. *Purity of ingredient*: The time to ignition is reduced with higher purity of ingredients.

j. *Container body material*: More heat loss is expected in compositions pressed in direct contact with metallic containers compared to composition filled in metallic containers with inner paper liner and hence higher the time to ignition.

k. *Ambient temperature*: Lower the ambient temperature, more time is required to attain ignition temperature.

l. *Ambient pressure*: Less ambient pressure would increase the time to ignition.

Ignition delays of the priming composition in inert countermeasure decoy flares [8] at very low pressures and at temperature extremes are shown in Table 3.7.

**TABLE 3.7**
**Ignition Delays at High Altitudes [8]**

| Altitude | Temperature ($^0$C) | Mean ignition delay (ms) |
|---|---|---|
| 0 feet | −40 | 115 |
| | 14 | 87.75 |
| | 100 | 42.75 |
| 20,000 ft | −40 | 440 |
| | 14 | 206.5 |
| | 100 | 84.2 |
| 40,000 ft | −40 | 333.5 |
| | 14 | 627 |
| | 100 | 499.6 |

The following approximate trend could be seen:

a. Increasing the temperature at constant altitude reduces the ignition delay
b. Increasing the altitude at constant temperature increases the ignition delay

It seems that pressure has a larger effect than temperature on the ignition delay; this would be expected as the gas generated at high altitude would diffuse and escape from the system quicker, resulting in less thermal impact.

Time to ignition of a pyrotechnic delay composition [2] is given as under.

$$t_{ig} = \frac{k\,\rho\,c\,A^2\left(T_{ig} - T_a\right)^2}{\pi e\,q_{in}^2} \tag{3.10}$$

where $k$ = Thermal conductivity, $\rho$ = Density
$c$ = Specific heat capacity of the sample
$T_0$ = Temperature, ($^0$K), the material starts at, $T_{ig}$ = Ignition temperature ($^0$K), $A$ = surface area, $q_{in}$ = Rate of heat input incident to the sample, i.e., the rate of heat energy transfer on the composition, $e$ = 2.71828

Since ignition energy $E = q_{in} \times t_{ig}$, (i.e., the product of rate of heat flow and time to ignition), the ignition energy requirement equation [2] is

$$E = \frac{k\,\rho\,c\,A^2\left(T_{ig} - T_a\right)^2}{\pi e\,q_{in}} \tag{3.11}$$

It could be seen from the above equation that the time to ignition and energy requirement increases with:

a. Increase in thermal conductivity
b. Increase in density
c. Increase in specific heat capacity
d. Increase of surface area
e. Increase in difference between ignition temperature and initial ambient temperature of the composition
f. Decrease of rate of heat input on the composition

The equation below is used to estimate ignition delay times caused by laser heating [9].

$$t_{ig} = \frac{\pi k\,\rho\,c\left(T_{ig} - T_0\right)^2}{4\,(AP/S)^2} \tag{3.12}$$

Where $k$ = Thermal conductivity, $\rho$ = Density, $c$ = Specific heat capacity, $T_{ig}$ = Ignition temperature, $T_0$ = Temperature at which the material starts at, $S$ = Area illuminated by the laser, $A$ = Absorptivity of the composition, $P$ = Specific ignition stimulus

This equation also explains that the time to ignition will increase with:

a. Increase in thermal conductivity (rapid dissipation of heat)
b. Increase in density (higher ratio of mass to heat)
c. Increase in specific heat capacity (more heat required to raise temperature)
d. Increase in surface area (large cross-sectional area illuminated by laser, leading to more heat loss)
e. Increase in difference between ignition temperature and initial temperature of the composition
f. Decrease of specific ignition stimulus
g. Decrease of absorptivity of the composition

Thus, for any composition, the time to ignition for a composition depends upon the initial start-up temperature, as all other factors referred above are predetermined in the ammunition. At very low start-up temperature (like in very cold environment), the time to ignition of composition is expected to be more since the external stimulus shall take more time to raise the temperature of the surface of the composition to its ignition temperature. At higher start-up temperature (like in a very hot environment), the time to ignition of composition would be less.

The above equations also reveal the ignition requirements of priming composition (transfer mix or first-fire composition) [2] on following aspects.

a. Ignition by caps or squibs produces intense flash for very short duration of around 250 milliseconds. Hence priming pyrotechnic should have low ignition temperature (low energy requirement for ignition, i.e., low activation energy) for quick ignition.
b. It is desirable that the priming composition should have as low thermal conductivity, low density, low specific heat and low ignition temperature values as possible, all leading to low time to ignition.

The factors affecting time to ignition are shown in Table 3.8.

## 3.7  HEAT OF COMBUSTION

All high energy materials have atoms bonded together. When these bonds are broken during reaction, energy is absorbed. Thus, bond-breaking is an endothermic process. However, energies are released during formation of new bonds of the product. Thus, bond-making is an exothermic process.

The new arrangement of bonds in the products does not have the same total energy as the bonds in the reactants. Since the bond energies of the products differ from the bond energies of the reactants, there is always a difference of energy when a reaction takes place. This difference in energy is termed as the heat of reaction.

**TABLE 3.8**
**Factors Affecting Time to Ignition**

| Parameters | Time to ignition |
|---|---|
| Higher ignition temperature | Higher |
| Higher thermal conductivity of composition | Higher |
| Higher density of composition | Higher |
| Higher specific heat capacity of composition | Higher |
| Higher cross section area of composition | Higher |
| Higher heat loss to surrounding | Higher |
| Higher conducting metal container body in contact with composition | Higher |
| Higher moisture | Higher |
| Higher initial start-up temperature | Lower |
| Higher ambient pressure | Lower |
| Higher ambient temperature | Lower |

*Heat of reaction*

= *Total energy absorbed when old bonds are broken in the reactants*

− *Total energy released when new bonds are formed in the products*

$$(3.13)$$

It is expressed in kilocalories per mole (or kilojoules per mole). Whether a reaction is endothermic or exothermic depends upon the difference between the energy required to break the bonds and the energy released on formation of new bonds.

a. If the total energy required to break bonds in the reactants is more than the total energy released when new bonds are formed in the products, it is an endothermic reaction.

b. If the total energy required to break bonds in the reactants is less than the total energy released when new bonds are formed in the products, it is an exothermic reaction.

An example would make it clear. Consider the simple reaction of hydrogen and iodine below

$$H_2(g) + I_2(g) \rightarrow 2HI(g) \qquad (3.14)$$

The average bond energies are H-H = 436 KJmol$^{-1}$ and I-I = 149 KJmol$^{-1}$. Hence the reactant side bond energy is 436 KJmol$^{-1}$ + 149 KJmol$^{-1}$ = 585 KJmol$^{-1}$. This amount of energy is needed to break the bonds of one mole of H-H and one mole of I-I.

Similarly, average bond energy H-I = 295 KJmol$^{-1}$. Hence bond energy released in formation of two H-I moles on product side shall be $2 \times 295$ KJmol$^{-1}$ = 590 KJmol$^{-1}$.

The energy difference $\Delta H$ = 585 KJmol$^{-1}$ −590 KJmol$^{-1}$ = (−) 5 KJmol$^{-1}$. (−) indicates that the reaction is exothermic with release of energy to surroundings.

We may understand exothermic and endothermic processes from simple activities surrounding us. The making of ice cubes and formation of snow in clouds is due exothermic process where heat is taken off the system while melting of ice cubes and conversion of frost to water vapour is an endothermic process where heat is absorbed by the system. Similarly, freezing, condensation and alkali–acid reactions are exothermic processes while melting, boiling and sublimation are endothermic processes where heat is absorbed.

The heat of combustion $\Delta H$ in (−) indicates an exothermic reaction where heat is produced during combustion, raising the surrounding temperature while the heat of combustion $\Delta H$ in (+) indicates endothermic reaction where heat is absorbed during combustion, thereby reducing the surrounding temperature.

In military pyrotechnic combustion, several endothermic and exothermic reactions take place. But the overall effect is the release of energy. In other words, in pyrotechnic compositions, the energy needed for the combustion to occur is less than the total energy released. As a result of this, the extra energy (−) $\Delta H$ is released and utilised partly as energy feedback for sustaining burning (i.e. ignition of next unburnt layer of composition), a part is lost to environment and a part energy is used up as special effects.

The heat of reaction may be calculated from the heat of formation of products and reactants as under, using enthalpy data when the reaction products are known or assumed (Hess law).

*Heat of reaction*

= *Sum of heat of formation of products (from its constituent elements)*

  − *Sum of heat of formation of reactants (from its constituent elements)*

(3.15)

The gross heat of reaction in terms of calories per gram is determined [6] by burning 1 to 2 g sample of pyrotechnic mixture in an inert atmosphere (nitrogen) in the standard bomb calorimeter submerged in water and recording the rise in water temperature.

The gross heat of combustion is measured [6] by burning 1 to 2 g samples of pyrotechnic mixture in an oxygen-filled (5 atmospheres) standard calorimeter bomb submerged in water and recording the rise in water temperature.

Heat of combustion of different pyrotechnic composition is different since the bond energies of reactants and products are different. Heat of combustion plays very important role in ignition and special effects of pyrotechnic compositions. Heat of combustion needs to be modified for various types of compositions by change of ingredient characteristics and processing. A higher heat of combustion needs special safety precautions in manufacture and handling of composition.

### 3.7.1 Factors Affecting Heat of Combustion in Pyrotechnic Compositions

The following factors affect the heat of combustion in pyrotechnic compositions:

a. *The type of ingredients and their proportions in pyrotechnic compositions* (e.g., flare composition and incendiary composition, containing metal fuels, release higher heat energy compared to signalling colour smoke composition containing lactose or sucrose as fuel. Similarly, compositions containing oxidisers like perchlorates give more heat of reaction than nitrates).

b. *The rate of combustion of pyrotechnic composition* (which determines the amount and rate of release of heat energy from the pyrotechnic composition).

c. *The combustion products of pyrotechnic composition and their quantities* (which may further react to produce heat or absorb heat and their temperatures)

It is desirable that the flow of heat is concentrated on a smaller area of the priming composition so that *"hotspots"* may form early, thereby reducing the energy requirement for combustion.

The summarised data for heat of combustion and heat of reaction of 180 pyrotechnic compositions are given in Table 3.9. The heat of reaction is generally lower than heat of combustion.

## 3.8 PROPAGATION OF COMBUSTION

Initially the atoms or molecules of ingredients in a composition having their own internal energy are in metastable form and do not react on their own. However, when an external ignition stimulus like flash or heat or friction is provided to the composition, their internal energy increases upwards till ignition temperature is achieved. At this point, the composition top fine layer gets ignited. The solid and liquid combustion products remain in the reaction zone while gaseous combustion

**TABLE 3.9**

**Heat of Combustion and Heat of Reaction [6]**

| Pyrotechnics | Heat of combustion (cal. $g^{-1}$) | Heat of reaction (cal. $g^{-1}$) |
|---|---|---|
| Initiatory | 2,619 ± 623 | - |
| Illuminants | 2,728 ± 1,514 | 1,475 ± 287 |
| Smoke | 2,794 ± 887 | 983 ± 319 |
| Gas | 2,261 ± 1,104 | - |
| Sound | 2,666 ± 789 | 933 ±112 |
| Heat | 1,746 ± 1,198 | 830 ± 495 |
| Time (Delay) | 682 ± 222 | 299 ± 101 |

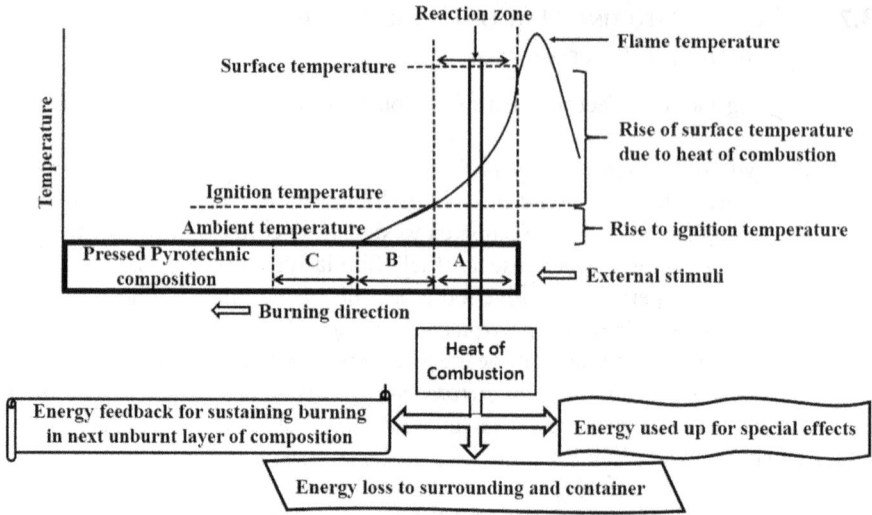

**FIGURE 3.10**   Burning of a Pressed Pyrotechnic Composition.

products move away from the combustion zone. The heat of combustion is used up as energy loss to surrounding and the container, energy used up for special effects and energy feedback for sustaining the burning process of the next layer.

Let us consider a pressed pyrotechnic composition which is theoretically sliced into fine layers "a," "b" and "c" where layers of a composition have been enlarged to show the combustion process in the layer (Figure 3.10).

a. Composition layer "a" where combustion is taking place showing energy distribution as energy being used for special effects, energy loss to surrounding and container and energy feedback for sustaining burning in next unburnt layer of composition.
b. Composition layer "b" where combustion is under initiation through "energy feedback for sustaining burning" by layer "a" by heat transfer through conduction, convection and radiation and formation of pre-ignition zone/hotspots.
c. Composition layer "c" which is unreacted consolidated composition awaiting "energy feedback for sustaining burning" by layer "b."

Thus, each new layer of composition passes through all the above three phases in reverse order (i.e., from layer "c" to layer "a") till entire composition is exhausted.

## 3.9   FLAME TEMPERATURE

The maximum flame temperatures of various pyrotechnic compositions are given in Table 3.10.

**TABLE 3.10**

**Function of a Composition and Maximum Temperature in the Flame [3]**

| Pyrotechnic composition | Maximum temperature in the flame (°C) |
|---|---|
| Smoke composition | 400–1,200 |
| Signal night | 1,200–2,000 |
| Illuminating and tracer | 2,000–2,500 |
| Rocket (solid fuel) | 2,000–2,900 |
| Incendiary (w/oxidiser) | 2,000–3,500 |
| Photo illuminating | 2,500–3,500 |

It could be seen that smoke composition has least flame temperature while incendiary and photoflash compositions have highest flame temperature.

The temperature of the pyrotechnic flame depends upon the distance of flame from the burning surface. The flame (due to hot gaseous products) nearer to burning surface will have a high flame temperature and the farthest from the burning surface will have the low flame temperature. The flame does not extend to a very long distance since the low temperature at far distance from the burning surface does not provide enough excitation energy to electrons for electromagnetic radiation in the visible range and, therefore, the hot gaseous products cool down. The movement of flame is due to gaseous products moving away from the burning surface.

Relationship between pressure and temperature for some typical pyrotechnic ignition compositions, ranging from black powder through percussion primer compositions (PA101, M42F1), igniter compositions (MTV, B/KNO3, Zr/KClO$_4$) and a delay train output composition (A IA) have been reported [10]. The approximate temperatures of some pyrotechnic ignition compositions, as computed from graph, at 40 atm. is given in Table 3.11. It could be seen that Zr/KClO$_4$ has the highest burning temperature while black powder the lowest.

## 3.10   TEMPERATURE DEPENDENCE OF RATE OF COMBUSTION

It could be seen that as the temperature of the composition is increased, a large number of atoms/molecules obtain energy higher than energy of activation. On achieving energy higher than the energy of activation of the composition by these atoms/molecules, these react to form combustion products. See Figure 3.11 for T1 where a few atoms/molecules (shown in shaded portion) have got energy of activation. If the temperature is further raised, say T2, then more atoms/molecules (shown under shaded portions) would gain energy higher than energy of activation and react, thereby rendering the combustion rapid [1].

Since pyrotechnic reactions are exothermic in nature, these produce heat of combustion and thus raise the temperature of the composition. An increase in thermal energy (temperature) further causes increase in rate of combustion as more

**TABLE 3.11**

**Temperature of Some Typical Pyrotechnic Ignition Compositions [10]**

| Composition | Ingredient | Percentage | Approximate temperature $^0C$ Computed at 40 atm. |
|---|---|---|---|
| MTV | Magnesium/ Teflon /Viton | 54 /30 /16 | 2,400 |
| A1A | ZirconiumIron/ Iron (III) oxide/ Diatomaceous earth | 65/ 25/ 10 | 4,400 |
| B/KNO$_3$ | Boron/ Potassium nitrate/ Laminac binder | 24/ 71/ 5 | 3,100 |
| Zr/KClO$_4$ | Zirconium/ Potassium perchlorate/ Graphite | 46/ 53/ 1 | 4,400 |
| Black powder | Potassium nitrate/ Charcoal/ Sulphur | 75/ 15/ 10 | 1,600 |
| PA 101 | Basic lead styphnate/ Barium nitrate/ Tetracene/ Aluminium powder/ Antimony sulphide | 53/ 22/ 5/ 10/ 10 | 2,500 |
| M42F1 | Boron/ Lead oxide/ Tetracene | 9.5/ 85.5/ 5 | 2,300 |

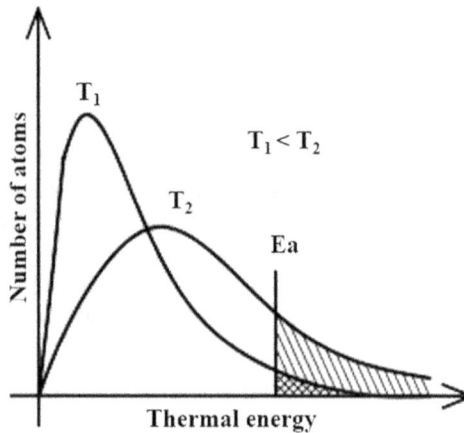

**FIGURE 3.11** Effect of Increase of Temperature on Thermal Energy Distribution of Atoms/Molecules (Reprinted from [1]).

atoms/molecules get activation energy though there is energy loss to the surroundings and energy used up for special effects.

Readers may refer [11] on increase in burn rate of titanium-based pyrotechnic composition with increase in burning front temperature.

a. *Arrhenius Equation:*

The temperature dependence of rate of reaction can also be understood by *Arrhenius equation*. Arrhenius equation is

$$k = A \ e^{-Ea/RT} \qquad (3.16)$$

Where $k$ = Rate constant ($s^{-1}$), $A$ = Pre-exponential constant or Arrhenius constant (Frequency factor), $Ea$ = Energy of activation ($Jmol^{-1}$), $R$ = Universal gas constant ($8.314 \times 10^{-3}$ $JK^{-1}mol^{-1}$), $T$ = Temperature in absolute $^0K$

As per Arrhenius equation on chemical kinetics [12], the rate of chemical reactions ($cm.sec^{-1}$) increases with rise in temperature since rise in temperature increases energy. It is therefore possible to derive the energy of activation from the rate of reaction at various temperatures.

The Arrhenius equation may be written in "*linearised*" form as

$$Ln \ k = Ln \ A - \frac{Ea}{RT} \qquad (3.17)$$

We may now deduce that the Arrhenius constant or pre-exponential constant ($k$) increases with increase in temperature ($T$) and hence the rate of reaction at higher temperatures will be high and for most chemical reactions, for every 10 $^0C$ rise in temperature, the value of $k$ gets almost doubled.

At extreme high temperature, extrapolation of the line on y-axis (Figure 3.12) gives the value of

$$Ln \ k = Ln \ A \ or \ k = A \qquad (3.18)$$

(The same ($k = A$) at higher temperature can be arrived at by considering the Arrhenius equation $k = A \ e^{-Ea/RT}$ where the value $Ea/RT$ approaches zero as temperature $T$ increases substantially and since $e^0 = 1$ and hence at extreme high temperature, $k = A$)

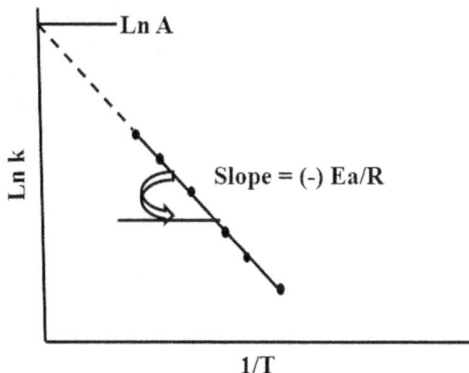

**FIGURE 3.12**   Energy of Activation Calculation Through Graph.

b. *Energy of Activation*: The above Arrhenius equation can also be written as

$$Ln\ k = (-)\frac{Ea}{RT} + Ln\ A \tag{3.19}$$

This is similar to standard equation $y = (-m)\ x + c$ which is a straight line with

$$Slope\ (m) = (-)\frac{Ea}{R} \tag{3.20}$$

Thus, an experimental graph may be plotted with $Ln\ k$ versus $1/T$ to determine the slope (Figure 3.12).

$$Slope\ in\ the\ graph = \frac{\Delta Ln\ k}{\Delta(1/T)}$$

$$Hence,\ Ea = (-)R\frac{\Delta Ln\ k}{\Delta(1/T)} \tag{3.21}$$

The energy of activation may also be calculated by knowing the Arrhenius constant or pre-exponential constant at two different temperatures, using Arrhenius integrated form

$$Lnk_2/k_1 = (-)Ea/R(1/T2-1/T1) \tag{3.22}$$

Another method is by the use of *"time to ignition"* (see Section 5.1.4) as the "time to ignition" ($t_{ign}$) may also be represented on the lines of Arrhenius equation as under [13].

$$t_{ign} = A.\ \exp(-Ea/RT) \tag{3.23}$$

where

$t_{ign}$ = Time to ignition at the temperature in degree absolute, $Ea$ = activation energy, a constant,

$E$ = universal gas constant, $A$ = a constant depending upon the material.

Hence it is possible to calculate the energy of activation on similar lines. It may be noted that time to ignition (sec.) becomes lower as the temperature ($^0K$) increases. However, the Arrhenius equation is not fully applicable for some delays at extreme temperatures.

## 3.11   THERMAL RUN-AWAY TEMPERATURE

A composition on heating causes rise in its temperature, i.e., a gain of heat energy. As the composition is heated further, it results in pre-ignition of a few grains of the composition at certain temperature and evolution of heat thereby gaining heat. However, a part of this heat energy is lost due to various factors. Let us understand this heat gain and loss which affects thermal run-away temperature.

### 3.11.1 Heat Loss

The heat loss is due to a number of factors like:

a. Heat loss to surrounding through convection and radiation, which will be more if the difference between pyrotechnic temperature and ambient temperature is high
b. Heat loss to other parts of the composition due to conduction for compositions containing metallic fuels and the increase will be more in compacted compositions
c. Heat loss to the metallic container body when in contact (hence paper tube inside metallic containers or suitable non-conductive coating of inner walls of metallic containers are given to prevent heat loss)
d. Heat loss due to very shape of the pressed composition, more surface area leading to more heat loss
e. Heat loss is more in pressed composition than in loose composition as heat may be easily conducted in pressed composition due to thermal conductivity.
f. Heat loss due to quantity of composition (higher the quantity, lower the heat loss since with large quantity of composition, the heat gained cannot escape easily to surrounding).

This heat loss is governed by the *Newton's Law of Cooling*.

Newton's Law of Cooling states *"the rate of cooling of any heated body at ambient temperature is proportional to the difference between the temperature of the body and that of the surrounding ambient temperature."*

Let us consider a pyrotechnic composition with ambient temperature $T_a$°C. Let the temperature of the pyrotechnic composition at a particular time "$t$" be $T_t$ ($T_t > T_a$). By Newton's Law of Cooling,

$$dT_t/dt = -k\,(T_t - T_a) \tag{3.24}$$

where $k$ is a positive proportionality constant and depends upon the type of composition and its volume, surface area, heat capacity, geometry and heat transfer coefficient, etc. The negative sign in the above equation may be noted since the temperature of the composition is decreasing (i.e., heat is removed) and thus the rate of change of temperature is negative.

The heat loss will be zero at the ambient temperature when $T_t = T_a$. The loss increases with increase in temperature of pyrotechnic composition as per above equation.

### 3.11.2 Heat Gain

The rate of heat gain can be arrived at from the Arrhenius equation on the rate of combustion of a pyrotechnic composition

$$k = A \ e^{-Ea/RT} \tag{3.25}$$

The rate of heat gained by the composition during combustion shall be the product of rate of combustion and the heat of combustion ($\Delta H$) as under.

$$\text{Rate of heat gained} = k \times \Delta H$$

$$\text{or Rate of heat gained} = A \ e^{-Ea/RT} \times \Delta H \tag{3.26}$$

Further, since heat of combustion ($\Delta H$) increases with increase in temperature of the composition, the rate of heat gained would therefore increase with increase in temperature. However, the graph is not a straight one due to exponential component in the equation.

Hence, if a graph is plotted for rate of heat gained against temperature, it would exhibit an exponential increase in heat gain with increase in temperature.

Both heat loss and gain occur simultaneously as the composition temperature is increased. But as the temperature of the composition is further raised, at a particular temperature (*thermal run-away temperature*), the heat loss would become lower than the heat gained by the composition, because, as the temperature increases, the rate of heat loss increases linearly but the rate at which heat is gained increases exponentially. Beyond this thermal run-away temperature, the composition will continue to increase in temperature due to self-sustained ignition. Furthermore, the rate of combustion will increase at higher temperature since heat gained would be substantially higher than the heat loss. Figure 3.13 [1] gives the rate of heat gain/loss against temperature.

Thermal run-away temperature $T_r$ may be defined as "*the minimum temperature to which the pyrotechnic composition must be raised at which the rate of heat gained is just more than the rate of heat loss to the surrounding*"

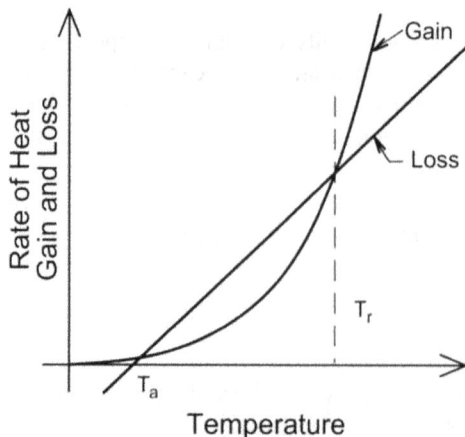

**FIGURE 3.13**   Thermal Run-Away Temperature (Reprinted from [1]).

It may thus be understood that the composition gains increase in temperature due to external stimuli leading to combustion of a few grains, but combustion becomes spontaneous at thermal run-away temperature (even after removal of external heat stimuli) where heat is continuously gained by the composition leading to sustained ignition.

Thermal run away depends upon:

a. Energy of activation (Lower energy of activation gives lower thermal run-away temperature)
b. Heat of combustion (Higher heat of combustion gives lower thermal run-away temperature)
c. Quantity of composition (Higher quantity of composition gives lower thermal run-away temperature)

The thermal run-away temperature also varies for the same composition when in loose form or when in pressed form and also depends on the ammunition system and the environment as heat loss/gain also depends upon these parameters.

It may be concluded that an exothermic combustion leads to thermal run-away, which starts when the heat produced by the pre-ignition combustion just exceeds the heat dissipated.

## 3.12  ESSENTIAL REQUIREMENT FOR PYROTECHNIC COMBUSTION

Depending upon the type of pyrotechnic composition, the combustion results in formation of various combustion products, in the form of solids, liquids, gases and vapours or a mixture of them, accompanied by the heat of combustion and are present on the burning surface or flame or nearby the flame. Combustion products may also react to produce heat of combustion. The heat of combustion ($\Delta H$) is used up as follows.

a. Some part of heat of combustion ($E_s$) is used up as energy in its special effects like flash, visible radiation, infrared radiation, sound, smoke and gas (subscript "s" signifies special effect).
b. Balance part of heat of combustion ($E_{fb}$) is used as feedback energy for sustaining the burning of the pyrotechnic composition from layer to layer (subscript " "fb" signifies energy feedback).
c. Some part of heat of combustion ($E_{loss}$) is lost (subscript "$loss$" signifies energy loss) to the atmosphere and metallic container body if in contact with the composition. The rate of conductive and convective heat loss is roughly proportional to the difference in temperature between the combustion temperature and the surroundings. Radiative heat loss is proportional to the difference in absolute temperature to the fourth power.

First two activities (a) and (b) are required for any pyrotechnic ammunition. However, this dissipation of energy to surroundings must be avoided by suitably

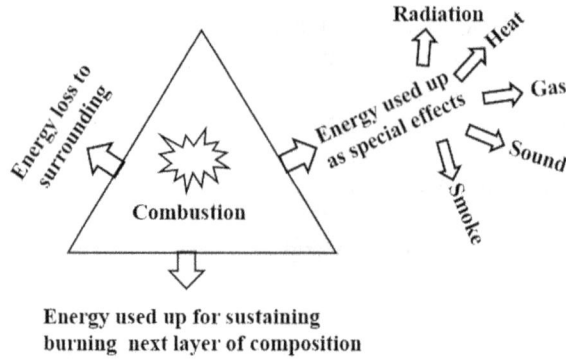

**FIGURE 3.14**   Distribution of Heat of Combustion.

designing the ammunition since higher dissipation of energy to surrounding may not allow sustaining the burning.

As an example, in illuminating compositions, composition is filled in a container assembled with paper liner or gap between candle and metallic container is filled with catalysed resin and dried to avoid dissipation of heat to surrounding through the metallic container and thus ensuring proper layer by layer combustion (see Section 4.3.2). Energy distribution of heat of combustion is shown in Figure 3.14

The energy distribution in combustion of a pressed pyrotechnic composition in a container is shown in Figure 3.15 with exaggerated layer of composition for ease of understanding.

The heat of combustion ($\Delta H$) may be written as

$$\Delta H = E_{loss} + E_s + E_{fb} \tag{3.27}$$

or $E_{fb} = \Delta H - (E_{loss} + E_s)$

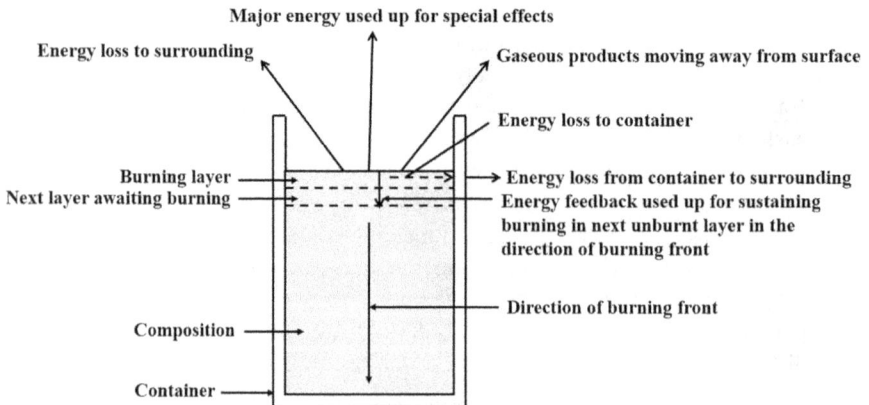

**FIGURE 3.15**   Energy Distribution in Combustion of Composition in a Container.

but since energy loss is the sum of energy loss due to conduction, convection and radiation, it can be written

$$E_{fb} = \Delta H - (E_{loss-cond} + E_{loss-conv} + E_{loss-rad} + E_s) \qquad (3.28)$$

Where $E_{loss-cond}$ = Energy loss due to conduction, $E_{loss-conv}$ = Energy loss due to convection and $E_{loss-rad}$ = Energy loss due to radiation.

This energy feedback for sustaining burning [ $E_{fb}$] must be more than the activation energy $E_a$ to raise the next layer of composition to its ignition temperature, i.e., $E_{fb} > E_a$.

In other words, even if the external ignition stimuli source is discontinued, the next layer would continue to burn only if this energy [ $E_{fb}$] is more than the energy of activation of the composition so that combustion takes place again in the next layer which will again produce combustion products and release heat of combustion and the process will go on till all the composition layers are exhausted. Figure 3.16 shows the process of combustion of pyrotechnic compositions.

Summing up, the sustained burning in a pressed pyrotechnic composition will proceed from layer to layer provided,

a. Combustion takes place in the layer of the composition generating sufficient energy and with its rate of combustion to such a level so as to offset the energy loss to the surrounding and energy loss to special effects during the burning of the composition.
b. The balance energy feedback for sustaining the burning process is more than the energy of activation of the next layer of composition.

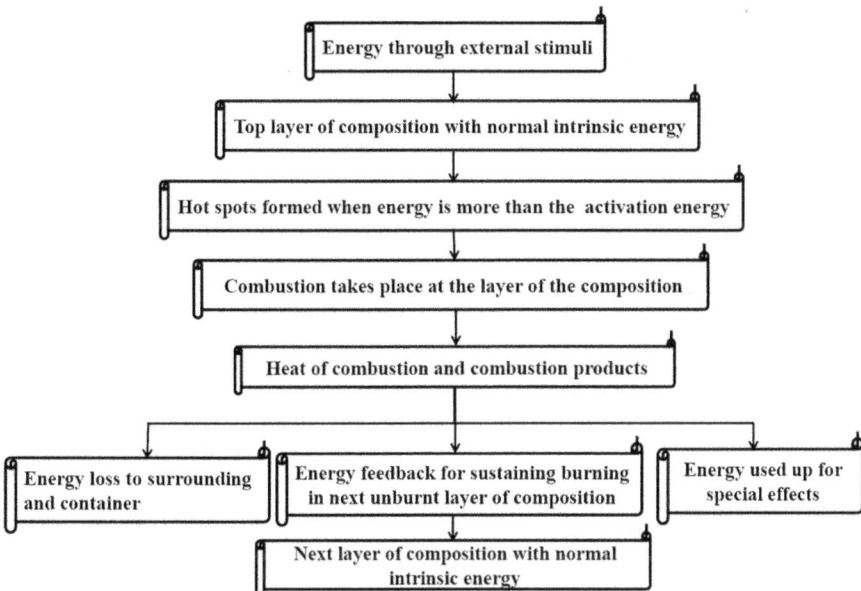

**FIGURE 3.16** Process of Combustion of Pyrotechnic Composition.

Rate of burning of pyrotechnic composition depends on three major factors:

a. Energy of activation (lower the better)
b. Heat of combustion (higher the better)
c. Energy feedback for sustaining burning to the next layer of composition (higher the better)

It may be stated that a composition having lower energy of activation, higher heat of combustion and higher feedback energy for sustaining burning to the next layer will have faster rate of combustion. Conversely, a composition having higher energy of activation, lower heat of combustion and lower feedback energy for sustaining the burning process to the next layer, shall have slower rate of combustion, subject to condition that $E_{fb} > E_a$.

Table 3.12 provides the role of these three parameters in the rate of combustion of a pyrotechnic composition.

## 3.13 IGNITION TRANSFER

*The route or path, through which ignition takes place, is termed as ignition train and the heat transfer is termed as ignition transfer.* Pressed pyrotechnic composition combustion works based on sequence of energy transfer from layer to layer of pressed pyrotechnic composition through any, some or all modes of heat transfer, i.e., conduction, convection and radiation. Any disruption in the energy transfer will lead to failure of the ignition train. Hence, the study of energy transfer for combustion is very important. The heat transfer to next layer may be provided by any, some or all sources, namely

a. Hot solid combustion products
b. Hot liquids/molten slag

**TABLE 3.12**
**Role of *Ea*, $\Delta H$ and $E_{fb}$ on Combustion Rate**

| Quantum of *Ea*, $\Delta H$ and $E_{fb}$ | Combustion rate |
|---|---|
| High energy of activation [$E_a$] | Low combustion rate |
| High heat of combustion [$\Delta H$] | High combustion rate |
| High energy feedback for sustaining burning [$E_{fb}$] | High combustion rate |
| Low energy of activation [$E_a$] | High combustion rate |
| Low heat of combustion [$\Delta H$] | Low combustion rate |
| Low energy feedback for sustaining burning [$E_{fb}$] | Low combustion rat |
| Energy feedback for sustaining burning [$E_{fb} < E_a$] | No combustion |

c. Hot gases/vapours

d. Hot flame

Let us discuss the three modes of heat transfer as under.

### 3.13.1 CONDUCTION

*"It is the transfer/flow of energy in a substance from a higher temperature region to a lower temperature region by means of atomic/molecular excitation without any net external motion."*

In a pyrotechnic composition, conduction occurs when a temperature gradient exists when the layer of composition is at a higher temperature to its next layer of composition.

Conduction takes place due to temperature gradient caused by the burning surface of pyrotechnic composition. In this process, energy is transferred from higher energy atoms/molecules to lower energy atoms/molecules in the compacted composition. Conduction is more if the fuel ingredient is metal and compaction is high and also through the metallic container used for filling/ pressing the composition provided the composition is in direct contact with the metal container.

Conduction follows the Fourier's law of conduction which states that *"the rate of heat flow (dq/dt) through a homogeneous solid is directly proportional to the area (A) of the section at right angles to the direction of heat flow, and to the negative temperature gradient (dT/dx) along the path of heat flow"* [14]. Fourier's law states

$$dq/dt = (-)kA(dT/dx)cal.\ sec^{-1} \tag{3.29}$$

where $q$ = Quantity of heat (cal), $t$ = time (sec), $dq/dt$ = Heat flow rate by conduction (cal.sec$^{-1}$), $k$ = Coefficient of thermal conductivity of the material (cal sec$^{-1}$ cm$^{-2}$) (°C/cm)$^{-1}$) , $A$ = Area for heat transfer (cm$^2$), $T$ = Temperature (°C), $x$ = Distance (cm), $dT/dx$ = temperature gradient across the material (°C cm$^{-1}$) (Figure 3.17).

In a steady state $dq/dt = q$ (heat flow rate)

Hence,

$$q = kA(T2 - T1)/x\ cal. \tag{3.30}$$

The factors affecting the heat transfer by conduction are:

a. *Thermal conductivity*: Higher for higher thermal conductivity of the ingredients

b. *Surface area*: Higher for higher surface area of composition containing conducting ingredients

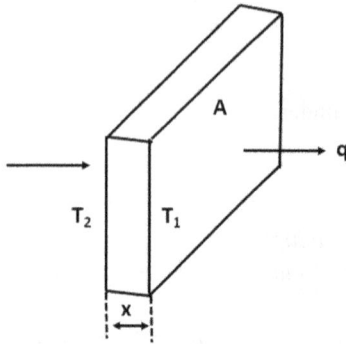

**FIGURE 3.17**   Conduction of Heat(based on [14]).

c. *Temperature gradient*: Higher for higher temperature gradient
d. *Thickness of composition*: Higher for lower thickness of composition
e. *Pressed composition*: Higher for pressed composition with metalic fuels since grains are more near each other

Conductivity varies from material to material, being higher for metals, lower for non-metals, very low for liquids, and extremely low for gases. The good metal thermal conductors used in pyrotechnics in the decreasing order are copper, aluminium, tungsten, magnesium, molybdenum and iron (see Table 2.2).

### 3.13.2   Convection

*"It is the transfer of energy by mass motions of hot fluids (gases/hot vapours/liquid) during combustion resulting in actual transport of energy in the pyrotechnic composition"*

The amount of heat transferred by convection is [14]

$$q = h_t A\ \Delta T,\ cal \tag{3.31}$$

Where $q$ = quantity of heat transferred (cal), $h_t$ = Heat transfer coefficient (cal cm$^{-2}$ $^0$C$^{-1}$), $A$ = Area for heat transfer (cm$^2$), $\Delta T$ = Temperature difference between the composition layers ($^0$C)

The rate of heat transfer by convection through hot molten slags, vapours, gaseous combustion products depends upon

a. *Fluid properties* such as temperature, density, viscosity, and thermal conductivity
b. *Cross-sectional area*: Higher rate of heat transfer for higher cross-sectional area

  c. *Temperature gradient*: Higher rate of heat transfer for higher temperature gradient
  d. *Composition density*: Uncompacted composition with higher porosity assist in rate of heat transfer while compacted compositions hinder

The higher the thermal temperature, higher the quantum of hot fluids (hot gases/vapours) that permeates through the pyrotechnic composition layers, the higher and faster shall be the transfer of energy. This permeability shall be more for the composition which are granulated or the composition is compacted with low pressure, i.e., porous. The pores in composition allow hot fluids (gases/vapours/fluids) to seep in for heating further layers in the composition. Organic fuels on combustion give more gases, and hence provide heat through convection of hot gaseous products.

### 3.13.3 RADIATION

It is the propagation of electromagnetic waves and is governed by Stefan–Boltzmann Law, which states that *"the energy q radiated by a black body radiator per unit time from an area A is proportional to the fourth power of the absolute temperature T"* and is given by [14]

$$q/A = \sigma T^4 \ or \ q = A \ \sigma T^4 \text{watt} \ m^{-2} \tag{3.32}$$

A body that does not absorb all incident radiation (also known as a gray body) emits less total energy than a black body and is characterised by an emissivity factor and is written as

$$q = \varepsilon A \ \sigma T^4 \tag{3.33}$$

where $\varepsilon$ = Emissivity of the surface, dependent upon material and temperature, dimensionless ($\varepsilon < 1$), $q$ = Radiation emitted or heat transfer per unit time (W), $\sigma$ = Stefan–Boltzmann constant, $5.6703 \times 10^{-8}$ (W $m^{-2}$ $^{0}K^{-4}$), $A$ = Area of the emitting surface ($m^2$), $T$ = Absolute temperature ($^{0}K$)

Thus, electromagnetic radiation due to black body radiation will be high at higher temperatures.

With the ambient temperature $T_a$, net energy radiated at temperature $T$ by an area $A$ per unit time will be

$$\Delta q = q - q_a = \varepsilon \ A \ \sigma [T^4 - T_a^4] \tag{3.34}$$

Readers may see reference [15] on review of methods to determine the ignitability of pyrotechnic compositions.

## 3.14  ENERGY PROPAGATION DIAGRAM

On providing ignition stimulus to the composition, the composition layer may or may not pick-up ignition; depending upon the intensity and duration of energy received from the external stimulus:

   a. If the energy from external stimulus is lower than the activation energy of composition, the composition would not ignite.
   b. If the energy from external stimulus is higher than the activation energy of composition, the composition would ignite.

Furthermore, once ignition takes place in the composition, the composition is capable of providing ignition to the next layer of composition or next pyrotechnic composition in the explosive train/chain provided its energy feedback for sustaining burning is more than the energy of activation of the composition or the next layer of another pyrotechnic composition.

Some pyrotechnic compositions are difficult to ignite. Such compositions are primed with priming compositions. Priming compositions have low energy of activation and a high heat of combustion. These priming compositions alone or in combination with booster composition help in igniting various pyrotechnic compositions (see Chapter 19).

The above concept of propagation of burning in a pyrotechnic composition (based on Shimuzi diagrams)   may be understood by *"Energy Propagation Diagram"* as below.

The following terms have been used:

Ea = Energy of activation of external stimulus

Efb = Energy feedback of external stimulus for self-sustaining ignition

Es = Energy for special effects of external stimulus (flame for ignition of composition)

Ea1, Ea2 and Ea3 = Energy of activation of compositions 1, 2, 3 for three compositions

Efb1, Efb2 and Efb3 = Energy feedback of composition 1, 2 and 3 for burning next unburnt layer of composition

Eap = Energy of activation of priming composition

Efbp = Energy feedback of priming composition for burning next unburnt layer of composition

It may be noted that the energy loss to surrounding ($E_{loss}$) and energy loss due to special effects (*Es*) have not been shown in the figure for the compositions since this energy goes out of the composition and does not play in ignition of the composition layers.

Let us consider combustion of three compositions 1, 2 and 3 by an external stimulus.

### 3.14.1 ENERGY PROVIDED FOR IGNITION BY EXTERNAL IGNITION STIMULI Es IS LESS THAN THE ENERGY OF ACTIVATION Ea1 OF THE COMPOSITION 1 (FIGURE 3.18)

Consider that an external stimulus (e.g., a match stick is rubbed against match box to produce flame) is allowed to ignite the top layer of a pressed pyrotechnic composition 1. The external stimulus has its own energy of activation $Ea$ (match gets it from friction induced energy) and burns and provides energy feedback $E_{fb}$ for sustaining its own burning and special effects energy Es for ignition to a layer of pressed pyrotechnic composition 1. It could be seen that the energy feedback for sustaining burning $E_{fb}$ is more than its own energy of activation $Ea$. Hence, external stimuli (i.e., match stick composition) continue to burn till exhausted.

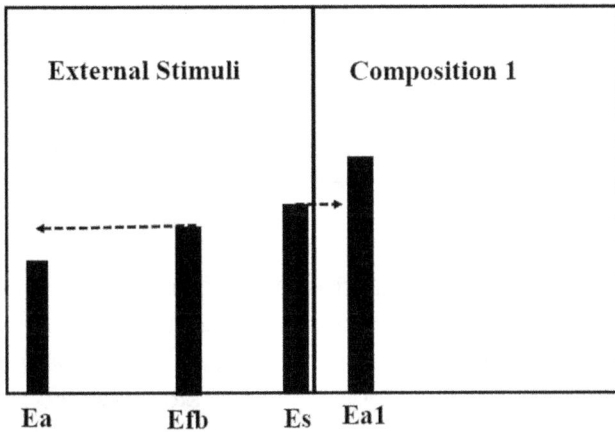

**FIGURE 3.18**  Composition Unable to Pick-Up Ignition.

However, the special effects energy Es for ignition is much lower than the activation energy of composition 1. Hence no ignition of composition 1 takes place. This may be due to composition configuration itself like being an incendiary composition having high energy of activation and hence the energy provided is not sufficient to achieve the ignition temperature of the composition or composition having high moisture content and heat energy is used up in evaporating the moisture rather than increasing the energy of the composition

### 3.14.2 ENERGY PROVIDED FOR IGNITION BY EXTERNAL IGNITION STIMULI Es IS MORE THAN THE ENERGY OF ACTIVATION Ea1 OF COMPOSITION 1 BUT ENERGY FEEDBACK FOR SUSTAINING BURNING Efb1 OF COMPOSITION 1 IS LESS THAN ITS OWN ENERGY OF ACTIVATION Ea1 (FIGURE 3.19)

Consider that above external stimuli is allowed to ignite the composition 1 where the energy from external stimuli Es is more than the energy of activation $Ea1$ of composition1. Hence composition 1 would pick up ignition. The burning shall continue till the external stimulus is available. However, ignition would stop after

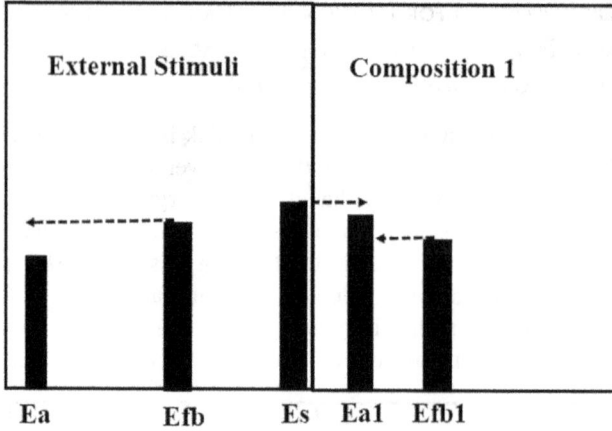

**FIGURE 3.19** Composition Unable to Self-Sustain Combustion.

removal of external ignition stimuli since composition's own energy for sustaining burning Efb1 is less than its own energy of activation *Ea1*. The composition 1 is therefore not a self-sustaining burning composition.

### 3.14.3 ENERGY PROVIDED FOR IGNITION BY EXTERNAL IGNITION STIMULI ES IS MORE THAN THE ENERGY OF ACTIVATION EA1 OF COMPOSITION 1 BUT ENERGY FEEDBACK FOR SUSTAINING BURNING EFB1 OF COMPOSITION 1 IS ONLY SLIGHTLY MORE THAN ITS OWN ENERGY OF ACTIVATION EA1 (FIGURE 3.20)

Ignition would take place in composition 1 as energy Es from external ignition stimuli is more than the energy of activation *Ea1* of the composition 1. The ignition would proceed slowly in the layer on its own (after removal of external stimuli)

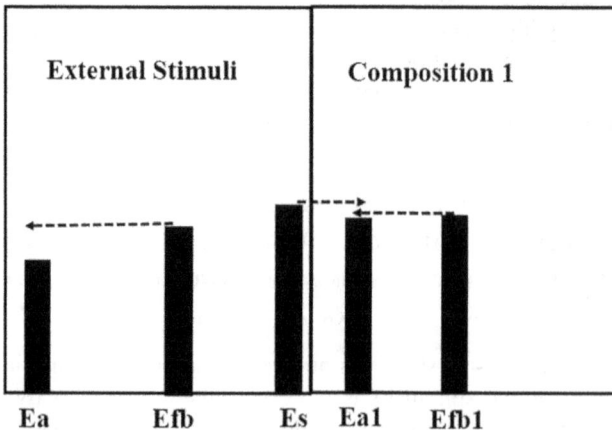

**FIGURE 3.20** Composition Unreliable for Self-Sustained Combustion.

since energy feedback for sustaining burning Efb1is just little more than the energy of activation *Ea1*.

However, the ignition may stop by external causes like low temperature, wind, etc. The composition is not reliable for consistent performance. As an example, it may show star "blind" when fired from the weapon as burning star immediately on ejection will have a high velocity and it will reduce the energy feedback Efb1for sustaining burning temporarily for some time.

### 3.14.4 ENERGY PROVIDED FOR IGNITION BY EXTERNAL IGNITION STIMULI Es IS MORE THAN THE ENERGY OF ACTIVATION Ea1OF COMPOSITION 1 AND THE ENERGY FEEDBACK FOR SUSTAINING BURNING Efb1FROM COMPOSITION 1 IS SUBSTANTIALLY MORE THAN ITS OWN ENERGY OF ACTIVATION Ea1 (FIGURE 3.21)

Ignition would take place in layer of composition 1 as energy Es from external ignition stimuli is more than the energy of activation Ea1 of the composition 1. The self-sustained ignition of composition layer would proceed (even after removal of external stimuli) as the energy feedback Efb1 for sustaining burning is much more than energy of activation Ea1 of layer of composition 1. The external factors like low temperature, wind etc. would have least impact on burning. The composition is a self-sustaining burning composition. Such pyrotechnic compositions are essentially required for military pyrotechnic ammunition filling.

### 3.14.5 ENERGY PROVIDED FOR IGNITION BY EXTERNAL IGNITION STIMULI Es IS LESS THAN THE ENERGY OF ACTIVATION Ea1 OF COMPOSITION 1 AND HENCE A PRIMING COMPOSITION IS PRESSED OVER COMPOSITION 1 (FIGURE 3.22)

The external ignition stimuli Es is less than the energy of activation *Ea1* of layer of composition 1. Hence ignition would not be picked by composition 1. To overcome

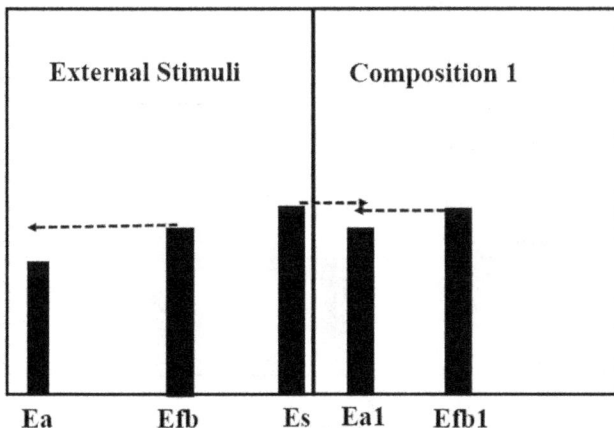

**FIGURE 3.21**  Composition Self-Sustained in Combustion.

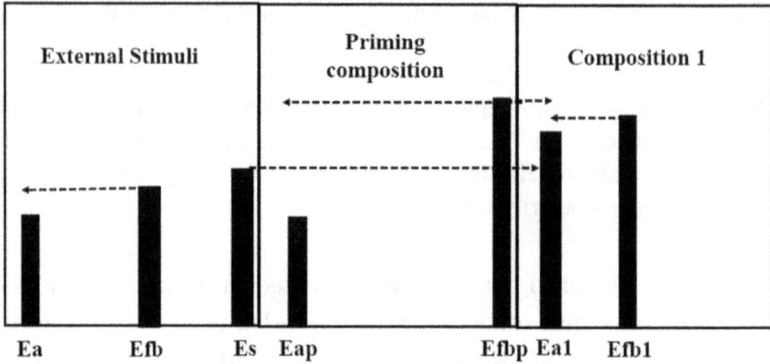

**FIGURE 3.22** Role of Priming Composition.

this situation, a priming composition is pressed over the composition 1. The priming composition having low energy of activation $Eap$ will pick up ignition from external stimuli's energy $Es$. Priming compositions have a very high energy feedback Efbp , which is more than the energy of activation $Ea1$ of composition 1 and hence composition 1 would pick up ignition. The energy feedback for sustaining burning Efb must be more than its own energy of activation $Ea1$ for self-sustaining burning of composition 1.

### 3.14.6  IGNITION OF PRESSED COMPOSITION CONSISTING OF THREE (OR MORE) COMPOSITIONS (FIGURE 3.23)

Consider three (or more) compositions pressed together. It is essential that the energy feedback from each composition should be higher than its own activation energy for self-sustained burning as well as more than the energy of activation of subsequent composition for ease of ignition.

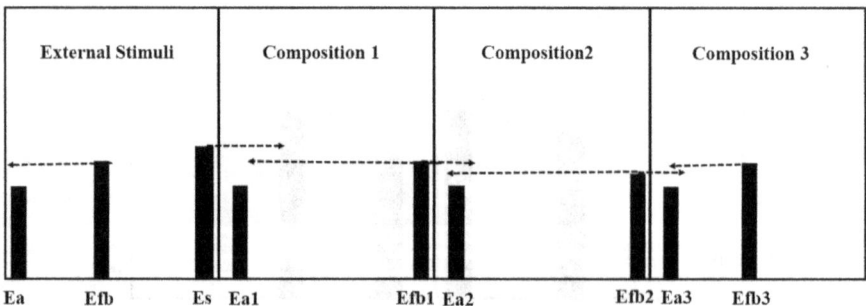

**FIGURE 3.23** Compositions Train with Three Compositions Capable of Self-Sustained Combustion.

## 3.15  PROPAGATION INDEX

This provides an assessment of suitability of a delay composition to sustain burning in a long small diameter tube where the heat losses to the surrounding are significant. It is defined as the ratio of heat of combustion to ignition temperature.

$$\text{Propagation Index} = \frac{\text{Heat of Combustion}}{\text{Ignition Temperature}} \tag{3.35}$$

A higher heat of combustion and lower ignition temperature and retaining the heat near or next to combustion zone is essential for proper propagation of combustion, giving higher propagation index. A higher propagation index means the composition will sustain the burning well. A lower heat of combustion with higher ignition temperature will result in combustion to propagate slowly or not propagate at all, i.e., low propagation index.

## 3.16  THERMAL ANALYSIS FOR UNDERSTANDING COMBUSTION PROCESS

Thermal analysis provides thermal behaviour and an understanding of combustion process. These tests detect exothermic or endothermic changes that occur in the specimen while it is being heated. These changes may be related to dehydration, decomposition, crystalline transition, melting, boiling, vaporisation, polymerisation, oxidation or reduction, etc.

A few techniques are given below:

a. Differential thermal analysis [DTA]
b. Differential scanning calorimetry [DSC]
c. Modulated temperature differential scanning calorimetry [MTDSC]
d. Thermogravimetric analysis [TGA]
e. Evolved gas analysis [EGA]
f. Combustion calorimetry
g. Temperature profile analysis
h. Thermal conductimetric analysis

The first two thermal analysis procedures have wide applications and are being used extensively. The detection of the type and character of the products of pyrotechnic combustion is a vital part of thermal analysis study. In addition, optical and scanning electron microscopy, infrared spectroscopy, mass spectrometry and X-ray diffraction are also used.

Readers may see the collaborative study data [16], which has been made on samples of pyrotechnic compositions (delays, coloured smokes and a percussion primer) by a number of laboratories. These laboratories have used a variety of techniques to characterise the ignition/ignition transfer behaviour of the compositions like temperature of ignition, heat of reaction, burning rate, time to ignition, safety tests (impact, electrostatic and thermal) and DSC/DTA analysis.

Readers may also see reference [17] for thermal methods of analysis with principles, applications and problems and reference [18] for characterisation tests for pyrotechnic ignition and ignition transfer.

## REFERENCES

[1] K.L. and B.J. Kosanke, *"Pyrotechnic Ignition and Propagation: A Review"*, Journal of Pyrotechnics, 1997, Issue No. 6.

[2] D.M. Johnson," *Ignition Theory Application to the Design of New Ignition Systems"*, RDTR No. 56, U.S. Naval Ammunition Depot, Crane, Indiana, 24 November 1965.

[3] A.A. Shidlovskiy, *"Principles of Pyrotechnics"*, 3rd Edition, Moscow, 1964. (Translated by Foreign Technology Division, Wright-Patterson Air Force Base, Ohio, 1974), American Fireworks News, 1 July 1997. ©1997, Rex E. &S.P., Inc.

[4] Chen-Guang Zhu, Hai-Zhen Wang, and Li Min," *Ignition Temperature of Magnesium Powder and Pyrotechnic Composition"*, Journal of Energetic Materials, 2014, Volume 32, Pages 219–226, ©Taylor and Francis.

[5] Dr. Herbert Ellern, *"Military and Civilian Pyrotechnics"*, © Chemical Publishing Co., New York, 1968.

[6] F.L. McIntyre, *"A Compilation of Hazard and Test Data for Pyrotechnic Compositions"*, Contractor Report ARLCD-CR-80047, October 1980. U.S. Army Research and Development Command, Large Caliber Weapon System Laboratory, Dover, New Jersey.

[7] Zaheer-Ud-din Babar and Abdul Qadeer Malik, *"Thermal Decomposition, Ignition and Kinetic Evaluation of Magnesium and Aluminium Fuelled Pyrotechnic Compositions"*, Central European Journal of Energetic Materials, 2015, Volume 12, Issue 3, Pages 579–592.

[8] Clive Woodley, R. Claridge, N. Johnson, and A. Jones, *"Ignition and Combustion of Pyrotechnics at Low Pressures and at Temperature Extremes"*, Defence Science Technology, 2017, Volume 13, Issue 3, Pages 119–126.

[9] Mirko Schoenitz and L. Dreizin Edward, "Consolidated *Energetic Nanocomposites: Mechanical and Reactive Properties"*, Proc. 36th International Annual Conference of ICT, Karlsruhe, Germany, 2005.

[10 Leo.V. De Yong and Frank.J. Valenta, *"Evaluation of Selected Computer Models for Modeling Pyrotechnic and Propellant Devices"*, Indian Head Technical Report 1279, Indian Head, MD, Naval Ordnance Station, 8 September 1979.

[11] A.G. Rajendran, C.B. Kartha, and V.V. Babu, *"Burn Rate Studies of a Titanium Based Pyrotechnic Composition"*, © Defence Science Journal, April 2000, Volume 50, Issue 2, Pages 199–206.

[12] S. Glasstone, K. J. Laidler, and H. Eyring, *"The Theory of Rate Processes"*, McGraw Hill Book Co.Inc., New York, 1941.

[13] *"Theory and Application"*, AMCP 706-185, Engineering Design Handbook Military Pyrotechnic Series, Part One, Headquarters, US Army Material Command, Washington DC, April 1967.

[14] *"Design of Ammunition for Pyrotechnic Effect"*, AMCP 706-188, Engineering Design Handbook Military Pyrotechnics Series, Part Four, Headquarters U S Army Material Command, 5001, Eisenhower Ave, Alexandria, VA 22304, March 1974.

[15] Leo de Yong, *"Prediction of Ignition Transfer Reliability in Pyrotechnic Systems using the VARICOMP Technique"*, Report MRL-R-898, February 1986, Department of Defence, Defence Science and Technology Organisation, Materials Research Laboratories, Melbourne, Victoria.

[16] P. Barnes, Leo. de Yong, J. Domanico, P, Twadawa, and F. Valenta," *A Comparison Between Several Standard Methods Used to Characterize the Ignition/Ignition Transfer of Pyrotechnic Compositions - A Collaborative Study"*, Editor L. de Yong, Part I data Report MRL-R-1043, Pt 1, February 1987, Department of Defence, Defence Science and Technology Organisation, Materials Research Laboratories, Melbourne, Victoria, Australia.

[17] P.J. Haines, *"Thermal Methods of Analysis: Principles, Applications and Problems"*, Chapman and Hall, 1995.

[18] R. Kelly, G.L. Lindsley, and N.R. Williams, *"Characterization Tests for Pyrotechnic Ignition and Ignition Transfer"*, Proceedings of the Seventh International Pyrotechnics Seminar of 14–18 July 1980, Volume 1.

# 4 Factors Affecting Pyrotechnic Performance

## 4.1 GENERAL

There are several factors affecting the performance of pyrotechnic composition. For example, a simple pyrotechnic composition gunpowder (with ingredients sulphur, charcoal and potassium nitrate) under following conditions shall behave differently under different conditions:

a. A moist gunpowder on ignition stimulus will not pick-up ignition at all
b. A pressed gunpowder (in a safety fuse) on ignition stimulus will burn slowly with low temperature and flame
c. Loose gunpowder spread on a platform on ignition stimulus will burn vigorously without exploding
d. A loose gunpowder in an enclosed container on ignition stimulus will explode with very high temperature and flame

Similarly, duration of burning varies for each type of composition, such as photoflash compositions in milliseconds, tracer compositions in few seconds and smoke and illuminating compositions in several seconds. The overall performance of pyrotechnic ammunition depends upon the inherent characteristics of pyrotechnic composition, design features of the ammunition and type of environment. Most of this combustion behaviour may be discussed under three main categories as follows.

a. *Characteristic of Composition* (Section 4.2)
   i. Chemical nature of ingredients
   ii. Heat of combustion of fuel
   iii. Exothermic/endothermic decomposition of oxidiser
   iv. Thermal conductivity
   v. Electromagnetic radiation from combustion products
   vi. Specific heat
   vii. Density
   viii. Percentage of ingredients and oxygen balance
   ix. Purity of ingredients
   x. Particle Size, Shape and Porosity and Surface Area
   xi. Type and quantity of binder
   xii. Type and quantity of burn rate modifier

DOI: 10.1201/9781003093404-4

  xiii.  Uniformity of mixing/blending
  xiv.  Loading pressure (degree of compaction)
   xv.  Moisture and or volatile matter
  xvi.  Exposed surface
  xvii.  Products of combustion


  b. *Design Features of Ammunitions* (Section 4.3)
    i.  Method of initiation
   ii.  Container material
  iii.  Column dimension
   iv.  Venting/confinement
    v.  Spin


  c. *Type of Environment* (Section 4.4)
    i.  Ambient temperature
   ii.  Ambient pressure
  iii.  Wind direction, velocity and background light
   iv.  Humidity, moisture, dust, smoke, heavy fog, snow and rain

The general trend of factors affecting pyrotechnic composition performance have been given below, *but the effects of various factors are not always linear and there may be some exceptions.*
Let us discuss each factors as under.


## 4.2  CHARACTERISTICS OF COMPOSITION

Composition is the main factor responsible for all the special effects of a pyrotechnic ammunition or device.


### 4.2.1  CHEMICAL NATURE OF COMPOSITION

The chemical nature of the main ingredients in the composition defines the energy available and the likely products. Change in ingredients, varying percentage of ingredients, changes in moisture content, use of burn rate modifiers, binders and dyes; varying particle size, shape, granulation, homogeneity of mixing/ blending, compaction density, etc. lead to changes in several parameters as listed below.

  a. Ignitability
  b. Energy of activation
  c. Ignition temperature
  d. Heat of combustion
  e. Combustion rate
  f. Combustion temperature
  g. Flame temperature

h. Heat capacity
i. Thermal conductivity
j. Hygroscopicity
k. Combustion products
l. Mechanical strength like compaction strength
m. Special effects
n. Safety characteristics like impact sensitivity, percussion sensitivity, friction sensitivity, thermal (flame) sensitivity, electrostatic discharge (spark) sensitivity

Some examples of change of ingredients resulting in different combustion effects are as follows:

a. Composition of different oxidisers like sodium nitrate, strontium nitrate and barium nitrate with magnesium would produce yellow flame, red flame and green flame, respectively.
b. Sodium nitrate has active oxygen content of 47% and barium peroxide has active oxygen content of 9.5%. Hence, the rate of combustion and flame temperature for same fuel magnesium shall be higher for the former.
c. Thermite compositions produce high heat of combustion with negligible amount of gas while smoke compositions with organic fuels containing carbon, hydrogen and oxygen produce low heat of combustion with copious amount of gas due to formation of gaseous $CO_2$ and $H_2O$.
d. Some fuels like gallic acid, potassium benzoate and potassium picrate are exclusively used in gas production in whistling compositions.
e. A fuel with potassium chlorate gives an ignition temperature much lower than that produced with potassium perchlorate, the latter being comparatively slow in burn rate.
f. Heat of combustion may vary from 200 cal.gm$^{-1}$ in delay composition to 2,500 cal.gm$^{-1}$ for photoflash composition.
g. The reaction temperature may vary from 200°C in smoke composition to 3,500°C for photoflash composition.

## 4.2.2   HEAT OF COMBUSTION OF FUELS

All pyrotechnic compositions on combustion produce heat. The extent of heat production depends upon the type of ingredients. The chemical nature of the fuel and oxidiser largely determines the energy of activation (*Ea*) required for breaking the existing bonds and forming combustion products with new bonds with heat of combustion (Δ*H*).

Metals like magnesium or aluminium as fuel give high heat of combustion compared to non-metals like sulphur and lactose. In case of illuminating, signalling flare, incendiary, infrared flare and tracer composition, enough heat and flame are required, which is possible only by having high heat of combustion by use of metallic fuels. Organic fuels cannot generate such high heat and flame. Therefore, to obtain high temperatures, one must use metallic fuels like magnesium, aluminium, titanium, zirconium with powerful oxidisers like potassium chlorate or potassium perchlorate.

Metallic fuels cannot be used for a signalling colour smoke composition in lieu of organic fuels, as extreme temperatures due to high heat of combustion during combustion would decompose the dye giving black smoke. Hence, organic fuels (like sucrose, fructose, etc.) having low heat of combustion producing high gaseous products (which assist in expelling dyes) are used for signalling colour smoke compositions. Therefore, to obtain low temperatures, one must use organic fuels like starch, lactose, etc. with moderate oxidisers .

The heat of combustion depends upon, amongst others, exothermicity of reactants, melting point of reactants and products and stoichiometric ratios of fuel and oxidiser.


### 4.2.3   Exothermic/Endothermic Decomposition of Oxidisers

In many pyrotechnic compositions, if the oxidiser heat of decomposition is higher, then more energy is required for its decomposition and hence with such an oxidiser in a composition, the linear burn rate would be slower.

For example, $KNO_3$ has endothermic decomposition ($\Delta H$ (+)75.5 kcal. $mole^{-1}$) and hence the composition containing $KNO_3$ will have higher energy of activation and hence slow combustion rate than composition containing $KClO_3$, which has an exothermic decomposition ($\Delta H$ (–)10.6 kcal. $mole^{-1}$) and hence will have lower energy of activation of the composition and fast combustion rate.


### 4.2.4   Thermal Conductivity

Thermal conductivity of composition affects the energy feedback for sustaining the burning (*Efb*) of the composition. During combustion of a pressed pyrotechnic composition column, heat travels from layer to layer increasing the temperature of the next composition layer to its ignition temperature. If this is not achieved, the propagation of combustion will not take place and combustion will stop.

The higher the thermal conductivity of the composition, the faster would be the rate of transfer of the heat produced during combustion to the next layer of the composition and hence there will be less loss of heat to the surrounding which would result in faster burning of the composition. However, this would make the ignition little difficult since the heat energy would be transmitted very fast from layer to layer and may make it difficult to achieve the ignition temperature of the first layer. Metals are best thermal conductors while organic compounds are worst. Addition of fine copper and silver particles increase the thermal conductivity of many gasless delay compositions.

Inhibitors/retardants like kaolin/clay/inert composition have a detrimental effect on burning of a column of composition. In general, thermal conductivity increases with increasing metal fuel content and loading pressure while it decreases with increasing particle size as this makes the composition porous.

## 4.2.5   ELECTROMAGNETIC RADIATION FROM COMBUSTION PRODUCTS

i. *Visible Radiation*: The composition for illuminating and coloured flare signals make use of ingredients which only yield the desired visible radiation. Thus, strontium nitrate is used as an oxidiser for red colour signals, barium nitrate or barium chlorate is used for green colour signals while sodium nitrate is used for yellow colour signals. Infusion of chlorine, using chlorine donors (like polyvinyl chloride, chlorinated rubber) assist in intensifying colour purity. Combustion products should also give maximum emission in the visible range.

ii. *Infrared Radiation*: The black body radiation depends upon the temperature. *Wien's Displacement Law* states that *"The black body radiation curve for different temperature peaks at a wavelength $\lambda_{max}$ (m) is inversely proportional to the temperature $T(^{0}K)$"* and given by equation

$$\lambda_{max} = 2.897756 \times 10^{-3} m.K.T^{-1} \qquad (4.1)$$

Hence as the temperature (T) increases, the wavelength ($\lambda max$) shifts towards shorter wavelength, i.e., hotter materials emit most of the radiations in shorter wavelengths. It may be noted that infrared wavelengths are higher with values 780 nm to 1 mm while visible wavelengths are lower at 380 nm to 780 nm. Hence in case of infrared compositions, the flame temperature should be lower than the illuminating flare compositions.

The infrared output depends upon the nature, concentration and emissivity of combustion products, amongst other factors. Carbon at high temperature is the ideal combustion product for infrared radiation since its emissivity is high, near 1.

## 4.2.6   SPECIFIC HEAT

Specific heat of the composition is the amount of heat required to raise the temperature of unit mass of substance by 1°C. Hence, if the specific heat of the composition is higher, it would require more heat energy to raise it to its ignition temperature and reduce the temperature of the combustion zone and consequently reduce the linear burn rate.

## 4.2.7   DENSITY

The density of ingredients in the composition affects the linear burn rate. Higher density of ingredients results in slower linear burn rate. This is because the same heat energy is distributed over a large mass in a high-density material. The density also determines the amount of composition that can be filled in an ammunition. The densities of a few fuel and oxidiser ingredients are given in Table 4.1.

**TABLE 4.1**
**Density of Ingredients**

| Material | Density $(g.cm^{-3})$ |
|---|---|
| Aluminium | 2.702 |
| Magnesium | 1.107 |
| Zirconium | 6.490 |
| Phosphorous | 2.340 |
| Sulphur | 2.070 |
| Potassium nitrate | 2.109 |
| Potassium chlorate | 2.320 |
| Potassium perchlorate | 2.520 |
| Barium peroxide | 4.96 |
| Barium nitrate | 3.24 |
| Barium chromate | 4.498 |
| Barium chlorate | 3.86 |
| Iron oxide (red) | 3.0 |
| Iron oxide (black) | 2.8 |
| Lead chromate | 6.3 |
| Sodium nitrate | 2.26 |
| Strontium nitrate | 2.99 |

### 4.2.8 Percentage of Ingredients and Oxygen Balance

Percentage change of ingredients brings changes in the nature/properties of the composition. These effects vary from composition to composition. Variation in the percentage of various ingredients like fuel, oxidiser, burn rate modifiers, binders and colour intensifiers leads to changes in thermal properties of the composition and combustion products. For example,

a. An increase in binder in an illuminating composition mostly results in slowing down the linear burn rate, i.e., increases the linear burn time and reduces the luminosity.

b. Use of burn rate modifiers like calcium oxalate or inert thermal insulators like kieselguhr, fuller's earth and fumed silica will result in slower linear burn rate, i.e., increase in linear burn time and decrease in luminosity.

c. An increase in fuel content will increase the linear burn rate up to a certain limit and then decrease. For example, the burn rate of magnesium–sodium nitrate increases with the increase in magnesium content from 30% to 70%, beyond which the linear burn rate reduces.

d. The heat of combustion is maximum at the stoichiometric proportion of fuel and oxidiser while rate of combustion is higher at much higher fuel content than the stoichiometric ratios of the fuel and oxidiser. When the fuel is a metal, there is an increased heat transfer to next layer of composition. In case of non-metals, like sulphur or carbon, the heat transfer is the lowest. It can be stated that below the stoichiometric level, the excess oxidiser, which is an inert diluent and heat insulator, retards burning. The maximum heat of combustion takes place at stoichiometric ratio of fuel to oxidiser and any further changes in fuel/oxidiser ratio brings down the combustion since excess fuel or excess oxidiser do not contribute to combustion and acts as heat sink material.

## Oxygen Balance

Pyrotechnic compositions have neutral/zero, positive or negative oxygen balance, depending upon the quantity of oxidiser and fuel. Oxygen balance indicates degree of oxidation of a high energetic material during combustion. It also indicates the number of grams of oxygen that are in excess or deficient for 100 grams of the composition. A variation in oxygen balance leads to change in thermal properties of composition like burn rate and heat of combustion and special effects as well as combustion products.

The oxygen balance or sufficiency of oxygen is categorised as under.

a. *Neutral/Zero Oxygen Balance*: The oxidiser present in the composition is just sufficient to oxidise the combustible fuel. Thus, both fuel and the oxidiser are consumed fully during combustion. This ratio is also termed as stoichiometric ratio. The sensitivity is maximum for such composition.

b. *Positive Oxygen Balance*: The oxidiser present in the composition is more than the required quantity needed to oxidise the combustible fuel. Thus, the fuel is consumed fully while some oxidiser is left. This is not desirable as composition is not utilised fully and occupies space in ammunition.

c. *Negative Oxygen Balance*: The oxidiser present in the composition is less than that required for complete oxidation of combustible fuel. This is an incomplete combustion where toxic gases like carbon monoxide and nitric oxide may be formed. Thus, the oxidiser is fully consumed and the remaining fuel may react with atmospheric oxygen.

The exact oxygen balance can only be obtained if the exact reaction equation is used, which requires a fundamental, experimentally based understanding of all of the chemical reactions taking place. To determine the fuel consumed by 1 g of oxygen as well as oxygen balance, reaction equations are required to be set up as under.

Consider the following equations where oxygen required for complete fuel consumption is present.

a. *Inorganic metal fuel aluminium* (Atomic weights Al = 26.98, O = 15.99):

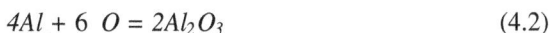

$$4Al + 6\ O = 2Al_2O_3 \tag{4.2}$$

Fuel consumed per gram of oxygen = 107.92/95.94 = 1.12 g. Further, aluminium does not have its own oxygen and there is a deficiency of six oxygen atoms for four aluminium atoms. Hence oxygen deficiency percentage for aluminium = $(-)100 \times 95.94/107.92 = (-)\ 89\%$.

b. *Inorganic non-metal fuel boron* (Atomic weight B = 10.8, O = 15.99):

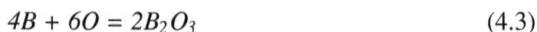

$$4B + 6O = 2B_2O_3 \qquad (4.3)$$

Fuel consumed per g of oxygen = $43.2/95.94 = 0.45$ g. Further, boron does not have its own oxygen and there is a deficiency of six oxygen atoms for four boron atoms. Hence oxygen deficiency percentage = $(-)100 \times 95.54/43.2 = (-)\ 222\%$

c. Organic fuel naphthalene: (Molecular weight naphthalene = 128.17, O = 15.99)

$$C_{10}H_8 + 24\ O = 10CO_2 + 4H_2O \qquad (4.4)$$

Fuel consumed per gram of oxygen = $128.17/383.76 = 0.33$ g. Further, naphthalene does not have its own oxygen atom and there is a deficiency of 24 oxygen atoms for one molecule of naphthalene. Hence oxygen deficiency percentage = $100 \times 383.76/128.17 = (-)\ 299.4\%$

Oxygen balances of some ingredients are given in Table 4.2.

---

**TABLE 4.2**
**Oxygen Balance of Some Ingredients**

| Ingredients | Oxygen balance (%) |
| --- | :---: |
| Aluminium | $(-)89$ |
| Anthracene | $(-)296.6$ |
| Ammonium nitrate | $(+)20$ |
| Ammonium perchlorate | $(+)34$ |
| Boron | $(-)222$ |
| Carbon | $(-)266.7$ |
| Epon 813/Versamid 140 $C_{83}H_{98}N_4O_{14}$ | $(-)\ 234$ |
| Lactose | $(-)106.7$ |
| Laminac/Lupersol $C_{42}H_{39}O_{16}$ | $(-)175$ |
| Magnesium | $(-)65.5$ |
| Naphthalene | $(-)299.4$ |
| Nitrocellulose | $(-)33.4$ |
| Phenolic resin | $(-)240.1$ |
| Sulphur | $(-)100$ |
| Potassium chlorate | $(+)39.2$ |
| Potassium nitrate | $(+)39.6$ |
| Sodium nitrate | $(+)47$ |
| Strontium nitrate | $(+)37.8$ |

It may be seen that most inorganic materials have a positive oxygen balance while organic materials have a negative oxygen balance. This is due to presence of more C and H atoms in the molecules of organic material requiring more oxygen to convert to their respective oxides. For this reason, a very common strategy to improve the oxygen balance of energetic materials is to blend inorganic and organic compounds together to compensate for the imbalance.

Table 4.3 depicts fuel consumption per gram of oxygen and grams of oxidiser required to release 1 g of oxygen.

Table 4.3 helps in matching the parts of oxidiser required by the part of fuel for stoichiometric proportions and computing the oxygen balance of composition as given below.

a. *Matching the Parts of Oxidiser Required by the Part of the Fuel for Stoichiometric Proportion:*

Let us consider the combustion between magnesium and sodium nitrate using Table 4.3. As 1 g of oxygen is liberated by 2.13 g of sodium nitrate while 1 g of

## TABLE 4.3
## Fuel Oxygen Correlation

| Fuels | Grams of fuel consumed/g of oxygen | Oxidiser | Grams of oxidiser required to release 1 g of oxygen |
|---|---|---|---|
| Aluminium | 1.12 | Barium chlorate | 3.12 |
| Anthracene | 0.44 | Barium chromate | 10.6 |
| Boron | 0.45 | Barium nitrate | 3.27 |
| Carbon | 0.38 | Barium peroxide | 10.6 |
| Iron | 2.32 | Iron oxide (red) | 3.33 |
| Lactose | 0.94 | Iron oxide (black) | 3.62 |
| Naphthalene | 0.33 | Lead chromate | 13.5 |
| Magnesium | 1.52 | Lead dioxide | 7.48 |
| Phosphorous (red) | 0.78 | Lead oxide (litharge) | 14.0 |
| Phosphorous (yellow) | 0.78 | Lead tetroxide (red lead) | 10.7 |
| Polyvinyl chloride | 0.78 | Potassium chlorate | 2.55 |
| Silicon | 0.88 | Potassium nitrate | 2.53 |
| Shellac | 0.47 | Potassium perchlorate | 2.17 |
| Sulphur | 1.00 | Sodium nitrate | 2.13 |
| Titanium | 1.50 | Strontium nitrate | 2.63 |
| Tungsten | 3.83 | | |
| Zinc | 4.09 | | |
| Zirconium | 2.85 | | |

oxygen consumes 1.52 g of magnesium, hence 1.52 g of magnesium requires 2.13 g of sodium nitrate. Hence, this ratio of magnesium fuel against sodium nitrate oxidiser is considered as stoichiometric ratio where oxidiser is just sufficient to consume all the fuel. This composition is termed as zero or neutral oxygen balance. The percent of magnesium will be $1.52 \times 100/(1.52 + 2.13)$, i.e., 41.6% and percentage of sodium nitrate will be $2.13 \times 100/(1.52 + 2.13)$, i.e., 58.4%.

Therefore, magnesium–sodium nitrate composition shall be

Magnesium 41.6%

Sodium nitrate 58.4%

The above can be obtained by 5 moles of Mg (121.5 g) reacting with 2 moles of $NaNO_3$ (170g), total reacting mass 291.5 g (where percentage of Mg is 41.6% and percentage of $NaNO_3$ is 58.4 %) and may be represented as

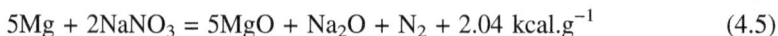

$$5Mg + 2NaNO_3 = 5MgO + Na_2O + N_2 + 2.04 \text{ kcal.g}^{-1} \qquad (4.5)$$

However, any excess of fuel over this stoichiometric composition shall result in negative oxygen balance. With Mg: $NaNO_3$ ratio of 46.6: 53.4, the combustion proceeds as

$$6Mg + 2NaNO_3 = 6MgO + 2Na + N_2 + 1. 9kcal.g^{-1} \qquad (4.6)$$

However, a very excess of magnesium with Mg:$NaNO_3$ ratio of 72:28, the combustion proceeds as follows where excess magnesium reacts with atmospheric oxygen (negative oxygen balance).

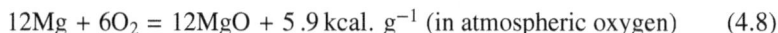

$$18Mg + 2NaNO_3 = 6MgO + 2Na + N_2 + 12Mg + 2.0 \text{ kcal. g}^{-1} \qquad (4.7)$$

$$12Mg + 6O_2 = 12MgO + 5.9 \text{ kcal. g}^{-1} \text{ (in atmospheric oxygen)} \qquad (4.8)$$

A large number of pyrotechnic compositions have negative oxygen balance as atmospheric oxygen also takes part in combustion, and thus receives more output from the composition.

b. *Computing the Oxygen Balance of Composition:*

To calculate the oxygen balance, one must divide the parts of oxidiser by the number of parts of the oxidiser that is required to yield one part of oxygen. This gives the total oxygen supplied. Similarly, divide the parts of fuel by the number of parts of that fuel that demand one part of oxygen to be burnt completely. This gives the total oxygen demand. Subtracting the oxygen demand from oxygen supplied, the resultant number gives the oxygen balance, whether positive or negative.

Let us calculate the oxygen balance of above magnesium sodium nitrate 41.6/58.4 mixture composition.

Here, oxygen supplied by sodium nitrate = 58.4/2.13 = 27.41

Oxygen demand by magnesium = 41.6/1.52 = 27.36

Oxygen balance = oxygen supply–oxygen demand = 27.41 – 27.36 = 0.05 g for 100 g of composition. Here, the oxidiser is just sufficient to meet the oxygen demand by magnesium. Hence this composition has neutral or zero oxygen balance.

Let us calculate the oxygen balance of a given yellow light composition [1] as follows where shellac acts as binder as well as a fuel.

| | |
|---|---|
| Potassium chlorate | 60 |
| Sodium oxalate | 25 |
| Shellac | 15 |

*Calculation of oxygen supplied by potassium chlorate* = 2.55 g of potassium chlorate liberates 1 g of oxygen

Hence 60 g of potassium chlorate shall liberate 60/2.55 g = 23.53 g of oxygen.

*Calculation of oxygen consumed by sodium oxalate* =

8.37 g of sodium oxalate consumes 1 g of oxygen

Hence 25 g of sodium oxalate shall consume 25/8.37 g = 2.98 g of oxygen.

*Calculation of oxygen consumed by shellac* =

0.47 g of shellac consumes 1 g of oxygen

Hence 15 g of shellac shall consume 15/0.47 = 31.91 g of oxygen.

Therefore,

oxygen supplied by potassium chlorate = 23.53

Cumulative oxygen demand = 2.98 + 31.91 = 34.89.

Oxygen balance = oxygen supply – oxygen demand = 23.53 – 34.89 = (–) 11.36 g of $O_2$ for 100 g of composition.

   c. *Formula for Oxygen Balance of an Explosive:*

The oxygen balance of an explosive with formula $C_aH_bN_cO_d$ and molecular weight M and considering complete conversion of C to $CO_2$ and H to $H_2O$ is given as below [2].

$$\text{Oxygen balance\%} = [d - 2a - b/2] \times (1600/M) \qquad (4.9)$$

*For metal [M] containing compound of empirical formula* $C_aH_bN_cO_dM_n$

$$\text{Oxygen Balance\%} = [d - 2a - b/2 - n] \times (1600/M) \qquad (4.10)$$

*Where n = number of atoms of metallic oxide formed.*

   Importance of oxygen balance can be understood from the fact that oxygen balance has been taken as a criterion for evaluation as a screening technique for thermal hazard assessment.

**TABLE 4.4**

**Influence of Oxidiser Ratio and Oxygen Balance on Flame Emission Spectra and Burning Properties** [3]

| Ingredients | Percentage | | | | | | | | |
|---|---|---|---|---|---|---|---|---|---|
| Ba (NO$_3$)$_2$ | 88.4 | 76.1 | 64.1 | 51.8 | 39.8 | 27.5 | 15.5 | 3.2 | 0.00 |
| NH$_4$ClO$_4$ | 0.00 | 12.1 | 24.0 | 36.1 | 48.0 | 60 | 72 | 84 | 87.2 |
| Shellac | 11.6 | 11.8 | 12.0 | 12.1 | 12.3 | 12.4 | 12.6 | 12.6 | 12.8 |
| | Oxygen balance | | | | | | | | |
| | 0 | 0 | 0 | 0 | 0 | 0 | 0 | 0 | 0 |
| | Pellet density (g cm$^{-3}$) | | | | | | | | |
| | 2.25 | 2.14 | 2.04 | 1.95 | 1.86 | 1.77 | 1.70 | 1.64 | 1.58 |
| | *Performance* | | | | | | | | |
| Combustion Rate (mm s$^{-1}$) | 0.38 | 0.42 | 0.62 | 0.73 | 0.79 | 0.85 | 0.96 | 1.19 | 1.24 |
| CIE X | 0.376 | 0.335 | 0,304 | 0.258 | 0.231 | 0.221 | 0.249 | 0.354 | 0.510 |
| CIE Y | 0.504 | 0.555 | 0.576 | 0.625 | 0.655 | 0.666 | 0.647 | 0.566 | 0.423 |
| Excitation purity (%) | 64.1 | 67.9 | 65.6 | 68.4 | 71.3 | 72.6 | 72.9 | 76.6 | 80.1 |
| DW (nm) | 564 | 555 | 548 | 539 | 534 | 533 | 538 | 558 | 586 |

DW = dominant wavelength, CIE refers to the 1931 International Commission on Illumination (CIE) colour space value

Apart from oxygen, chlorine also acts as an oxidiser for metal fuels like those liberated from hexachloroethane, hexachlorobenzene, polyvinylchloride, etc.

   d. *Effect of Oxidiser Ratio and Oxygen Balance on Burn Rate:*

The influence of oxidiser ratio and oxygen balance on flame emission spectra and burning properties through spectrophotometer [3] have been shown in Table 4.4 and Table 4.5.

   In Table 4.4, oxidiser ratio was varied, keeping oxygen balance zero. In this way, different amounts of barium and chlorine were introduced. Increasing NH$_4$Cl increased the combustion rate as well as colour purity due to formation of BaCl emitter and quenching of BaO continuum.

   In Table 4.5, oxidiser ratio was kept constant (NH$_4$Cl was 1.24 times higher than Ba (NO$_3$)$_2$) and the oxygen balance was varied. The emission intensities of the BaO and BaCl emitters were fixed due to the constant ratio of oxidisers, which provide barium and chlorine separately. Incandescence was found to be a dominating emission when oxygen balance was less than 20%. The best green colour was obtained with 37.5% of Ba (NO$_3$)$_2$, 46.5% of NH$_4$Cl and 16% shellac.

   Report on oxygen balance for a few pyrotechnic compositions for a gas generator of base bleed projectile is shown in Table 4.6.

   The effect of oxygen balance on burn rate for some of the compositions in Table 4.6 may be seen in Table 4.7 with its variation coefficient (VC) (see Section 4.2.13).

**TABLE 4.5**

**Influence of Oxidiser Ratio and Oxygen Balance on Flame Emission Spectra and Burning Properties** [3]

| Ingredient | Percentage | | | | | | | | | |
|---|---|---|---|---|---|---|---|---|---|---|
| Ba (NO$_3$)$_2$ | 42.0 | 41.1 | 40.2 | 39.3 | 38.4 | 37.5 | 35.7 | 33.9 | 32.1 | 30.4 |
| NH$_4$ClO$_4$ | 52.0 | 50.9 | 49.8 | 48.7 | 47.6 | 46.5 | 44.3 | 42.1 | 39.9 | 37.6 |
| Shellac | 6.0 | 8.0 | 10.0 | 12.0 | 14.0 | 16.0 | 20.0 | 24.0 | 28.0 | 32.0 |
| | Oxygen balance | | | | | | | | | |
| | 18.6 | 12.7 | 6.7 | 0.8 | −5.1 | −10.9 | −22.5 | −34.0 | −45.4 | −56.6 |
| | Pellet density (g cm$^{-3}$) | | | | | | | | | |
| | 2.06 | 2.07 | 2.03 | 1.98 | 1.96 | 1.94 | 1.85 | 1.80 | 1.75 | 1.71 |
| | Performance | | | | | | | | | |
| Combustion rate (mm s$^{-1}$) | – | – | 0.72 | 0.77 | 0.73 | 0.74 | 0.71 | 0.72 | 0.69 | 0.71 |
| CIE X | – | – | 0.289 | 0.229 | 0.214 | 0.206 | 0.270 | 0.408 | 0.493 | 0.521 |
| CIE Y | – | – | 0.638 | 0.654 | 0.665 | 0.671 | 0.624 | 0.517 | 0.459 | 0.439 |
| Excitation Purity (%) | – | – | 80.1 | 70.9 | 71.2 | 71.5 | 70.9 | 77.6 | 85.5 | 88.4 |
| DW (nm) | – | – | 547 | 534 | 531 | 530 | 542 | 568 | 581 | 585 |

– = Combusted with no visible flame, DW = dominant Wavelength, CIE refers to the 1931 International Commission on Illumination (CIE) colour space value.

It could be seen that lactose-based composition with variation in oxygen balance from (+)15% to (−)14 % results in substantial decrease of linear burn rate by 21.4% while ascorbic acid (C$_6$H$_8$O$_6$) based composition with variation in oxygen balance from neutral to positive oxygen balance +15.52% exhibits only 5.83% increase.

Although the oxygen balance of a pyrotechnic redox system is a useful parameter for predicting performance of pyrotechnic composition, it is often masked by other decomposition reactions as well.

## 4.2.9 PURITY OF INGREDIENTS

The higher the purity of the ingredients, faster would be the combustion since more reactants are available. The time to ignition is reduced with higher purity of ingredients. It is essential to check purity and other relevant test criteria of each batch of ingredients before taking them into use for manufacture of composition. Impurities of ingredients affect the purity of coloured flames. Impurities may pose problems during storage and affect shelf life. There are certain impurities in ingredients that are not desirable like presence of copper or ammonia in red phosphorous or ground glass or sand in composition, since the former two reduce the stability of red phosphorous while the latter increases the sensitivity of the composition. In fact, some initiatory compositions deliberately use ground glass in definite proportions as an ingredient to improve sensitivity of the composition. Fuels forming oxide layer during storage and oxidisers absorbing moisture affect the burn rate.

**TABLE 4.6**

**Pyrotechnic Compositions with Oxygen Balance [4]**

| Composition no. | Ingredients | | | Oxygen balance percentage (%) |
|---|---|---|---|---|
| | Fuel | Oxidiser | Binder | |
| | Neutral oxygen balance | | | |
| 010/12 | Lactose/27.6 | KClO₄/69.4 | Viton A/3.0 | −0.06 |
| 006/12 | Lactose/26.8 | KClO₄/69.2 | Viton A/4.0 | −0.02 |
| 001/12 | Lactose /26.1 | KClO₄/68.9 | Viton A/5.0 | −0.70 |
| 004/12 | Lactose /25.2 | KClO₄/68.8 | Viton A/6.0 | −0.06 |
| 005/12 | Lactose /24.4 | KClO₄/68.6 | Viton A/7.0 | −0.02 |
| 013/12 | Ascorbic acid /30.9 | KClO₄/66.1 | Viton A/3.0 | −0.26 |
| 012/12 | Ascorbic acid /30.0 | KClO₄/66.0 | Viton A/4.0 | −0.33 |
| 007/12 | Ascorbic acid /29.1 | KClO₄/65.9 | Viton A/5.0 | −0.46 |
| 011/12 | Ascorbic acid /28.0 | KClO₄/66.0 | Viton A/6.0 | −0.46 |
| | Positive oxygen balance | | | |
| 002/12 | Lactose 22.9 | KClO₄71.8 | 5.3 | +15.24 |
| 008/12 | Ascorbic acid 25.9 | KclO₄69.1 | 5.0 | +15.52 |
| | Negative oxygen balance | | | |
| 003/12 | Lactose /29.3 | KclO₄/65.7 | Viton A/5.0 | −14.39 |
| 009/12 | Ascorbic acid /32.5 | KclO₄/62.5 | Viton A/5.0 | −14.26 |

### 4.2.10 PARTICLE SIZE, SHAPE, POROSITY AND SURFACE AREA

Particle size, shape, porosity and surface area play an important role in pyrotechnic composition performance. Their effects can be seen in bulk density, particle flow characteristics, hygroscopicity, compactibility, homogeneity in blending, aggregation/de-aggregation during blending or storage and above all reactivity. The effect of each is as under.

a. *Size:* Decrease in particle size
   i. Increases surface area to volume
   ii. Increases bulk density
   iii. Decreases porosity and pore size
   iv. Gives higher compaction density
   v. Increases more surface contact in the composition with other ingredients and thus improves homogeneity of the composition
   vi. It gets more affected by moisture content and causes adhesion and cohesion leading to poor flow characteristics

**TABLE 4.7**

**Effect of Oxygen Balance on Burn Rate [4]**

| Composition no. | Oxygen balance (%) | Burning rate (mm/sec) | | | | BR$_{av}$ | SD | VC |
|---|---|---|---|---|---|---|---|---|
| | | Lactose-based composition | | | | | | |
| 001/12 | −0.69 | 0.70 | 0.66 | 0.70 | 0.71 | 0.69 | 0.02 | 2.98 |
| 002/12 | +15.24 | 0.69 | 0.71 | 0.69 | 0.69 | 0.70 | 0.01 | 1.56 |
| 003/12 | −14.39 | 0.55 | 0.55 | 0.54 | 0.55 | 0.55 | 0.00 | 0.88 |
| | | Ascorbic acid–based composition | | | | | | |
| 007/12 | −0.46 | 1.21 | 1.21 | 1.19 | 1.20 | 1.20 | 0.01 | 1.01 |
| 008/12 | +15.52 | 1.24 | 1.23 | 1.20 | 1.26 | 1.23 | 0.03 | 2.03 |
| 009/12 | −14.26 | 1.28 | 1.27 | 1.26 | 1.28 | 1.27 | 0.01 | 0.93 |

vii.  Reduces energy of activation and thus decreases ignition temperature, thereby making ignition easy

viii. Reduces ignition delays

ix.   Increases rates of combustion

x.    Increases sensitivity to impact

xi.   Increases sensitivity to friction

xii.  Increases sensitivity by fine metallic fuels to electrostatic discharge due its ability to develop static charges

xiii. Higher reactivity makes it susceptible to compatibility problems and shelf-life problems

xiv.  Affects free flow ability as large surface area causes more friction in free flow

xv.   Affects aggregation/de-aggregation of particles during blending

b. *Shape*: The ingredients of a pyrotechnic composition may be of *spherical type, spheroid type, needle type, flake type, crystalline type* or a combination thereof.

   i.   The more spherical a particle, the better is the free flow ability due to friction at one point only. Other shapes like needles or flakes do not possess free flow ability.

   ii.  Spherical particles settle faster than non-spherical particles during storage and transportation.

   iii. The reactivity of spherical particles is less compared to flakes since area per unit mass is less than flakes.

c. *Porosity:* Decrease in particle size decreases porosity and pore size. Distribution of pores, dimension of pores (i.e., surface area of pores) and whether the pores are closed or connected is important. Increase in porosity has following effects:

   i.   Reduces bulk density.

    ii.   Increases in exposed surface area inside composition.

   iii.   Increases permeability (allows hot gases/vapours/liquids to move freely inside the pores of the composition for heat transmission).

   iv.   Reduces drying time

    v.   Increases hygroscopicity of hygroscopic ingredients.

As burning is a surface phenomenon, a higher surface area results in faster burning as more area of the ingredient particles take part in the combustion due to much more intimate contact with each other. As the fuel particle size decreases, the surface to volume ratio increases and it increases the number of contact points with the oxidiser and improves composition homogeneity. Thus, fine particles having higher surface area burn faster than big particles. It may be concluded that if linear burn rate is required to be increased, fine particles (apart from other avenues) are required to be used. Hence, in many pyrotechnic compositions, fine as well as coarse fuel is used for adjustment of burning time. To improve free flow properties of the composition, wet granulation or flow modifiers like aerosil or zinc stearate may be added.

It must be understood that a compromise is to be achieved for the particle size since extremely fine particles, especially dust, would be hazardous to handle as burn rate would be very fast. This is particularly important for metallic fuels (see Table 4.8) which are prone to combustion in air since a little stimulus in the form of static charge, heat, compaction or friction induced energy may cause ignition. It may be noted that even human body can accumulate static charge up to 25 mJ.

All the ingredients used in pyrotechnic composition are required to have a desired particle size of the ingredient for consistency of performance. It may also be stated that homogeneity of a composition is better if the particle size variation amongst ingredients is least. A reproducible burn rate may be obtained only if the particle size distribution is uniform in the composition.

   d. *Surface Area*:

The chemical reactivity of metal fuel depends upon its shape, which in turn depends upon its method of manufacture as explained below.

    i.   *Atomisation*, wherein fluid metal is sprayed into an atmosphere (generally nitrogen) where the fluid cools and solidifies as *spherical particles*. In case of presence of oxygen, a crust of oxide is formed, resulting into *spheroids*.

   ii.   *Grinding* of metals, where *granular particles* are formed with sharp angular features.

   iii.   *Milling or stamping*, where flakes are formed.

The surface to mass ratio differs in each of these particles. The spherical shape has the lowest surface area and hence is the least reactive compared to other shapes. An uneven and porous material shall have more surface area. Following is the order of reactivity from lower to higher reactive particle.

   *Spherical <spheroid <granular <flake*

**TABLE 4.8**
**Ignition and Explosibility of Metal Powders [5]**

| Material | Ignition temperature °C | | Minimum explosive concentration oz/cu ft | Min. igniting energy for dust cloud mJ | Maximum pressure psig | Maximum rate of pressure rise Psi.sec$^{-1}$ | Index of explosibility |
|---|---|---|---|---|---|---|---|
| | Cloud | Layer | | | | | |
| Aluminium atomised | 650 | 760 | 0.045 | 50 | 73 | 20,000+ | >10 |
| Magnesium -aluminium alloy | 430 | 480 | .020 | 80 | 86 | 10,000 | >10 |
| Magnesium | 620 | 490 | .040 | 40 | 90 | 9,000 | >10 |
| Zirconium | 20 | 190 | .045 | *15 | 55 | 6,500 | >10 |
| Titanium | 330 | 510 | .045 | 25 | 70 | 5,500 | >10 |
| Aluminium -silicon alloy | 670 | - | .040 | 60 | 74 | 7,500 | 3.6 |
| Iron, carbonyl | 320 | 310 | .105 | 20 | 41 | 2,400 | 1.6 |
| Boron | 470 | 400 | <.100 | 60 | 90 | 2,400 | 0.8 |
| Zirconium alloy | 420 | 340 | - | 30 | 43 | 300 | - |

These data apply to relatively coarse dust (through a No. 200 sieve) but not to submicron powder.
*In this test less than 1 gram of powder was used. Larger quantities ignited spontaneously. Electrical spark source used for ignition of metals.

It may be noted that specific surface area (surface area associated with one gram of composition) is more important than particle size.

In case of hygroscopic ingredients having lumps; the lumps are broken in a crusher mill and sieved and dried to obtain smaller grains with more surface area than the original grain.

The process of milling, pulverising, atomising or grinding of ingredients enhances the reactivity or sensitivity by producing new surfaces, new edges and new corners where atoms are not so well bonded than those deep insides. Milling also causes lattice deformation and distortions leading to higher reactivity as particle size is reduced.

*Crystal defects* in the oxidiser crystal structure like dislocations, cracks and other discontinuities play an important role as they provide sites favourable to the formation of *"hotspots"*. Smaller crystals have more of these imperfections. Therefore, as the oxidiser particle size is decreased, reactivity increases and the ignition temperature decreases. Impurity and presence of foreign matter in the lattice also affects burning characteristics. However, the surface area of fuel is more important

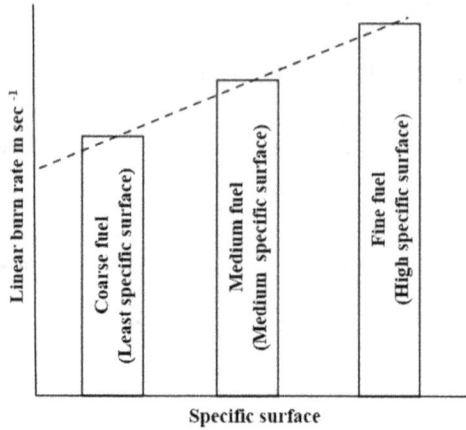

FIGURE 4.1   Effect of Specific Surface on Linear Burn Rate.

than oxidiser for its effect on burn rate since oxidiser in most cases melts at or before ignition temperature and hence the role of its particle shape and size is less.

Figure 4.1 shows the rates of burning of a fuel having coarse particles (least specific surface), medium particles (medium specific surface) and fine particles (high specific surface), with all other ingredients unchanged. It could be seen that rate of burning is slow with fuel having coarse particles (least specific surface) while it is fast with fuel having fine particles (high specific surface).

Table 4.9 displays the effect of change in specific surface of magnesium particle on burning characteristics.

Table 4.10 provides effect of specific surface on rate of burning for composition SR 563 using cut magnesium powder grade "O."

## TABLE 4.9
## Effect of Change in Specific Surface of Magnesium on Burning Characteristics [6]

| Composition | Ground magnesium | Atomised magnesium |
|---|---|---|
| Ground magnesium | 66.6 | - |
| Atomised magnesium | - | 66.6 |
| Sodium nitrate | 28.6 | 28.6 |
| Resin | 4.8 | 4.8 |
| | Characteristics | |
| Candles per sq. In. | 200,000 | 178,000 |
| Burning rate, in./min. | 9.4 | 57 |
| Density | 1.56 | 1.65 |
| Candle-seconds/g | 50,000 | 69,200 |

**TABLE 4.10**
**Effect of Specific Surface on Inverse Burn Rate [7]**

| Specific surface (cm². g⁻¹) | Rate of burning (seconds. inch⁻¹) |
|---|---|
| 225 | 15.0 (slower) |
| 245 | 14.2 |
| 275 | 12.5 |
| 344 | 11.3 |
| 403 | 10.0 |
| 570 | 8.1(Faster) |

Table 4.11 shows the effect of particle size on burning rate and candle power for atomised magnesium–sodium nitrate–polyvinylchloride–laminac composition.

i. *Mathematical Expression on Surface Area and Particle Size:*

Consider spherical particles for ease of understanding. Let $N_1$ particles of radius $R_1$ be broken to obtain $N_2$ particles of radius $R_2$. A mathematical derivation is as follows.

Obviously $R_1 > R_2$ and $N_2 > N_1$
The total surface area ($S_1$) of $N_1$ particles = $4\pi R_1^2 N_1$
The total surface area ($S_2$) of $N_2$ particles = $4\pi R_2^2 N_2$
The ratio of surfaces $S_2/S_1 = (N_2/N_1)(R_2/R_1)^2$

**TABLE 4.11**
**Effect of Particle Size on Linear Burn Rate [6]**

| Ingredients | Average particle size (microns) | Percentage | | | |
|---|---|---|---|---|---|
| Magnesium At. 20/50 | 437 | 48 | - | - | - |
| Magnesium At. 30/50 | 322 | - | 48 | - | - |
| Magnesium At. 50/100 | 168 | - | - | 48 | - |
| Magnesium At. 100/200 | 110 | - | - | - | 48 |
| Sodium nitrate | 34 | 42 | 42 | 42 | 42 |
| Polyvinyl chloride | 27 | 2 | 2 | 2 | 2 |
| Laminac resin 4116 | - | 8 | 8 | 8 | 8 |
| Time–Intensity Data | | | | | |
| Candle power, 10³ Candles | - | 130 | 154 | 293 | 285 |
| Burning rate in.min⁻¹ | - | 2.62 | 3.01 | 5.66 | 5.84 |

The volume being same,

$$4/3\pi(R_1)^3N_1 = 4/3\pi(R_2)^3N_2 \text{ or } N_2/N_1 = (R_1/R_2)^3$$

Hence

$$S_2/S_1 = R_1/R_2 \tag{4.11}$$

Since $R_1 > R_2$ and hence $S_2 > S_1$

Thus, surface area increases as the particle size is reduced.

Further, let us consider how much surface area increases with the breaking of an ingredient grain.

Consider again the particles as spheres of radius 1 cm or cubes of side 1 cm being broken to smaller spheres and cubes, which consequently increases the particle numbers as shown in Figure 4.2. The surface area ratio is the ratio of new surface area of all new particles to the original surface area of the unbroken particle.

As per Figure 4.2, in case of a sphere or cube, for every 10-fold decrease in radius of sphere or side of cube, the total surface area increases by 10-fold. A 100 nm particle causes increase in total surface area by 100,000 times.

## 4.2.11  QUALITY AND QUANTITY OF BINDERS

Binder has a strong role in performance of the pyrotechnic composition. Table 4.12 provides the binder influence on burning rate of compositions (composition details at Table 4.6) as below.

| Sphere | Radius | Numbers | Total surface area | Surface area ratio |
|--------|--------|---------|--------------------|--------------------|
|   | 1 cm | 1 | $4\pi$ cm$^2$ | 1 |
|  | 0.1 cm | $10^3$ | $40\pi$ cm$^2$ | 10 |
|  | 0.01 cm | $10^6$ | $400\pi$ cm$^2$ | 100 |
|  | 0.001 cm | $10^9$ | $4,000\pi$ cm$^2$ | 1,000 |
|  | 0.0001 cm | $10^{12}$ | $40,000\pi$ cm$^2$ | 10,000 |
|  | 0.00001 cm (100 nm) | $10^{15}$ | $4,00,000\pi$ cm$^2$ | 1,00,000 |
| **Cube** | **Side** | **Numbers** | **Total surface area** | **Surface area ratio** |
|   | 1 cm | 1 | 6 cm$^2$ | 1 |
|  | 0.1 cm | $10^3$ | 60 cm$^2$ | 10 |
|  | 0.01 cm | $10^6$ | 600 cm$^2$ | 100 |
|  | 0.001 cm | $10^9$ | 6,000 cm$^2$ | 1,000 |
|  | 0.0001 cm | $10^{12}$ | 60,000 cm$^2$ | 10,000 |
|  | 0.00001 cm (100 nm) | $10^{15}$ | 6,00,000 cm$^2$ | 1,00,000 |

FIGURE 4.2  Increase in Surface Area with Breaking of Ingredient Grain.

**TABLE 4.12**

**Binder Influence on Linear Burning Rate [4]**

| Composition | Binder Viton A % | Density $(g.cm^{-3})$ | Rate of burning $(mm.sec^{-1})$ | | | | $BR_{av}$ | SD | VC |
|---|---|---|---|---|---|---|---|---|---|
| | | | Lactose-based compositions | | | | | | |
| 010/12 | 3 | 1.73 | 0.73 | 0.70 | 0.68 | 0.74 | 0.71 | 0.03 | 3.94 |
| 006/12 | 4 | | 0.69 | 0.67 | 0.66 | 0.70 | 0.68 | 0.02 | 2.40 |
| 001/12 | 5 | | 0.70 | 0.66 | 0.70 | 0.71 | 0.69 | 0.02 | 2.98 |
| 004/12 | 6 | | 0.68 | 0.67 | 0.64 | 0.67 | 0.67 | 0.02 | 2.30 |
| 005/12 | 7 | | 0.64 | 0.63 | 0.63 | 0.64 | 0.64 | 0.00 | 0.71 |
| | | | Ascorbic acid-based compositions | | | | | | |
| 013/12 | 3 | 1.73 | 2.25 | 2.23 | 2.23 | 2.18 | 2.22 | 0.03 | 1.30 |
| 012/12 | 4 | | 1.60 | 1.66 | 1.64 | 1.64 | 1.64 | 0.02 | 1.47 |
| 007/12 | 5 | | 1.21 | 1.21 | 1.19 | 1.20 | 1.20 | 0.01 | 1.01 |
| 011/12 | 6 | | 1.13 | 1.13 | 1.14 | 1.13 | 1.14 | 0.01 | 0.52 |

Binder reduces the linear burn rate (i.e., increase burn time), reduces the burn temperature, reduces the luminosity and affects sensitivity to impact, friction and spark. It also acts as secondary fuel. Table 5.4 may also be referred for effect of binder on performance of $Mg/NaNO_3$/binder composition. Table 6.12 may also be seen for binder effect on inverse burn rate, ignition temperature, breaking strength, density and sensitivity.

### 4.2.12 QUALITY AND QUANTITY OF BURN RATE MODIFIERS

The burn rate modifiers affect the burn rate by modifying the thermal properties of the composition.

Some burn rate modifiers like sodium bicarbonate increase energy of activation by absorbing heat energy for its decomposition at a certain temperature ($270^0C$). Thus, initially on external stimulus, there is rise in the temperature of composition and as the temperature approaches the decomposition temperature of sodium bicarbonate, the energy is absorbed causing decomposition of the sodium bicarbonate and thus the composition needs higher input energy for ignition. In addition, the gaseous products dilute the gaseous combustion products and thus retard the combustion.

The burn rate modifiers, eventually, also lead to lowering of luminosity, which is measured in candela or candle power. The quality of flame colour is also affected by burn rate modifiers. Table 4.13 provides the effect of burn rate modifiers in flare composition.

**TABLE 4.13**

**Effect of Burn Rate Modifiers in Flare Compositions** [7]

| | SR 563 | | | | SR 562 |
|---|---|---|---|---|---|
| Magnesium Gr 0 | 50 | 50 | 50 | 50 | 50 |
| Sodium nitrate | 46 | 43 | 40 | 37 | 35 |
| Lithographic varnish | 4 | 4 | 4 | 4 | 4 |
| Calcium oxalate | - | 3 | 6 | 9 | 11 |
| | Characteristics | | | | |
| Rate of burning (sec. inch$^{-1}$) | 11 | 13.1 | 14.8 | 16.0 | 17.3 |
| Specific intensity (candles.sq. inch$^{-1}$) | 112,000 | 106,000 | 77,500 | 68,400 | 64,200 |
| Specific light flux (candles-second. g$^{-1}$) | 40,800 | 45,000 | 36,600 | 35,100 | 35,200 |

It could be seen that burn rate modifier calcium oxalate has reduced the linear burn rate (inch.sec$^{-1}$) and luminosity substantially. The slight increase in light flux by 3% addition of calcium oxalate is due to thermal radiation of decomposition product calcium oxide but its further addition lowers the light flux.

Similar effect can be seen in a delay composition where ferric oxide acts as a burn rate modifier (retarder) [8]. Readers may refer Section 6.5 for more details on burn rate modifiers.

## 4.2.13 Uniformity of Mixing/Blending

Suitable blenders of appropriate size for blending are important as some blenders are meant for dry blending while others are meant for wet blending. Ingredients vastly differing in density and size make it necessary to use binders.

Density, volume, size, shape, specific surface of ingredients affects blending. A reduced particle size of ingredients improves the uniformity in blending. The intimacy of ingredients increases the reactivity since pyrotechnic compositions are largely solid–solid reaction.

A composition may undergo segregation of ingredients during storage and hence it is desirable that the composition kept in storage before use is blended so that the composition becomes uniform. However, wherever possible, granulation with binders helps in avoiding segregation of the ingredients as binders bind them well forming granules in the composition.

A uniformly blended composition mix is always desirable. A non-uniform mix would lead to variation in burn rate and consequently give variation in performance since the ratio of fuel to oxidiser would vary from place to place in the same composition and hence the uniformity of rate of combustion may not be achieved and the heat of combustion would vary from place to place. Samples of composition are therefore required to be checked, if feasible, for correctness of mixing of ingredients.

Uniformity/homogeneity of pyrotechnic compositions can be determined by their *Variation Coefficient (VC) [4]*. Every pyrotechnic composition has a variation coefficient which is defined as *"the standard deviation to average burning rate ratio, multiplied by 100"* to obtain some meaningful and comparable data.

$$VC = SD/BR_{av} \times 100 \qquad\qquad (4.12)$$

Where $VC$ = variation coefficient $SD$ = standard deviation

$BR_{av}$ = burning rate average.

Variation coefficient is the measure of the homogeneity of the pyrotechnic mixture. A variation coefficient below 10 indicates good uniformity of the composition while a value of the variation coefficient below 5 indicates very high uniformity of the composition. The Table 4.7 provides typical data for variation coefficient of composition.

## 4.2.14  LOADING PRESSURE (DEGREE OF COMPACTION)

The burn rate is affected by degree of compaction of the composition as explained below.

  a. *A composition filled in loose form like those in photoflash compositions or loose gunpowder in cartridges*:

Such composition would have very high combustion rate as entire mass would burn in few milliseconds as hot gaseous products and/or hot fluid products and/or hot solid products will have easy access to the whole composition.

  b. *Pressed composition where propagation of burning in layers is through hot gaseous or hot fluid products*:

Most of the pyrotechnic compositions are porous. A higher loading pressure gives better consolidation/density but reduces the porosity and permeability. Finer ingredients also give better compaction and reduction in porosity and permeability. It gives lower linear burn rates as hot combustion products (hot gaseous products or hot fluid products) are unable to move to the next layer in the composition. However, the mass burn rate would be high since the pressed composition would have higher density.

For example, in a boron–barium chromate composition, where inverse burn rate increases (i.e., linear burn rate decreases) with increase in loading pressure is shown in Table 4.14 and Table 4.15 for composition a and b respectively.

  a. *Barium Chromate–Boron 95/5 composition*:

  b. *Barium Chromate–Boron 90/10 composition*:

**TABLE 4.14**
**Effect of Loading Pressure on Burn Rate [9]**

| Loading pressure ($10^3$ psi) | 36 | 18 | 9 | 3.6 | 1.3 | 0.5 |
|---|---|---|---|---|---|---|
| Mean burning rate (sec.in$^{-1}$) | 1.69 | 1.60 | 1.49 | 1.39 | 1.29 | 1.21 |
| Mean burning rate (sec. g$^{-1}$) | 0.648 | 0.655 | 0.645 | 0.642 | 0.646 | 0.693 |
| % coefficient of variation | 1.2 | 0.6 | 0.7 | 0.7 | 0.8 | 0.8 |

**TABLE 4.15**
**Effect of Loading Pressure on Burn Rate [9]**

| Loading pressure ($10^3$ psi) | 36 | 18 | 9 | 3.6 | 1.3 | 0.5 |
|---|---|---|---|---|---|---|
| Mean burning rate (sec.in$^{-1}$) | 0.070 | 0.653 | 0.619 | 0.586 | 0.558 | 0.544 |
| Mean burning rate (sec. g$^{-1}$) | 0.272 | 0.276 | 0.280 | 0.287 | 0.297 | 0.309 |
| % Coefficient of variation | 1.5 | 0.9 | 1.1 | 1.6 | 2.0 | 1.8 |

Readers may see Table 24.2 regarding the effects of compression on density, porosity and permeability of a coarse fraction of atomized magnesium powder.

   c. *Pressed composition where propagation of burning is through hot solid products (gasless):*

It has been observed that sometimes the linear burn rate increases with increasing compaction pressure (especially with metallic fuels) due to increased conduction but subsequently the linear burn rate reduces with further increase in compaction pressure. An explanation to this phenomenon is that on increasing compaction pressure, the reactants come closer and hence the linear burn rate is increased. However, at further increased pressure, the hot products are unable to move inside the pores due to which the linear burn rate reduces. Another cause is that pores get easily clogged due to combustion products due to its smaller size under higher compaction.

The compaction of pyrotechnic composition is linked to sensitivity as well as quantum of composition that may be filled in an ammunition as explained below.

   i. Since the porosity is reduced during higher pressing load, thereby not allowing heat to penetrate further inside the composition, pyrotechnic composition in compact form are less sensitive to flash/flame than in loose form.
   ii. A higher density of pellet needs more energy input from external stimuli for formation of hotspot as more heat is required to heat more quantum of composition in a higher compacted composition.

iii. The compaction load is important for ensuring that the pressed composition in the form of a star or candle or a delay element does not break during handling, transportation and firing.

iv. The density of the pressed composition is governed by the densities of the individual ingredients like fuel, oxidiser and other ingredients. Generally, higher the density achieved, the higher the amount of composition that can be packed in a container having fixed volume. Hence, a higher output in terms of burning time or delay time is expected. This is important for ammunition since space in ammunition is always critical.

The compaction load is more pronounced in delay composition. The rate of pressing and dwell time also has an effect on burn rate. Tracers require pressing load 25% more than the *"setback pressure."* Hence, tracers are pressed at a very high pressure at 586–862 MPa (85,000–125,000 psi). Some igniter compositions are filled with manual ramming or tamping so that the composition is thus neither loosely filled nor is highly pressed enabling long flame from the igniter for igniting incendiary gel compositions.Thus, a judicious choice is required in deciding the degree of compaction while designing the ammunition.

Fred. L. McIntyre [10] has reported (Table 4.16) the bulk density and loading density of 180 pyrotechnic compositions.

Cracks in pressed compositions like stars or flare candles are undesirable. Cracks provide space for combustion products (hot gases, hot solid and fluid products) to move inside beyond the next layer and may cause a chunk of composition to be thrown away, thereby reducing the burn time with the possibility of a ground fire with chunks falling on ground. This typical effect of crack in pressed candle for illuminating ammunition is shown in the Figure 4.3 where it could be seen that a chunk of pressed composition falls from the candle when the burning layer reaches to the cracked area. This, therefore, does not give proper burning time. Similarly, cracks in delay compositions may give erratic burn time.

## TABLE 4.16
## Bulk Density and Loading Density of Pyrotechnic Compositions [10]

| Pyrotechnics | Bulk density | Loading density |
| --- | --- | --- |
| Initiatory | - | 1.71 ± 0.55 |
| Illuminants | 0.98 ± 0.31 | 2.21 ± 0.59 |
| Smoke | 0.85 ± 0.23 | 1.61 ± 0.27 |
| Gas | 1.39 ± 0.42 | 1.48 ± 0.27 |
| Sound | 0.98 ± 0.42 | - |
| Heat | 1.31 ± 0.49 | - |
| Time | 2.02 ± 0.45 | 3.62 ± 0.82 |

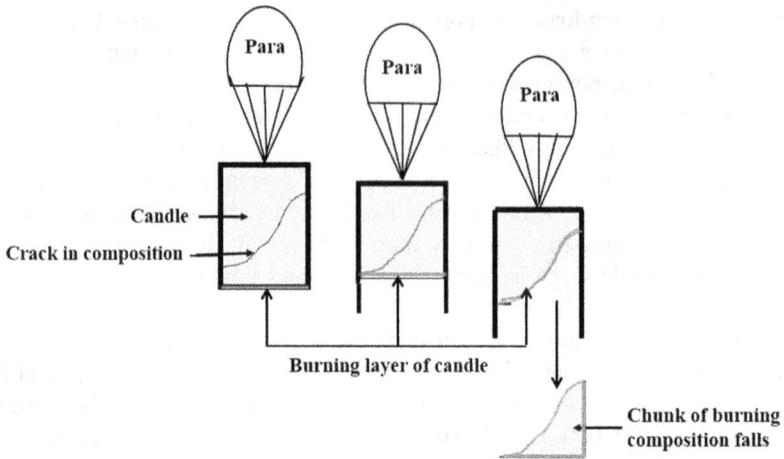

**FIGURE 4.3**   Effect of Crack in Illuminating Candle.

### 4.2.15   MOISTURE AND OR VOLATILE MATTER

Some ingredients in pyrotechnic compositions are hygroscopic and absorb moisture from the surrounding air. Moisture and/or volatile matter is an important factor as a part of the ignition stimuli is used up for vaporising the moisture or volatile matter and balance heat may not be sufficient to bring the composition to its ignition temperature. An analogy may be drawn to more time consumed in burning a moist wood since initial heat is used up in evaporating the moisture in the wood. The energy of activation (*Ea*) is affected due to moisture or volatile matter.

Table 4.17 [11] indicates the moisture absorbing tendency [water absorbed (g) by 2.0 g at 25°C (40–80 mesh)] of some of the oxidisers.

The Figure 4.4 shows that a pellet of composition on absorption of moisture increases its energy of activation and if kept in dryer for specified time and temperature, the original energy of activation may be obtained again.

Moisture has severe effects as listed below.

a. Causes failure to pick-up .ignition
b. Affects chemical stability of composition.
c. Reacts with metallic fuels, forming a layer of non-reactive oxide/hydroxide. This non-reactive layer affects the ignition and propagation of combustion resulting into non-ignition or less intense burning of the composition (thereby affecting the luminosity or the burn time or the heat output).
d. Reacts with metallic fuels like magnesium and produces gas which causes bulging and or cracking of paper cartridge cases during long storage.
e. Granulated composition with extra volatile matter during drying or storage causes cracking of granules as the volatile matter vaporises and leaves the granules and thus exposes the ingredients. Such composition shall lead to erratic results.

**TABLE 4.17**

**Water Absorbed by 2 g Oxidisers at 25 °C [11]**

| Size | Sodium nitrate | | Potassium nitrate | | Potassium chlorate | |
|---|---|---|---|---|---|---|
| (40-80 mesh) | Hrs. | $H_2O$ absorbed (g) | Hrs. | $H_2O$ absorbed (g) | Hrs. | $H_2O$ absorbed (g) |
| | 3 | 0.0713 | 2.75 | 0.0147 | 45.25 | 0.0660 |
| | 5.5 | 0.1355 | 18.75 | 0.1136 | 69.75 | 0.1055 |
| | 7.5 | 0.1970 | 25.25 | 0.1527 | 93.5 | 0.1392 |
| | 16 | 0.3294 | 42.75 | 0.2687 | 122.25 | 0.2340 |
| | - | - | 47 | 0.2896 | - | - |
| Ground | - | - | 7.25 | 0.0453 | 16 | 0.0248 |
| Very fine | - | - | 16.5 | 0.1056 | 69 | 0.1062 |
| | - | - | 40.5 | 0.2580 | 88.75 | 0.1488 |

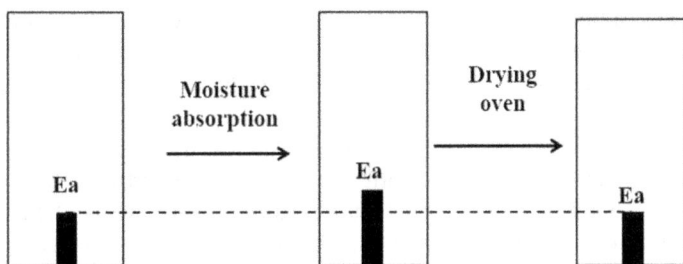

**FIGURE 4.4**    Effect of Moisture on Energy of Activation.

f. The composition with higher moisture/volatile matter takes more time to dry.

g. Reduces flowability of composition.

h. Reduces burn rate drastically by altering and loosening the composition

i. Affects compaction of the composition. The crushing load of pellets gets reduced considerably on exposure to moisture (see Table 24.1).

Following precautions are taken for moisture/volatile matter.

a. A humidity of 55% to 65% in the process (composition manufacture, filling, assembly and packing) is ideal for safety in manufacturing of pyrotechnic ammunitions.

b. Special precautions are required during composition manufacture in rainy season. All hygroscopic ingredients are required to be grinded and dried and sieved properly and then only to be mixed in an air-conditioned building with humidity of 55% to 65%.

c. All pyrotechnic composition batches are required to be tested for moisture content barring a few exceptions. The composition batches taken from storage rooms are required to be checked for moisture before use. If the moisture content is found more than specified, then the composition needs drying and blending before filling.

d. If the pellet with moisture content is dried for a short time, all the moisture will not be evaporated. Moisture gradient would be observed wherein maximum moisture will be in the centre of the pellet and the outer surface shall have least moisture. However, this moisture will get evenly distributed, if kept for long time, in a closed space. Only long duration of drying will expel all the moisture. However, some small percentage of moisture is considered necessary to be retained in the pellets for better controlled burning of pellets.

e. Signalling paper cartridges are invariably subjected to varnish or lacquer or NC dope to avoid moisture effect and packed in heat sealed polythene tubes to avoid ingress of moisture during long storage. Some naval ammunitions are packed in metallic containers by soldering the lid of the container.

It should be noted that absorption of moisture by the ingredients would be more if the particle size of ingredient is very fine, since more surface area is exposed. Similarly, absorption of moisture by the ingredients will be more if the composition is in loose form due its porosity while it will be less in pressed form.

A typical rubber valve is used in some cartridge cases to avoid ingress of moisture where the rubber valve is assembled in the central hole of a plastic disc having grooves over which a rubber ring is assembled. This plastic disc is assembled at the top of the metallic cartridge case and the cartridge case is turned over. If any gaseous product is formed during storage of the ammunition, the same is vented out through the rubber valve. This avoids any bulging of the cartridge case since it allows gaseous products to move out of the cartridge case and at the same time does not allow any ingress of moisture as shown in Figure 4.5.

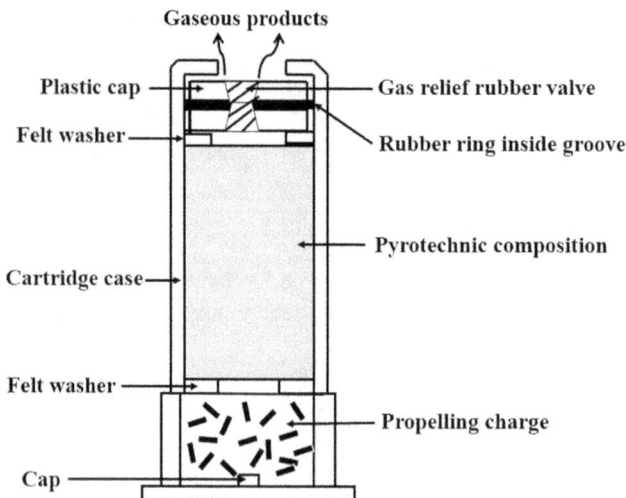

**FIGURE 4.5**  Cartridge with Gas Relief Rubber Valve.

## 4.2.16  EXPOSED SURFACE

Exposed surface of pyrotechnics is required for ignition to obtain the desired special effects. It must positively pick-up ignition from the flash source. The following may be noted for a pressed composition:

a. Burning is a surface phenomenon and burning takes place layer by layer, perpendicular to the layer in a pressed composition.
b. The more the surface area of layer exposed, the higher would be the mass burn rate.
c. Serration or holes in pellets increase the exposed surface area and for faster pick up of ignition.
d. In a loose composition, the particle surfaces are exposed due to porosity, resulting into faster combustion. In a photoflash composition, the composition is not compacted and remains in loose porous form in a closed system so that mass burning in milliseconds takes place (due to large exposed surfaces of ingredients) for better flash output for aerial night photography. Similarly, gunpowder used as propelling charge in several ammunitions is loosely filled and burns in milliseconds.
e. Burn time of pellets depends upon the web size of the pellet. *The web is the least distance between outer burning layer and the last burning layer in a pellet.* The burn rate is given by

$$\text{Linear Burn Rate} = \frac{\text{Web Thickness}}{\text{Time of Burning}} \tag{4.13}$$

The smaller the web, the lesser would be the burning time. Similarly, the higher the web, the more would be the burning time (Figure 4.6).

The mass burning of a pellet is termed as degressive, neutral or progressive, based upon its rate of burning with time as given below.

a. *Solid Pellet of Any Shape:* On burning, this type of pellet will burn with a constant diminishing surface and therefore will have a decreasing burn rate with time. This is known as *degressive burning.*
b. *Solid Pellet with a Central Hole:* On burning, the outer surface will be decreasing while inner surface will be increasing. These two phenomena will almost offset each other and the grain will have almost same surface area and therefore would burn with same burn rate. This is known as *neutral burning.*
c. *Solid Pellet Made with More than One Hole:* On burning, the increase of inner surface will be more than the decrease in the outer surface. Such pellets will burn with increasing burn rate and is known as *progressive burning.*

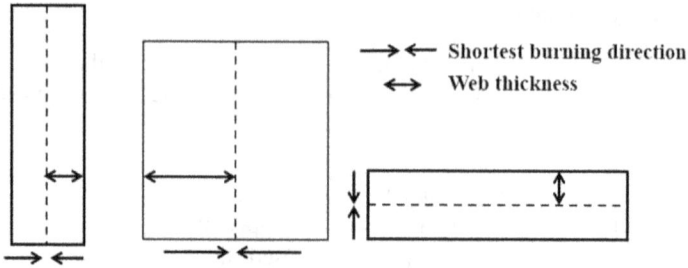

**FIGURE 4.6**  Burning Time Dependence on Web Thickness.

Some infrared pyrotechnic composition, after formation into pellets, are required to be grooved with sawing and milling machines (with water cooling arrangement or under water) for increasing the surface area of the pellet so that desired infrared radiation output and burning time are achieved. A higher surface area would give fast mass burning (low burn time) with increased output of infrared radiation.

In the smoke canister, containing annular pellets, the burning is not like a *cigarette type* burning (as in candles, delays, etc.). The burning of layers in smoke canisters moves *radially* (Figure 4.7) as it is ignited from the centre and the burning surface area increases as the burning radius increases. This causes emission of more quantum of smoke with time. This central burning also allows fluid combustion products (tar material) to settle at the bottom and does not inhibit escape of smoke.

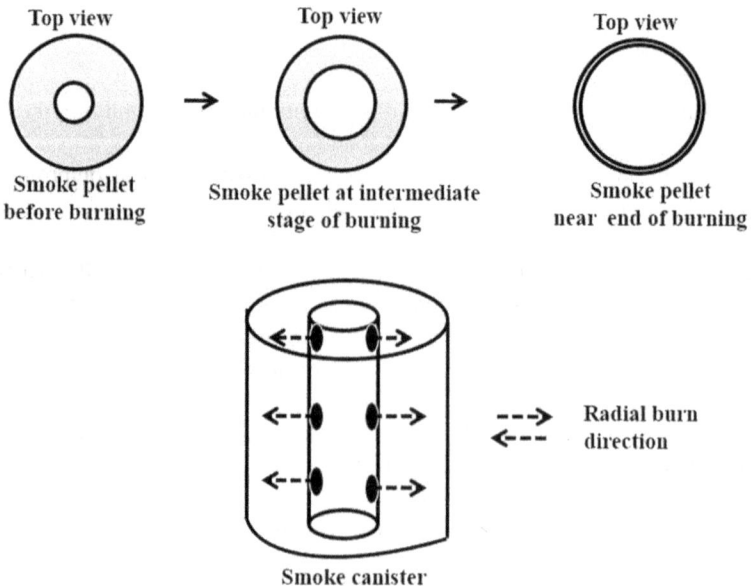

**FIGURE 4.7**  Radial Burning of Smoke Canister.

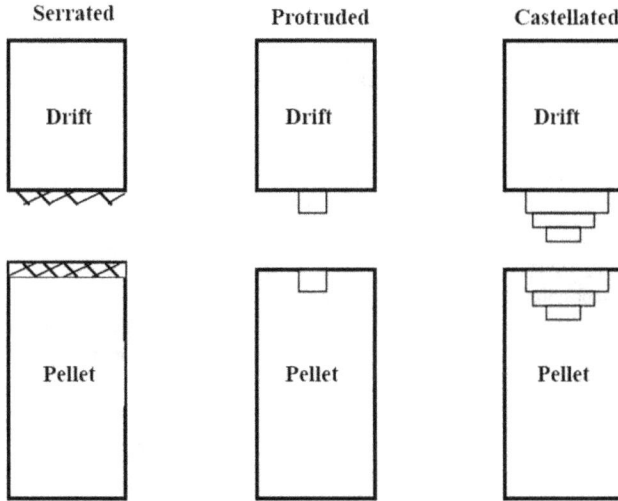

**FIGURE 4.8**   Methods to Increase Surface Area of Composition.

To increase the ease of ignition pick-up, punches with serrated marks or protrusion or castellated (step ladder) shape, shown in Figure 4.8, are used. Several shapes of pellets like circular, rectangular, square, etc., may be made. The shapes with cavity on exposed surface area (for retaining flash inside the cavity for ease of picking ignition) are possible by suitably designing a punch and mould used for pressing or by milling operation. The shape of a pellet may be varied as per design requirement of ammunition like cylindrical pellet for cartridges or triangle shaped for 155-mm red phosphorous or rectangular serrated pellet for infrared flares, each having different exposed surface area. Exposed surface of the pellets has strong bearing on burning output characteristics like luminosity, infrared radiation, burn time, etc.

Illuminating candles, which are pressed in a container, burn only from the exposed top surface linearly and hence provide more duration of burning. However, a signalling star (a pressed pellet) while ejected out from weapon burns from all sides, since all its sides are exposed and hence gives low burn time. Similarly, infrared flares, when ejected out from dispenser unit of aircraft burn from all sides since it is exposed from all sides and gives low burn time. Table 4.18 gives the comparative effect of surface exposure of some ammunitions on burn time.

## 4.2.17   PRODUCTS OF COMBUSTION

The pyrotechnic composition with variation in ingredient proportions gives different combustion products (and in different proportions) and heat of combustion as shown in Table 4.19.

These products of composition affect the performance of the composition. Every pyrotechnic composition for ammunition is suitably designed to provide useful products of composition. A few examples are given below.

**TABLE 4.18**

**Comparative Effect of Surface Exposure on Burn Time**

| Ammunition type | Exposure of surface | Burn time |
|---|---|---|
| Illuminating ammunition (composition pressed in candles) | Access of flame/heat to only top exposed surface of pressed candles | Burning layer by layer and hence gives long burn time for observation (several seconds) |
| Illuminating and signalling ammunition (composition pressed as star) and infrared flare decoys (pressed as flares) | Access of flame/heat to all outer exposed surfaces of pressed star or flare | Medium-fast burn time through all exposed surfaces and hence gives less burn time (few seconds) |
| Photoflash ammunition (composition loosely filled) | Access of flame/heat to whole mass of composition | Quick en masse burning and hence least burn time (milliseconds) |

**TABLE 4.19**

**Proposed Reactions for the $Si/PbO_2$ System [12]**

| Si (wt. %) | Reactions | Products | $-\Delta H$ (MJKg$^{-1}$) |
|---|---|---|---|
| 2.85 | $Si + 4PbO_2$ | $Pb_4SiO_6 + O_2$ | 0.71 |
| 3.77 | $Si + 3PbO_2$ | $SiO_2 + 3PbO + 1/2\ O_2$ | 1.00 |
| 5.55 | $Si + 2PbO_2$ | $Pb_2SiO_4$ | 1.64 |
| 7.26 | $2Si + 3\ PbO_2$ | $2PbSiO_3 + Pb$ | 1.89 |
| 10.5 | $Si + PbO_2$ | $SiO_2 + Pb$ | 2.38 |
| 15.0 | $3Si + 2PbO_2$ | $SiO_2 + 2SiO + 2Pb$ | 1.00 |
| 19.0 | $2Si + PbO_2$ | $SiO_2 + Pb + Si$ | 2.15 |
| 26.1 | $3Si + PbO_2$ | $Pb + Si + 2SiO$ | −.23 |

a. Priming compositions on combustion provide hot slag products for heating the next layer of composition so that ignition takes place.

b. Delay composition ingredients are so chosen as to provide gasless products for ammunition without vent or ammunition meant for high altitudes (low atmospheric pressure) so that no pressure is allowed to build up inside to avoid variation in burn rate.

c. *The combustion products provide the colour of flare signal* and hence right choice of ingredients is essential. For example, in colour signal composition, strontium monochloride gives red signal while barium monochloride provides green signal. Hence strontium nitrate with a chlorinating agent is used for production of strontium monochloride for red signal while barium nitrate with chlorinating agents is used for production of barium monochloride for green signals.

d. In illuminating composition containing magnesium, magnesium forms MgO as a combustion product and improves luminosity. Some part of magnesium as vapour reacts with atmospheric air to produce light, thereby enabling loading of excess magnesium to improve efficiency of the composition.
e. Presence of hydrogen, nitrogen, sulphur and chlorine element in the ingredients lead to corresponding gaseous products. These gaseous products have a special role as under.
    i. In case of illuminating, signalling stars and tracers, the composition provides little gaseous products (15%–20%) for the flame and increase the luminosity.
    ii. Large amount of gaseous products of combustion is required for screening since a large amount of combustion products are required to be expelled. Similarly, dyes (40%–50%) are required to be expelled from the colour smoke canisters. Hence organic fuel like lactose is largely used in signalling colour smoke composition so as to obtain maximum amount of $H_2O$, CO and $CO_2$ at low combustion temperatures.
    iii. Incendiary compositions have very less gas output compared to smoke and solid rocket fuel. In case of incendiary compositions, large gaseous combustion products are not desirable as otherwise gaseous products would take the heat away from the surface of combustion. Combustion products of incendiary composition are required in hot fluid state for required incendiary effect. However, a small number of gaseous products is essential in expanding the fire spot for destruction.

## 4.3  DESIGN FEATURES OF AMMUNITION

Characteristics of ammunition affect pyrotechnic performance as given below.

### 4.3.1  METHOD OF INITIATION

An external stimulus like heat, spark, flame or friction, etc. is essential for initiating the combustion of pyrotechnic composition. Such external initiation must be steady to raise the temperature of the pyrotechnic composition to its ignition temperature. To ensure ignition, in many compositions, certain composition known as priming composition or first fire charge with booster charges are used over pyrotechnic composition. The details are given in Chapter 19.

A spark or flame of high magnitude from an explosive is not desirable in pyrotechnic composition, as it may result in burning failure. This may be explained by the fact that on providing sudden, extreme high energy, the internal energy of all the atoms/molecules in a small part of the composition present on the surface rises much above the energy of activation level and causes a very high increase in rate of combustion leading to minor explosion. This results in part of composition being blown off and further propagation of combustion is stopped.

## 4.3.2  CONTAINER MATERIAL

Container material has an effect on the energy feedback (Efb) for sustaining the burning of the composition. Metallic containers are good conductors of heat. Container material has two distinct effects on burn rate.

a. A metallic container would transmit some heat during the burning of the composition to atmosphere through the container body, thereby taking away heat from combustion zone and thus slowing down the combustion rate since only a fraction of the heat would be carried to the next layer of composition for burning.
b. The heated metallic container would slowly heat the full composition in the container, while requirement is to heat the composition layer by layer.

This may be understood by heat absorption of metallic container as heat is absorbed Q (Joules) by the metallic container as per equation

$$Q = mc\Delta t \tag{4.14}$$

where Q = heat absorbed (Joules), $m$ = mass (kg), $c$ = specific heat of container material (J. (kg. $K^0)^{-1}$) and $\Delta t$ = temperature rise ($K^0$)

Therefore, for same quantum of heat provided by the composition during combustion, rise in temperature in thinner material (for mass being less and $c$ being constant) will be more and will affect burning properties of even much lower layers in the column due to conduction of heat downwards. If the thickness of the metallic container is increased, then it will have more outer surface area resulting in more heat loss to atmosphere as well as it will absorb more heat from the composition due to higher mass. The problem is more acute for illuminating compositions as metal powders are used, which generates high temperature. To avoid this problem, three practices are followed for such high temperature compositions.

a. *Use laminated paper liner:* These are used inside metal containers for holding the composition so that metal container is not in direct contact with the composition. This also avoids friction during consolidation with metallic containers and also avoids compatibility problems associated with pyrotechnic composition and the metallic container. Generally, laminated paper liners are kept longer to accommodate composition during filling/pressing due to low bulk density of composition and after pressing is over; the excess length of paper liner is trimmed to metallic container size.
b. *Coat the internal surface of the metallic container with a catalysed resin paste:* A coating of catalysed resin is given enough time to dry to form a protective coating and then the composition is pressed in the container, thus avoiding conduction of heat to metallic container.
c. *Pour catalysed resin around the pressed pellet:* The pressed pellet is kept centrally in a plastic liner (with an outside metal liner support) of diameter little bigger than the pressed candle so as to have a small gap between the

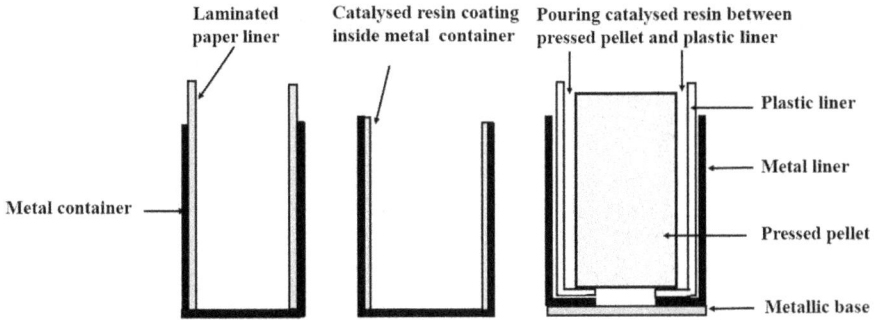

**FIGURE 4.9** Methods to Avoid Heat Loss in Pressed Metallic Container.

pellet outer diameter and the plastic liner internal diameter. Catalysed resin is poured in the gap from the sides and the resin is allowed to dry. The dried resin forms a thick coat of resin around the pellet and acts as an inhibitor and allows layer by layer burning of the pellet.

All the three practices are shown in Figure 4.9.

Tracers mostly use waxed paper sleeve in the tracer capsule to avoid heat loss.

However, the above methods cannot be followed with delay tubes due to their lower diameter. Delay housing material affects the burn rate of a gasless combustion in a delay tube where heat transfer is mainly through conduction. One may find variations in linear burn rate using the same composition in different metal housing tubes like those made up of aluminium, lead, copper or steel. Thermal effusivity and thermal diffusivity are important aspect in heat transfer in gasless combustion of delay pyrotechnic compositions and are dependent on density, specific heat and thermal conductivity of the housing tube.

*(a) Thermal Effusivity (e):* Thermal effusivity is related to the speed at which thermal equilibrium can be reached. In other words, a material's thermal effusivity is the measure of its ability to exchange thermal energy with its surroundings. Effusivity (sometimes called the heat penetration coefficient) is the rate at which a material can absorb heat. It is the property that determines the contact temperature of two bodies that touch each other. It is defined as *"the square root of the product of the material's thermal conductivity and its volumetric heat capacity (ρ Cp)"* and is mathematically expressed as

$$e = (\lambda \rho Cp)^{1/2} \qquad (4.15)$$

where $\lambda$ = thermal conductivity of the material ($W.m.^{-1}K^{-1}$), $\rho$ = density of the material ($Kg.m^{-3}$). $Cp$ = specific heat capacity of the material ($J. Kg^{-1}K^{-1}$).

The effusivity of materials varies due to their differing abilities to transfer heat. This is due to the differences in heat transfer through and between particles, and is therefore a function of particle size, particle shape, density, morphology, crystallinity and moisture content.

*(b) Thermal Diffusivity (α):* It is the thermal conductivity divided by density and specific heat capacity at constant pressure. It measures the ability of a material to conduct thermal energy relative to its ability to store thermal energy. It is mathematically expressed as

$$\alpha = \frac{\lambda}{\rho Cp} \tag{4.16}$$

A higher thermal effusivity and thermal diffusivity will yield rapid transmission of heat from the composition to the walls of the tube, while a lower thermal effusivity and diffusivity would result in a large part of the heat being retained by the delay body and only a small part would be conducted through. The densities and thermal properties of some materials are shown in Table 4.20.

Table 4.21 shows the inverse burn rates of five samples each in both 2024-T3 aluminium and 304 L stainless-steel tubes. Partial propagation was observed in the composition containing 25% $Sb_2O_3$. In this case, three out of the five aluminium tube items functioned properly, allowing an average time to be determined. All five stainless-steel tubes containing this composition failed to propagate completely.

Table 4.21 shows faster linear burn rates in aluminium tubes than in stainless-steel tubes. It has been explained [13] that this can be attributed to differences in the thermal effusivity of aluminium and stainless steel, which changed the reaction front heat loss and ultimately influenced the degree of sample pre-heating. Since aluminium has both higher thermal effusivity and thermal diffusivity than stainless

## TABLE 4.20
### Densities and Thermal Properties of Materials [13 ]

| Material | Density ($\rho$) (Kg.m$^{-3}$) | Specific heat capacity (Cp) (J. Kg. $^{-1}$K$^{-1}$) | Thermal conductivity ($\lambda$) (W.m.$^{-1}$K$^{-1}$) | Thermal effusivity (e) (W. s.$^{1/2}$ m.$^{-2}$K$^{-1}$) | Thermal diffusivity ($\alpha$) (m$^2$.s$^{-1}$) |
|---|---|---|---|---|---|
| Copper (pure) | 8,930 | 385 | 398 | 36,991 | $1.16 \times 10^{-4}$ |
| Aluminium (pure) | 2,700 | 900 | 210 | 22,590 | $8.64 \times 10^{-5}$ |
| Silicon (pure) | 2,329 | 713 | 124 | 14,350 | $7.47 \times 10^{-5}$ |
| Lead (pure) | 11,350 | 129 | 33 | 6,951 | $2.25 \times 10^{-5}$ |
| Aluminium alloy 2024–T3 | 2,780 | 875 | 121 | 17,156 | $4.97 \times 10^{-5}$ |
| Stainless steel 304 L | 8,030 | 500 | 16 | 8,015 | $3.99 \times 10^{-6}$ |

Data obtained from Matweb (www.matweb.com).

**TABLE 4.21**

**Average Inverse Burn Rate for Each Composition in Aluminium and Stainless-Steel Tubes [13]**

| $Si/Bi_2O_3/Sb_2O_3$ | Aluminium tubes (s.cm$^{-1}$) | Stainless-steel tubes (s.cm$^{-1}$) |
|---|---|---|
| 30/70/0 | 0.54 (0.0061) | 0.68 (0.073) |
| 30/65/5 | 0.74 (0.053) | 0.92 (0.077) |
| 30/60/10 | 0.91 (0.012) | 1.12 (0.079) |
| 30/55/15 | 1.04 (0.051) | 1.51 (0.10) |
| 30/50/20 | 1.41 (0.12) | 2.25 (0.13) |
| 30/45/25 | 1.82 (0.22) only 3 burnt properly | All 5 partial propagation |

(Standard deviations in parentheses).

steel, the heat generated by the reaction, and consequently transferred to the inner walls of the tube, is transmitted more rapidly down the walls in the aluminium tubes. It is hypothesised that the compositions inside the aluminium tubes experience more pre-heating than those in stainless-steel tubes, therefore enhancing reactivity by transmitting heat to unburnt layers more effectively. In the stainless-steel tubes, more of the thermal energy is confined to the composition, with less propagation through the tube walls. The burn times in these tubes are therefore longer.

However, there are reports where the above analogy does not apply. Hence, more research work is needed to determine the mechanism of burn rates in delay tubes.

It has been found [13] that in addition to tube material, the tube length, inner diameter, and outer diameter can have a large effect on delay performance. Changes in the surface/volume ratio of the tubes can influence burn rate, or even quench the reactions in some cases.

### 4.3.3 COLUMN DIMENSION

The combustion output increases with an increase in the surface area of burning since the volume of pyrotechnic composition generating heat is proportional to the square of the diameter of the column while heat loss through surface is proportional to the diameter of the column to a certain extent. Thus, the ratio of heat generated to the heat loss will be proportional to the diameter ($D^2/D = D$) and will be more if the diameter of the column of pressed pyrotechnic composition is more, and hence the combustion output would increase up to a certain value of diameter.

Thus, a composition pressed in a 10 mm diameter container would burn slower than the same composition pressed in a 50 mm diameter container.

Table 4.22 [14] gives the effect of flare diameter on luminosity and luminous efficiency of a flare composition containing 52% magnesium, 40% sodium nitrate and 8% polyester resin (a combination of 97.5% unsaturated polyester, 2% methyl ethyl ketone peroxide and 0.5% cobalt naphthenate).

**TABLE 4.22**

**Effect of Flare Diameter on Luminosity [14]**

| Diameter (mm) | Luminosity (cd) | Burning time (s) | Luminous efficiency (cd.s.g$^{-1}$) |
|---|---|---|---|
| 10 | 5,200 | 13.5 | 17,100 |
| 17 | 30,400 | 12.7 | 32,100 |
| 23 | 51,000 | 14.3 | 36,500 |
| 30 | 94,800 | 14.9 | 37,200 |

**TABLE 4.23**

**Effect of Aperture Diameter of a Tracer on Luminosity [15]**

| Tracer aperture diameter (mm) | Luminous intensity in perpendicular direction (Kilo Candles) |
|---|---|
| 6.0 | 2.6 |
| 8.0 | 3.7 |
| 11.0 | 7.2 |
| 14.0 | 13.2 |
| 16.0 | 26.4 |

Similar inference can be drawn from Table 30.4 where luminosity increases with increase in bore diameter (also candle diameter) of ammunition though compositions used may differ.Similarly, the effect of aperture diameter of a typical tracer on luminosity has been reported in Table 4.23.

Column dimension has more pronounced effect if the ambient temperature is low since heat loss is significant compared to the heat loss when the ambient temperature is high. The low diameter of delay columns and low ambient temperature results in significant heat loss to surrounding during combustion and may even stop further combustion. This is important for small diameter delay columns where the heat of combustion is low.

It may therefore be concluded that the heat losses are more predominant with reduction of column diameter and reduction in ambient temperature, which result in low burning rate. Failures in delay columns of lower dimensions at low temperatures are therefore expected for such pyrotechnic composition, which have very low burning rate.

### 4.3.4   VENTING /CONFINEMENT

When a pressed composition is enclosed in a container, its rate of burning would depend upon the extent of enclosed volume.

a. If the container has a large volume and the amount of burning composition is less, the confinement may not have any appreciable change in burning rate as there would not be any appreciable change in temperature and pressure inside the container.

b. If the container enclosed volume is less but has a vent, then also the burn rate would not vary much since venting would release heat and pressure.

c. If the container enclosed volume is less and without any vent, then the rate of burning would increase rapidly due to sudden increase in temperature and pressure inside the container.

In case of loose pyrotechnic composition in a closed container, the burning would not take place layer by layer but would take place in the whole mass since the combustion products like flame, hot gases and hot slag (solid and fluid products) will have easy access through the pores between particles, resulting the entire mass of composition being heated by convection and radiation of heat transfer, thereby increasing the pressure inside the container leading to faster burning of entire mass of composition. This would lead to sudden, steep rise in temperature and pressure and may even result in explosion (Figure 4.10).

Thus, an increase in charge/volume ratio shall result in higher burn rate in a closed container. This is very important for all gas-producing pyrotechnic compositions. To control the burn rate, the excess pressure may be reduced by suitable venting in the ammunition and reducing the burn rate. Venting of gases thus lowers the burn rate more, in contrast to those cases where the venting is not allowed.

Venting in one typical fuze having gaseous delays is accomplished by providing holes of a component closed by wax or aluminium foils to act as sealant against moisture, as shown in Figure 4.11. The wax or foil prevents the composition from external moisture during storage. These wax or aluminium foils are thrown out

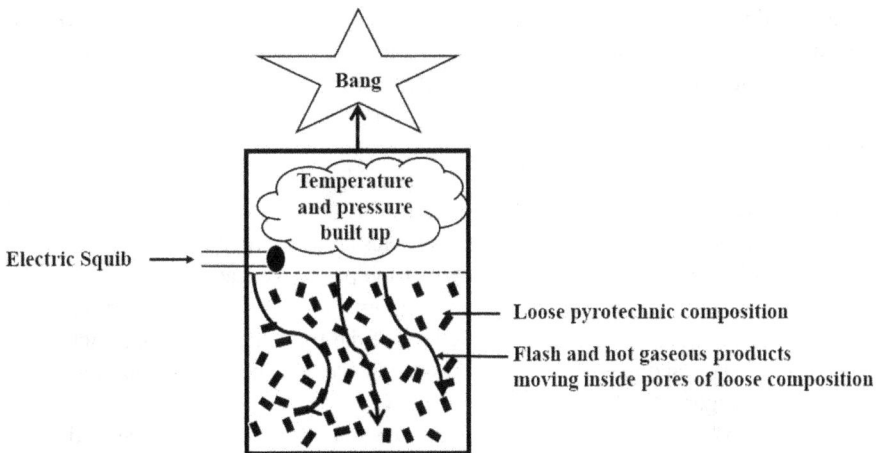

**FIGURE 4.10**  Rapid Combustion in a Closed Container.

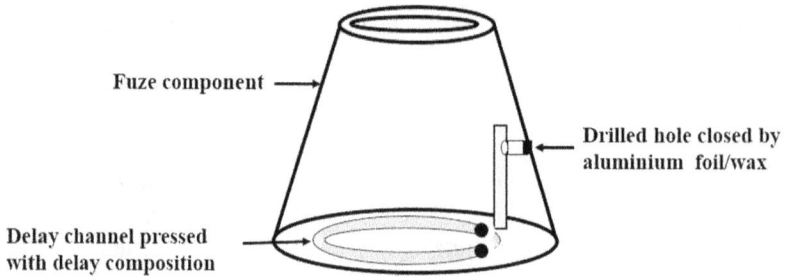

Fuze component ⟶

Drilled hole closed by
aluminium foil/wax

Delay channel pressed
with delay composition

**FIGURE 4.11**   Venting in a Typical Fuze.

when the pressure inside increases, and this does not allow the pressure built inside
to affect delay burn time.

### 4.3.5  SPIN

Spin increases the linear burn rates and thus reduces the burn time of illuminating
candles and tracers. Several artillery illuminating ammunitions have a system to
reduce the spin effect on the candle by providing break flaps (fins) in the canister
that open and reduce the spin effect.

In some high-calibre illuminating ammunition, an auxiliary parachute (or brake
parachute) with delay element is introduced so that after fuze functioning, the
canister (containing candle) with auxiliary parachute ejects out and hangs in the air
with auxiliary parachute. The break flaps (fins) of canister are opened, thereby
reducing the spin effect on the canister and the delay in canister then ignites and
ejects the main parachute with candle, thus allowing the main parachute to be
deployed with a candle.

Tracer compositions produce hot slag and hot gases. Linear burn times of tracers are
severely reduced due to high burn rates during spinning of the ammunition. Hence it is
desirable to ensure the burning time of pressed tracer in static proof much higher than
actually required in the ammunition during dynamic proof to compensate for the fast
burning under spin conditions. Tracer burning time reduces 50% to 33% of the static
burning time, depending upon the tracer composition and the spin rate. The cause for
this effect is attributed to less loss of slag during higher spin rate. As the centrifugal
force increases with radial distance, the burn rates shall be more with higher diameter of
the tracers. At higher spin of the tracer, the hot slags are centrifuged on the side walls of
the tracer and thus cause increased heat flow at the periphery leading to faster burning
than the composition at the centre, thereby forming a convex shape of the tracer pellet
composition. The slag reduces the open surface of the tracer pellet and does not allow
hot gases to escape and thus builds extra pressure inside, leading to faster burning of the
composition (Figure 4.12).

This convex/oval surface has a higher surface area than the flat surface, and leads
to faster burning. Thus, the major effect of spin is a change from a *"cigarette-type,"*
i.e., straight layer to layer burning at static condition to a *convex/oval* type burning
during spinning.

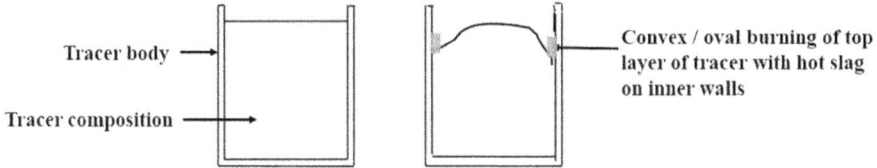

**FIGURE 4.12** Tracer Burning with Spinning.

Spin effect on linear burn rate, percent slag and candle power of pyrotechnic compositions [16] containing magnesium/strontium nitrate may be seen as given below. The stoichiometric equation is

$$5Mg + Sr(NO_3)_2 = 5MgO + SrO + N_2 \qquad (4.17)$$

The fuel oxidiser composition is given in Table 4.24.

The burn rate and percentage change from static conditions are in Table 4.25.

The slag percent and percent change from static condition are given in Table 4.26.

The candle power versus spin rate are at Table 4.27.

The apparatus setup [16] is shown in Figure 4.13. The major components consisted of a spinner (to hold both the sample and impart spin); a photocell (to detect and convert light output to measurable electrical signals) and an oscilloscope (to display the output). Auxiliary equipment included a photometer, a pyrotechnic ignition system, a stroboscope and an oscilloscope recording camera.

In case of delays [7], high rates of spin cause gassy fillings to burn faster than they do at rest and gasless compositions burn slower

## 4.4 TYPE OF ENVIRONMENT

The parameters of environment affect the pyrotechnic performance as explained below.

### 4.4.1 Atmospheric Pressure

An increase in burn rate is mostly observed with increase in the atmospheric pressure. The phenomenon is more pronounced when the composition gives more

**TABLE 4.24**
**Fuel Oxidiser Composition [16]**

| Composition | Magnesium | Strontium nitrate |
|---|---|---|
| Stoichiometric | 36.3 | 63.7 |
| Fuel deficient ((-) 30 mole %) | 28.8 | 71.2 |
| Fuel rich ((+) 30 mole %) | 42.8 | 57.2 |

**TABLE 4.25**

**Linear Burn Rate and Percentage Change from Static Condition [16]**

| Spin rate (rpm) | Deficient Mg | | Stoichiometric Mg | | Excess Mg | |
|---|---|---|---|---|---|---|
| | Burn rate (in.sec$^{-1}$) | % Change | Burn rate (in.sec$^{-1}$) | % Change | Burn rate (in.sec$^{-1}$) | % Change |
| 0 | 0.059 | - | 0.120 | - | 0.160 | - |
| 20,000 | 0.094 | 59.3 | 0.159 | 32.5 | 0.195 | 21.8 |
| 28,000 | 117 | 101.8 | 0.160 | 33.3 | 0.195 | 21.8 |
| 35,000 | 118 | 103.5 | 0.179 | 49.1 | 0.208 | 30.0 |
| 43,000 | 139 | 135.6 | 0.186 | 54.9 | 0.227 | 41.9 |

**TABLE 4.26**

**Slag Percent and Percent Change from Static Condition [16]**

| Spin Rate (rpm) | Deficient Mg | | Stoichiometric Mg | | Excess Mg | |
|---|---|---|---|---|---|---|
| | Slag % | % Change | Slag % | % Change | Slag % | % Change |
| 0 | 69.5 | - | 58.3 | - | 49.1 | - |
| 20,000 | 65.6 | (-)5.6 | 63.0 | 8.1 | 57.4 | 16.9 |
| 28,000 | 72.4 | 4.2 | 64.5 | 10.6 | 62.2 | 26.6 |
| 35,000 | 66.7 | (-)4.1 | 60.7 | 4.2 | 63.7 | 29.6 |
| 43,000 | 70.0 | 0.1 | 66.7 | 13.1 | 62.5 | 27.2 |

**TABLE 4.27**

**Candle Power Versus Spin Rate [16]**

| Spin rate (rpm) | Maximum candlepower | | |
|---|---|---|---|
| | Deficient magnesium | Stoichiometric | Excess magnesium |
| 0 | 160 | 1,400 | 5,100 |
| 20,000 | 700 | 4,600 | 4,800 |
| 28,000 | 1,350 | 3,200 | 6,200 |
| 35,000 | 1,250 | 3,750 | 5,750 |
| 43,000 | 950 | 4,550 | 9,050 |

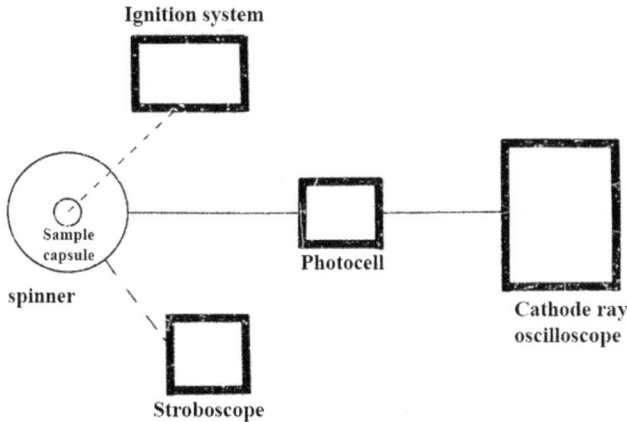

**FIGURE 4.13**   Pyrotechnic Spin Performance Test Apparatus (Reprinted from [16]).

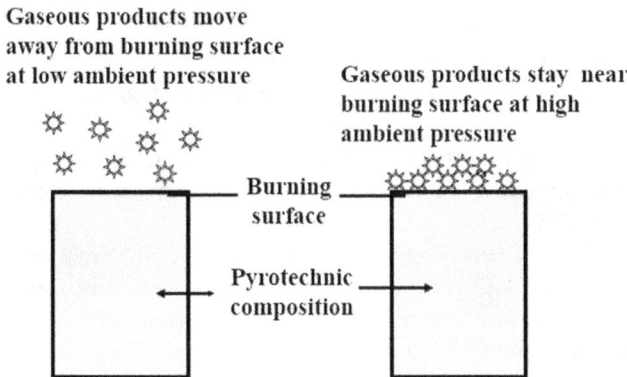

**FIGURE 4.14**   Effect of Pressure on Combustion.

gaseous products. At low pressures in high altitudes, the hot combustion products move away from the burning layer and hence do not contribute in the ignition of next layer. At higher pressures, the hot combustion products remain near the burning layer and hence contribute in the ignition of next layer and raise the temperature of reaction resulting in fast burning (Figure 4.14).

Therefore, a composition producing gas will be affected by external pressure. Hence, if there is no variation in combustion rate of a composition under varied external pressures, it may be concluded that the composition combustion is "gasless" combustion like thermite compositions that produce almost no gas. The organic fuels like tetra nitro oxanilide and some binders like shellac produce large amount of gases and hence burn rate of compositions containing such ingredients is dependent on pressure.

Also, the burn rate would decrease at high altitudes due to low atmospheric pressure for such compositions, which depends upon the atmospheric oxygen

**TABLE 4.28**

**Effect of Altitude on the Linear Burn Rate of Al-oxidant and Mg-Oxidant Compositions [17]**

| Fuel oxidant | Burning rate (cm.s$^{-1}$) | | | Percent decrease at altitude |
|---|---|---|---|---|
| | Ambient | Altitude | | |
| | | 20,000 ft | 40,000 ft | |
| Al-NaNO$_3$ | 0.318 | 0.168 | - | 47 |
| Al-KClO$_4$ | 0,210 | - | 0.101 | 52 |
| Al-Teflon | 0.151 | - | 0.081 | 47 |
| Mg-NaNO$_3$ | 0.143 | 0.125 | - | 12 |
| Mg-Teflon | 0.218 | - | 0.201 | 8 |

available for burning. For example, illuminating composition containing magnesium and sodium nitrate where magnesium is used more than the stoichiometric ratio, the excess magnesium will react with atmospheric oxygen and contribute to illumination though at a slower rate at high altitudes, since oxygen availability is low. This is important for use of illuminating ammunition at high altitudes or hilly terrains.

The effect of altitude on linear burn rates of some pyrotechnic compositions is shown in Table 4.28.

Signalling red and yellow flares are found to have increased colour value at simulated higher altitudes, while no changes are observed in green signal flares [6].

At a very low pressure, the combustion becomes difficult and at extreme low pressures, the combustion may stop. A decrease in atmospheric pressure results in increased candle flame size (Figure 4.15). It can be said that candle flame size is inversely proportional to air pressure.

The increased flame (movement of hot burning gaseous products) at lower atmospheric pressure is due to hot gaseous products that have a tendency to move away from the burning surface taking along with it some of the hot combustible products. On the other hand, the decreased flame at higher atmospheric pressure is

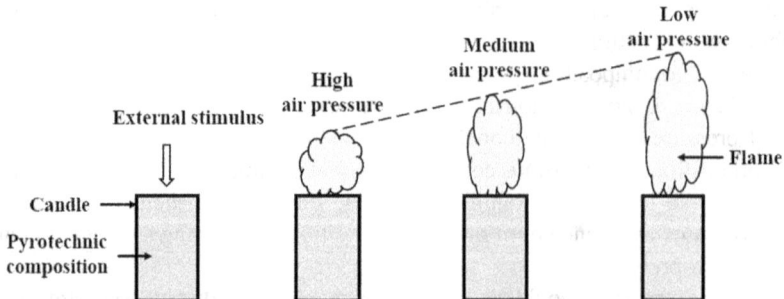

**FIGURE 4.15** Effect of Pressure on Candle Flame Size.

due to the fact that hot gaseous products have a tendency to remain near the burning surface along with some of the hot combustible products.

Burn rates are usually fit to the empirical de St. Robert or Vieille's equation

$$R = ap^n \qquad (4.18)$$

where $R$ = linear burn rate (cm. $\sec^{-1}$), $p$ = gas pressure on composition (atmosphere), $a$ = burning rate coefficient and $n$ = pressure index, $a$ and $n$ determined by best fit. The value on $n$ is pressure dependent and more the gas produced during combustion, the higher shall be the value of $n$. Therefore, higher the gassy composition, the higher is the value of $n$. The value of $n$ is therefore more for gas-producing composition than non-gassy composition [18].

A report [18] on the effect of atmospheric pressure on burn rates mentions the following procedure used to generate the pressure pulse from the burning sample. The bomb and the pressure system are evacuated with a mechanical vacuum pump and then pressurised with nitrogen gas to the desired pressure. The sample is ignited, and the change in pressure is detected by the pressure transducer, amplified by a Kistler 504E amplifier, recorded on an Analogic Data Precision 6000A waveform recorder and stored on a floppy disk.

The composition used was $TiH_2/KClO_4/Viton$ with a mass ratio of 30/65/5. The composition was pressed into small 316 stainless-steel charge holders in two increments with a pressing pressure of 42.34 MPa (6,140 psi), the same as used in the NASA cartridge.

The burn rates of the pyrotechnic $TiH_2/KClO_4/Viton$ with a mass ratio of 30/65/5 have been measured [18] as a function of pressure in nitrogen up to 312 MPa (45 Kpsi). The burn rates were fit to $R = a\,p^n$, with $a$ = 2.055 cm/sec/MPa$^n$ and $n$ = 0.472 between 0.15 MPa (22 psi) and 21.6 MPa (3.13 Kpsi) and $a$ = 4.38 cm/sec/MPa$^n$ and $n$ = 0.266 between 70 MPa (10.15 Kpsi) and 312 MPa (45.25 Kpsi). The results are given in Table 4.29.

The results indicate that an increase in atmospheric pressure increases the linear burn rate.

The inverse burn rate of pyrotechnics at particular ambient pressure is related to burn rate at normal atmosphere by empirical formula [7]

$$R_n = R_a p^n \qquad (4.19)$$

where $R_a$ = inverse rate of burning (sec. $\text{inch}^{-1}$) at particular atmospheric pressure, $R_n$ = inverse rate of burning (sec. $\text{inch}^{-1}$) at normal atmosphere, $p$ = pressure in atmospheres, $n$ = fractional index (to be determined by generating data on rate of burning at two different pressures).

The variation of burn time of 1 g charge weight pellets of some flare compositions under vacuum are as shown in Table 4.30.

This shows that increase in vacuum level decreases the burn time. Such data helps in arriving at suitable composition for functioning at higher altitudes where air pressure is very low.

**TABLE 4.29**

**The Initial Pressure, Average Pressure, and Burn Rate Values of TiH$_2$/KC10$_4$/Viton, OEA Lot TR 120591, in Nitrogen [18]**

| Initial pressure (MPa abs.) | Average pressure (MPa abs.) | Burn rate (cm.sec$^{-1}$) | Initial pressure (MPa abs.) | Average pressure (MPa abs.) | Burn rate (cm.sec$^{-1}$) |
|---|---|---|---|---|---|
| 0. 100 | 0.155 | 0.759 | 69.119 | 70.332 | 13.833 |
| 0.183 | 0.279 | 1.166 | 82.909 | 84.205 | 13.622 |
| 0.410 | 0.679 | 1.773 | 103.525 | 104.801 | 14.960 |
| 0.803 | 1.217 | 2.198 | 117.384 | 118.701 | 15.732 |
| 1.479 | 1.975 | 2.852 | 138.000 | 139.386 | 16.817 |
| 2.100 | 2.624 | 3.230 | 138.276 | 139.662 | 15.937 |
| 2.879 | 3.437 | 3.654 | 143.171 | 144.488 | 16.215 |
| 3.610 | 4.203 | 4.047 | 151.790 | 152.921 | 16.875 |
| 4.727 | 5.361 | 4.603 | 172.475 | 173.895 | 17.294 |
| 11.132 | 11.884 | 6.015 | 206.950 | 208.267 | 18.487 |
| 13.890 | 14.683 | 7.257 | 245.562 | 246.665 | 18.403 |
| 20.785 | 21.592 | 8.771 | 278.520 | 279.596 | 20.644 |
| 34.644 | 35.609 | 10.742 | 310.375 | 311.540 | 19.279 |
| 48.365 | 49.420 | 12.242 | | | |

**TABLE 4.30**

**Variation of Burn Time Under Vacuum [19]**

| Composition | Vacuum level (mm of Hg) | Burn time (seconds) |
|---|---|---|
| Mg/NaNO$_3$/Resin | 710 | 3.66 |
| 56/37.5/6.5 | 610 | 3.99 |
| | 510 | 4.28 |
| | 410 | 4.34 |
| | 385 | 4.88 |
| Mg/NaNO$_3$/E605 | 760 | 4.65 |
| 42/54/4 | 610 | 5.47 |
| | 460 | 6.40 |
| | 310 | 7.62 |
| Mg/NaNO$_3$/Calcium Oxalate/Resin | 760 | 4.87 |
| 56/37.5/3*/6.5 | 660 | 5.36 |
| | 560 | 6.52 |
| | 410 | 6.86 |

* over and above.

For a pyrotechnic composition to burn at high altitudes (i.e., low atmospheric pressures) above 30,000 feet, it is preferable to use composition producing less gaseous products during combustion so that the heat is not carried away by gas and the energy feedback *(Efb)* for sustaining the burning is available for further combustion. Table 8.5 may be referred for performance of illuminating flares in air, argon and nitrogen where the luminosity is increased in oxygen atmosphere while decreases in argon and nitrogen atmosphere.

### 4.4.2  AMBIENT TEMPERATURE

Table 4.31 shows the effect of low, normal and high temperatures on the burning rate of the $B/BaCrO_4/FG$ delay mixture.

It could be seen that the burn rate is temperature dependent. The effect of temperature is more pronounced at low temperatures as the delay body acts as a heat sink; heat being taken away and given to the surroundings. At low temperatures, the difference of temperature between burning composition and surrounding is high, hence more heat is conducted away from the burning pyrotechnic composition, resulting in lower combustion rate.

### 4.4.3  WIND DIRECTION, WIND VELOCITY AND NATURE'S BACKGROUND LIGHT

Wind direction velocity and nature's background light as contrast is relevant for visibility of screening smoke, signalling colour smoke, signalling flare and tracer compositions.

Screening smoke compositions, signalling colour smoke compositions and day tracers with smoke composition may be visible during day time while in case of coloured flare signalling and night tracer composition, the same are more visible during nights.

During day time, the smoke cloud is required to stay for some time for its intended purpose of screening the movement of personnel and equipment or infrared

---

**TABLE 4.31**

**Test Results of $B/BaCrO_4/FG = 15/84/1$ Delay Mixture at Different Operating Temperatures [20]**

| Operating temperature | Mean delay time (s) | Mean burning rate (mms$^{-1}$) | Mean standard deviation in delay time | Mean standard deviation in burning rate |
|---|---|---|---|---|
| - 40 | 0.310 | 48 | 0.008 | 1.14 |
| + 21 | 0.295 | 51 | 0.005 | 1.00 |
| + 70 | 0.273 | 55 | 0.003 | 0.71 |

FG = Fish Glue (Binder)

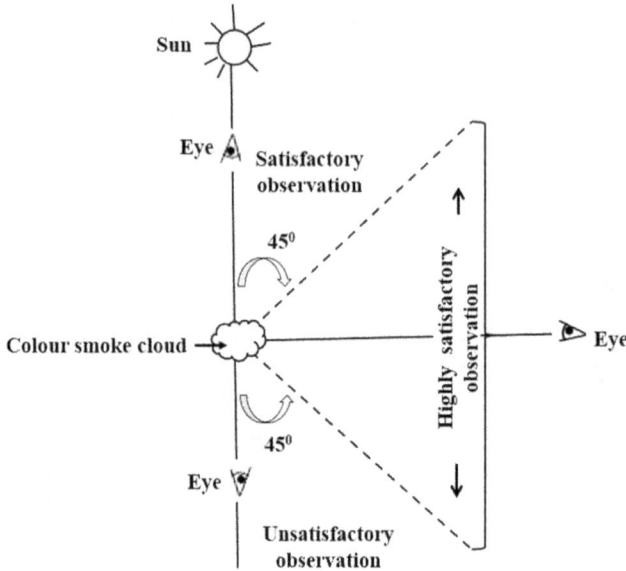

**FIGURE 4.16**   Observation of Coloured Smoke (Modified from [1]).

attenuation of radiation from tanks etc, but high wind velocity may cause the smoke of cloud to disappear. Cloud of smoke dissipates faster with faster wind velocity.

Further, observation of coloured smoke is affected by day time background if the observer's eye and smoke and the sun are in the same line. The ideal angle is $45^0$–$135^0$ for the angle formed between the sun, smoke and observer's eye as shown in Figure 4.16.

During night time, in case of illuminating ammunition, the candle descends in the air at designed rates and its burning rate and luminosity are affected by the wind at the particular height. It also causes shift in the parachute descend due to wind velocity. The extent and direction of the flame of the candle depends upon air pressure, wind velocity and direction.

### 4.4.4   HUMIDITY, MOISTURE, DUST, SMOKE, HEAVY FOG, SNOW, AND RAIN

Humidity and moisture have severe effect on pyrotechnic compositions as explained earlier. Dust and smoke obscure visibility of pyrotechnic illumination, signals and smoke. The effect is less predominant for red and yellow signals than the green and blue signals. However, heavy fog, snow, rain would make them totally obscure and hence pyrotechnic ammunition would not serve the desired purpose. This is a serious limitation of some of the pyrotechnic compositions.

## 4.5   EFFECT OF VARIOUS PARAMETERS ON RATE OF COMBUSTION

The burning rate of a pyrotechnic delay composition depends upon the rate at which the next composition layer is raised to its ignition temperature. It could be seen that the parameters affecting combustion of a pyrotechnic composition are very large and the relative magnitude of all the parameters on combustion characteristics varies with every pyrotechnic composition. Hence it is very difficult to make a model including all parameters of pyrotechnic delay composition for correctly predicting its combustion characteristics. Despite these drawbacks, certain equations for delay compositions have been developed for specific function.

a. Based on Homogeneous Chemical Kinetic Theory [12]:

The simplest chemical kinetic theory assumes the following:

i. Composition- and temperature-independent physical properties
ii. Absence of phase transitions
iii. A thin reaction zone
iv. Postulates a gasless exothermic $n^{th}$ order solid-state reaction
v. An Arrhenius-type temperature dependence for the rate constant

This yields the following expression for the linear burning rate under adiabatic reaction conditions:

$$u^2 = \frac{g(n)RT_c^2}{E_a(T_c - T_o)}(\alpha k_o) \; exp \; (-Ea/RT_c) \tag{4.20}$$

where $u$ = combustion wave velocity (m s$^{-1}$); $g(n)$ = a dimensionless function of the reaction order $n$ that assumes values between 0.5 and 2; $R$ = gas constant (8.314 J mol$^{-1}$K$^{-1}$); $T_o$ = initial temperature; $T_c$ = the maximum temperature of the burning column ($^0$K); $k_o$ = Arrhenius pre-exponential factor of the reaction rate constant (s$^{-1}$); $E_a$ = apparent Arrhenius activation energy (J mol$^{-1}$); $\alpha$ = the effective thermal diffusivity (m$^2$ s$^{-1}$).

It could be inferred that linear burn rate increases with an increase in reaction temperature, increase in effective thermal diffusivity, decrease in activation energy and decrease in temperature difference between initial temperature and maximum temperature of the burning column.

b. Based on Heterogeneous Mass Transport (i.e., diffusion) [12]:

When the condensed phase reactions are mass and energy transport limited, the reaction rate is determined by the thermal and mass diffusivities together with the reactant particle sizes. The expression for linear burning is

$$u^2 = \frac{6RT_c^2}{E_D(T_c-T_0)}\frac{(\alpha D_0)}{d^2}exp\,(E_D/RT_c) \tag{4.21}$$

where $D_0$ [m$^2$ s$^{-1}$] and $E_D$ [J mol$^{-1}$] are the effective pre-exponential and apparent activation energy for the diffusion coefficient, $\alpha$ is thermal conductivity and $d$ is the measure of the particle size distribution of the reactants expressed in $m$. The interpretation of the other variables is similar to that in equation under (4.20 ).

It could be inferred that linear burn rate increases with the increase in thermal conductivity, increase in reaction temperature, decrease in temperature difference between initial temperature and maximum temperature of the burning column, increase in thermal diffusivity, decrease in apparent activation energy for diffusion coefficient and decrease in particle size.

Figure 4.17 shows the summarised factors affecting pyrotechnic composition performance.

Some parameters that affect the pyrotechnic composition performance predominantly are given in Table 4.32.

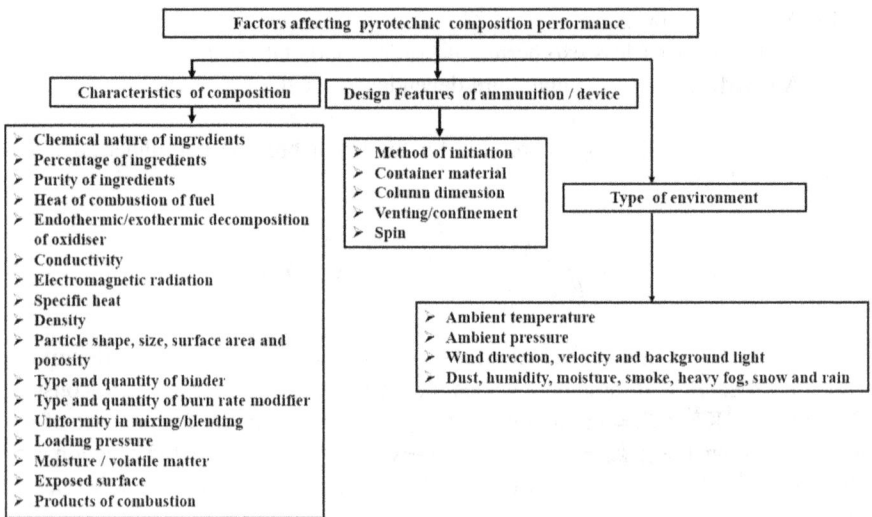

**FIGURE 4.17**   Summary of Factors Affecting Pyrotechnic Composition Performance.

**TABLE 4.32**

## Effect of Various Parameters on Rate of Combustion

| Parameters | Rate of combustion |
| --- | --- |
| Higher conductivity of fuel/composition | Faster |
| Higher heat of combustion of fuel | Faster |
| Higher exothermic decomposition of oxidiser | Faster |
| Higher heat of combustion of composition | Faster |
| Higher surface area of ingredients | Faster |
| Higher exposed surface area of pressed pellets | Faster |
| Loose composition | Faster |
| Spinning of pressed composition | Faster |

**TABLE 4.32** *(Continued)*
**Effect of Various Parameters on Rate of Combustion**

| Parameters | Rate of combustion |
| --- | --- |
| Higher ambient temperature | Faster |
| Higher ambient pressure | Faster |
| Higher loading pressure of composition with metallic fuels | Faster |
| Closed system | Faster |
| Higher loading pressure (heat transfer through gaseous products) | Slower |
| Higher density of ingredients | Slower |
| Addition of inhibitor, retardant and inert material | Slower |
| Higher specific heat of composition | Slower |
| Higher binder percentage | Slower |
| Higher moisture/volatile percentage | Slower |
| Higher energy of activation of composition | Slower |
| Higher ignition temperature of composition | Slower |
| Lower energy feedback for sustaining burning | Slower |

## REFERENCES

[1] A.A. Shidlovskiy, *"Principles of Pyrotechnics"*, 3rd Edition, Moscow, 1964. (Translated by Foreign Technology Division, Wright-Patterson Air Force Base, Ohio, 1974), American Fireworks News, 1 July 1997. ©1997, Rex E. &S.P., Inc.

[2] J.P. Agrawal, *"High Energy Materials: Propellants, Explosives and Pyrotechnics"*, Page 17–18, 2010 © Wiley-VCH Verlag GmbH & Co. KgaA, Weinheim. ISBN:978-3-527-32610-5. Reproduced with permission.

[3] Dominykas Juknelevicius, Rytis Kubilius, and Aruna Ramanavicius, *"Oxidiser Ratio and Oxygen Balance Influence on the Emission Spectra of Green Colored Pyrotechnic Flame"*, European Journal of Inorganic Chemistry, 2015, Pages 5511–5515. https://www.researchgate.net/publication/283732243_Oxidizer_Ratio_and_Oxygen_Balance_Influence_on_the_Emission_Spectra_of_Green-Colored_Pyrotechnic_Flames.

[4] Sinisa Pasagic, Dusan Antonovic, and Sasa Brzic, *"Influence of Technological Parameters on the Combustion Velocity of Pyrotechnic Compositions for Gas Generator of Base Bleed Projectile"*, Central European Journal of Energetic Materials, 2015, Volume 12, Issue 2, Pages 331–346.

[5] Murray Jacobson, Austin R. Cooper, and John Nagy "Explosibility of Metal Powders,"Report of investigation 6516, U. S. Dept. of the Interior, Bureau of Mines 1964.

[6] *"Theory and Application"*, AMCP 706-185, Engineering Design Handbook Military Pyrotechnic series, Part One, Headquarters, US Army Material Command, Washington D.C., April 1967.

[7] J.C. Cackett *"Monograph on Pyrotechnic Compositions"*, Royal Armament Research and Development Establishment, Fort, Halstead, Sevenoaks, Kent 1965.

[8] S.R. Yoganarasimhan and O.S. Josyulu, *"Reactivity of the Ternary Pyrotechnic System Red Lead – Silicon – Ferric Oxide"*, Defence Science Journal, January 1987, Volume 37, Issue 1, Pages 73–83.

[9] *"Explosive Train"*, AMCP 706-179, Engineering Design Handbook Explosive Series, Explosive Trains, Headquarters, U S Army Material Command, 5001, Eisenhower Ave, Alexandria, VA 22304, January 1974.

[10] F. L. McIntyre, *"A Compilation of Hazard and Test Data for Pyrotechnic Compositions"*, Contractor Report ARLCD-CR-80047, October 1980. U.S. Army Research and Development Command, Large Caliber Weapon System Laboratory, Dover, New Jersey.

[11] *"Properties of Materials Used in Pyrotechnic Compositions"*, Engineering Design Handbook-Military Pyrotechnic Series, Part Three, Headquarters, US Army Material Command, Washington D.C., October 1963.

[12] Walter W. Focke, Shepherd M. Tichapondwa, Yolandi C. Montgomery, Johannes M. Grobler, and Michel L. Kalombo, *"Review of Gasless Pyrotechnic Time Delays"*, Propellants, Explosives, Pyrotechnics, 30 May 2018, Volume 44, Issue 1. ©WILEY ONLINE LIBRARY, Reproduced with permission.

[13] Jay C. Poret, Anthony P. Shaw, Lori J. Groven, Gary Chen, and Karl D. Oyler, *"Environmentally Benign Pyrotechnic Delays"*, Pyrotechnics Technology and Prototyping Division, US Army RDECOM-ARDEC, Bldg 1515, Picatinny Arsenal, New Jersey, USA. 38th International Pyrotechnics Seminar, Denver, Colorado, 10–15 June 2012, Pages 494–500.

[14] Saeid Bagherpour, Mohammad Mahdavi, and Ebrahim Abedini, *"Investigation of Thermal Behavior of Energetic and Non-Energetic Binders on Luminous Efficiency of High-Performance Miniature Flares"*, Defence Technology, April 2019, Volume 15, Issue 2, Pages 193–197.

[15] Dr M. R. Somayajulu, *"Pyrotechnic-An Overview"*, Pyrotechnic Seminar Proceedings, 2nd April 2005, Ordnance Factory Dehuroad, Pune, India.

[16] Walter J. Puchalski, *"The Effect of Angular Velocity and Composition on Pyrotechnic Performance"*, Report No. FA-YR-74011, Frankford Arsenal Philadelphia, Pennsylvania, August 1974. Distributed by National Technical Information Service US Department of Commerce, Springfield, VA 22151.

[17] Patricia L. Farnell, Clement Campbell, and Francis R. Taylor, *"Effect of Reduced Atmospheric Pressure on the Performance Characteristics of Pyrotechnic Compositions containing Aluminum"*, Technical Report ARAED-TR-89024, November 1989, U.S. Army Armament Research, Development and Engineering Center Armament Engineering Directorate, Picatinny Arsenal, New Jersey.

[18] John A. Holy, *"Burn Rates of $TiH_2/KClO_4$ /Viton and Output Testing of NASA SKD26100098-301 Pressure Cartridges"* NASA Contractor Report 188357, National Aeronautics and Space Administration, 1993.

[19] S. M. Panday, *"Light Intensity Measurement"*, Continued Education Programme on Instrumentation for Testing of Propellants, Pyrotechnics and Allied Devices 03–07 November 2003, High Energy Materials Research Laboratory, Pune.

[20] Azizullah Khan, Abdul Qadeer Malik, Zulfiqar Hameed Lodhi, and Syed Ammar Hussain (2017), *"Development and Parametric Study of $B/BaCrO_4/FG$ Pyrotechnic Delay Composition"*, Combustion Science and Technology, DOI: 10.1080/001022 02.2017.1410800

# 5 Determination of Performance Parameters

## 5.1 GENERAL

The performance parameters of a pyrotechnic composition may be determined through various tests. These tests provide vital information on characteristics of pyrotechnic compositions, which allows a judicious choice of composition for a pyrotechnic ammunition.

### 5.1.1 SENSITIVITY TO EXTERNAL STIMULI

The measurement of sensitivity to external stimuli has been discussed in detail under Chapter 2 as this is considered most vital. Unless the pyrotechnic composition is sensitive to external stimuli, no combustion would take place to observe the pyrotechnic effects. Further, this sensitivity also allows to choose a suitable firing device and the safety precautions required during its use.

### 5.1.2 MOISTURE ABSORPTION

A composition is considered better if its absorption of moisture is less. However, many ingredients prone to moisture absorption are yet found to be very useful. Hence, suitable precautions are taken to ensure to carry out processes within the defined humidity range by providing air-conditioned buildings and suitable dryers for drying of ingredients/compositions.

The moisture content is determined by heating a small sample of composition at specified temperature for specified duration and recording the weight loss and is expressed as percentage. Moisture percent acceptance limit is usually specified for the composition.

### 5.1.3 APPARENT BULK DENSITY AND LOADING DENSITY

The *apparent bulk density* is expressed as grams per millilitre ($g.ml^{-1}$) or grams per cubic centimetre ($g.cm^{-3}$) or kilogram per cubic metre ($kg.m^{-3}$).

*Bulk density* of a pyrotechnic dry composition is expressed as the ratio of the mass of dry composition and its corresponding untapped volume (including the void). Bulk density depends on the following factors:

a. Type of the composition
b. Density and porosity of dry pyrotechnic composition
c. Spatial arrangement of particles in the composition

d. Particle size and distribution
e. Shape and surface roughness of the particles
f. Method of storage and external jerks/impact/vibrations

The bulk density of a composition is determined by:

a. Measuring the volume of a known mass of sieved composition
b. Measuring the mass of a known volume of composition

*Loading density* of pyrotechnic composition is the ratio of the mass of pyrotechnic composition loaded per unit volume. Pyrotechnic compositions are filled in ammunition or compacted as pellet and then assembled. A higher loading density permits higher amount of pyrotechnic composition in the pyrotechnic ammunition. However, certain initiatory compositions may lead to *dead pressed* condition at very high compaction.However, higher loading density reduces porosity of the composition.

### 5.1.4   IGNITION TEMPERATURE

Generally, three methods are used for determination of ignition temperature.

a. Hot Bath Immersion Technique
b. Differential Thermal Analysis (DTA)
c. Thermogravimetric analysis (TGA)

The method for arriving at the ignition temperature by hot bath immersion technique (Julius Peter's method) is shown in Figure 5.1.

This is measured in a temperature bath (woods metal alloy of bismuth 50%, lead 25%, tin 12.5% and cadmium 12.5% having 70°C melting point), which makes use

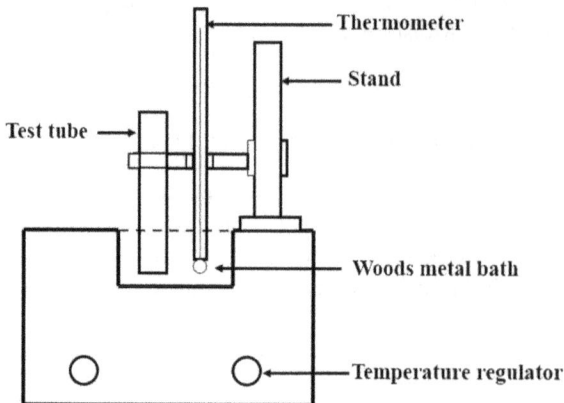

**FIGURE 5.1**   Temperature Bath.

of a thermometer to measure the temperature of the bath. The pyrotechnic sample 10–15 mg (5 mg for initiatory composition) is taken in a 10–12 mm test tube and held in position in the bath. The bath temperature is kept approximately 20°C less than the likely ignition temperature. The temperature of the bath is increased at the rate of 5°C per minute. The temperature at which the ignition takes place is taken as the ignition temperature of the composition. The test is repeated several times to arrive at the average value of ignition temperature.

Another method is by using the Henkin and McGill [1] method, where the composition is heated at a temperature and time taken for its ignition is noted. Obviously, the higher the input temperature, shorter would be the time for ignition. A graph is plotted with temperature on x-axis and time to ignition on y-axis. The temperature at which the ignition occurs within 5 seconds is taken as its ignition temperature. Figure 5.2 shows the time to explosion for nitrocellulose.

The temperature at which the ignition takes place after infinite time (taken by extrapolating the curve for ignition on the graph) is known as the minimum spontaneous ignition temperature. The self-ignition temperatures (within 5 seconds of exposure) for pyrocellulose, guncotton and nitroglycerine have been reported as 170°C–230°C.

Another method [2] is to obtain the ignition temperature by determining the bath temperature and corresponding time to ignition for a given sample weight and composition, using the equation

$$t_i = \frac{(t_1^2 \times Tb_1 - t_i^2 \times Tb_i^2)}{t_1^2 - t_i^2} \ldots \tag{5.1}$$

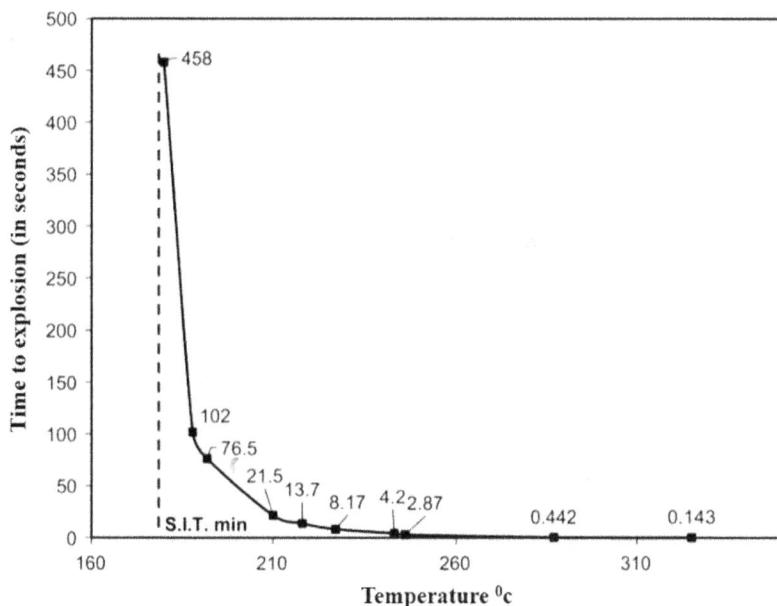

**FIGURE 5.2**   Time to Ignition Versus Temperature (Reprinted with permission from [1]).

**TABLE 5.1**

**Time to Ignition at Various Bath Temperatures [2]**

| Bath temperature(°C) | Time to ignition (seconds) |
|---|---|
| 702 (say $Tb_1$) | 25.6 (say $t_1$) |
| 682 (say $Tb_2$) | 35.6 (say $t_2$) |
| 673 (say $Tb_3$) | 39.4 (say $t_3$) |
| 659 (say $Tb_4$) | 73.5 (say $t_4$) |

where first bath temperature is shown as $Tb_1$ and other bath temperatures as $Tb_i$ and time to ignition in first bath $t_1$ and $t_i$ for other three baths where $i = 2$, 3 and 4.

A typical data for time to ignition of 10% boron and 90% barium chromate is given in Table 5.1.

The time to ignition calculated using above equation gives:

$$t_{i2} = 660.5 \,°C, \, t_{i3} = 651.8 \,°C \text{ and } t_{i4} = 653.0 \,°C, \text{ average being } t_i = 655.1 \pm 4.7 \,°C$$

The variation in results of ignition temperature at various laboratories is due to

(a) Variation in the rate of heating the composition
(b) Difference in equipment and procedure used in various laboratories
(c) Conditions of environment (ambient pressure, temperature and humidity) during measurements
(d) Variation in the composition characteristics (size, shape, surface area, purity, proportion, moisture, homogeneity of composition etc.).

The DTA technique provide not only the ignition temperature but also the heat of reaction and activation energy.

## 5.1.5   HEAT OF COMBUSTION AND HEAT OF REACTION

Each pyrotechnic composition has its own *heat of combustion* due to various ingredients and their varying proportions that are different from other compositions. Thus, thermites and illuminating compositions produce a very high heat output and signalling colour smoke compositions produce very low heat output. This may be measured using bomb calorimeter. Combustion releases heat energy causing rise in temperature of bomb calorimeter containing some water. Noting the initial temperature and final temperature after combustion and heat capacity of calorimeter and specific heat of water, heat of combustion or heat of reaction is calculated using the standard equations. This is expressed as calories per mole or kilocalories per mole or joules per mole.

The gross *heat of combustion* is measured by burning 1 to 2 g samples of pyrotechnic mixture in an oxygen-filled (5 atmospheres) standard calorimeter bomb submerged in water and recording the rise in water temperature.

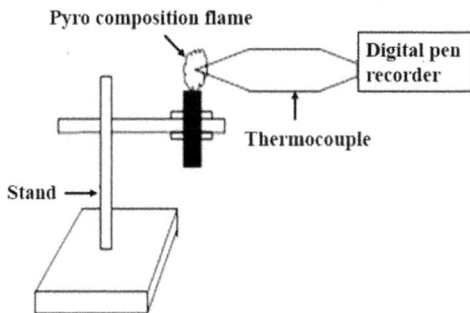

**FiGURE 5.3**   Flame Temperature Measurement by Thermocouple (Reprinted from [3]).

The gross *heat of reaction* in terms of calories per gram is determined by burning 1 to 2 g sample of pyrotechnic mixture in an inert atmosphere (nitrogen) in the standard bomb calorimeter submerged in water and recording the rise in water temperature.

### 5.1.6   FLAME TEMPERATURE

Each pyrotechnic composition will have different flame temperatures. This is measured using thermocouples or optical pyrometers. In case of thermocouple, a suitable thermocouple [3] for the desired temperature range is used. The temperature is recorded by thermocouple on digital pen recorder in millivolts signal and the same is converted into temperature by using the conversion table (Figure 5.3).

The values obtained are always less than the actual values as some part of the heat is always dissipated through conduction, convection and radiation. This is expressed as °C or °K.

### 5.1.7   LUMINOSITY AND BURN TIME

The illuminating compositions are required to provide a desired luminosity and burn time. The illuminating compositions are tested for luminosity (in candela or candle power) and burning time in a luminosity tunnel. The luminosity tunnel also has an air exhaust system that, to some extent, simulates the air drift which the star or candle (attached to parachute) encounters during descent in air (Figure 5.4).

The illuminating composition in pressed form as candle is mounted on a stand in the front of the tunnel. A squib is attached to the flare so that it may be ignited using a remote control. A detector camera is placed at a specified distance from the illuminating candle.

The fan is switched on and when the fan has achieved the set speed, the flare is ignited from remote control.

The light emanating from the candle is detected by the camera that converts it into electrical signal. The electrical signal is amplified by the photometer and the analogue output of the photometer is transmitted to computer. The computer screen exhibits the intensity of light in the form of a graph as the burning proceeds till it ends. The

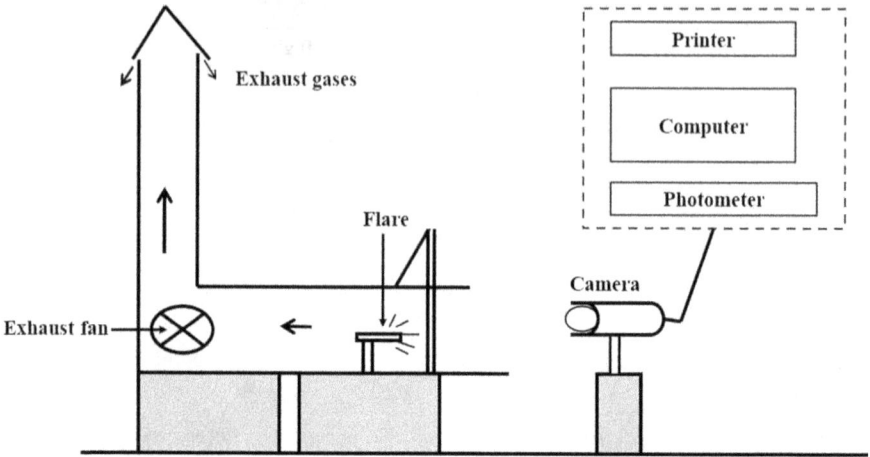

FIGURE 5.4   Luminosity Test Measuring Setup.

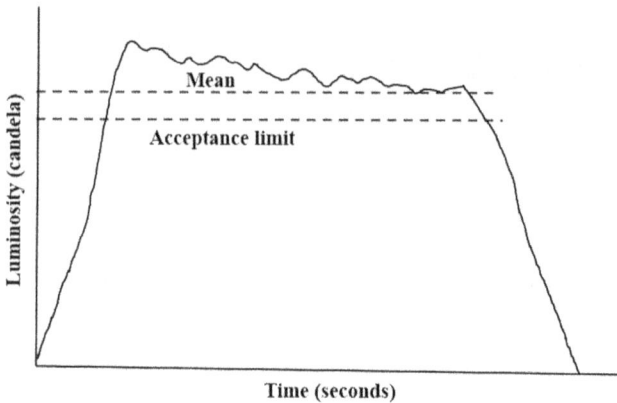

FIGURE 5.5   Luminosity Time Curve.

computer also computes the average value of light intensity and burn time and displays it in the form of a graph. A typical luminosity with burn time curve is shown in Figure 5.5.

## 5.1.8   INFRARED INTENSITY AND EMISSION TIME

The infrared radiation is measured on a radiometer using various filters for determining the infrared radiation in each filter bands. The setup is similar to the above used for light intensity for illuminating composition except that a radiometer is used instead of a photometer.

The radiometer measures [4] the percentage attenuation of radiation energy emitted by the infrared source at desired discrete regions of 2–2.4, 3–5 and 4–18 microns or continuous filter 2.5–14.5 microns. The radiometer collects the radiation through mirrors and lenses and projects on the infrared detector. This infrared detector converts the radiation into electrical signal and is processed in the electronic system and displayed on the monitor as IR intensity and burn time and is also stored in the hard disc. It is expressed as kw.sr$^{-1}$ (kilowatt per steradian).

## 5.1.9  Flame Length of Composition

Primers, pressed compositions and igniters are sometimes required to be tested for the minimum flame length. A thin wire at specified minimum distance from the primer or pressed composition is kept, which melts if the candle flame length is as specified (Figure 5.6).

## 5.1.10  Ignition Delay

The time required between the application of firing pulse and functioning of the composition is termed as ignition delay. A counter is used to measure the time interval between the instant voltage pulse applied to the functioning of the device/ squib. The counter starts due to the pulse derived suitably from firing voltage and stops due to pulse from photo-sensing unit.

On actuation of the firing switch, the power supply unit supplies necessary firing current to the power device [4]. Signal pulse of suitable amplitude and polarity is derived from firing pulse and is applied to START the electronic counter. Firing of electro-explosive device like squib produces flash, which is sensed by a phototransistor unit and suitably amplified, shaped and used to STOP the electronic counter. The digital electronic counter displays the delay in ignition device (Figure 5.7).

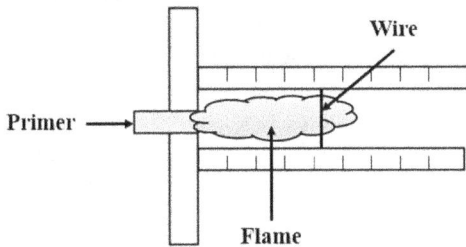

**FIGURE 5.6**  Flame Length of Pressed Composition.

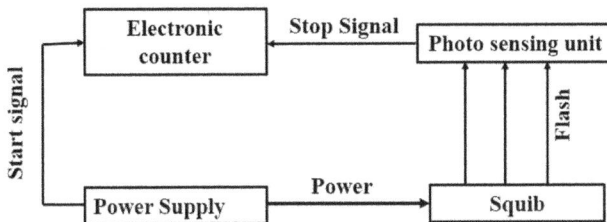

**FIGURE 5.7**  Ignition Delay Measurement (Reprinted from [4]).

## 5.1.11  BURN RATE

The burn rates of each pyrotechnic composition are different and may vary from a few milliseconds to several minutes. Compositions in loose form burn vigorously while those in pressed form like stars, candles, tracers or delay component burn slowly.

The burn time of a pressed composition is measured as the linear burn rate, i.e., the length of the composition burnt per unit time. Alternatively, this can also be measured as the inverse burn rate, i.e., the time taken to burn a unit distance. Both burn rates are inverse of each other. Electronic counter starts on ignition of composition at a specified distance and stops when the ignition reaches another specified distance. The counter shows the burn time for the composition for specified distance.

Other methods are to measure the mass burnt per unit time or volume burnt per unit time.

Spinning causes fast burning of tracers. Without spin, the tracer burning time would be much more. Since tracers are assembled to spin-stabilised projectiles, hence it is essential to record the tracer burn time with simulated spin so that the tracers give proper burn time expected during actual use of the ammunition. The tracers are tested in an apparatus with spinning facility. The spin rate is made similar to the ammunition in which tracers are to be assembled. The spin is imparted to the tracer and then the tracer is ignited through a squib and its burn time recorded.

Delay compositions are generally developed and tested for delay time in lead rolling tubes. These are subsequently tested using suitable measuring instruments.

The method [5] consists of rigid delay body, percussion primer, pyrotechnic delay composition and flame composition. Percussion cap of this mechanically initiated delay element is hit with a striking pin. The firing system consists of electro-mechanical switch. Burning time starts when the pin hits the percussion primer, and stops when the light dependent resistor (sensor) detects the flame of the delay composition. To protect the sensor from the slag, a perspex disc is assembled in front of it. The burning time is measured with digital oscilloscope. Measured results are in milliseconds, therefore the sensitivity of the instruments used are in microsecond to avoid motion blur. A customised chronometer is used for the measurement of the burning time simultaneously. Schematic diagram of the delay burning time measurement system is shown in Figure 5.8.

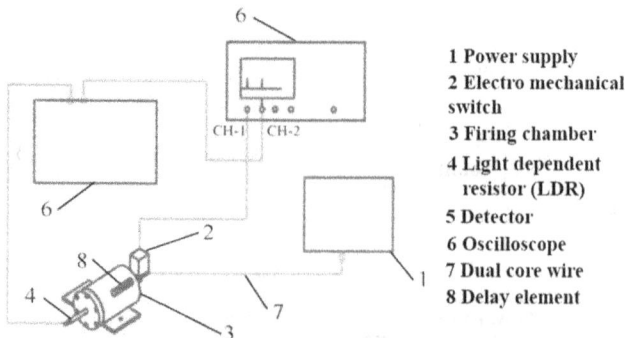

1 Power supply
2 Electro mechanical switch
3 Firing chamber
4 Light dependent resistor (LDR)
5 Detector
6 Oscilloscope
7 Dual core wire
8 Delay element

FIGURE 5.8  Delay Measurement System for Pyrotechnic Delay (Reprinted from [5]).

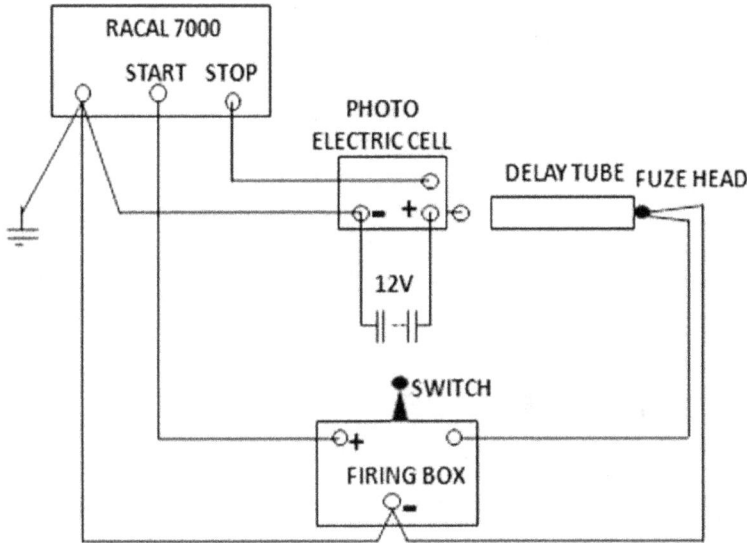

**FIGURE 5.9**  Determination of Burn Rates of Delay Compositions (Reprinted from [6]).

The burning times of the various delay compositions may also be determined using a Racal SA 7000 interval counter [6]. The firing pulse required to ignite the match head also triggers the start mode of the counter. The counter is stopped by a pulse generated by a photoelectric cell placed at the opposite end of the delay tube. A schematic diagram of the circuitry used is shown in Figure 5.9.

### 5.1.12  Ignition Energy

A joulemeter [4] is used for determining the minimum energy required to fire squibs or electro-explosive devices. Measurement of minimum energy required to fire such devices is an important acceptance criterion with respect to safety.

The joulemeter basically consists of a precisely measured capacitor that can be charged to precisely known voltages, and then switched into the EED. The energy delivered to EED is the ½ CV$^2$, where "C" is the value of capacitor in farads and "V" is the charging voltage in volts. The arrangement is shown in Figure 5.10.

Another method of determination of minimum electrical energy required for functioning is by using the equation

$$\text{Electrical energy} = i^2 \times R \times t \tag{5.2}$$

where $i$ = the current passing through EED measured using hall sensor; $R$ = the EED bridge resistance measured with safety ohmmeter and $t$ = the time for which electrical energy is supplied. The test setup is shown in Figure 5.11 [7].

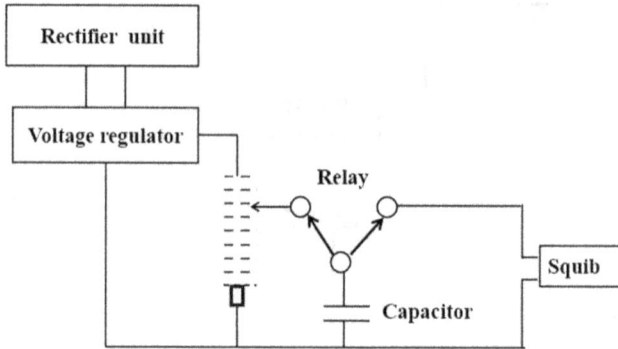

**FIGURE 5.10**  Ignition Energy Measurement (Reprinted from [4]).

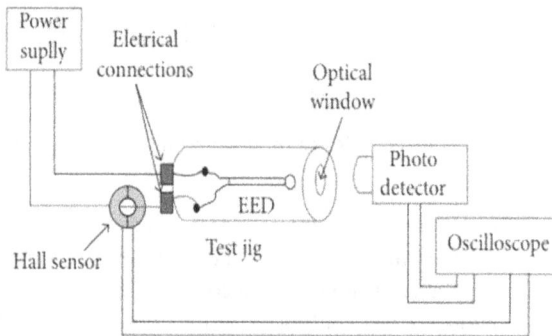

**FIGURE 5.11**  Experimental Setup to Determine the Electrical Energy for Ignition of EEDs (Reprinted from [7]).

If $t1$ is the time at which the current supply started, $t2$ is the time at which bridge wire in EEDs breaks (denoting that current levels through EED become zero) and $t3$ is the photodetector pulse indicating the start time of EED ignition. Two cases arise.

*Case 1: Bridge wire breaks before the ignition ($t2 < t3$):* in this case, the time for which the actual energy is supplied to EED is the difference between the time at which current supply started and the time at which the bridge wire in EEDs breaks. For this case, time $t$ in Equation (5.1) is $t2 - t1$.

*Case 2: Bridge wire breaks after the ignition ($t2 > t3$):* in this case, the time for actual energy supplied to EED is the difference between the time at which the current supply started and the time at which EED ignition started. For this case, time $t$ in Equation (5.1) is $t3 - t1$.

Using measured EED bridge resistance ($R$) and current ($i$) supplied through EED, the energy for EED ignition is calculated through Equation (5.1) by considering time $t$ as $t2 - t1$ for Case 1 and $t3 - t1$ for Case 2. The summarised results are at Table 5.2.

It shows that higher values of supplied current reduce electrical energy for functioning and also reduce the time to ignition.

**TABLE 5.2**
**Electrical Energy Levels for Ignition of EEDs at Different Current Levels [7]**

| Current i (A) | Resistance (Ω) | Time at which bridge wire breaks t2 (ms) | Time at which photodetector signal is received t3 (ms) | Δt (ms) | Electrical energy (mJ) |
|---|---|---|---|---|---|
| 1.05 | 2.135 | 45.188 | 23.708 | 23.708 | 55.80 |
| 1.32 | 2.05 | 9.98 | 8.678 | 8.678 | 31.00 |
| 2.04 | 2.2 | 2.102 | 5.19 | 2.102 | 19.25 |
| 2.74 | 2.0 | 1.132 | 6.58 | 1.132 | 16.99 |
| 3.44 | 2.18 | 0.572 | 1.852 | 0.572 | 14.76 |

## 5.1.13 COMPACTION STRENGTH OF PRESSED COMPOSITION

This is measured on a compact strength testing machine. The compacted composition is kept on the machine and the load is increased slowly. A graph is obtained as the article experiences the force till it starts crumbling. The pressure at crumbling point is taken as the strength of the compacted article. This is very important for such pressed articles which experience heavy stresses on firing like those experienced in heavy calibre ammunitions. The designer specifies the values for such composition which may be several tons. Compaction strength of pellets may be seen in Table 24.1.

## 5.1.14 INFRARED ATTENUATION BY SMOKE

The attenuation power of smoke is measured through a smoke chamber of volume 1.8 m$^3$. The distance between infrared source and radiometer is kept at 12 meters while the distance between smoke chamber and the radiometer is kept at 10 meters. The infrared source is a lamp, 150-watt hot plate with filters [4].

The efficacy of the composition for the attenuation for different wavelengths may be ascertained by changing filters (discrete filters 2, 2.4, 3–5 and 8–14 microns while continuous filter 2.5–14.5 microns). (Figure 5.12).

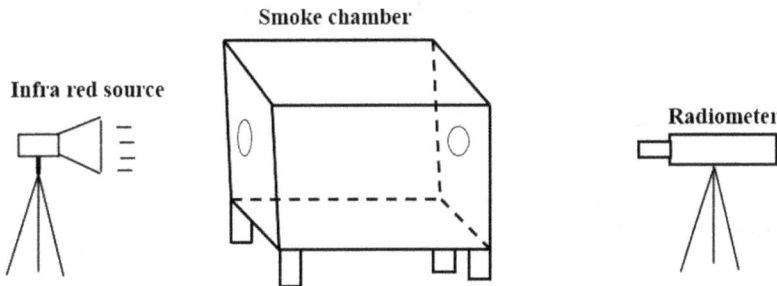

**FIGURE 5.12** Measurement of Infrared Attenuation by Screening Smoke (Reprinted from [4]).

The infrared radiation output of the infrared source radiation is measured through the windows of the smoke chamber by the radiometer. The infrared attenuating smoke composition pellet is then burnt inside the smoke chamber and the smoke chamber is filled with infrared attenuating smoke. Transmitted infrared radiation output through smoke is again measured. Percent attenuation is given as

$$\text{Percent attenuation} = \frac{(\text{Initial} - \text{transmitted}) \text{ radiant intensity} \times 100}{\text{Initial radiant intensity}} \quad \dots \quad (5.3)$$

The extinction coefficient may also be calculated using Beer–Lambert Law (Section 12.2.2). Readers may refer Table 12.17, which shows the percentage attenuation by pyrotechnic smoke compositions.

### 5.1.15  OBSCURATION POWER OF SCREENING SMOKE

This is measured in similar way using a smoke chamber described in infrared attenuation by smoke method except that a photometer is used instead of a radiometer.
    The quality of smoke is expressed as *Total Obscuration Power.*

Total Obscuration Power

$$= \frac{\text{Volume of smoke}(\text{m}^3) \text{ produced per kg}}{\text{Smoke thickness}(\text{metres}) \text{ required to obscure a filament lamp of 40W}} \quad \dots$$

$$(5.4)$$

A higher concentration of smoke results in higher total obscuration power.

### 5.1.16  SOUND INTENSITY

Sound is created due to pressure waves in the air. The sound intensity is measured [4] in decibels. A high-quality microphone converts the air pressure of the sound into an electronic signal with voltage proportional to the sound pressure variations. After amplification and filtering, the signal is fed to rectifier through amplifier. The averaged DC level signals are converted to DC output proportional to its r.m.s. level, which is fed to the soundmeter to give direct indication of the exact sound pressure level in decibels as shown in Figure 5.13.

### 5.1.17  ELECTRICAL CONDUCTIVITY

Electrical conductivity of electrical sensitive compositions can be measured. Readers may see the detailed method for electrical conductivity measurement reported under reference [8].

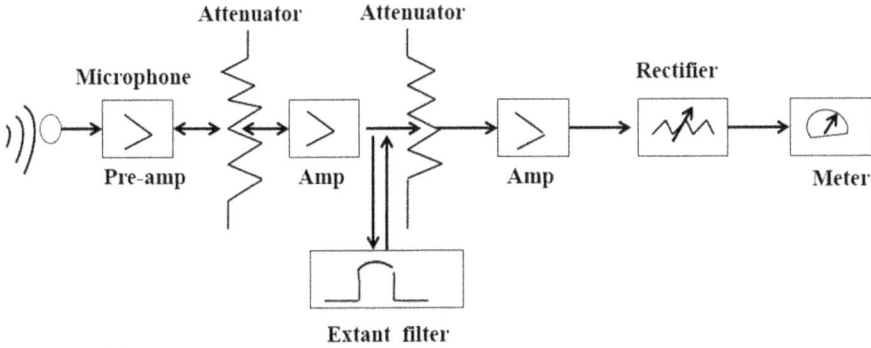

**FIGURE 5.13** Sound Level Meter (Reprinted from [4]).

## 5.1.18 THERMAL CONDUCTIVITY

A report [9] on the thermal conductivity of the photoflash composition (consisting of 30% potassium perchlorate, 30% barium nitrate and 40% aluminium) desensitised (using carbon as coatant) mentions use of instrument called a thermal comparator (manufactured by the Thermal Properties Research Co., Lafayette, Indiana) since the thermal conductivity on the test sample is obtained from a calibration curve derived from reference standards.

The instrument registers the rate of cooling experienced by a heated probe upon contact with the surface of the test material. The probe assembly, consists of a thermocouple sensing tip, heater and a thermal reservoir, held at a temperature above the room temperature. Upon contact of the probe tip of thermal conductivity X1, with the surface of the material at room temperature and having thermal conductivity X2, the tip immediately drops to an intermediate temperature, which registers either digitally or on a strip chart recorder as an emf. A plateau is reached that represents a near steady state temperature level and it is this level that relates to the thermal conductivity of the material. Table 5.3 shows values of thermal conductivity of a photoflash composition.

**TABLE 5.3**
**Thermal Conductivity Measurement of Photoflash Composition [9]**

| Photoflash designation | Carbon content of photoflash mixture (%) | Thermal conductivity $(W. cm^{-1} {}^{\circ}C^{-1})$ |
|---|---|---|
| 555M | 0 | 0.0051 |
| 555B | 0 | 0.0053 |
| 161M | 0.5 | 0.0048 |
| 59M | 1.0 | 0.0035 |
| 62M | 1.5 | 0.0026 |

It is noted that there is about a 10% drop in conductivity when 0.5% carbon is added to the mixture. The value decreases 35% at 1% carbon level and by 50% at 1.5% carbon level. Thus, the carbon concentration appears to have a pronounced effect on conductivity. However, the absolute value of conductivity is in all cases extremely low, which is normal for loose or near loose powders. Readers may see more details in the reference [9].

### 5.1.19  VOLUME OF GASEOUS COMBUSTION PRODUCTS

The volume of gas produced is important for ejecting the star or the parachute–candle system in the ammunition. It is also required for screening smoke and colour smoke compositions and is useful for producing sound as well as acting as a gas generator for missile systems. The gases are mostly water vapour, carbon monoxide, carbon dioxide and nitrogen. It is also important to know the quantum of gases produced by delay composition to know if the same are to be treated as gaseous or gasless (low gas) for appropriate use in ammunitions. A method to measure the same in a closed bomb by measuring transient pressure [10] is given below.

The volume of gases produced by 1.0 g of composition during pyrotechnic combustion through electrical firing was determined by measuring the transient pressure (through a pressure transducer with sensitivity of 143.9 mV/MPa) in the 50-ml closed bomb. The signals were processed on DEWEsoft. Then pressure–time (P–t curves) were drawn. The test system is shown in Figure 5.14.

Since the addition of composition is small, suppose the temperature of the gaseous products in the closed bomb was close to that of gaseous products under atmospheric pressure. Thus, according to the equation of state for the gas, the volume of the composition under atmospheric pressure is:

$$V_0 = P_1 V_1 / P_0 \ldots \tag{5.5}$$

where

$V_0$ = Volume of gaseous products in combustion of compositions under atmospheric pressure, L

$P_0$ = Atmospheric pressure (1 atm.), Pa

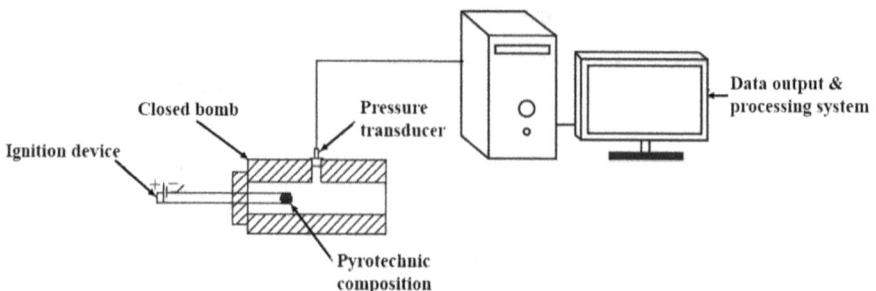

**FIGURE 5.14**  Arrangement for Determination of Volume of Combustion Products (Reprinted from [10]).

$V_1$ = Volume of the closed bomb, L
$P_1$ = Peak pressure measured by the pressure transducer, Pa

## 5.1.20 EFFICIENCY OF COMPOSITION

The efficiency per unit mass or per unit volume in producing the desired pyrotechnic effect can be evaluated. It is of importance since the space in ammunitions is generally a constraint. Further, it also has a bearing on the cost of the ammunition. As an example, luminous efficiency of flare compositions has been reported [11] where 8 flares of 17 mm diameter and 20 mm height were pressed with same loading pressure using following compositions.

  i. Magnesium 52.0%, sodium nitrate 40.0%, binder 8.0%
 ii. Magnesium 54.3%, sodium nitrate 41.7%, binder 4%

The test results are given in Table 5.4.

## TABLE 5.4
## Effect of Various Binders on Luminous Efficiency [11]

| Performance | None | Polyester[a] | | Epoxy[b] | | NC | | CR | |
|---|---|---|---|---|---|---|---|---|---|
| | | 8% | 4% | 8% | 4% | 8% | 4% | 8% | 4% |
| Density (g.cm$^{-3}$) | 2.10 | 1.86 | 2.00 | 1.86 | 2.00 | 1.86 | 2.00 | 1.86 | 2.00 |
| Luminous intensity (cd) | 51,600 | 29,500 | 36,700 | 26,000 | 32,600 | 40,200 | 45,800 | 21,400 | 26,500 |
| Burn time (s) | 3.6 | 8.4 | 6.7 | 8.9 | 7.0 | 5.0 | 4.2 | 10.4 | 8.5 |
| Burn rate (mm.s$^{-1}$) | 4.44 | 2,26 | 2.69 | 2.13 | 2.57 | 3.80 | 4.52 | 1.83 | 2.07 |
| Luminous efficiency (cd.s.g$^{-1}$) | 23,200 | 31,000 | 31,200 | 28,900 | 28,500 | 25,100 | 24,000 | 27,800 | 28,200 |

*Note*: a = polyester resin (97.5%), methyl ethyl ketone peroxide (2%) and cobalt naphtenate (0.5%), b = bisphenol-A epoxy resin (66.66%), curing agent (33.33%)
NC = nitrocellulose, CR = calcium resinate

It could be seen that the luminous efficiency, including luminosity and burn rate are more with use of polyester-based flares presumably due to its higher oxygen content.

## 5.1.21 COLOUR QUALITY

Colour in signalling flare compositions can be measured as a wavelength of the light emitted by the burning composition. The Commission International de L'Eclairage (CIE) developed the chromaticity curve that is currently used in the determination of the dominant wavelength and spectral purity of luminous sources. The signalling compositions are checked for the "visible range" of radiation 0.380–0.780 microns.

## REFERENCES

[1] Hyman Henkin and Russell McGill, *"Rates of Explosive Decomposition of Explosives, Experimental and Theoretical Kinetic Study as a Function of Temperature"*, Industrial and Engineering Chemistry, 1952, Volume 44, Pages 1391–1395. Copyright © 1952 American Chemical Society.

[2] Duane M. Johnson, *"Ignition Theory: Application to the Design of New Ignition System"*, RTDR No. 56, 24 November 1965, US Naval Ammunition Depot, Crane, Indiana.

[3] D. K. Jawale, *"Temperature Measurement"*, Continued Education Programme on Instrumentation for Testing of Propellants, Pyrotechnics and Allied Devices, November 2003, High Energy materials Research Laboratory, Pune, India.

[4] S. B. R. Thakur, *"Performance Evaluation of Various Effects Produced by Pyrotechnics"*, Continued Education Programme Course on Pyrotechnics, August, 2003, High Energy Materials Research Laboratory, Pune, India.

[5] Azizullah Khan, Abdul Qadeer Malik, Zulfiqar Hameed Lodhi, and Zain Ul Abdin, *"Effect of Body Material and Temperature Variation on the Performance of the Time Delay Pyrotechnic Compositions"*, © Defence Technology, 2018, Volume 14, Pages 261–265.

[6] John R. Bentley and Paul P. Elischer, *"Development of a Gasless Pyrotechnic Cap"*, REPORT MRL-R-776, May 1980, Materials Research Laboratories, Ascot Vale, Victoria 3032, Australia.

[7] Kameshwar Kumar Mishra, Aleti Sudheer Babu, Chandrakala P. Shetty, and Himanshu Shekhar, *"Method to Determine the Electrical Energy for Ignition of Electro-Explosive Devices"*, Journal of Aerospace Technology and Management, 2015, Volume 7, Issue 3, Pages 285–288.

[8] Eric Collins, *"Ignition Sensitivity of Composite Energetic Materials to Electrostatic Discharge"*, A Dissertation in Mechanical Engineering Submitted to the Graduate Faculty of Texas Tech University in Partial Fulfilment of the Requirements for the Degree of Doctor of Philosophy, Texas Tech University, © Eric Collins, August 2013.

[9] S. Dallman, S. Werbel and F.R. Taylor, *"Photoflash Compositions Desensitized by Coatants"*, Technical Report ARLCD-TR-80043, December 1980. US Army Armament Research and Development Command, Large Caliber Weapon Systems Laboratory, Dover, New Jersey.

[10] Jie Li, Hua Guan, Dongming Song, Qi Wang and Jun Du, *"Effects of Gas Production on Acoustic Radiation Characteristics of Underwater Pyrotechnic Combustion"*, Journal of Low Frequency Noise, Vibration and Active Control, 2015, Volume 34, Issue 1. pages 1–8.

[11] Saeid Bagherpour, Mohammad Mahdavi, and Ebrahim Abedini, *"Investigation of Thermal Behavior of Energetic and Non-Energetic Binders on Luminous Efficiency of High-Performance Miniature Flares"*, © Defence Technology, April 2019, Volume 15, Issue 2, Pages 193–197.

# 6 Ingredients of Pyrotechnic Compositions

## 6.1 GENERAL

Various ingredients used in pyrotechnic compositions are classified as:

a. Fuels
b. Oxidisers
c. Binders
d. Burn rate modifiers
e. Colour intensifiers
f. Colour dyes
g. Special additives

Fuel and the oxidiser are the main ingredients of a composition. Binders bind these two ingredients for better homogeneity. Burn rate modifiers are used to modify the burn rate. Colour intensifiers are used in signalling flare compositions, while colour dyes are used in signalling colour smoke compositions. In addition, special additives are required for a variety of purposes. We now discuss each of these ingredients below.

## 6.2 FUELS

Fuel is the most vital ingredient in a pyrotechnic composition.

### 6.2.1 Role of Fuels

The various roles of a fuel in a pyrotechnic composition are as follows:

a. It reacts with the oxidiser to produce combustion products and special effects like heat, flash, screening smoke, etc.
b. It acts as means of modifying the burn rate by varying fuel percentage and/or size of the fuel particles.
c. It affects the ignition temperature of the composition. For example, low melting fuels (like sulphur (m.p. 119°C) and lactose (m.p. 202°C)) with potassium chlorate give low ignition temperatures of 220°C and 195°C, respectively while magnesium (m. p. 649°C) and aluminium dust (m.p. 660°C)

DOI: 10.1201/9781003093404-6

with potassium chlorate give higher ignition temperatures of 540°C and 785°C, respectively.

d. It contributes to the heat of combustion of the composition.

e. Some fuels provide hot slags for positive and sustained pick-up of ignition.

## 6.2.2 PARAMETERS FOR CHOICE OF FUELS

The choice of fuel for use in pyrotechnic composition depends on the following:

### 6.2.2.1 Heat of Combustion

All illuminating compositions, incendiary compositions, tracer compositions, photoflash compositions require very high heat of combustion to produce the desired special effects such as luminosity, heat and visibility of coloured light. Hence, fuels for such compositions should be chosen such that they yield higher heat of combustion. Metallic fuels like magnesium and aluminium have very high heat of combustion and hence are preferred in the above compositions. Magnesium is mostly used in illuminating and tracer compositions while aluminium is mostly preferred in photoflash and incendiary compositions.

In white/grey screen smoke ammunition and coloured smoke ammunition, large amount of gases are required for ejection of smoke particles. A high heat of combustion in smoke composition would result in flaming of dye, giving grey/black smoke, which is not desirable. Hence, organic fuels that produce less heat of combustion like lactose, sucrose, etc. are used, which release copious amounts of gaseous products due to presence of large number of carbon, hydrogen and oxygen atoms in their molecules for ejection of combustion products or dye particles. Table 6.1 provides the heat of combustion of some fuels.

*The heat of combustion is maximum at stoichiometric proportion of fuel and oxidiser while rate of reaction is higher at a little higher fuel content than the stoichiometric ratios of the fuel and oxidiser.*

### 6.2.2.2 Grinding and Sieving

The fuel must be solid (with a few exceptions of liquid fuels used in incendiary compositions). The fuel is required to be grinded and sieved for proper sieve size to ensure desired burning rate of the composition by suitable choice of sieve spectrum. This is possible only when the fuel is in solid form. The entire operation of grinding and sieving must be safe. The manufacturer of fuel may also supply fuel within the desired sieve spectrum as desired by the ammunition manufacturer.

It may be mentioned that magnesium is manufactured in several shapes and surface areas and the performance of these varies drastically. Suitable trials are required to arrive at the sieve size, proportion of various ingredients, pressing load and dwell time, etc., to obtain the desired burning time and special effects.

### 6.2.2.3 Ease of Oxidisability

The space available in any ammunition is limited, and hence the maximum output is possible only if the oxygen requirement by the fuel for combustion is low so that the composition may have higher quantity of fuel for exhibiting special effects for a

**TABLE 6.1**
**Heat of Combustion of Fuels**

| Fuel | Heat of combustion (kcal. $g^{-1}$) |
|------|-------------------------------------|
| Aluminium | 7.4 |
| Boron | 14.0 |
| Carbon | 7.8 |
| Lactose | 3.94 |
| Magnesium | 5.9 |
| Phosphorous red | 5.8 |
| Phosphorous white | 5.9 |
| Silicon | 7.4 |
| Sulphur | 2.2 |
| Titanium | 4.7 |
| Tungsten | 1.1 |
| Zinc | 1.3 |
| Zirconium | 2.9 |

prolonged duration. The fuel must be moderately oxidisable with oxidisers as well as with atmospheric oxygen.

This allows the combustion to take place easily; for instance, magnesium with melting point 649°C and low boiling point 1,107°C allows it to vaporise quickly and burn with atmospheric oxygen. Hence, magnesium is often added over and above the stoichiometric ratio in the pyrotechnic composition. However, the same is not possible with aluminium, although it has a melting point of 660°C but a boiling point of 2,500°C, which is very high.

Some fuels like white phosphorous or red phosphorous are easily oxidisable and form extreme sensitive compositions. Readers may see Table 4.3 on fuel oxygen correlation.

The higher the fuel consumption per gram of oxidiser, the better is the oxidisabilty of the fuel. However, in fuels, a thin layer of oxide film is formed on the surface of a metal due to an oxidation reaction from air or moisture surrounding the material. This oxide coating over metal fuels reduces the oxidisabilty of the metal fuel.

The metal oxide ratio, also known as N. Pilling and E. Bedworth ratio (or P–B ratio), is the ratio of the volume of the elementary cell of a metal oxide to the volume of the elementary cell of the corresponding metal (from which the oxide is created). In other words, it is the ratio of volume of the oxide formed to the volume of the metal consumed. This can be shown in the equation:

$$R_{PB} = Voxide/Vmetal = (Moxide \times \rho metal)/(n \times Mmetal \times \rho oxide) \quad (6.1)$$

where $R_{PB}$ is the Pilling–Bedworth Ratio, $M$ is the molecular or atomic mass, $n$ is the metal atom per oxide molecule (Example $n = 1$ for MgO, $n = 2$ for $V_2O_5$ and $Cr_2O_3$), $V$ is the molar volume, $\rho$ is the density. P–B ratio provides an idea about the type of such films.

   i. A thin and porous oxide film would have the ratio less than 1. Example, magnesium, lithium, sodium, potassium.
   ii. A thick and non-porous oxide film would have a ratio greater than 1. Example aluminium, antimony, lead, copper, titanium and chromium. A ratio greater than 2 would result in chipping off the film. Example iron.

The Pilling–Bedworth ratio of some of the metals are magnesium to magnesium oxide MgO (0.81), aluminium to aluminium oxide $Al_2O_3$ (1.28), zirconium to zirconium (iv) oxide $ZrO_2$ (1.56), titanium to titanium (iv) oxide $TiO_2$ (1.73), iron to iron (III) oxide $Fe_2O_3$ (2.14) and chromium to chromium (iii) oxide (2.07).

To obtain the Pilling–Bedworth ratio, one needs to make the equation of metal-to-metal oxide and put the appropriate values. For example, 24.3 g of magnesium reacting to form 40.3 g of oxide as given below.

$$Mg + \tfrac{1}{2}\, O_2 = MgO$$

Magnesium metal density = 1.74 $gcm^{-3}$, magnesium oxide density = 3.58 $gcm^{-3}$
   Hence, Pilling–Bedworth ratio = Volume of oxide/volume of metal

$$= \frac{40.3 \times 1.74}{3.58 \times 24.3} = 0.806(\text{say } 0.81)$$

### 6.2.2.4 Combustion Products

The combustion products must ensure the following on as required basis.

   a. Large volumes of gases are required to be produced (in case of screening smoke ammunition and signalling colour smoke ammunition) to eject particles of combustion products or dyes into air
   b. Large volume of gases are required for whistling composition. Fuels Gallic Acid, Potassium benzoate, Sodium benzoate, Potassium picrate, Sodium salicylate, Potassium dinitrophenolate, Potassium hydrogen phthalate are used for whistling composition.
   c. High luminosity is required (in case of illuminating ammunition) where combustion products of fuel like MgO act as incandescent to provide luminosity.
   d. The quantum of gaseous products and solid and fluid particles of the emitters should be optimum, since an increase or decrease in these would affect the

**TABLE 6.2**
**Combustion Products and their Characteristics of Some Fuels**

| Fuel | Combustion product | Combustion product Melting Point(°C) | Combustion product Boiling Point(°C) |
|------|--------------------|--------------------------------------|--------------------------------------|
| Aluminium | $Al_2O_3$ | 2072 | 2980 |
| Boron | $B_2O_3$ | 450 | 1860 |
| Carbon | $CO_2$ | – | – |
| Magnesium | $MgO$ | 2852 | 3600 |
| Phosphorous red | $P_2O_5$ | – | – |
| Phosphorous yellow | $P_2O_5$ | – | – |
| Silicon | $SiO_2$ | 1610 | 2230 |
| Sulphur | $SO_2$ | – | – |
| Titanium | $TiO_2$ | 1830 -1850 | 2500-3000 |
| Tungsten | $WO_3$ | – | 1800 |
| Zinc | $ZnO$ | 1800 | 1950 |
| Zirconium | $ZrO_2$ | 2700 | 5000 |

    flame dimension (consequently the luminous area). This is important for il-luminating composition, tracers and flare signal composition.

e. Hot fluid decomposition products are required for incendiary and thermite compositions to achieve incendiary/thermite effect as well as for priming and booster effect.

f. Boron based compositions are good priming compositions since boron has a very high heat of combustion as well as low melting point of its combustion product $B_2O_3$ which allows molten hot slag to move a greater area for quick ignition of composition, compared to other fuels.

g. Priming composition SR 252 on combustion provides fusible silicates in hot molten state to provide heat to main composition

Table 6.2 provides the combustion products and the melting and boiling points of some fuels.

### 6.2.2.5 Thermal Conductivity of Fuel

A higher thermal conductive fuel transmits the conduction of heat to next layer of composition, especially for gasless delay compositions.Addition of fine copper and silver particles increase the thermal conductivity of many gasless delay compositions.

### 6.2.2.6 Melting Point of Fuel

A low melting point fuel like sulphur or lactose lowers the ignition temperature of the composition. A high melting point fuel like magnesium or aluminium gives higher ignition temperature. Hence, the ease of ignitability varies (see Section 6.2.2.3) with type of fuel.

### 6.2.2.7   Stability

The fuel must be stable at extreme temperatures of -40°C and +60°C. This is considered necessary since the ammunition is required to function at extreme low temperatures in snow-bound areas as well as at high altitudes in air and also at extreme high temperatures at deserts. It should also be stable over the years during storage, adding to the shelf life.

### 6.2.2.8   Hygroscopicity

The fuel should not be hygroscopic as this would jeopardise the functioning and storage shelf life of the ammunition.

### 6.2.2.9   Purity of Fuels

The purity of fuels affects the combustion properties. Impurities reduce the actual fuel content and also have their own combustion properties. Presence of impurities like copper or ammonia in red phosphorous affects its stability.

### 6.2.2.10   Toxicity

Toxicity is the inherent ability of a chemical substance to produce injury through skin contact, ingestion and inhalation. The fuel must not be toxic and also should not produce toxic products during combustion so as to affect the personnel handling the fuel as well as during manufacture and use of the composition.

However, riot control agents are meant to affect physiologically on human beings. Therefore, during manufacture of compositions with such ingredients, the personnel are required to wear protective clothing and equipment like masks, goggles, use of skin creams, ventilation/suction of buildings, etc.

### 6.2.2.11   Sensitivity

Pyrotechnic compositions are sensitive to mechanical energy (includes impact, percussion and stab energy), frictional energy, electrical energy (includes electrostatic discharge, i.e., spark energy) and thermal energy (primer, igniter, laser beam, fuze), etc. The fuel should not make the composition extra sensitive so as to prohibit its use. This is possible by avoiding dusts and fines and regulating sieve spectrum of fuel. Metal powders in fine form and dust are hazardous (see Table 4.8).

### 6.2.2.11   Availability and Cost

The fuel must be readily available at reasonable cost.

### 6.2.3   Classification of Fuels Used in Pyrotechnic Compositions

Table 6.3 shows the classification of fuels used in pyrotechnic ammunitions (the list is not exhaustive).

**TABLE 6.3**
**Classification of Fuels Used in Pyrotechnic Compositions**

| | |
|---|---|
| Metallic fuels (high calorific value) | Magnesium [Mg], Aluminium [Al], Zirconium [Zr], Nickel [Ni], Zinc [Zn], Iron [Fe], Antimony [Sb], Titanium [Ti], Molybdenum [Mo], Zirconium–nickel alloy [Zr-Ni], Magnesium–Aluminium Alloy [Mg–Al] |
| Non-metallic fuels (medium calorific value) | Sulphur [S], Boron [B], Silicone [Si], Phosphorous yellow [P], Phosphorous red [$P_4$], Charcoal [$C_7H_4O$] |
| Inorganic compounds | Antimony sulphide [$Sb_2S_3$], Calcium silicide [$CaSi_2$], Phosphorous sesquisulphide [$P_4S_3$] |
| Organic compounds | Anthracene [$C_{14}H_{10}$], Naphthalene [$C_{10}H_8$], Glucose [$C_6H_{12}O_6$], Lactose monohydrate [$C_{12}H_{22}O_{11}.H_2O$], Fructose [$C_{12}H_{22}O_{11}$], Sucrose [$C_{12}H_{22}O_{11}$], Tetranitrocarbazole [$C_{12}H_5N_5O_8$], Tetra nitro oxanilide [$C_{14}H_8N_6O_{10}$], Liquid petroleum products like kerosene, naphtha, and benzene |
| Exclusively used in whistling composition | Gallic acid [$C_7H_6\,O_5$], Potassium benzoate [$C_7H_5KO_2$], Sodium benzoate [$C_7H_5NaO_2$], Potassium picrate [$C_6\,H_2KN_3O_7$], Sodium salicylate[$C_7H_5NaO_3$], Potassium dinitrophenate [$C_6H_3KN_2O_5$], Potassium hydrogen phthalate [$C_8H_5KO_4$] |

## 6.3 OXIDISERS

Oxidiser is one of the vital ingredients of a pyrotechnic composition.

### 6.3.1 ROLE OF OXIDISERS

The roles of an oxidiser are as follows:

a. It oxidises the fuel.
b. It determines the rate-determining step in most pyrotechnic compositions.
c. It also acts as means of modifying the burn rate by varying oxidiser percentage.
d. It also acts to provide colour to the flame during combustion of illuminating and signalling flare composition.
e. It modifies the products of combustion

Oxidisers mostly release oxygen to the reducing fuel by melting and thermal decomposition. In addition to ionic solid oxidisers, metal and halogenated organic compounds (like hexachloroethane) or polymers like polytetrafluoroethylene (PTFE) are also capable of undergoing redox reactions even though they do not have an oxygen.

Oxidisers having high hygroscopicity, scarce availability, high cost and lower oxygen content are not suitable for pyrotechnic compositions. Therefore, a lot of oxidisers are not useful and only compounds like sodium, potassium, strontium and barium are practically useful under controlled conditions.

## 6.3.2   PARAMETERS FOR CHOICE OF OXIDISER

The following parameters determine the choice of oxidiser.

### 6.3.2.1   Heat of Decomposition of Oxidiser

The oxidiser must have a moderate heat of decomposition. The oxidiser having high exothermic heat of decomposition would lead to low ignition temperature of the composition, thereby making the composition highly reactive while an endothermic heat of decomposition would lead to a high ignition temperature of the composition and may make ignition difficult. The heat of decomposition of oxidisers is given in Table 6.4.

### 6.3.2.2   Grinding and Sieving

The oxidiser must be solid. The oxidiser is required to be grinded and sieved for proper sieve size to ensure proper mixing/blending with consistent burning of the composition and to vary burning rate by suitable choice of sieve spectrum of the oxidiser. This is possible only when the oxidiser is in solid form. The entire operation of grinding and sieving must be safe. However, between fuel and oxidiser, the fuel sieve size is predominant in deciding the burn rate since oxidiser, in most cases, melts near the ignition temperature of the composition.

### 6.3.2.3   Active Oxygen Content

The oxidiser must possess high active oxygen content and spontaneously give oxygen during pyrotechnic composition combustion. Active oxygen content is the amount of oxygen liberated during combustion of oxidiser. This is different from

**TABLE 6.4**

**Heat of Decomposition of Oxidisers**

| Oxidiser | Heat of decomposition (kcal. mole$^{-1}$) |
|---|---|
| Sodium nitrate | 60.5 |
| Potassium nitrate | 75.5 |
| Potassium chlorate | (–)10.6 |
| Potassium perchlorate | (–)0.68 |
| Barium peroxide | 17 |
| Barium nitrate | 104 |
| Barium chlorate | (–)28 |
| Strontium nitrate | 92 |

(–) indicates that the decomposition is exothermic

the amount of oxygen available in the oxidiser. This can be understood by the decomposition of sodium nitrate as under.

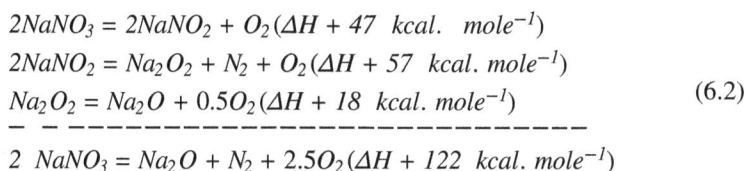

$$2NaNO_3 = 2NaNO_2 + O_2 (\Delta H + 47 \ \ kcal. \ \ mole^{-1})$$
$$2NaNO_2 = Na_2O_2 + N_2 + O_2 (\Delta H + 57 \ \ kcal. \ mole^{-1})$$
$$Na_2O_2 = Na_2O + 0.5O_2 (\Delta H + 18 \ \ kcal. \ mole^{-1})$$ 
$$\overline{- \ - \ -\ - - - - - - - - - - - - - - - - - - - - - - - - - -}$$
$$2 \ \ NaNO_3 = Na_2O + N_2 + 2.5O_2 (\Delta H + 122 \ \ kcal. \ mole^{-1})$$

(6.2)

The atomic weights of atoms are sodium (Na) = 23, nitrogen (N) = 14 and oxygen (O) = 16. The molecular weight of sodium nitrate is 85 {23 + 14 + (3 × 16) = 85}).

The oxygen content in sodium nitrate is 48 g out of total 85 g or 1 g of sodium nitrate has 0.57 g (48/85 or 57%) of oxygen.

However, on decomposition, 2 moles of sodium nitrate ($NaNO_3$) having molecular weight 170 give 5 moles of oxygen having molecular weight 80 {5 × 16 = 80}. In other words, 170 g of sodium nitrate releases 80 g of oxygen. The oxygen liberated by 1 g of sodium nitrate is 0.47 g (80/170 or 47%), which is the active oxygen content. Table 6.5 gives the active oxygen content per gram of the oxidiser in percentage.

### 6.3.2.4   Melting Point of Oxidiser

As explained earlier (see Section 3.5.1.2,), melting of oxidiser causes formation of more intimate mixture of the fuel and the oxidiser, thereby resulting in initiation of ignition. Thus, the lower the melting point of oxidiser, the lower would be the ignition temperature of the composition and faster would be the combustion rate. Sodium nitrate and potassium nitrate-containing pyrotechnic compositions have a low ignition temperature than compositions containing barium nitrate due to low melting point of both sodium nitrate (308°C) and potassium nitrate (334°C) compared to barium nitrate (592°C).

The above oxidisers also act to ease the ignition of pyrotechnic composition containing other oxidisers. Thus, $KNO_3$ having low melting and decomposition temperature is added in some compositions containing Ba ($NO_3$)$_2$, the latter having high melting and decomposition temperature. This leads to ease of ignition and smooth combustion of the composition.

### 6.3.2.5   Purity of Oxidiser

Highly pure oxidisers cause rapid combustion of the composition. Foreign matters in the oxidisers are not desirable as they reduce the stability of the pyrotechnic composition. It should also not contain ingredients like fuel and sensitising ingredients like sand, fine glass, etc., which would make the composition highly sensitive.

### 6.3.2.6   Stability

The oxidiser must be stable at −40°C to +60°C. The oxidiser should withstand extreme climatic temperatures in snow-bound areas, at high altitudes as well as deserts. It should also be stable over the years giving adequate storage life.

**TABLE 6.5**

**Quantum of Oxygen Release and Active Oxygen Percent of Oxidiser**

| Oxidiser | Grams of oxygen released/gram of oxidiser | Active oxygen content (%) |
|---|---|---|
| Barium chlorate | 0.32 | 32 |
| Barium chromate | 0.095 | 9.5 |
| Barium nitrate | 0.31 | 31 |
| Barium peroxide | 0.09 | 9 |
| Iron oxide (red) | 0.30 | 30 |
| Iron oxide (black) | 0.28 | 28 |
| Lead chromate | 0.074 | 7.4 |
| Lead tetroxide | 0.093 | 9.3 |
| Lead peroxide | 0.13 | 13 |
| Potassium chlorate | 0.39 | 39 |
| Potassium nitrate | 0.40 | 40 |
| Potassium perchlorate | 0.46 | 46 |
| Sodium nitrate | 0.47 | 47 |
| Strontium nitrate | 0.38 | 38 |

### 6.3.2.7  Hygroscopicity

The moisture content in the oxidiser depends upon the humidity and the temperature of the air as well as the property of the oxidiser and its surface area in contact with the moist air. The oxidiser should preferably be non-hygroscopic. In other words, an oxidiser, preferably, should not have affinity towards water, as the absorbed water may react with other ingredients, making the composition unusable. This non-hygroscopicity is desirable for proper pick-up of ignition and continuity of ignition. The relative hygroscopicity of some oxidisers in descending order are as under.

*Potassium Nitrate (most hygroscopic) > Sodium Nitrate > Strontium Nitrate > Potassium Chlorate > Potassium Perchlorate > Barium Nitrate (least hygroscopic).*

However, oxidisers like sodium nitrate and potassium nitrate, though highly hygroscopic, are still used in various ammunitions due to release of high quantum of oxygen during combustion.

The hygroscopicity is taken care of by the use of suitable dryers for drying oxidiser and operations like mixing, pressing and assembly and packing carried out in suitable controlled air-conditioned buildings, maintaining the humidity between 55% and 65%.

### 6.3.2.8  Availability and Cost

The oxidiser must be readily available at reasonable cost.

#### 6.3.2.9   Toxicity

The oxidiser must not be toxic and also should not produce toxic products during combustion so as to affect personnel handling the oxidiser as well as during manufacture of the composition and also during the use of the ammunition. Oxidisers containing barium, lead, chromium are toxic in nature. Efforts are on to use green environmentally friendly oxidisers.

#### 6.3.2.10   Flame Colour

The oxidiser should not affect/mask the colour of the flame in signalling or illuminating ammunitions. For example, during combustion of illuminating compositions containing sodium nitrate, sodium atoms give yellow flame which is suitable for illumination. But it is not suitable for green or red signalling compositions as yellow flame would mask the green or red colour.

Hence, the oxidiser barium nitrate forming barium monochloride in presence of chlorine donor and oxidiser strontium nitrate forming strontium monochloride in presence of chlorine donor are used in green and red signalling composition, respectively.

Examples of some oxidisers imparting colour to the flame are given in Table 6.6.

#### 6.3.2.11   Sensitivity

Pyrotechnic compositions are sensitive to mechanical energy (includes impact, percussion and stab energy), frictional energy, electrical energy (includes electrostatic discharge, i.e., spark energy) and thermal energy (primer, igniter, laser beam, fuze), etc. The oxidiser should not make the composition extra-sensitive so as to pose difficulty in its use.

**TABLE 6.6**
**Oxidisers Imparting Colour to Flame**

| Ingredient | Colour |
|---|---|
| Strontium nitrate, strontium oxalate | Red |
| Sodium nitrate, sodium oxalate | Yellow |
| Barium nitrate, barium chlorate | Green |
| Copper chloride | Blue |
| Potassium nitrate | Violet |

#### 6.3.3   Classification of Oxidisers Used in Pyrotechnic Compositions

Table 6.7 shows the classification of oxidisers used in pyrotechnic ammunitions.

**TABLE 6.7**

## Classification of Oxidisers Used in Pyrotechnic Compositions

| | |
|---|---|
| Nitrates | Sodium nitrate [NaNO$_3$], strontium nitrate [Sr (NO$_3$)$_2$], barium nitrate [Ba (No$_3$)$_2$], potassium nitrate [KNO$_3$] |
| Chlorates | Potassium chlorate [KClO$_3$], barium chlorate [Ba (ClO$_3$)$_2$} |
| Perchlorate | Potassium perchlorate [KClO$_4$], ammonium perchlorate [NH$_4$ClO$_4$] |
| Chromates | Barium chromate [BaCrO$_4$], lead chromate [PbCrO$_4$] |
| Oxides | Lead oxide [PbO], lead tetroxide [Pb$_3$O$_4$], zinc oxide [ZnO], manganese dioxide [MnO$_2$], iron (III) oxide [Fe$_2$O$_3$], iron (II, III) oxide [Fe$_3$O$_4$] |
| Peroxides | Lead peroxide [PbO$_2$], barium peroxide [BaO$_2$] |
| Polymers | Polytetrafluoroethylene/PTFE/Teflon/Fluon [(-CF$_2$-CF$_2$-)$_n$] |

### 6.3.4 SOME TYPICAL DECOMPOSITION OF OXIDISERS

Some typical decomposition of oxidisers with active oxygen content and heat of decomposition ($\Delta$H) values and some end uses are at Table 6.8.

**TABLE 6.8**

## Decomposition of Oxidisers

| Oxidiser/ Molecular wt./ Active oxygen% | Decomposition | M.P.°C | $\Delta$H kcal. mole$^{-1}$ | Some Uses |
|---|---|---|---|---|
| NaNO$_3$/85/ (+)47 | 2NaNO$_3$ = Na$_2$O + N$_2$ +2.5O$_2$ | 308 | (+) 60.5 | Illuminating |
| KNO$_3$/101.1/ (+)39.6 | 2 KNO$_3$ = K$_2$O + N$_2$ +2.5O$_2$ | 334 | (+) 75.5 | Priming |
| KClO$_3$/122.6/ (+)39.2 | KClO$_3$ = KCl +3/2O$_2$ | 356 | (−)10.6 | Smoke, Signal |
| KClO$_4$/138.6/ (+)46.2 | KClO$_4$ = KCl + 2O$_2$ | 525 | (−)0.68 | Incendiary |
| Sr (NO$_3$)$_2$/212/ (+)37.7 | Sr (NO$_3$)$_2$ = SrO + N$_2$ + 2.5 O2 | 570 | (+)92 | Tracer, signal |
| Ba (NO$_3$)$_2$/261.3/ (+)30.6 | Ba (NO$_3$)$_2$ = BaO + N$_2$ + 2.5O$_2$ | 592 | (+)104 | Signal, incendiary |
| BaO$_2$/169.3/ (+)9.5 | BaO$_2$ = BaO + 0.5O$_2$ | 450 | (+)17 | Tracer priming |
| MnO$_2$/87/ (+)36.8 | MnO$_2$ = Mn + O$_2$ | 535 | (+)125 | Incendiary |

(+)$\Delta$H shows endothermic decomposition, (-)$\Delta$H shows exothermic decomposition

## 6.4 ROLE OF BINDERS

The main role of a binder is to improve process ability and mechanical strength. However, it also affects the thermal properties of the composition.

a. *Process Ability:*
  i. Improves the cohesion of ingredient particles for manufacture of granules for ease of pelleting
  ii. Avoids segregation due to difference in the density and sieve size of the ingredients during processing, transit and storage
  iii. Provides homogeneity of composition
  iv. Provides a coat to fuel particles (waterproofing) to avoid corrosion/aerial oxidation
  v. Provides bonding to the empty hardware
  vi. Improves flow properties of some of the compositions
  vii. Reduces dusting in composition manufacturing process

b. *Mechanical Behaviour:*
  i. Modifies compaction strength
  ii. Makes more rigid pellets, thereby enhancing the mechanical strength and rigidity of the pellet with less pressing load (prevents the pressed pellets from getting cracked or broken or crushed due to stresses of processing, handling and transportation of ammunitions as well as the force of setback, set forward, spin, mechanical shock, etc. during dynamic firing)

c. *Thermal Properties:*
  i. Modifies the sensitivity of the composition to ignition
  ii. Modifies burn temperature since binder reduces the surface temperature, thereby reducing the rate of burning. In case of illuminating compositions, it reduces luminosity

Some binders act as plasticisers and some as oxidisers, while organic binders also act as a secondary fuel. Plasticiser binders like gilsonite, asphaltum, viton and sulphur flow in between the pores of ingredients under compaction. The binders acting as a secondary fuel consume oxygen during combustion of the composition. The binder thus reduces the linear burn rate and minimises the temperature of the flame.

Most compositions are pressed in the form of a pellet or a star or a candle. The quantity of the binder, among other things, plays an important role in consolidation of the pellet and an increase in binder percentage increases the strength (crush load) of the pellet. The binders are not used more than 10%–12% in a pyrotechnic composition as the binding effect becomes insignificant at higher binder percentage. Binders work better when dissolved in a suitable solvent and then mixed in the composition. These solvents are required to be evaporated in subsequent operations.

A polyester resin-based composition is required to be pressed within a short but appropriate time. If pressed early, the gases entrapped inside may come out forming cracks in the pellet. However, if it is delayed, the binding property of the resin is affected severally, and adhesion property may be lost (it may show very low compaction strength in load testing equipment). The pellets are required to be dried for proper curing and then packed in hermetically sealed boxes. Polyester resins make the pressed pellet very hard.

Thiokol rubber makes the extruded pellet soft, flexible and resilient. Thiokol-based compositions typically need not be pressed/extruded and may be poured into the container and cured in a heating chamber. The proportion of Thiokol rubber is much higher in the composition than normal binders and hence sensitivity of such compositions is much lower.

The presence of Thiokol rubber makes these composition waterproof and in-sensitive to impact, shock and friction and hence is safe to handle. These rubber polymers thus act in the following three ways:

a. Avoid ingress of moisture in the composition
b. Are less sensitive and hence safe to handle
c. They do not need hydraulic or air or manual presses and heavy moulds and punches for pressing

Table 6.9 provides the effect of binders on pyrotechnic composition.

Table 6.9 show that the use of binder reduces the burn rate, reduce the burn temperature and reduce the luminosity. Table 6.10 gives the effect of binder on inverse burn rate, sensitivity to impact, sensitivity to friction, sensitivity to spark and breaking strength of igniter compositions.

Table 5.4 may also be referred for effect of binder on performance of Mg/NaNO3/binder composition.

### 6.4.1   CLASSIFICATION OF BINDERS USED IN PYROTECHNIC COMPOSITIONS

Binders can be broadly classified as natural and synthetic polymers, organic and inorganic polymers, thermoplastic and thermosetting polymers etc. However, in common terminology and understanding, binders may be classified as in Table 6.11.

---

**TABLE 6.9**
**Effect of Binder on Pyrotechnic Compositions [1]**

| Composition | Composition I | Composition II | Composition III |
|---|---|---|---|
| Barium nitrate | 59.5 | 56.5 | 56.5 |
| Magnesium | 40.5 | 38.5 | 38.5 |
| Phenolic resin | | 5 | - |
| Nitrocellulose | - | - | 5 |
| | Characteristics | | |
| Burn temperature $^0$c | 1658.1 | 1594.2 | 1432.6 |
| Average luminous intensity(cd) | 65741.2 | 47235.6 | 32439.8 |
| Average burn rate (m.s$^{-1}$) | 16.8 | 12.5 | 9.0 |

**TABLE 6.10**

**Effect of Binder on Inverse Burn Rate and Sensitivity [2]**

| Composition | Inverse burn rate (s.cm$^{-1}$) | Sensitivity to impact, height of 50% explosion (cm) | Sensitivity to friction (kg) | Spark sensitivity | Breaking strength (N) |
|---|---|---|---|---|---|
| | Igniter composition based on boron | | | | |
| B/KNO$_3$/PMMA 30:70:5 | 0.76 | 85 | >36 | 5J-NF | 451 |
| B/KNO$_3$/PMMA 30:70:10 | 0.86 | 91 | >36 | 5J-NF | 645 |
| B/KNO$_3$/PEC 30:70:10 | 1.35 | 84 | >36 | 5J-NF | 575 |
| | Igniter composition based on magnesium | | | | |
| Mg/KNO$_3$/PMMA 45:50:5 | - | 100 | 28.8 | 5J-NF | 207 |
| Mg/KNO$_3$/PMMA 42:50:8 | - | 164 | >36 | 5J-NF | 226 |
| Mg/KNO$_3$/PEC 42:50:8 | - | 100 | 28.5 | 5J-NF | 230 |
| | Delay composition based on ferrosilicon and red lead | | | | |
| FeSi/Pb$_3$O$_4$/PMMA 25:75:1 | 0.35 | >175 | 24 | 6.0mJ-NF 9.4mJ-F | - |
| FeSi/Pb$_3$O$_4$/NC 25:75:1 | 0.65 | 103 | 24 | 18.4mJ-NF 20.5mJ-F | - |

PMMA = polymethyl methacrylate, PEC = plasticised ethyl cellulose, F = function, NF = not function

## 6.4.2   PARAMETERS FOR CHOICE OF BINDERS

Each binder has its own characteristics like reactivity, melting point, heat of combustion, combustion products, quantum of gas evolved, acting as fuel to some extent and affecting sensitivity. Hence a suitable choice of binder for a particular composition is required.

The binders are polymers having high molecular weight. It is necessary that these polymers have very high oxygen content (more than 40%) so that all the carbon is used up for combustion. In case the oxygen content is less than 40%, the carbon which has not been oxidised, would become incandescent and degrade both the colour and the light output. In addition, the following parameters are important while selecting a binder.

a. It should be neutral and compatible with ingredients and the metallic/paper components. It should not react with fuel, oxidiser and other ingredients.
b. It should be non-hygroscopic.

**TABLE 6.11**

**Classification of Binders Used in Pyrotechnic Compositions**

| Type | Binders |
|---|---|
| Synthetic resins | 1. Bakelite resin [HSR 8111], accelerator [1021] & catalyst [Q 8013] |
| | 2. Epoxy resins like [dobecot 520 F & hardener EH 408] |
| | 3. Calcium resinate [Ca $(C_{19}H_{29}COO)_2$] |
| | 4. Vinyl acetate-alcohol resin (VAAR) |
| | 5. Viton (vinylidene fluoride/hexafluoropropylene copolymer) [(-$CH_2$-$CF_2$-)$_n$-$CF_2$-$CF(CF_3$)-]$_m$ |
| Natural resins | 1. Bees wax [$C_{15}H_{31}COOC_{30}H_{61}$] |
| | 2. Paraffin wax [$C_nH_{2n+2}$ n=24-36] |
| | 3. Starch [$C_6H_{10}O_5)_n$] |
| | 4. Dextrin [(-$C_6H_{10}O_{5-})_n$.$H_2O$)] |
| | 5. Shellac |
| | 6. Acaroid resin, etc. |
| Oils | 1. Boiled linseed oil (used mostly in signal composition) |
| | 2. Lithographic varnish (used mostly in illuminating composition) |
| | 3. Castor oil, etc. |
| Lacquer/ varnish/dope | 1. Nitrocellulose with low nitrogen content |
| | Note: Thinner consistency material is termed "*Lacquer*", moderately thicker consistency material is termed "*Varnish*" and thicker consistency material is termed "*Dope*" |
| | 2. Ethyl cellulose dissolved in Hercules solvent (consisting of amyl acetate 40%, acetone 40% and toluene 20%) with other stabilisers |
| | 3. Shellac dissolved in ethyl alcohol, etc. |
| Rubber | 1. Chlorinated rubber (other names alloprene, parlon) |
| | 2. Thiokol rubber |
| | 3. Silicone rubber, etc. |

c. It should have a high softening point and low glass transition temperature.

d. It should cure/mature in a reasonable time, providing enough time for processing a composition.

e. It should be able to form a thin film of the binder on drying up.

f. It should not be toxic and its products of decomposition should not be toxic.

g. It should be readily available in the market at a reasonable cost.

Synthetic resins are preferred over natural resins as characteristics of natural resins change due to variation in source, processing methods and weather conditions leading to variation in composition performance from batch-to-batch while the synthetic binders give better performance. Waxes are used after melting for coating fuel. Water-soluble binders are not preferred as moisture affects pyrotechnic composition ingredients like fuels magnesium and aluminium and hygroscopic oxidisers like nitrates and peroxides.

Lithographic varnish gets oxidised in the initial stages generating heat which is hazardous. If a heap of composition is used, then the polymerisation of varnish or boiled linseed oil would cause a dry and hard film on the outer side and the heat generated during the process would not be allowed to be released, causing auto-ignition. Hence, raking during drying is a must. Thus, as a safety measure, a composition containing lithographic varnish or boiled linseed oil is dried at an ambient temperature by spreading thin layer of composition in a tray in a dry building for 48 hours with occasional raking, using wooden spatula.

### 6.4.3  SOLVENTS FOR BINDERS

The solvents used for some of the binders are given in Table 6.12.

## 6.5  BURN RATE MODIFIERS

Burn rate modifiers play vital role in controlling the burn rate of pyrotechnic compositions.

### 6.5.1  ROLE OF BURN RATE MODIFIERS

Burn rate modifiers act as cooling agents or catalysts or retardants or sensitivity modifiers that affect the burn rate by altering the activation energy, heat of reaction and energy feedback for sustaining burning of the pyrotechnic composition. They act in such a way so as to either reduce the burn rate, or enhance the burn rate or reduce the sensitivity or enhance the sensitivity of the composition with minimal

**TABLE 6.12**
**Solvents for Some Binders**

| Binder | Solvent |
| --- | --- |
| Carboxymethyl cellulose | Water |
| Chlorinated rubber | Acetone or methyl ethyl ketone |
| Dextrin[(-$C_6H_{10}O_{5-}$)$_n$.$H_2O$)] | Water |
| Gum arabic or acacia gum | Water |
| Nitrocellulose (low nitrogen content) | Acetone or amyl acetate |
| Ethyl cellulose | Ethyl alcohol |
| Plasticised ethyl cellulose | Toluene |
| Red gum or acaroid resin | Alcohol |
| Starch (rice) [$C_6H_{10}O_5$)$_n$] | Water |
| Shellac | Ethyl alcohol |
| Viton (vinylidene fluoride/hexafluoropropylene copolymer)[(-$CH_2$-$CF_2$-)$_n$-$CF_2$-$CF(CF_3)$-]$_m$ | Acetone |

**TABLE 6.13**

**Effect of SiO$_2$ on the Linear Burn Rate of a 30:70 Sb:KMnO$_4$ Composition [3]**

| SiO$_2$ percentage | Linear burn rate (mm. s$^{-1}$) |
|:---:|:---:|
| 0 | 1.9 |
| 7 | 1.6 |
| 19 | 1.3 |

effect on special effects output. Burn rate modifiers may be classified into four groups as under.

a. *As Cooling and Retarding Agent*:
   A pyrotechnic composition having higher linear burn rate may be slowed down by the addition of a burn rate modifier as retardant. These acts by:
   i. Cooling the system by absorption of heat of combustion, also known as heat sink materials. Inert materials and compositions like clay, kaolin, kieselguhr, chromic oxide and magnesium oxide are used as heat sink materials to retard the rate of burning as these absorb the heat of combustion and occupy a position in the mass of composition so as to keep the fuel and oxidant separated and thus reduce the rate of burning. Table 6.13 shows an example of use of the inert additive SiO$_2$ for reducing the linear burn rate in the Sb/KMnO$_4$ system.
   ii. Cooling the system by producing gas that dilutes the air, thereby separating the reactants and reducing the concentration of oxygen, both reducing the probability of reaction, also known as gas-producing burn rate modifiers.
   iii. Cooling the system by endothermic decomposition to absorb heat and thus reduce the burn rate.

Similarly, magnesium carbonates effectively control the flaming of signalling colour smoke compositions by reducing the combustion rate by reducing the heat of combustion, and thus avoiding higher temperatures that lead to flaming of dye in the compositions. However, a higher percentage of burn rate modifier may reduce the density of smoke from dense to rare, which is not desirable. The decomposition of some burn rate modifiers as retardants producing carbon dioxide are as follows:

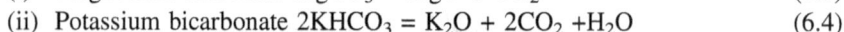

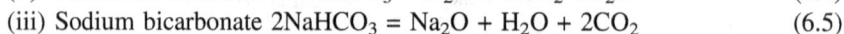

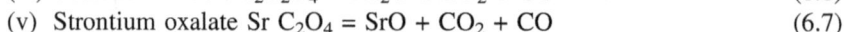

(i)   Magnesium carbonate $MgCO_3 = MgO + CO_2$ $\hspace{2cm}$ (6.3)
(ii)  Potassium bicarbonate $2KHCO_3 = K_2O + 2CO_2 + H_2O$ $\hspace{1cm}$ (6.4)
(iii) Sodium bicarbonate $2NaHCO_3 = Na_2O + H_2O + 2CO_2$ $\hspace{0.5cm}$ (6.5)
(iv)  Sodium oxalate $Na_2C_2O_4 = Na_2O + CO_2 + CO$ $\hspace{1.5cm}$ (6.6)
(v)   Strontium oxalate $Sr\ C_2O_4 = SrO + CO_2 + CO$ $\hspace{1.5cm}$ (6.7)

**TABLE 6.14**
**Burn Rate Modifiers**

| Nomenclature | Molecular mass | M.P. (°C) | Density (g.cm$^{-3}$) |
|---|---|---|---|
| Magnesium carbonate (MgCO$_3$) | 84 | 540 decompose | 2.958 |
| Potassium bicarbonate (KHCO$_3$) | 100 | Decompose | 2.17 |
| Sodium carbonate (Na$_2$CO$_3$) | 106 | 851 | 2.54 |
| Sodium oxalate (Na$_2$C$_2$O$_4$) | 134 | 250 decompose | 2.34 |
| Strontium oxalate (SrC$_2$O$_4$) | 176 | 150 decompose | 2.08 |

These burn rate modifiers also take away heat for their own decomposition (Table 6.14) and thus cool the system and modify the burn rate of composition.

Further, ferric oxide (Fe$_2$O$_3$) is used as retardant for many delay compositions, though it is an oxidiser. However, since it has less oxidative properties than red lead and hence acts as retardant in delay compositions containing red lead, silicon and ferric oxide.

b. *As Catalysts*:

A pyrotechnic composition's lower burn rate may be increased by addition of burn rate modifiers acting as catalysts. These act by increasing the heat in the system and thermal conductivity. For example, linear burn rate in illuminating and tracer composition may be increased by addition of some burn rate modifiers as catalyst like zirconium, which increases the heat output and also being a conductive ingredient aids in energy feedback for sustaining burning through conduction in the composition. Similarly, some gasless delay compositions linear burn rates may be increased by addition of fine conductive copper or silver.

c. *As Sensitivity Modifiers:*

Some burn rate modifiers act as sensitivity modifiers like paraffin wax (also acts as binder), zinc stearate (reduces the sensitivity to friction and flame). However, some burn rate modifiers increase the sensitivity of the composition like lead thiocyanate, ferrosilicon, sulphur, sand, glass powder, etc. and are deliberately used in cap compositions.

Similarly, some burn rate modifiers like manganese dioxide (MnO$_2$) and potassium dichromate (K$_2$Cr$_2$O$_7$) reduce the decomposition temperature of oxidisers like potassium chlorate and potassium perchlorate, respectively, and hence increase the sensitivity of the composition and reduce the ignition temperature.

### 6.5.2 PARAMETERS FOR CHOICE OF BURN RATE MODIFIER

The following parameters are important while selecting a burn rate modifier.

a. It should be neutral and compatible with ingredients and the metallic/paper containers
b. It should be non-hygroscopic
c. It should not be toxic and its products of decomposition should not be toxic
d. It should be readily available in the market at reasonable cost
e. Its combustion products should not mask the effect on the colour of signal composition

These burn rate modifiers are required to be used judiciously, especially in signal composition since decomposition products of these modifiers will have a masking effect on the colour of the signal. Hence, for signal yellow, sodium oxalate is used (sodium atom provide in yellow spectral emission) and for signal green, barium oxalate (forming barium monochloride with colour intensifiers provide green spectral emission) is used while for signal red, strontium oxalate (forming strontium monochloride with colour intensifier provide red spectral emission) is used.

## 6.6   COLOUR INTENSIFIER

Colour intensifiers play a special role in imparting colour during the combustion process.

### 6.6.1   ROLE OF COLOUR INTENSIFIERS

To obtain flame colour, certain ingredients are necessary, which on combustion, produce species capable of emitting desired radiation in the visible spectrum. Flame colour is related to type of oxidisers (metal salts) that impart colour. However, chlorine is required to be infused in the composition to enhance colour purity (except yellow) by addition of colour intensifiers that are rich in chlorine and act as chlorine donor. Thus, colour intensifiers intensify the colour for better visibility. The mechanism of flame colour has been explained in Chapter 9.

### 6.6.2   TYPES OF COLOUR INTENSIFIERS (AS CHLORINE DONOR)

Since the oxidisers are generally strontium nitrate, strontium oxalate, barium nitrate, etc., and do not have any chlorine content, the monochloride species are produced by addition of chlorine donor chemicals. Although some oxidisers like chlorates and perchlorates do have chlorine, chlorine-rich ingredients are still added as chlorine donors. A list of chlorine donors with chlorine content is given in Table 6.15.

These ingredients are added to intensify the colour of the flame. Polyvinyl chloride $[-CH_2.CHCl.]_n$, where n is about 4,000 and contains approximately 56% chlorine and chlorinated rubber $[C_{10}H_{11}Cl_7]_n$ contains approximately 67% chlorine, is used in red star signal composition to intensify the colour. However, in addition, these also act as bonding and retarding agent. An optimum quantity of colour intensifier is required as very low or very high percentage shall affect colour (faint colour) or flame temperature (dim flame).

**TABLE 6.15**
**Chlorine Content of Chlorine Donors**

| Chlorine donor ingredient | Chlorine content (%) |
|---|---|
| Chlorinated rubber [$C_{10}H_{11}Cl_7$]$_n$ | 67 |
| Dechlorane (hexachlorocyclopentadiene dimer) [$C_{10}Cl_{12}$] | 78.3 |
| Hexachlorobenzene [$C_6Cl_6$] | 74.7 |
| Hexachloroethane [$C_2Cl_6$] | 89.9 |
| Chlorinated polyisopropylene "Parlon" [$C_4H_6Cl_2$]$_n$ | 66 |
| Polyvinyl chloride [-$CH_2.CHCl$-]$_n$ | 56 |
| Polyvinylidene chloride -[($C_2H_2Cl_2$) $_n$]- | 73 |

## 6.7  DYES

Dyes play special role in imparting colour during colour smoke combustion process.

### 6.7.1  ROLE OF DYES

Dyes play a role in signalling colour smoke composition by exhibiting the colour of the dye when expelled from the ammunition system. The dye particles absorb some wavelength of the incident radiation and scatter and reflect the rest of the radiation. The human eye perceives the complimentary colour. The details are given under signalling colour smoke composition Chapter 13.

### 6.7.2  PARAMETERS FOR CHOICE OF DYES

These dyes should have following properties:

a. The dyes should sublimate at low temperatures (around 450–500°C).
b. Sublimation should take place without decomposition of the dye.
c. The smoke colour should be of high purity and distinct and stable in air for better colour visibility (i.e., retain in air for more duration).
d. The dye molecular weight should be lower than 450.
e. The dye should have thermal stability so that it does not decompose easily. Thus, the dye should sublime very fast so that it does not decompose while under exposure of heat of combustion.
f. The dye should have high flash point so that it does not burn at low temperatures.

Dye series anthraquinone, azine, azo, quinoline and xanthene are most preferred. However, anthraquinone dyes are superior to all other dyes for colour smoke as these are stable and include all shades of red, orange, yellow, blue and violet.

Readers may see reference [4] for various dyes for red, green, orange, orange-red, blue and violet smoke.

It is necessary that the signalling colour smoke composition does not have high heat of combustion as dyes may decompose at higher temperatures. Hence, fuels like sucrose and lactose are used, which give low combustion temperatures as well as give high quantity of gaseous products to expel the dye in air. Metallic fuels are not used as they give higher combustion temperature. Sodium or potassium bicarbonates may be used in the composition to control the burn rate, thereby preventing flaming of dye. It should be ensured that other ingredients of the composition should not make its own smoke, which could alter the colour shade of the dye. Lactose and sugar are preferred fuels while potassium chlorate is a preferred and suitable oxidant for colour signalling compositions.

In signalling colour smoke ammunition, extra dye in loose form or in a pellet form may be added, which takes away excess heat and gets vaporised in air to give bright smoke of the dye, thereby preventing flaming of the dye .

## 6.8   SPECIAL ADDITIVES

These additives play an important part in the manufacture of pyrotechnic compositions. These do not affect special effects of composition.

### 6.8.1   ROLE OF SPECIAL ADDITIVES

Some typical roles played by special additives in processing of pyrotechnic compositions are as follows:

a. zinc stearate in the composition helps to
   i. Inhibit agglomeration of ingredients
   ii. Improve flowability of ingredients
   iii. Reduce the attractive forces between the ingredient particles, by forming a solid barrier
   iv. Reduce the friction between the ingredient particles, by lubricating the solid ingredient
   v. Neutralise electrostatic charges
   vi. Help in safety aspect of processing pyrotechnic compositions
b. *Acting as conducting medium* for pyrotechnic compositions, e.g., graphite is used as a conducting medium for electric cap composition or fine copper or silver in gasless delay compositions
c. *Acting as solvents* like acetone, ethyl alcohol, amyl acetate, methyl ethyl ketone, dibutyl phthalate, etc. for binders like shellac, nitrocellulose, ethyl cellulose, Viton, etc.
d. *Assist in avoiding friction to punch and mould* by use of zinc stearate (0.25%–2.0%) in composition to prevent friction and sticking of composition on punches and also graphite as a lubricant for punches and moulds during pressing, thus avoiding friction between the punch and the mould.

## 6.9   MATERIAL SAFETY DATA SHEET

Material safety data sheet is an important document with respect to the ingredients. Safety data sheets provide information about personal hazards associated with the material and its environmental hazards. It provides guidelines with respect to safety and emergency measures to be adopted. It has 16 sections as mentioned below:

Section 1: Product and company identification, Section 2: Hazards identification, Section 3: Composition and information on ingredients, Section 4: First aid measures, Section 5: Firefighting measures, Section 6: Accidental release measures, Section 7: Handling and storage, Section 8: Exposure controls and personal protection, Section 9: Physical and chemical properties, Section 10: Stability and reactivity, Section 11: Toxicological information, Section 12: Ecological information, Section 13: Disposal information, Section 14: Transport information, Section 15: Regulatory information, Section 16: Other information.

A material safety data sheet for ingredients is a must for all production organisations as it enhances safety of the personnel.

## 6.10   FACTORS GOVERNING THE CHOICE OF INGREDIENTS FOR COMPOSITION

The ingredients used in pyrotechnic compositions are mostly solids with a few liquids. Each of these ingredients are required to be chosen judiciously to get the desired performance from the composition. Ingredients should be chosen for manufacture of composition for desired special effects based upon some of the properties of the ingredients, such as:

a. Heat of combustion of fuel
b. Active oxygen percent in oxidiser
c. Melting point and boiling point of fuel
d. Density and purity of ingredients
e. Particle size, shape, porosity, surface area and crystal structure of ingredient
f. Hygroscopicity of ingredients
g. Safety in handling of ingredients
h. Toxicity of ingredients
i. Combustion products and their toxicity
j. Availability and cost of ingredients
k. Compatibility of ingredients with other ingredients/components during prolonged storage

There are a large number of ingredients used in pyrotechnic compositions for a variety of ammunitions. These have been provided with their common names, other names, molecular formula, molecular weight and use in reference [5]. Readers may also refer a book [6] on chemicals used in military pyrotechnic compositions with emphasis on the occurrence, commercial methods of production and physical properties of the chemicals.

## REFERENCES

[1] Di-hua Ouyang, *"Effect of Different Binders on the Combustion Characteristics of Barium Nitrate/Magnesium containing Pyrotechnic Mixtures"*, **Central European Journal of Energetic materials**, 2013, Volume 10, Issue 2, Pages 209–215.

[2] S.P. Sontakke, S.D. Kakade, R.M. Wagh, A.G. Dugam and P.P. Sane, *"Polymethyl Methacrylate as a Binder for Pyrotechnic Compositions"*, © *Defence Science Journal*, October 1995, Volume 45, Issue 4, Pages 349–352.

[3] Walter W. Focke, Shepherd M. Tichapondwa, Yolandi C. Montgomery, Johannes M. Grobler, and Michel L. Kalombo, *"Review of Gasless Pyrotechnic Time Delays"*, *Propellants, Explosives, Pyrotechnics*, 30 May 2018, Volume 44, Issue 1. ©WILEY Online Library, Reproduced with permission.

[4] *"Theory and Application"*, AMCP 706-185, Engineering Design Handbook Military Pyrotechnic Series, Part One, Headquarters, US Army Material Command, Washington D.C., April 1967.

[5] *"Properties of Materials used in Pyrotechnic Compositions"*, AMCP 706-187, Engineering Design Handbook-Military Pyrotechnic Series, Part Three, Headquarters, US Army Material Command, Washington D.C., October 1963.

[6] Henry Burnel, *"Military Pyrotechnics, Vol. 3 of 3: A Study of the Chemicals Used in the Manufacture of Military Pyrotechnics"*, Forgotten Books Publishers, 28 July 2018 (Classic Reprint).

# 7 Manufacture of Pyrotechnic Compositions

## 7.1 GENERAL

The manufacture of a pyrotechnic composition involves the handling of various ingredients to form the desired composition through various stages. Some of the ingredients used in the composition are hazardous or may make the composition hazardous. Hence, all safety precautions are required to be taken.

There are several types of pyrotechnic compositions used in military pyrotechnic ammunitions/devices like containers/pots, cartridges, grenades, bombs, shells and decoys. The special effects of the ammunition are mainly based on the performance of the main composition. Main compositions are classified on the basis of the special effects they produce on combustion.

Other pyrotechnic compositions like tracers, delays, caps, squibs, igniters, priming, booster, relay and fuze compositions, etc. assist in performance of the ammunition but do not exhibit special effects of the pyrotechnic ammunition. For example,

a. Delays do not give visible special effects but provide time interval for next ignition and assist in achieving range, safety, etc.
b. Tracers trace the path of the projectile by providing visible or infrared radiation and assist in firing the ammunition but do not provide special effects as special effects are provided by the main composition in the projectile.
c. Initiatory compositions do not give visible special effects but are important in assisting ignition of the composition.

Amongst many pyrotechnic compositions, some major military pyrotechnic compositions are given in Figure 7.1.

## 7.2 NOMENCLATURE OF PYROTECHNIC COMPOSITIONS

There are a very large number of pyrotechnic compositions developed by pyrotechnicians over the years for variety of ammunitions. All these compositions have a suitable nomenclature so that there is no ambiguity with regard to its use as a composition. The compositions are mostly designated as per the initials of the laboratory where they were developed or finally accepted. Every country has its own method of nomenclature. This nomenclature helps in proper identification and use.

DOI: 10.1201/9781003093404-7

| Military pyrotechnic compositions | |
|---|---|
| **Other compositions** | **Main compositions** |
| Priming compositions | Illuminating compositions |
| Booster compositions | Signalling flare compositions |
| Initiatory compositions | Photoflash composition |
| Friction compositions | Screening smoke compositions |
| Electrical squib compositions | Infrared smoke compositions |
| Cap conducting compositions | Signalling colour smoke compositions |
| Igniter compositions | Riot control compositions |
| Base bleed igniter composition | Incendiary compositions |
| Tracer compositions | Simulating compositions |
| Infrared tracer compositions | Infrared flare compositions (decoys) |
| Delay compositions | Infrared illuminating compositions |
| Miscellaneous compositions | Flame and smoke composition |
| | Gun powder compositions |

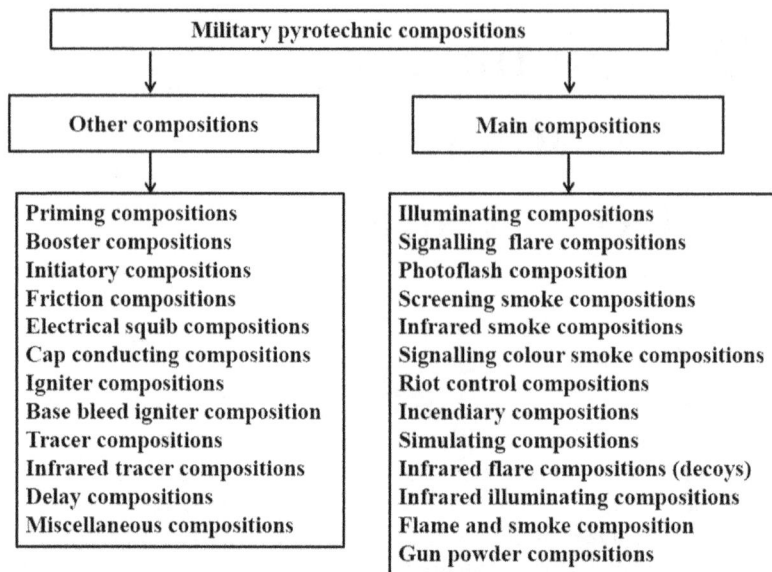

**FIGURE 7.1** Military Pyrotechnic Compositions.

A few such nomenclatures of compositions are given below, where the first few letters indicate the source.

a. SR stands for compositions developed at Royal Armament Research and Development Establishment, Fort Halstead, UK.
b. RD stands for compositions developed at old Research Department, Woolwich or Explosive Research and Development Establishment, Waltham, Abbey or Royal Armament Research and Development Establishment, Fort Holstead.
c. IND/ME stands for compositions developed by HEMRL (High Energy Materials Research Laboratory, Pune, India) (old name ERDL Explosive Research Development Laboratory, Pune, India) and authorised for use by Controller of Quality Assurance Military Explosives, India and signifies as Indian military explosive.
d. JSS stands for Joint Service Specification compositions, which are common in the Indian Army, Air Force and Navy.
e. PN stands for compositions developed by Chemical Defence Establishment (CDE) Porton, a UK research laboratory.
f. D followed by a number stands for compositions developed by Denel, South Africa.
g. NOL stands for compositions developed by Naval Ordnance Laboratory, USA.
h. PA stands for compositions developed by Picatinny Arsenal, New Jersey, USA.
i. FA stands for compositions developed by Frankford Arsenal, Pennsylvania, USA.

**TABLE 7.1**
**Composition SR 264A(M) and SR 269**

| Ingredient | SR264A(M) | SR269 |
|---|---|---|
| Hexachloroethane | 44.5 | 44 |
| Zinc oxide | 44.5 | 40.5 |
| Calcium silicide | 9 | 13.5 |
| Potassium nitrate | 2 | 2 |

The digits in the composition number stand for the particular composition mix. Sometimes the composition is modified from the original one by changing the sieve size of ingredients and/or proportion of ingredients or by deletion of an ingredient and/or inclusion of another ingredient for better special effects. Nomenclature of such compositions includes additional suffix, which indicates that the composition has been modified/changed or with a new number. This may be amplified by an example of composition SR264A(M) and composition SR269 (Table 7.1).

There are same ingredients in both the compositions but there is a variation in the composition percentage between SR264A(M) and SR269. The suffix (M) in the composition SR264A(M) indicates that the original composition has been modified.

It may be noted that there are several compositions known by other names as well. Examples are

a. Priming compositions are elsewhere also referred as ignition composition, quick composition, ignition charge, quick composition, first fire charge or starter mixture.

b. Initiatory compositions are elsewhere also referred as priming mixtures.

c. Booster Compositions are elsewhere also referred as intermediate composition or transitional composition or intermediate fire.

In addition to coloured smoke compositions with dye colour and a few others like SR 214 which is white in colour, most of the pyrotechnic compositions are greyish or black and hence compositions after manufacture are required to be kept in suitable containers duly labelled with composition designation and the batch number and quantity with date of manufacture to avoid ambiguity.

Let us now discuss the sieves.

## 7.3  SIEVES

Ingredients used in pyrotechnics are of various forms, shapes and sizes, which are distributed in the whole mass. However, the burning characteristics are different for different particle sizes. Hence, to achieve uniform and consistent burning, use of ingredients in proper sieve size is essential.

Sieving is a process of separating particles of varying sizes. The particle bigger than the hole in the sieves are retained while the smaller ones pass through the sieves. The sieves assist in removal of unwanted foreign matter like nails, screws, etc. as well as removal of oversize particles. Using multiple sieves results into gradation of the material into several particle sizes. These particle sizes have severe impact on the pyrotechnic behaviour (see Section 4.2.10).

## 7.3.1 STANDARD SIEVES

There are several types of standard sieves used across the globe. Table 7.2 gives the standard sieve numbers as per B.S.S., A.S.T.M and I.S. with approximate sieve sizes in microns.

**TABLE 7.2**
**Standard Sieve Sizes**

| B.S.S(410/1969) | A.S.T.M(11-70) | I.S. (469/1972) | Microns |
|---|---|---|---|
| 4 | 5 | 4.00 mm | 4000 |
| 5 | 6 | 3.35 mm | 3353 |
| 6 | 7 | 2.80 mm | 2812 |
| 7 | 8 | 2.36 mm | 2411 |
| 8 | 10 | 2.00 mm | 2057 |
| 10 | 12 | 1.70 mm | 1680 |
| 12 | 14 | 1.40 mm | 1405 |
| 14 | 16 | 1.18 mm | 1204 |
| 16 | 18 | 1.00 mm | 1003 |
| 18 | 20 | .850 mm | 850 |
| 22 | 25 | .710 mm | 710 |
| 25 | 30 | .600 mm | 600 |
| 30 | 35 | .500 mm | 500 |
| 36 | 40 | .425 mm | 420 |
| 44 | 45 | .355 mm | 355 |
| 52 | 50 | .300 mm | 300 |
| 60 | 60 | .250 mm | 250 |
| 72 | 70 | .212 mm | 210 |
| 85 | 80 | .180 mm | 180 |
| 100 | 100 | .150 mm | 150 |
| 120 | 120 | .125 mm | 120 |
| 150 | 140 | .106 mm | 105 |
| 170 | 170 | .090 mm | 90 |
| 200 | 200 | .075 mm | 75 |
| 240 | 230 | .063mm | 63 |
| 300 | 270 | .053mmm | 53 |
| 350 | 325 | .045 mm | 45 |
| 400 | 400 | .037 mm | 37 |
| 500 | 500 | -.025 mm | 25 |

Thus, as sieve size (number of openings in one square inch of a screen) increases, the hole size in the sieve decreases (i.e., a greater number of holes in the sieve). For example, a sieve size of 4 BSS means a bigger hole of 4,000 micron while 400 BSS means a smaller hole size of 37 microns.

## 7.3.2 WORKING OF SIEVES

A gradation of material into various sizes is done through a column of sieves with the bigger sieve holes on the top and smallest sieve hole size at the bottom. The sieves are provided with shaker arrangement with automatic digital timer to stop after pre-set timings.

A fixed quantity of the material is poured over the top sieve. Some material is retained on the top sieves, which are bigger than the sieve hole. The rest of the material passes through the first sieve and falls over the second sieve.

Here again the material with particle size bigger than the sieve hole of second sieve is retained while the rest of the material passes through the sieve and falls over to the third sieve and so on.

The percentage retained on a sieve is calculated by

$$\% \text{ retained} = \frac{\text{Weight of material retained on the sieve} \times 100}{\text{Weight of the fixed material}} \quad (7.1)$$

The percentage of passing through the sieves may be calculated by adding all the quantities retained on various sieves (cumulative retained) and finding the percentage as

$$\% \text{ cumulative retained} = \frac{\text{Weight of cumulative material retained} \times 100}{\text{Weight of fixed material}}$$
$$(7.2)$$

The percentage of cumulative passing is calculated as

$$\text{Percent cumulative passing} = 100 - \% \text{ cumulative retained.} \quad (7.3)$$

A small quantity of the material may be sieved through a single mesh by hand and the material passed through the sieve may be taken in a bowl for retention. The results of sieving mainly depend upon comparative size of the ingredient particle and hole size in the sieve, the orientation of the particles, sieving duration and extent of sieve movement. Sieving gives a quick mass distribution of particles of varied sizes but some of its drawbacks are:

  a. It works with only dry ingredient particles.
  b. It does not segregate the particles based upon density, porosity or surface features.

c. The ingredient particles may be spherical, elongated or flat and as such would behave differently when shaken or vibrated.

d. It is possible for a few elongated particles to jump and pass through the sieve through its nose. There are chances that a few particles may agglomerate due to electrostatic charges and humidity and may give wrong particle size.

e. There is a lower limit of sieve hole size

It is essential to clean the sieves after each use to ensure that the holes in the sieves are not clogged due to the composition.

## 7.4   MANUFACTURE OF PYROTECHNIC COMPOSITIONS

Manufacture of compositions mostly involves the following three major stages:

a. *Input*: Composition technology documents, process buildings, process equipment, storage buildings, process machinery/equipment, process tools, storage buildings, utilities (electrical power, water, steam, etc.), manpower, ingredients, General Safety Directives/Work Instructions/ Standing Instructions, safety gadgets, earthing of plant and machineries etc.

b. *Process*: Crushing lumps, sieving, drying, coating, premixing, final mixing (hand/mechanical mixing), granulation, maturation/drying, final sieving, storage

c. *Output*: Final composition, sampling and analysis with analysis report, test/ proof results, remixing/re-drying, waste disposal

Manufacture of a new composition involves the following steps:

a. Decide the objective (special effects) of the composition like illuminating, infrared illuminating, photoflash, incendiary, signaling flare, screening smoke, color smoke, infrared smoke, incendiary, sound and flash etc.

b. Understand the material Safety Data Sheet (MSDS) of ingredients and it should be available. Understand the role played by each ingredient in the composition and classify them as fuel, oxidiser, binder, burn rate modifier (sensitiser or catalyst or retarder), colour intensifier, dye, additives, etc. and choose the appropriate ingredients suitable for special effects (see section 6.10 for factors governing choice of ingredients for composition).

c. Choose right combination of fuel and oxidiser with other ingredients of suitable sieve spectrum ensuring compatibility and less toxic ingredients and less toxic combustion products

d. Ensure the composition is sensitive to external stimuli for ignition but not too sensitive to be hazardous. A higher sensitive composition needs to be made by hand in small quantities with special precautions during manufacturing. The most hazardous compositions are initiatory compositions, photoflash compositions, whistle compositions, chlorate-based compositions, zirconium-based compositions etc. Initiatory compositions are made in jelly mould.

e. Make experimental General Safety Directives/Work Instructions/Standing Instructions, duly approved by the competent authority.

f. Choose appropriate weight of mixing composition to be manufactured, based on sensitivity of composition

g. Choose appropriate process of mixing of composition, i.e., dry mixing, wet mixing, hand mixing, machine mixing, etc.

h. Choose appropriate granulation process.

i. Choose appropriate drying process, i.e., ambient drying or hot oven drying,

j. Choose appropriate blenders for blending composition batches

k. Ensure initial quantity to be made in small amounts and the batch quantity may be increased depending upon the sensitivity of the composition.

l. Test composition batches in laboratories and proof of filled components or ammunitions at proof ranges.

m. Decide whether binder is to be used for fuel coating, whether granulation is required (granulated dough to be sieved generally at 600 microns passing and 300 microns retaining) and whether premixing of ingredients is required and whether maturation of composition is required. In case of large number of ingredients, some of them are 'premixed' for better homogeneity. Also, if the composition is to be granulated with addition of binder, some ingredients are required to be 'premixed'. The dry compositions (without binders) are generally allowed to pass through either 300 or 600 or 850 μm IS

n. Measure various parameters of composition like bulk density, sensitivity, burn rate, ignition temperature and energy of activation etc.

o. Safety certificate for composition to be made and approved by competent authority.

p. Suitable priming and booster compositions should also be made for ease of pick-up of ignition.

q. Ensure composition is used before expiry of shelf life (Generally, shelf life of a composition batches is taken as 21 days, while initiatory compositions have very less shelf life of 72 hours only. These may be remixed after expiry of shelf life and used. Some composition batches using epoxy resins are to be used up within hours to avoid hardening of the composition

r. Ensure disposal of waste as per provisions of disposal instructions

A pyrotechnic composition must meet the design requirements of following major parameters.

a. Sensitivity to external stimulus (like flame, flash, friction, spark depending upon energy of activation or ignition temperature)

b. Special effects of composition

c. Thermal properties like combustion rate, combustion temperature, heat of combustion, heat capacity, thermal conductivity

d. Desired combustion products and their characteristics

e. Chemical stability and compatibility

f. Physical properties of the composition (like bulk density and loading density, ingredient particle size distribution, moisture content etc.)

g. Low toxicity of composition and low toxicity of combustion products

## 7.5    Transportation and Storage of Ingredients

The ingredients are transported from manufacturing companies in trucks or rail wagons, packed suitably in boxes, drums or bags, each ensuring that the ingredients are not exposed to atmosphere and that the containers are unbreakable. The ingredients are required to be stored in storage house as per standard practice. Further, ingredients are transported from storage house to grinding and or sieving as well as to mixing houses inside the factory. The following precautions are required.

a. The ingredients are required to be stored as per compatibility. The incompatible ingredients like fuel and oxidiser should not be kept in the same storage bays. All volatile chemicals like acetone, alcohol, varnishes/lacquers are required to be kept in separate building.

b. All the chemicals are required to be suitably stored with specified packages/ containers like drums, bags, etc. with correct label details like nomenclature, specification, sieve size, batch no. and date, quantity and initials of manufacturer address and corresponding supply orders.

c. There should be adequate space between the two ingredients in the storage bay for ease of handling of individual ingredients like sampling as well as bulk drawl for mixing.

d. The ingredients should preferably be kept on wooden dunnage and the bays must be cleaned periodically.

e. It should be ensured that no spillage takes place and the ingredients are not allowed to mix in the bays.

### 7.5.1    Testing of Ingredients

All the ingredients are required to be tested by the companies supplying ingredients. However, as a measure of caution, sometimes these are tested at manufacturer's analytical laboratory to ensure that they conform to the specification. It must be ensured that that the bulk does not contain grit/foreign matter.

Proper sampling procedure is to be followed so that the sample represents the bulk. The tests include purity, moisture and sieve size, amongst others. All ingredients are required to be in accordance with the specification. However, there are cases where particle size variation is observed, especially for metallic powders like magnesium, zirconium, etc. These may be considered for use only after testing by actual practical end product performance.

### 7.5.2    Preparation of Ingredients

Some ingredients especially those which are hygroscopic in nature like sodium nitrate and potassium nitrate form lumps during storage and are required to be broken using suitable crushers/grinders and then dried and sieved to appropriate sieves.

The machines are to be ensured for proper electrical earthing and a dry run to be made. Each time only a small amount is to be pulverised, ensuring avoiding extreme dust particles. These should be remotely operated, and personnel may enter the room, with protective clothing and nose piece over nose, after stopping the machine and allowing the dust to settle down for a while. A lot of dust and powder gets deposited on the machine and inside the walls of the room. These are required to be cleaned on a daily basis.

These grinded ingredients are to be sieved to obtain desired particle sieve size as well as to remove any extraneous material and lumps, if any. These sieved materials are further dried, as required, just before mixing of compositions. This is done in suitable dryers with suitable thermostat controls, to ensure proper temperature and avoid any overheating in the dryer. The ingredients are laid over trays and are suitably raked with wooden spatula to open out next layer of composition for drying. The temperature for drying varies from ingredient to ingredient. Special care should be taken to ensure that no mix up of different ingredients takes place inside the oven ensuring that the dryers are properly earthed.

Metallic powders (like magnesium, aluminium, etc.) having a tendency to segregate are required to be blended for uniform sieve spectrum.

The blender should be properly earthed. The same blender should never be used for the fuel and the oxidiser for obvious reasons. Proper blending avoids variation in combustion rate of composition from one point to another in the same batch and from batch to batch.

Some binders having restricted shelf life, like nitrocellulose varnish, epoxy resins, etc. are to be consumed within shelf life.

### 7.5.3   WEIGHING OF INGREDIENTS

The weighing machines are to be suitably earthed, and dust if any, be cleaned. Special care must be taken to ensure that the fuel and oxidiser are not weighed on the same weighing machine. Accuracy in weighing is important as it affects the proportion of the ingredients in the composition, thereby affecting the performance.

Depending upon the quantity to be weighed, electronic type balances (small quantity in milligrams), single or double pan balance (small quantity in grams) and platform type weighing machine (large quantity in kilograms) may be used.

Ingredients of specified sieve size duly dried, wherever required, are weighed. The composition specification may give the ingredients as a percentage or in parts. In case the composition is given in parts, the same may also be converted to percentage.

The conversion from parts into percentage may be made by adding all the parts and then calculating the percentage of each part in the composition mix. The formula for percentage for any ingredient would be the ratio of specific ingredient part to the total ingredient part multiplied by 100 as

$$\text{Percentage of ingredient} = \frac{\text{Specific ingredient part } \times 100}{\text{Total part}} \quad (7.4)$$

A simple example with an imaginary composition is given in Table 7.3

**TABLE 7.3**
**Conversion of Parts to Percentage**

| Ingredients | Parts | Percentage | Percentage (rounded off) |
|---|---|---|---|
| Fuel (A) | 55 | 55 × 100/108 = 50.9245 | 51 |
| Oxidiser (B) | 24 | 24 × 100/108 = 22.2216 | 22 |
| Oxidiser (C) | 16 | 16 × 100/108 = 14.8144 | 15 |
| Burn rate modifiers (D) | 8 | 8 × 100/108 = 7.4072 | 7.5 |
| Binder (E) | 5 | 5 × 100/108 = 4.6295 | 4.5 |
| Total | 108 | - | 100 |

## 7.5.4 MIXING

*Mixing pyrotechnic ingredients is the process of thoroughly combining different ingredients to produce a composition of a desired quantity with consistent particle size distribution, colour, texture, reactivity and other required attributes so as to provide consistent performance.*

Hence, a standard procedure is required to be followed during mixing to get consistent performance. Therefore, general safety directives or work instructions or standing instructions should be made and approved by the competent authority. General safety directives for composition manufacture includes, composition nomenclature, general safety directives number and its approval reference, safety certificate of the composition, and its approval reference, building/room number authorised for the work, man limit, explosives and other materials limit, firefighting classification, process details for composition manufacture and special precautions etc.

Due to safety reasons, pyrotechnic compositions are mixed in batches and not continuous mixing. For a new mix to be processed, initially, a very small composition batch should be made in presence of supervising personnel using all safety precautions like using gloves, safety shield, earthing of bowls, humidity of process building, etc. and should be tested for performance. If satisfactory, the mix quantity may be increased, depending upon sensitivity and requirement of composition.

Generally, the maximum quantity for hand mixing is about 1–2 kg, extendable to 5 kg. Priming compositions, igniter compositions and delay compositions are mostly made by hand mixing as the requirement is less. Illuminating compositions, coloured smoke compositions and screening smoke compositions are mixed in a planetary mixer or sigma mixer with quantity around 20–25 kg, extendable to 50 kg. The relative humidity of the process building should be between 55% and 65%.

In case of large number of ingredients, some of them are "premixed" for better homogeneity. Also, if the composition is to be granulated with addition of binder, some ingredients are required to be *"premixed."*

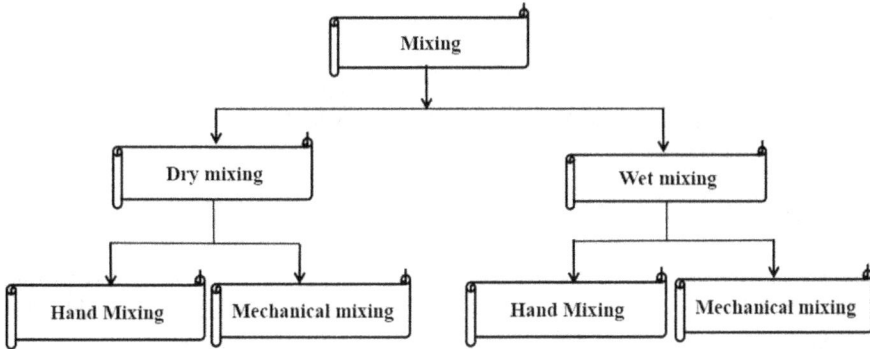

**FIGURE 7.2** Types of Mixing.

It may be mentioned that each composition has its own method of mixing. Mixing may be dry mixing or wet mixing, and further as hand mixing and mechanical mixing, depending upon the ingredients and the requirement (Figure 7.2).

### 7.5.4.1 Dry Mixing

There are two methods as hand mixing (dry) and mechanical mixing(dry) as under.

i. *Hand Mixing (Dry)*

Hand mixing is done when

a. Quantity required is very less (like priming, delays and fuze compositions) compared to main pyrotechnic compositions which are required in bulk quantity.
b. Composition is of high sensitivity, i.e., highly reactive/hazardous
c. There is a wide variation in the density of ingredients like SR 252 (see Table 19.1).composition containing ferrosilicon having wide variation in densities with other ingredients potassium nitrate and SMP.

The following methods are used.

a. *Diapering Method:*
   This is the oldest method for dry mixing of a very small quantity. In this method, the ingredients, duly sieved, are kept in the centre of a thick paper. The one corner end of the paper is raised so that the ingredients roll down. When it reaches the end of the paper, the paper is rolled down from the other corner end. This is repeated several times for proper homogeneity. This is suitable for compositions which are friction-sensitive or shock-sensitive and the required quantity is less.

b. *Screening Method:*

The ingredients are mixed in a wooden or paper trough through wooden spatula and then allowed to pass through a sieve. The process is repeated till a homogenous composition is obtained.

ii. *Mechanical Mixing (Dry):*

The initiatory compositions, which are highly sensitive, mixed in small jelly mould (ribbed cylindrical container, made of antistatic material) behind the shield. The ingredients are added in the jelly mould and are allowed to rotate at an inclined position slowly for specified period. The drive is generally given by simple water turbine, in which the water is fed through an overhead tank. The initiatory composition is unloaded behind the shield by tipping the composition in the jelly mould on to a small antistatic material container. This method of dry mixing of initiatory compositions is associated with high sensitivity and hazards and hence extreme safety measures are required to be followed.

Screening smoke compositions and colour smoke compositions are generally made in sigma mixer as dry compositions.

### 7.5.4.2   Wet Mixing

Following considerations are important in wet mixing.

a. Pot life of the mix, i.e., the period for which wet compositions can be used for granulation and pressing
b. Use of suitable suction and ventilation for the likely gas/vapours /fumes of the solvents (like xylene, acetone, etc.) which are to be driven out from the mixing process
c. Relative viscosity of the ingredients like binders

Generally, wet mixing is safer than dry mixing. There are two methods as hand mixing(wet) and mechanical mixing (wet) as under.

i. *Hand Mixing (Wet):*

The ingredients are hand mixed in a bowl. The bowl, the table top with aluminium sheet and other equipment are earthed. The bowl for mixing is smooth and polished, without any burrs, as often friction due to the burr may cause spark and ignition of the composition. The face shield is required to be used using perspex sheets in front of the operator (Figure 7.3).

The quantity (1 kg to 2 kg) to be mixed should be as per the general safety directives or work instructions or standing instructions made for the purpose duly approved by the competent authority. The mixing bays should have adequate ventilation for vapours when using wet mixing. The relative humidity of the bay should be 55%–65%.

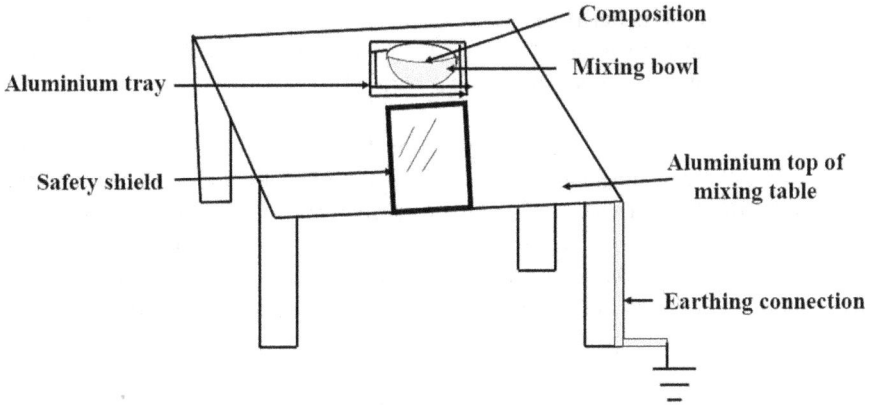

**FIGURE 7.3** Hand Mixing.

ii. *Mechanical Mixing (Wet):*

Large quantities of pyrotechnic compositions are mixed in planetary mixers or sigma mixers through a remote control. As an example, a typical procedure followed for illuminating compositions is as follows:

a. Magnesium powder is first coated with a binder, like synthetic resin, in a planetary mixer (Figure 7.4). The addition of oxidiser and other ingredients is done afterwards. The composition, after mixing in a planetary mixer, is laid on a trough and raked at regular intervals with a wooden spatula for proper maturation. The composition is to be taken for pressing neither too early nor too late. This comes with the experience. If the composition is taken early for compacting/pressing, cracks develop on the pellets/candles in a short time. However, if the composition is taken for pressing after a prolonged time, the

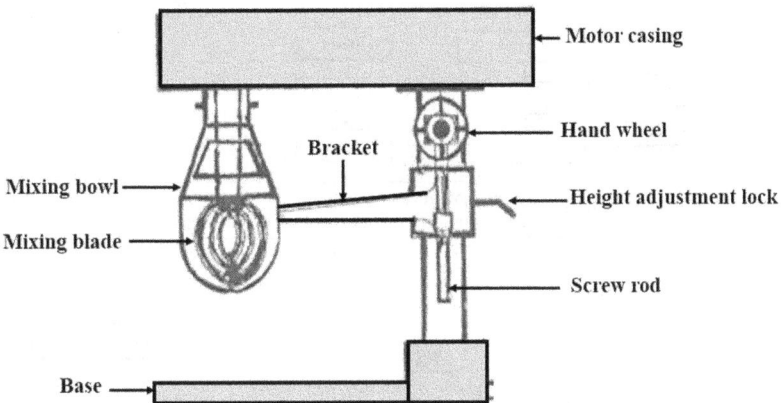

**FIGURE 7.4** Planetary Mixer.

strength of the pellet is reduced, which may not comply with the crush load test and cause cracking of pellet during stress from firing or may cause detachment of the composition during burning.

b. In case of sensitive compositions using boiled linseed oil/lithographic varnish, coated magnesium requires maturation for 48 hours by keeping the composition in a shallow tray, with periodic raking by wooden spatula to avoid auto ignition due to formation of a hard film over the composition. Raking helps in opening the composition, thereby allowing heat to be dissipated to the atmosphere. Then, further ingredients are mixed using a wooden spatula.

In case of not so sensitive compositions, other ingredients are added to coated magnesium powder in the planetary mixer. After completion of mixing, the composition is laid on aluminium tray for maturation for specified hours with periodic raking and sieved.

In case of a power failure, the power is switched off. If power resumes within 15 minutes, the process is resumed. If not, composition is taken out and in case of boiled linseed oil, mix thoroughly by hands wearing gloves and then sieve through specified sieve; however, in case of resins, the composition must be disposed. An overall approximate mode of mechanical wet mixing process for typical pyrotechnic composition is shown in Figure 7.5

There are several types of mechanical mixers used in pyrotechnic composition manufacture. The type of composition decides the type of mixer to be used and is included in the general safety directives.Two examples of typical mixers are

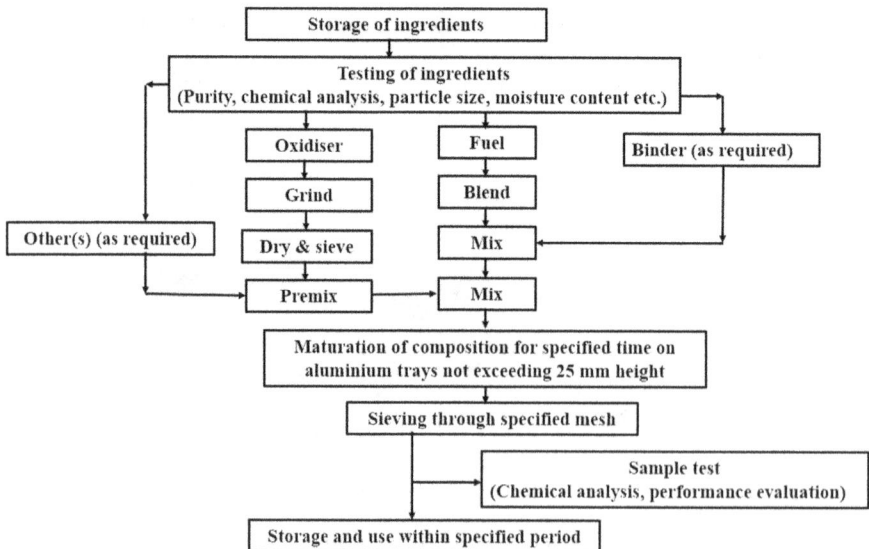

FIGURE 7.5  Flow Chart of Mixing of Composition.

a. *Planetary mixer*: A planetary mixture is a bowl with a stirrer which moves around its own axis as well around the circular path. The gap between the stirrer and the bowl is the most important. It is customary to take out the composition sticking to the walls of the bowl periodically and mix with the bulk.
b. *Sigma mixer*: These are mechanical mixers with jackets for keeping uniform temperature during mixing operation.

These machines are installed in cubicles and earthed. The doors are interlocked in such a way that the machine cannot start until the doors are locked. Similarly, unless the power to machine is cut off, the door cannot be opened. This ensures safety of the personnel when the machine is on. A board "*MIXING IN PROGRESS*" is displayed. It is necessary to ensure that the mixer is clean. A dry run is to be made every day to ensure proper working of the mixer.

## 7.5   TYPES OF MIXING OF COMPOSITION

Types of mixing of compositions is based on the following:

a. Full dry powder pyrotechnic compositions which do not have any binder
b. Dry powder pyrotechnic compositions which may have little binder for proper adhesion and consolidation under press
c. Compositions which use resins as binders which have only a few hours shelf/pot life
d. Compositions which use oil as binder which has long shelf life of 1 month minimum
e. Elastic pyrotechnic compositions which use uncured fluid plastic material (like polysulphides, unsaturated polyesters, polypropylene glycol) which are extruded or moulded and kept in oven at appropriate temperature and duration for maturation.

The elastic pyrotechnic compositions have an advantage over conventional pyrotechnic composition:

a. *Waterproof*: The composition becomes waterproof due to higher quantity of elastic material.
b. *Better shelf life*: The composition can retain property in long storage as it is not affected by humidity.
c. *Filling flexibility*: The composition may be filled in any shape or extruded to any desired shape.
d. *Less sensitive*: The composition is less sensitive and hence safe for use in manufacture

e. *No machinery required*: The hydraulic, pneumatic or fly presses are not required for pressing the composition as the composition may be directly poured into the container and cured.

However, since the proportion of plastic material is much higher than conventional binders, its pyrotechnic efficiency is less. Hence to improve pyrotechnic efficiency, more fine ingredients are used in such composition.

## 7.6  GRANULATION OF COMPOSITIONS

*Granulation is a process in which individual ingredients are made to adhere to form bigger multi- ingredient entities, called granules. The granulation process eliminates segregation feature from the composition and gives uniformity to the composition for its regular and consistent performance.*

Non-aqueous binders are preferred since many ingredients like oxidisers in the composition are hygroscopic and metals in the composition react with water. Therefore, a non-aqueous solvent is used for granulating hygroscopic compositions. In granulation process, some binder like NC-based lacquer, bonosol and plasticised ethyl cellulose are used.

The composition is made in a dough form and is allowed to dry for approximately 30 minutes to remove volatile contents. Granulation is done through sieves which are clean and are suitably earthed during processing so as to dissipate static electricity. The granulation of composition is done on two sieves, one for passing and the other for retention, (e.g., 25 BSS (600 micron) for passing and 52 BSS (300 micron) for retention). The fines particles passing through 52 BSS sieve may again be made into dough form by use of solvents (example amyl acetate 40%, acetone 40% and toluene 20%) and granulated again using the above process.Granulation process requires that the dough should not be too wet thereby causing agglomeration of granules or too dry so as to form dust.

These are dried at ambient or hot condition in a dryer as per stipulation of general safety directives/work instructions/ standing instructions, so that the solvent is evaporated. If the drying is done fast at high temperature for short time, then the composition gets a dry coat of binder over it much earlier and further volatile matter may not escape.

If the solvent is not evaporated properly during the drying stage, then the vapour may escape during storage in hot places or during tests at hot conditions, by opening the binder layer over the ingredient, exposing the ingredient. Such compositions are expected to give erratic results.

Suitable provision for escape of solvent vapour from the bay is required to be made. The operator must have an anti fume respirator to avoid inhalation of solvent vapours as well as use antistatic rubber hand gloves.

Similarly, many ingredients are of small size, irregular shape or having such surface characteristics which do not allow it to have proper flow characteristics. Granules produced from such ingredients shall be larger and more isodiametric, both these factors leading to improved flow properties.

An ideal granulation will contain all the constituents of the mixture in the correct proportion in each granule. The granulation process has the following advantages:

a. Improves uniformity/homogeneity of ingredients in the composition
b. Improves their free-flowing properties
c. Improves controlled burning by providing granular shape
d. Reduces the tendency of dusting and segregation
e. Gives a good bonding between the ingredients
f. Avoids absorption of moisture
g. Does not stick to tools like scoops, pressing punch, etc.

Sieving of the composition varies as per the design of the composition. Generally, compositions with binders are allowed to pass through 600 μm and retained at 300 μm IS sieve during granulation. Sieves should not be damaged or have sharp edges or rough surfaces to avoid friction during sieving. Automatic granulation machines are also now available.

## 7.7   DRYING AND MATURATION OF COMPOSITIONS

The ambient drying consists of the composition spread evenly over aluminium trays with the thickness of composition not exceeding 25 mm. The composition is raked periodically, if required.

Consider a fresh wet pyrotechnic composition requiring removal of volatile matter is dried in a dryer. Initially, the composition temperature would increase with time till the boiling point (say T1) of the volatile material is achieved. The temperature (T1) would remain constant for some time till all volatile matter is expelled out. Then the composition temperature would again rise till it achieves the set temperature of the dryer (say T2). It may be noted that the dryer temperature is much below the ignition temperature of the composition for safety reasons (Figure 7.6).

The drying of composition by removal of moisture makes the composition free flowing for subsequent operations of filling and pressing as well as avoids effect of moisture on hygroscopic ingredients and compaction strength of pellets.

These dryers work on certain parameters given in Table 7.4.

Following precautions are necessary:

a. Earthing of the dryer.
b. The dryer is clean before and after use.
c. In no case, two incompatible compositions are dried together or allowed to mix.
d. Overheating is avoided (by installing suitable double thermostatic controls in the oven. These thermostatic controls are required to be checked at periodic intervals.
e. Solvent vapours are removed from the area suitably.
f. The dryers are not left with sensitive explosives overnight or unmanned.
g. No material is dried which contains volatile ingredients like hexachloroethane or camphor

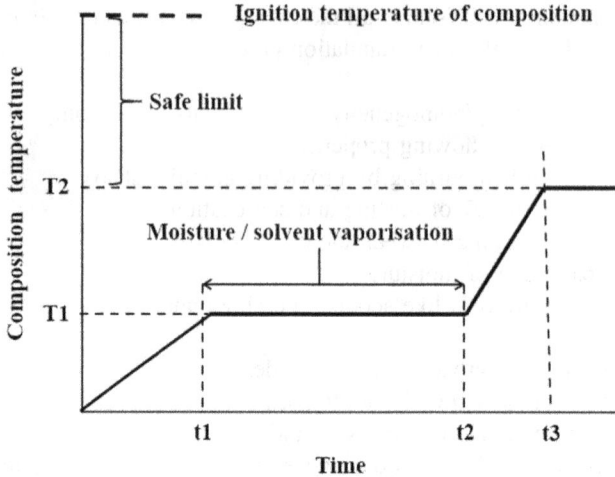

**FIGURE 7.6** Effect of Moisture on Drying Time.

**TABLE 7.4**
**Working of Dryers**

| Parameters | Type |
|---|---|
| Mode of Operation | Batch |
| Heat input | Conduction, convection and radiation |
| State of material in dryer | Stationary on trays |
| Quantity of material | As specified in GSD/SI/WI |
| Drying medium | Mostly hot air (by hot water or steam) |
| Drying temperature | Above boiling temperature of volatile matter and as specified in GSD/SI/WI |
| Relative motion between drying medium and drying composition | Static composition |
| Drying duration | As specified in GSD/SI/WI |

### 7.7.1 Factors Affecting Drying of Composition

The following factors affect the drying of the composition (also ingredients) having moisture and or solvent content.

a. *Quantity of solvent in the composition*: The higher the solvent content, the more the time would be required for drying
b. *Solvent boiling point in the composition*: The higher the boiling point of the solvent, the more the time would be required for drying

c. *Particle size of the granules*: The higher the particle size, the more the time would be required for the drying. The smaller the particle size, the faster would be the drying time.

d. *Porosity of the composition*: The higher the porosity of the composition, the lower would be the drying time.

e. *Thickness of the composition layer under drying*: The higher thickness of the composition would require more time for drying. Hence raking of the composition, using wooden spatula is required to expose new surfaces of composition at periodic intervals for uniform and early drying.

f. *Dryer temperature*: The higher the dryer temperature, the lower would be the drying time. However, there is an upper limit to the dryer temperature for the composition to avoid ignition of composition.

g. *Rate of air flow in the oven*: The faster the rate of flow of the air in the oven, lower would be the drying time.

## 7.8   BLENDING OF COMPOSITIONS

Sometimes segregation of ingredients takes place in few dry mixing compositions during storage. Segregation is more if the density variation and or particle size variation is more between the ingredients and hence such composition is required to be mixed again before use, either by hand mixing or by blenders (Figure 7.7). Blending cycle time is important as short blending cycle time may result in non-intimate composition while too long blending a cycle time shall result in stratification or separation of ingredients. In pyrotechnic composition manufacture, uniformity of the composition is important for consistency in results.

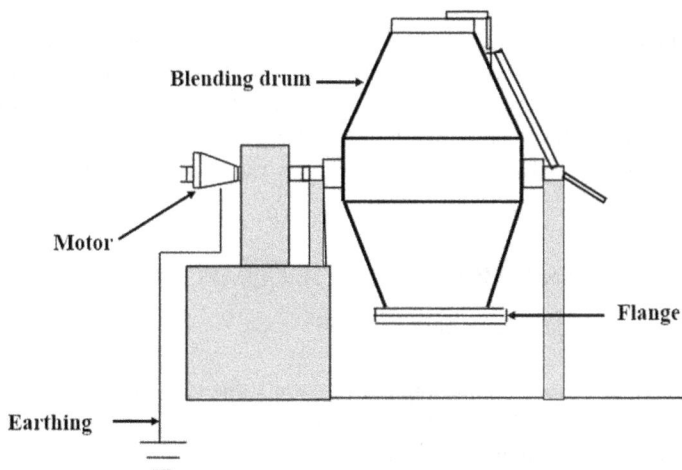

**FIGURE 7.7**   Blending Machine.

### 7.8.1 Factors Affecting Blending

Blending is expected to produce uniform mixing of a composition and the composition is also expected to retain its uniformity in storage. However, this is seldom achieved due factors causing non-uniformity of composition as under.

a. *Ingredient particle characteristics*: Particle size, shape, density, size distribution, particle flow characteristics, cohesivity, hygroscopicity of ingredients, agglomeration
b. *Blending Process*: speed, duration of blending, quantity of composition.
c. *Segregation:* of individual ingredients due to density difference.
d. *Vibrations:* during handling and storage
e. *Free fall:* A free fall of container with composition shall induce shock and cause segregation

Ingredients having lower size and higher density tend to go down in the container containing composition during storage. Similarly, higher size and lower density ingredients tend to concentrate on the top. This causes a variation in the uniformity of mixing of the composition. This demixing is the unintended separation of ingredients and it must be minimised for better performance of the composition.

The main cause for segregation or demixing of ingredients in composition are:

i. *Vibration and percolation*: Smaller mass or size ingredients try to occupy the voids between bigger particles causing a separation of the ingredients with different mass or size.
ii. *Differential motion during transportation*: Particles due to variation in mass or size shall have different trajectories and thus get segregated during transportation.

Compositions without binders are prone to segregation. However, compositions with binders (granulation) using oil or wax are prone to aggregation. The problem is compounded with hygroscopic ingredients leading to caking of composition. The finer ingredients are more prone to agglomeration. This may be checked before use of the composition and agglomeration may be removed by breaking lumps and sieving.

## 7.9 STORAGE AND TRANSPORT OF COMPOSITIONS

The following procedure is followed.

a. Pyrotechnic compositions are kept in compatible containers on dunnage duly earthed for dissipation of static electricity. Many ingredients of pyrotechnic composition are hygroscopic in nature and needs adequate care by proper cover.

b. It is necessary to label each composition container for its identity like no-menclature of the composition, batch number, quantity, date of production etc. since many compositions look alike.

c. It is essential that such storage houses are opened every day for ventilation of any released gaseous products.

d. Sometimes segregation of ingredients takes place for some composition (without a binder) during storage and hence such compositions are required to be re-mixed and sieved again before use. Similarly, some compositions with binder are prone to aggregation. In this case, the composition is required to be re-sieved before use.

e. Transportation of highly sensitive composition should be done by hand by slow walk while others may be carried through hand trolley.

f. No dust or sunlight should be allowed to fall on the composition.

## 7.10  TESTING/EVALUATION OF COMPOSITIONS

The drawl of sample for tests should ensure the following:

a. The sample drawn for test must be a true representation of the bulk.

b. The quantity should be sufficient to provide material for all the tests.

c. A part of the sample, as master sample or reference sample may be kept for future reference, if required.

d. The container used for keeping sample should be compatible with the composition.

e. The container should be labelled with composition nomenclature, batch no., batch qty., sample qty., date of sampling

Most of the compositions are tested in laboratory to check the correct percentage of the composition, moisture and other specified parameters of the composition like calorific value.

Compositions may be sentenced by laboratory as re-dry, remix or reject. Composition failing for moisture or volatile matter content may be rectified by re-drying and sieving the composition and subjecting the same for retest. Ungranulated compositions failing for percentage of ingredients can be remixed with fresh in-gredients and subjected to test again. Granulated compositions are difficult to rectify and rejected compositions may be disposed as per procedure.

Some compositions are required to be pressed and assembled as sub-assemblies, and proved in proof range to confirm that the satisfactory performance of the composition. For example, signal stars, smoke canisters or delays are fired statically to see the performance before bulk assembly in ammunition.

Crush load tests are checked for some composition pellets to ensure that the pellet has specified crush load to withstand the stresses during proof.

However, some illuminating composition, containing epoxy-based binders, which get dried very fast, are not tested in laboratory since composition cannot remain unused for long as it becomes hardened in a short time. Hence these compositions are checked for the performance like burn time and luminosity for illuminating composition after pressing as stars and candles.

Similarly, tracer composition pressed in tracer capsules are tested under spinning for the burn time.

In any case, the performance of the composition is more important than the once-approved specification of the composition. This is important since large number of factors affect the manufacture of a composition. Hence, minor variation in laboratory test results may be ignored subject to condition that the composition is satisfactory in all other parameters and ammunition performance using the composition is satisfactory.

A general safety directives/ standing instructions/work Instruction are required to be made for these laboratory and proof tests for clarity of the tests as well as safety of the personnel.

## 7.11   CHECKLIST FOR PYROTECHNIC COMPOSITION MIXING

The following checklist may be used for pyrotechnic composition mixing:

a. The production building has approved general safety directives/work instructions/standing instructions and approved layout of the building for the mixing job
b. Explosive limit and man limit are within authorised limits
c. All the machines/equipment/tables are earthed and cleaned before and after mixing. Static earthing test done on monthly basis and lightning protector test done every six-monthly basis.
d. Machine/equipment has a dry run
e. Personnel are with approved danger building dress and approved personnel. Protective appliances are available.
f. All personnel are well trained in the job and read safety data sheet of ingredients and safety certificate of composition.
g. Humidity of building is within authorised limits (55% to 65%) and air conditioning duct and filters cleaned every three moths or as required.
h. Static charges of personnel dissipated by personnel tester
i. No prohibited article is kept inside the room
j. All tools are non-sparking tools
k. No other mixing process is carried out in the building
l. Logbook with activities and timing being recorded
m. Building surrounding and fire hydrant surroundings are kept clean.
n. Firefighting equipment and first aid, as approved, are available in serviceable condition and easily accessible.
o. Disposal of waste in oil is being done periodically

The most hazardous compositions are photoflash compositions, whistle compositions, chlorate-based compositions, zirconium-based compositions etc. Disposal of waste composition and fired ammunitions are required to be made as per the provisions of disposal instructions.

## 7.12 DRAWBACKS IN LITERATURE ON PYROTECHNIC COMPOSITIONS

It may be mentioned that the pyrotechnic composition given in literature is as good as an advice data. The data has several drawbacks as it does not provide full status of all the ingredients and activities related to production of compositions, due to space crunch in literature like

a. Purity and other chemical characteristics of each ingredient
b. Sieve size distribution
c. Mode of mixing ingredients in composition
d. Time schedule for mixing
e. Time schedule for drying
f. Permissible moisture content in composition
g. Granulation sieve sizes, if any
h. Ambient conditions of tests/proof, etc.
i. Performance characteristics of filled/pressed compositions (pressing load and dwell time, compaction strength and area exposed of composition)

Characteristics and general essential requirements of pyrotechnic compositions of all pyrotechnic compositions have been given in Sections 1.5 and 1.6, respectively. Essential requirements of individual pyrotechnic compositions are given in the relevant composition Chapters 8–22.

Readers may see reference [1] on how physical mixing of ingredients poses inconsistencies in performance of batches.

## REFERENCE

[1] Lisa H. Humphreys, Simon J. Coles, Ian Sinclair, and Ranko Vrcelj, "*Mixing and Quantifying Multi-Component Materials for Pyrotechnic Applications*", *Acta Crystallographica*, 2014, Volume A70, Page C1233.

# 8 Illuminating Compositions

## 8.1 ROLE

Warfare at night demands visibility of the terrain for tactical needs like target detection (locating object), recognition (type of object) and identification (exact object description) at night by all personnels to enable to carry out any of the following:

a. Navigate and engage targets like enemy troops/equipment like tanks, vehicles and materials
b. Target location for hitting the target
c. Landing of war planes and materials at night
d. Making target for night practice fire
e. Tracking flare for missiles

However, the following problems are faced in the night in identifying the enemy target.

a. The normal human eye or high-power binoculars or telescope or image magnifiers are unable to see the target during night.
b. Search lights, if used, have a limited range and that too would reveal the location of the search light.

The infrared instruments may reveal the target since some equipment like tanks and human body radiate infrared radiation. However, these have limitations of range and area of search. Further, this is limited to only a few people and it is not possible to communicate the target type and location to all crew members. If the distance is large, it is difficult to identify the target. Thus, there are limitations in observing the target at night with above means.

The illumination from illuminating composition provides means to identify the target area. It provides

a. Larger target area of observation
b. Visibility to large number of fighting personnel
c. Target to light source distance is adjustable (i.e., with adjustable fuze or delay)
d. Sufficient illumination like daytime (i.e., illumination of desired value in visible light)

DOI: 10.1201/9781003093404-8

e. Sufficient duration for identification of target (i.e., burn time sufficient for observation)

f. Does not reveal the location of source of firing

Thus, illuminating compositions and infrared flare compositions are used to illuminate the warfare area at night with visible/infrared radiation and are known as night warfare compositions. Table 8.1 provides the general illumination required for a battlefield.

**TABLE 8.1**
**General Illumination Required in Battlefield [1]**

| Type of requirement | Illumination |
|---|---|
| Illumination sufficient for reconnaissance of large stationary target or small moving objects | 1 lux |
| Illumination for distinguishing small stationary objects | 2 lux |
| Illumination necessary to achieve well-aimed direct fire against most type of objects | 5 lux |

## 8.2 CHARACTERISTICS OF ILLUMINATING COMPOSITIONS

The illuminating composition provides luminosity of specified candela for a specified time (burn time). Luminosity and burn time are important characteristics of illuminating compositions. However, both are required to be considered together since linear burn time and luminosity are reciprocal to each other. Thus, within a given space in an ammunition and specified composition ingredients, if flare linear burn time is required to be high, the luminosity would be low and vice versa.

### 8.2.1 LUMINOSITY

Illumination is based on a high flame temperature, which produces large quantities of excited sodium atoms in vapour state together with combustion product (incandescent solid magnesium oxide particles). Electrons in sodium atoms get excited and move to higher energy levels and then come back to original state, thereby emitting light energy (photons) in the bright yellow band of visible range (see Section 9.2).

Since loose illuminating composition would burn very fast, these compositions are used in the pressed form as stars or candles to increase the burn time. To ensure that these compositions as stars or candle give desired illumination and burn time, these are tested in the luminosity tunnel before assembly of the same in the ammunition.

The rate of burning versus luminosity may be plotted on a graph for a series of composition with variations in fuel oxidiser ratio. The composition having desired luminosity (with corresponding burn time) may be selected. Addition of a small amount of sodium fluoride or cryolite in some illuminating compositions have been found to increase the luminosity.

Small-calibre illuminating ammunitions capable of firing at a small range eject burning star and provide illumination. These pressed stars are fired from the ground, aircraft, ship or raft and are usually not supported by parachute except a few. Stars of small size used in small calibre ammunitions burn from all sides and hence give low burn time and luminosity. The burn time is so adjusted that the burning is complete in the air itself and does not burn further on the ground to avoid ground fire.

High-calibre illuminating ammunition capable of firing at a long range ejects the burning candle in the air with parachute(s) to provide illumination. The parachute retards the rate of fall of the candle. The following points may be noted:

a. High-calibre candles burn unidirectional and layer by layer and hence give more burn time.

b. The higher the luminosity of the composition in candles, the better is the visibility. This requirement of higher luminosity makes it necessary to have a larger burning surface of the candle with appropriate composition. Hence, the diameter of the candle is preferably kept large.

c. The formation of gaseous substances in illuminating composition produces flame during burning of the composition ensuring increased luminous energy.

d. The requirement of candlepower increases with increase in height of deployment of candle. Hence, the aircraft-dropped illuminating ammunitions require very high luminosity followed by artillery, mortar and hand-held ammunitions.

e. Rate of descent of candle with parachute should provide adequate burn time to the candle in air.

Table 8.2 provides candlepower requirement necessary to provide specified illumination on the ground from given heights above the ground.

In general, for magnesium–sodium nitrate illuminating compositions:

a. Linear burn rate and luminosity increase with an increase in magnesium content (between 32% and 60%)

b. Linear burn rate and luminosity decease with an increase in percentage of binder (2%–16%)

c. Linear burn rate and luminosity increase with a decrease in size of magnesium powder (from coarser to finer)

d. Mass burn rate and luminosity increase with increase in pressing load

e. Luminosity increases with an increase in candle diameter

Table 8.3 shows the burning characteristics for stoichiometric mixtures of various oxidants with atomised magnesium. It could be seen that as the molecular weight of oxidiser increases, there is corresponding decrease in the fuel content. A decrease in

TABLE 8.2

**Candlepower Requirement Versus Height of Illuminating Source** [2]

| Height above ground | Ground brightness | |
|---|---|---|
| | 0.1 (foot-lambert) | 1.0 (foot-lambert) |
| 100 feet | 1,000 | 10,000 |
| 500 feet | 25,000 | 250,000 |
| 1,000 feet | 100,000 | 1,000,000 |

TABLE 8.3

**Physical Data and Burning Characteristics for Stoichiometric Mixtures of Various Oxidants with Atomised Magnesium** [2]

| Oxidant | Stoichiometric ratio (oxidiser/fuel) | Calculated heat of reaction | | Luminous intensity (candle.sq.in$^{-1}$) | Burning rate (in. min$^{-1}$) | Efficiency candle. (sec. g$^{-1}$) |
|---|---|---|---|---|---|---|
| | | k cal | cal. g$^{-1}$ | | | |
| Lithium nitrate | 53.2/46.8 | 631.0 | 2,430 | 109,000 | 13.9 | 17,500 |
| Sodium nitrate | 58.3/41.7 | 595.4 | 2,060 | 102,000 | 13.1 | 15,500 |
| Potassium nitrate | 62.5/37.5 | 569.8 | 1,760 | 27,500 | 6.9 | 8,000 |
| Calcium nitrate | 57.5/42.5 | 647.1 | 2,260 | 64,000 | 6.8 | 18,000 |
| Strontium nitrate | 63.5/36.5 | 626.8 | 1,881 | 50,500 | 7.7 | 12,500 |
| Barium nitrate | 68.8/31.8 | 615.5 | 1,606 | 45,000 | 5.1 | 14,000 |
| Potassium perchlorate | 58.8/41.2 | 515.8 | 2,441 | 37,000 | 5.2 | 15,000 |

fuel content leads to decrease in the heat of reaction, thus affecting linear burn rate, luminosity and efficiency.

For higher illumination, lower molecular weight oxidisers are preferred, ensuring the control over humidity as these are more hygroscopic. Among various oxidisers, sodium nitrate is the only preferred oxidiser giving yellow light for illumination.

## 8.2.2 Burn Rate

It is desirable to have higher duration of burning so as to have a prolonged duration of visibility. This makes it mandatory to have longer length of the candle since burning of candle takes place from layer to layer. But the space for composition is restricted in illuminating ammunitions and the rate of fall of candle even with parachute due to gravity is inevitable. Higher duration of burning time means slower linear burn rate which, in turn, gives lower luminosity. Thus, it is not possible to have both higher luminosity and higher burn time. Hence a compromise is required between burning time and luminosity. As a rough estimate, for typical illuminating composition, a reduction in 30% burn time has shown sevenfold increase in luminosity. Burning time can be controlled by various methods (see Chapter 4), such as

a. Varying the ingredients and their ratio (fuel, oxidiser, binder, burn rate modifier, catalyst, etc.)
b. Varying sieve sizes of ingredients
c. Varying pressing load etc.

## 8.2.3 Factors Affecting Burn Time and Illumination

The target illumination depends on several factors, including:

a. *Constituents of pyrotechnic composition*
   i. Properties of ingredients in composition
   ii. Heat of combustion, burn rate and flare intensity
   iii. Temperature of incandescent particles in the flame
   iv. Shape, size and spectral radiation of the flame
   v. Combustion-produced smoke as aerosol moves upwards initially as burning candle with parachute descends and does not affect luminosity. However, it attenuates the luminosity at the end of the burning of composition.
b. Design features of ammunition
   i. Compaction density
   ii. Diameter of candle (higher diameter gives higher luminosity to a certain extent)
   iii. Height of deployment of candle above ground (higher illumination achieved by use of higher diameter of candle for higher heights)
   iv. Rate of descent of candle (depends upon candle mass, surface area of parachute, mass of parachute, height of deployment, porosity of parachute cloth and type of parachute like umbrella type or unicross type etc.)
c. *The prevailing atmospheric conditions and target*
   i. Ambient temperature
   ii. Ambient pressure

iii. Ambient humidity
iv. Wind drift
v. Degree of natural illumination
vi. Presence of smoke, dust, fog, snow, storm in air
vii. Target radiation
viii. Target inherent contrast
ix. Target area
x. Target background radiation (colour and reflective characteristics) of other materials
xi. Target to observer distance
xii. Distance(range) of its functioning relative to the observer and target

## 8.3 PREFERENCE OF MAGNESIUM AS FUEL

White light is produced by fuels like aluminium, magnesium, titanium, tungsten and zirconium. However, preference of magnesium fuel is based upon its properties (Table 8.4).

Magnesium is preferred over aluminium for illuminating composition due to

a. Low melting point (649°C compared to 660°C of aluminium) causes magnesium to burn with more ease.
b. Low boiling point (1,107°C compared to 2,467°C of aluminium) allows excess magnesium with its low heat of vaporisation (128 kJ/mole compared to 294 kJ/mole of aluminium) to vaporise and burn with atmospheric oxygen to provide additional input of heat and light.
c. Combustion product magnesium oxide (MgO) is a white light emitter having high melting point and boiling point.
d. Magnesium forms a thin and porous oxide film while aluminium forms a thick and non-porous oxide film in contact with air. This ensures the ease of oxidisability of magnesium more than aluminium as oxygen diffuses through the porous oxide layer on magnesium and reacts with magnesium but is difficult in case of aluminium with thick non porous oxide film. Pilling Bedworth ratio for magnesium is 0.81 while it is 1.28 for aluminium (see Section 6.2.2.3.).

Magnesium is ahead of aluminium in all aspects which makes it more suitable for increasing the flame temperature.The only drawback with magnesium is that it gets oxidised if kept open and also gets easily affected by moisture. Aluminium scores over magnesium for cost, ready availability and heat of combustion (7.4 $Kcal.g^{-1}$ compared to 5.9 $Kcal.g^{-1}$ of magnesium).

## 8.4 ILLUMINATING COMPOSITIONS WITH PARACHUTE

The choice of fuel is restricted to mainly magnesium (due to low melting and boiling points of magnesium) and to some extent aluminium. Preferred oxidisers are sodium nitrate (due to low melting point of sodium nitrate and strong radiation in

**TABLE 8.4**
**Comparative Data of Some Fuels for White Light [3]**

| Fuel | Melting point (°C) | Boiling point (°C) | Heat of combustion (kcal. g$^{-1}$) | Heat of vaporisation (cal. mole$^{-1}$) | Combustion products | Melting points of oxides | Boiling points of oxides |
|---|---|---|---|---|---|---|---|
| Aluminium | 660 | 2,467 | 7.4 | 70,200 | $Al_2O_3$ | 2,072 | 29,80 |
| Magnesium | 649 | 1,107 | 5.9 | 30,750 | $MgO$ | 2,852 | 3,600 |
| Titanium | 1,660 | 3,287 | 4.7 | 102,500 | $TiO_2$ | 1,830–1,850 | 2,500–3,000 |
| Tungsten | 3,410 | 5,660 | 1.1 | 191,000 | $WO_3$ | – | 1,800 |
| Zirconium | 1,852 | 4,377 | 2.9 | 139,000 | $ZrO_2$ | 2,700 | 5,000 |

yellow band, despite higher hygroscopicity) and sometimes addition of barium nitrate.

When an illuminating composition for high-calibre ammunition is to be considered, it should be mandatory to attach a parachute attachment to the candle for prolonged burning time in air. Parachute allows it to descend slowly while pressed candle allows the pyrotechnic composition to burn layer by layer giving more burn time. If the desired flare burn time is 30 seconds above the ground, a suitable composition giving more than 30 seconds is required to be developed. Left over spin and parachute descent, both, enhance the linear burn rate and thus reduce the burn time in air.

An illuminating composition is required to be pressed in the candle with paper liner to reduce energy loss to container. When the ammunition is fired with an electronic or mechanical time fuze setting so as to ensure functioning at desired distance in air, say at 400 m above the ground (with a designed rate of parachute descent, say approximately 5 m/seconds), the candle with parachute would be ejected out at 400 m above the ground (Figure 8.1). Presuming no loss in mass of candle during burning, the parachute will hold the candle in the air and descend down with approximately 5 m/seconds. The radius of light falling over the area would go on reducing. The candle is expected to extinguish in 30 seconds. During this period, the parachute descent would be 150 m (i.e., 250 m above ground) and hence will illuminate the ground by slow descent from 400 m to 250 m in the air, above the ground. The burnt candle with parachute then touches the ground after 50 seconds. However, in actual, the rate of candle descent is not constant as there is loss of mass of candle during burning and hence the descent rate decreases with time. The illumination on ground increases as the flare descends with parachute.

In some illuminating shells, an auxiliary parachute is used which initially holds a canister (containing the candle and the main parachute). The canister flaps are opened during the spin, thereby reducing the spin of the canister. The main parachute along with the candle is then allowed to be deployed after some delay through a delay element, so that there is no spin effect on candle.

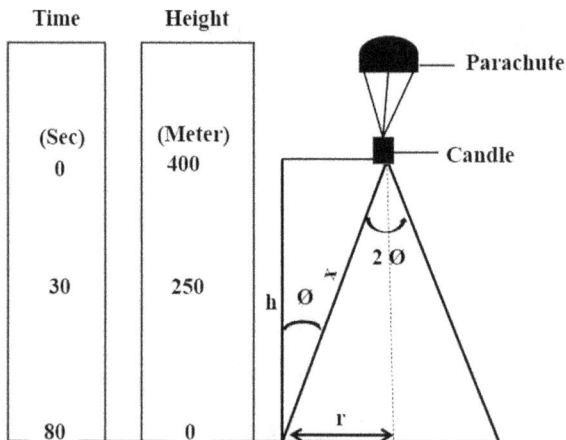

**FIGURE 8.1**  Illumination of Target Area.

## 8.5  RELATIONSHIP BETWEEN GROUND ILLUMINATION AND CANDLE ALTITUDE

The relationship between ground illumination at a particular distance from the candle height and candle's luminosity can be understood as follows.

Consider a sphere of radius $r$ (metres) with a point light source at the centre. It is known that a light source with luminous intensity 1 candela gives a luminous flux of $4\pi$ lumens. Hence, if the light source has luminous intensity $L$ candela, then the luminous flux shall be $4\pi L$ lumens. The illumination $E$ (lm m$^{-2}$) on the inside surface of the sphere (spherical surface) having surface area $4\pi r^2$ shall be

$$E = \text{Luminous flux/Area, (lm m}^{-2}) = 4\pi L/4\pi r^2 \,(\text{lux}) = L/r^2 \,(\text{lux}) \quad (8.1)$$

This shows that the illumination will decrease with increase in radius. However, if the candle or flare (light source) illuminates a plane surface at a distance x metres from the candle and Ø be the included angle between the line from the source of light to the target and the normal to the target (Figure 8.1), then illumination $E$ at the surface shall be (Lambert's Cosine Law)

$$E = L \, \cos \, \emptyset/x^2, \quad (\text{lux}) \quad (8.2)$$

If $r$ is the distance or radius of the target from the vertical ground point and $h$ the altitude of the candle or flare, then illumination at the target can be calculated as under.

Since $\cos \emptyset = h/x = h/(r^2 + h^2)^{1/2}$ since $x^2 = (r^2 + h^2)$

Hence, $E = Lh/(r^2 + h^2)^{3/2}$, (lux)

$$L = \frac{E}{h}(r^2 + h^2)^{3/2}, \,(\text{lux}) \quad (8.3)$$

Thus, required minimum luminosity $L$ within a radius $r$ of the candle may be arrived at with the desired illumination $E$ with the specified candle deployment height. The flare height should be such that it not only gives sufficient burn time and avoids ground burning but also provides the required illumination on the ground. However, there is an upper limit to luminosity $L$ of the candle due to design constraints.

The area illuminated on the ground, the candle height and its illumination are interrelated. For example, as the height of candle deployed is increased, the illuminated area on ground increase but illumination decreases. Readers may see reference [4] for ground area-flare altitude relationship. Readers may refer [5] for better understanding of deployment of illuminating ammunitions and reference [6] on aircraft parachute flare simulation.

It is desirable to have the candle burning half way by the time candle reaches its optimum height. The height of deployment of illuminating ammunitions and rate of descent of candle with parachute and the luminosity of candle are designed such that the illumination achieved over a desired target area is for maximum time.

Multiple or successive firing of illuminating ammunitions is an option for better illumination, longer duration and larger area of combat field.

## 8.6 PERFORMANCE OF ILLUMINATING COMPOSITION IN AIR, ARGON AND NITROGEN

Table 8.5 provides the performance of an illuminating composition in air, argon and nitrogen.

**TABLE 8.5**

**Illuminant Composition Burn Time and Luminosity in Air, Argon and Nitrogen [7]**

| Composition Mg/NaNO₃ /binder | Air | | Argon | | Nitrogen | |
|---|---|---|---|---|---|---|
| | Burn time (in.sec⁻¹) | Luminosity (kilo candela) | Burn time (in.sec⁻¹) | Luminosity (kilo candela) | Burn time (in.sec⁻¹) | Luminosity (kilo candela) |
| 40/52/8 | 0.052 | 82 | 0.048 | 26 | 0.048 | 23.0 |
| 47.5/44.5/8 | 0.060 | 133 | 0.064 | 14.7 | 0.063 | 10.7 |
| 55/37/8 | 0.076 | 177 | 0.075 | 5.0 | 0.075 | 4.8 |
| 60/32/8 | 0.076 | 160 | 0.084 | 2.4 | 0.098 | 2.7 |
| 40/55/5 | 0.062 | 80 | 0.061 | 50.0 | 0.063 | 37.8 |
| 55/40/5 | 0.074 | 137 | 0.071 | 41.5 | 0.076 | 28.0 |
| 47.5/47.5/5 | 0.086 | 178 | 0.088 | 25.8 | 0.094 | 20.0 |
| 44/51.5/4.5 | 0.042 | 60 | 0.045 | 22.0 | 0.044 | 20.0 |
| 58/37.5/4.5 | 0.060 | 140 | 0.062 | 12.0 | 0.061 | 13.3 |

The following could be seen:

a. Marked decrease in luminosity in nitrogen and argon though burn rates not significantly affected
b. Oxygen in air contributes significantly in luminosity

## 8.7 TYPICAL ILLUMINATING COMPOSITIONS

Some of the typical illuminating compositions are given in Table 8.6.

Cackett [9] has reported elastic-illuminating compositions using Thiokol rubber (Table 8.7).

Readers may see reference [10] on a tamp castable illuminating flares and reference [11] on substitution of aluminum for magnesium as a fuel in flares. Readers may see reference [12] on factors affecting the stoichiometry of the Magnesium-Sodium nitrate combustion reaction.

**TABLE 8.6**
**Illuminating Compositions [8]**

| Ingredient | Percentage | | | |
|---|---|---|---|---|
| | SR343A | SR 562 | SR563 | SR580 |
| Magnesium powder Gr O | 50 | 49 | 50 | – |
| Magnesium powder Gr IV | – | – | – | 60 |
| Sodium nitrate | – | 39 | 46 | 35.5 |
| Barium nitrate | 40 | – | – | |
| Binder (acaroid resin) | – | – | | 4.5 |
| Lithographic varnish | – | 5 | 4 | – |
| Calcium oxalate | 5 | 7 | – | – |
| Paraffin wax | 5 | – | – | – |
| Burn rate (sec. in$^{-1}$) | 11 | 18–20 | 12 | 2 |
| Intensity (candela in$^{-2}$) | 46,000 | 50,000-70,000 | 100,000 | 350,000-400,000 |

**TABLE 8.7**
**Typical Elastic-Illuminating Compositions [9]**

| Ingredients | Slow burning | Fast burning |
|---|---|---|
| Thiokol rubber L.P.2 | 15.0 | 14.0 |
| Magnesium powder Gr.3 | 54.0 | – |
| Magnesium powder Gr.5 | – | 57.3 |
| Sodium nitrate | 27.0 | 28.7 |
| Oxamide | 4.0 | – |
| Performance | | |
| Rate of burning (sec. inch$^{-1}$) | 19.8 | 6.2 |
| Specific intensity candles. (sq. inch)$^{-1}$ | 49,000 | 137,000 |
| Specific luminous flux (candles-sec. g$^{-1}$) | 33,600 | 35,900 |

## 8.8   SPECIAL REQUIREMENTS OF ILLUMINATING COMPOSITIONS

Apart from general essential requirements of compositions as given in Section 1.6, illuminating composition should preferably possess the following special requirements.

   a. The composition should give maximum emission in the desired visible solar radiation with optimum amount of composition for desired duration.
   b. The fuel and oxidiser present in the pressed composition should produce extreme heat output per unit mass of composition.

c. The excess fuel should react with the atmospheric oxygen to produce high heat of combustion.

d. The combustion product (e.g., MgO) should also give incandescence in the visible range

Infrared illuminating compositions are also used during night warfare and cannot be seen through naked eyes but only through infrared devices. Readers may see illuminating infrared compositions in Section 18.10.

## REFERENCES

[1] S. M. Nirgude, *"Design Philosophy of Illuminating Ammunitions"*, Continued Education Programme Course on Pyrotechnics, High Energy Materials Research Laboratory, Pune, India, October 2005.

[2] *"Theory and Application"*, AMCP 706-185, Engineering Design Handbook Military Pyrotechnic series, Part One, Headquarters, US Army Material Command, Washington D.C., April 1967.

[3] Reuben Daniel, *"Choice of Materials for Pyrotechnic Compositions"*, Continued Education Programme Course on Pyrotechnics, High Energy Materials Research Laboratory, Pune, India, August 2003.

[4] Maj Robert L. Hilgendorf, S/SGT Robert G. Searle, and Ronald A. Erickson, *"The Effect of Hovering Flares on Visual Target Acquisition"*, Report Number NWC TP 5722, January 1975, Naval Weapon Center, China Lake, California.

[5] *"Flare Effectiveness Factors: A Guide to Improved Utilization for Visual Target Acquisition"*, Target Acquisition Working Group Report 61 JTCG/ME-74-10, November 1973.

[6] Joseph J. Angoti, *"Aircraft Parachute Flare Simulation"*, RDTR No. 157, 1 October 1969, Research and Development Department, Naval Ammunition Depot, Crane, Indiana.

[7] David R. Dillehay, *"Illuminant Performance in Inert Atmospheres"*, Proceedings of the Fourth International Pyrotechnic Seminar, 22–26 July 1974. Colorado 80210.

[8] *"Service Textbook of Explosives"*, 1972, Ministry of Defence, UK, Reprinted in India in 1976.

[9] J. C. Cackett, *"Monograph on Pyrotechnic Compositions"*, Royal Armament Research and Development Establishment, Fort, Halstead, Sevenoaks, Kent 1965.

[10] R. F. Eather, *"A Tamp Castable Composition for Illuminating Flares"*, Proceedings Fourth International Pyrotechnics Seminar, Steamboat Village Inn Steamboat Village, Colorado, July 1974, Hosted by Denver Research Institute /University Of Denver.

[11] B. Jackson, Jr., F. R. Taylor, R. Motto, and S. M. Kaye, *"Substitution of Aluminum for Magnesium as a Fuel in Flares"*, Technical Report 4704, January 1975, Picatinny Arsenal, Dover, New Jersey.

[12] A. J. Beardell and D. A. Anderson, *"Factors Affecting the Stoichiometry of The Magnesium-Sodium Nitrate Combustion Reaction"*, Proceedings Third International Pyrotechnics Seminar Colorado Springs, Colorado, August 1972.

# 9 Signalling Flare Compositions

## 9.1 ROLE

Signalling flare compositions make use of coloured flare to communicate various signals. These compositions are used for:

a. Ground-to-ground, ground-to-air or vice versa communication
b. Sea-to-sea, sea-to-ground and sea-to-air communication
c. Submarine to sea surface or air communication
d. Distress (search and rescue) communication

These compositions produce light with different colours and intensities. It has been observed that flame colours, namely red, orange, yellow, green and white are easily identifiable from a long distance and can be distinguished from each other. Hence these colours are most widely used in signalling ammunitions.

In warfare, coordinated code signals are used. This is achieved through signal flare compositions of various colours. These signals are changed from time to time and place to place for tactical reasons.

The colour dimensions are:

a. Hue (like red, yellow, green and white, etc.)
b. Saturation (like pastel, deep and rich)
c. Brightness (like dim or bright)

The signal compositions on burning are required to provide the following:

a. They should provide the required colour signal with high colour saturation.
b. The intensity should be as specified so that it is visible from long distance. The visibility depends on the distance, candlepower, colour and weather conditions.
c. They should burn for the specified time so that it may be noticed or identified properly.

Human eye can perceive the light in the visible range (0.380 to 0.780 μm) of electromagnetic spectrum while night vision devices can see in the infrared radiations (780 to 1,000 μm).

DOI: 10.1201/9781003093404-9

## 9.2   MECHANISM OF PRODUCTION OF FLAME COLOUR

The emission of radiation from a signaling flare composition is mainly through three sources, as mentioned below.

  a. Atomic excitation (e.g., sodium atom) giving yellow line spectrum
  b. Molecular excitation (e.g., strontium monochloride) giving red band spectrum
  c. Incandescence (e.g., magnesium oxide) giving continuous spectrum

The visible range of spectrum wavelength in the increasing order of wavelengths is:

*Violet < Blue < Green < Yellow < Orange < Red*

Violet colour has the shortest wavelength but the highest photon energy while red has the highest wavelength but lowest photon energy. The human eye colour sensitivity varies among individuals and so does the colour observation. Table 9.1 provides the approximate colour spectrum range as the colour spectrum is continuous with no distinct colour boundaries between a colour and the others.

Signalling flare compositions on combustion are capable of producing some of these colours. The colour of the flame is obtained by using suitable oxidisers (metal salts) which are responsible for producing specific colours. However, it is necessary to use colour intensifiers for better visibility from even long distances.

It may be seen that except sodium, which produces yellow light, all other colours are produced by metal chlorides formed during the combustion process. The following monochlorides of metals produce colours:

---

**TABLE 9.1**
**Approximate Visible Spectrum Range**

| Perceived by human eye | Emitted wavelengths | |
|---|---|---|
| | microns (μm) | nanometres (nm) |
| Violet | 0.380 to 0.435 | 380-435 |
| Blue | 0.435 to 0.480 | 435-480 |
| Greenish blue | 0.480 to 0.490 | 480-490 |
| Bluish green | 0.490 to 0.500 | 490-500 |
| Green | 0.500 to 0.560 | 500-560 |
| Yellowish green | 0.560 to 0.580 | 560-580 |
| Yellow | 0.580 to 0.595 | 580-595 |
| Orange | 0.595 to 0.650 | 595-650 |
| Red | 0.650 to 0.780 | 650-780 |

1 micrometre (μm) = $10^{-6}$ meter = 1,000 nanometre;
1 nanometre = $10^{-9}$ meter

---

    a. Strontium monochloride – Red
    b. Barium monochloride – Green
    c. Copper monochloride – Blue

Most of these metal chlorides are not stable and thus get converted into metal oxides that give different colour spectrum. To stabilise these metal chlorides, chlorine is infused through colour intensifiers (chlorine donors) to obtain an intense colour spectrum. Chlorine creates molecular emitting species and serves the desired purpose during combustion.

    a. Forms chlorine-containing volatile molecular species with the colour-producing metal
    b. Produces sufficient concentration of emitting species in vapour phase
    c. These emitting species are excellent emitters of narrow bands of visible light spectrum

A compromise is necessary for addition of colour intensifier as it consumes heat to vaporise itself and thus reduces the flame temperature, which in turn causes flame colour dim. The colour obtained is governed by

    a. Quantity of metal salts and concentration of colour-producing atoms/molecules
    b. Type and quantity of colour intensifier (chlorine donor)
    c. The ratio of fuel and oxidiser
    d. Flame temperature
    e. Presence of oxygen

A good-quality coloured flame is possible by suitable choice of ingredients, especially ensuring that the composition

    a. Produces just sufficient heat to produce desired emitting species capable of emitting desired wavelength, but not very high so as to dissociate emitting species
    b. Produces just sufficient heat to ensure emitting species to remain in vapour state
    c. Produces emitting species in vapour state to be sufficiently in excited state to provide desired wavelength
    d. Does not produce any incandescent species in solid or vapour state which may affect or mask flame colour

The intensity of the light emission is dependent on the quantity of atoms and molecules that are excited during combustion. Table 9.2 shows some of the oxidisers with its combustion species emitting wavelengths (variations in wavelength reported in literature) in the visible range and seen by human eye as flame colour. It may be seen that only a few elements like sodium, barium, strontium and copper are preferred.

**TABLE 9.2**

**Combustion Species of Oxidiser Emitting Wavelength in Visible Range**

| Ingredients | Flame colour | Approximate wavelength (nm) |
|---|---|---|
| Sodium atom | Very strong yellow | 588.9950, 589.5924 |
| Barium monochloride | Strong green | 507, 513.8, 516.2, 524.1, 532.1 |
| Strontium chloride | Strong red | 661.4, 662.0, 674.5, 675.6 |
| Copper (I)chloride | Blue/violet | 420–460 |

Other flame colours may be obtained through a mix of various flame colours. Metals like lithium (red), boron (green), thallium (green), rubidium (red) and caesium (blue) also produce strong colours but their use is not practical because of cost, toxicity or the nature of the compounds.

The heat of the combustion along with formation of colour-emitting products cause emission of light in the narrow zone of visible electromagnetic spectrum as specific colour of that zone and is perceived by the eyes. The reason for this emission of light is that the electrons in the atoms or molecules are excited during heat of combustion and move from ground level to higher orbital and then come back to the ground level. During this act, energy in the form of light (photon) is released.

The atomic spectrum is produced by atomic vapours in the form of a line spectrum, while molecular spectrum is produced by the molecules in gaseous state in the form of a band spectrum.

Electrons move in orbits around the nucleus at specified fixed energy levels. These energy levels are different for each element or in other words each element has different energy level structures. Thus, each element has its own individual spectral fingerprint due to different energy level structures.

Electrons in the farther orbits from the nucleus have more energy than those near the nucleus.

A ground state electron is at the lowest energy level possible. As each energy level of an electron has a fixed energy, an electron from a lower energy level can go to higher level only when it receives the right quantum of energy equivalent to the higher energy level. Thus, energy added by stimuli like heat can move the electron to a higher energy level, called an excited state of electron. Atomic vapours and gases and molecules in gaseous state give out heat energy during combustion; some electrons in the orbit absorb this additional energy and move to a higher energy level. This movement to higher energy level is with a fixed quantum of the energy (absorption of energy).

The atom is not stable when the electrons are in higher energy levels. Therefore, the electron reverts to its original state with low energy level and thus stabilises the atom. During this process, for each different quantum jump, light energy (photon energy) corresponding to the energy difference between the two different energy levels is released in the form of electromagnetic radiation.

The differences are unique for each element. The colour is determined by the energy of the emitted photons $\Delta E$, which determines their characteristic wavelength ($\lambda$). This is represented by formula

$$\Delta E \, (\text{Energy}) = h \, (\text{Planck's constant}) x f \, (\text{Frequency}), \qquad (9.1)$$

Since $f$(Frequency) = c(speed of light)/$\lambda$(Wavelength), hence

$$\Delta E = hc/\lambda \qquad (9.2)$$

where $\Delta E$ = energy of photon, i.e., energy released by electron on coming back from excited state orbital to the ground state (joules), $h$ = Planck's constant ($6.62606896 \times 10^{-34}$ joules-seconds) (or $4.1357 \times 10^{-15}$ ev second) $f$ = frequency of light (cycles.second$^{-1}$), $\lambda$ = wavelength of light (metres) $c$ = speed of light ($2.99792458 \times 10^8$ m.second$^{-1}$)

It may be noted that since $hc = 1,240$ ev. nm, hence photon energy (ev) = 1240/$\lambda$

$$(9.3)$$

From the above equations, it may be concluded that the higher the energy of photon released, the lower would be the wavelength ($\lambda$) of light. The values approximately vary between 1.59 ev and 3.26 ev for visible radiation.

As an example, sodium has 11 electrons, 11 protons and 12 neutrons with atomic mass 23 (Figure 9.1).

Of these 11 electrons, 2 are in the first energy level, 8 are in the second energy level and 1 is in third energy level with configuration $1s^2 2S^2 2p^6 3s^1$. The energy difference between the 3p and 3s state for sodium is $3.37 \times 10^{-19}$ J.

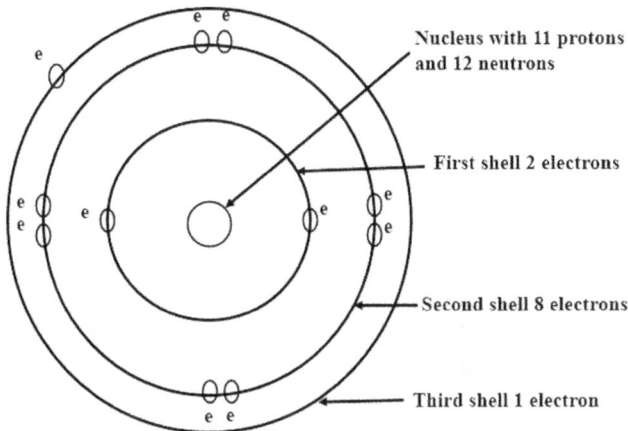

**FIGURE 9.1** Sodium Atom.

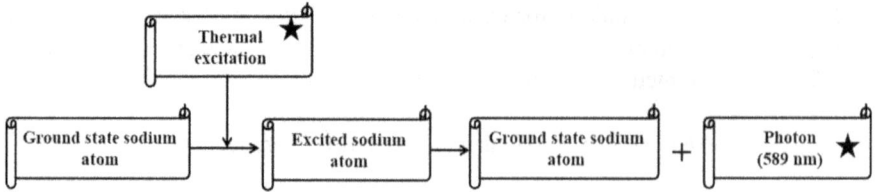

FIGURE 9.2    Electromagnetic Radiation by Sodium Atom.

So, a photon with this energy is released when an electron moves from the higher to the lower energy level in the form of electromagnetic radiation. The wavelength of radiation may be calculated using the above formula by taking the nearest values,

$$\lambda = ch/E$$

$$\lambda = \frac{2.99792458 \times 10^8 \text{ m. second}^{-1} \times 6.63 \times 10^{-34} \text{ joules} - \text{second}}{3.37 \times 10^{-19} \text{ Joules}} \quad (9.4)$$

$= 5.89.7 \times 10^{-7}$ $m$ = 589.7 nm which is in the visible segment of the electro-magnetic yellow radiation. This is shown in Figure 9.2.

Actually, two sodium lines are visible with values 589.593 nm (D1 line) and 588.996 nm (D2 lines), the difference between the two lines is 0.597 nm.

Similarly, in case of a photon with $3.84 \times 10^{-19}$ joules, the corresponding value of wavelength is 517.6 nm, which corresponds to visible segment of the electro-magnetic green radiation.

Some salient points about photons are that photons

a. Behave like a particle and a wave simultaneously
b. Exist as moving particles with constant velocity of light in vacuum
c. Have zero mass and rest energy
d. Carry energy and momentum dependent on frequency, but do not have electric charge
e. Are destroyed/created when radiation is absorbed/emitted

## 9.3   COLOUR PURITY

In a signalling composition, it is nearly impossible to have radiation in a single part of the spectrum (i.e., the flame colour purity as 100%, termed as monochromatic). Thus, if colour purity of a yellow radiation signal is 65%, it may be understood that the yellow signal is equivalent to a mixed signal of 65% monochromatic radiation of yellow wavelength and 35% radiation of fully white source. It is measured based on the chromaticity curve developed by Commission International de L'Eclairage (CIE) to determine the dominant wavelength and spectral purity of luminous sources (see Section 5.1.21).

## 9.4 FACTORS AFFECTING PERCEPTION OF FLAME COLOUR

Perception of flame colour depends upon the following factors:

a. Radiation of composition in visible electromagnetic spectrum
b. Flare height and distance from the observer
c. Luminous intensity of flare
d. Colour saturation value
e. Background illumination (background with contrast gives better visibility)
f. Atmospheric conditions like dust, smoke, haze, fog rain etc.
g. Degree of separation from other colours
h. Absorption of colour radiation by atmosphere
i. Limitation of human eye sensitivity for discriminating colours

Table 9.3 gives the candlepower of coloured light necessary for visibility at a distance of 5,000 yards.

Distances at which, under average weather conditions, various types of signals may be recognised are governed by the following aspects [1]:

a. *Factors of design*: This includes colour and candlepower of the flare as higher candlepower flares may be seen from farther distances than lower candlepower flares.
b. *Factors of position*: This includes height at which flares function. Signals may be seen from greater distances but, due to the tendency of colours to change with distance and the tendency of several lights to merge into one, reliable recognition of the type of signal should not be expected at distances notably greater than 1,500 yards (0.85227 miles) in the daytime, or 2 miles at night.
c. *Factors of environment*: The signal visibility is affected by environmental factors such as dust, humidity, moisture, rain, fog, snow, clouds and sky background. The colour and its sharpness are lost and looks hazy. The green and blue signals are affected more severally than red and yellow signals as longer wavelengths red and yellow can penetrate haze and fog.
d. *Location of observer*: The location of the observer and the signal with respect to the sun plays an important role in visibility of colour.

**TABLE 9.3**
**Candlepower of Coloured Light Necessary for Visibility at 5,000 Yards [1]**

| Condition of environment | Red | Amber | White | Green |
|---|---|---|---|---|
| Night, clear | 1.0 | 2.0 | 2.5 | 2.8 |
| Night, rain light | 1.2 | 2.1 | 3.0 | 3.2 |
| Night, overcast and haze | 3.2 | 4.1 | 3.1 | 5.9 |
| Night, rain heavy | 8.9 | 33.5 | 132.0 | 33.5 |
| Night, snow light | 222.0 | 835.0 | 1,556.0 | 567.0 |
| Day, overcast and haze | 2,000.0 | 2,111.0 | 3,222.0 | 4,000.0 |
| Day, clear | 4,778.0 | 7,556.0 | 11,111.0 | 10,000.0 |

The luminosity of signal composition depends on similar lines of illuminating composition given under Section 8.2.1.

## 9.5   RED SIGNALLING FLARE COMPOSITIONS

Magnesium is the best choice of fuel for red signalling composition. A strong red signal is obtained from strontium monochloride. Hence, to have strontium monochloride, strontium oxalate and strontium nitrate are used as oxidisers and chlorine is infused into the composition by use of potassium chlorate or potassium perchlorate or chlorinated organic compounds like chlorinated rubber, dechlorane or polyvinylchloride.

Infusing chlorine also allows more magnesium to react with chlorine to form MgCl, which is invisible and does not affect colour formed due to strontium monochloride. Further, presence of carbon provided by chlorine donors also allows any MgO formed to react and form Mg vapour and CO which does not affect colour. These chlorine donors also cool the flame, so that white light does not overpower the red light due to strontium monochloride. It also reduces the reaction rate for proper burn rate. The chlorine formed during combustion at high temperature also reacts with any strontium oxide (formed, if any) to form strontium monochloride. Hence, these signal compositions are required to be made in negative oxygen balance to avoid formation of strontium oxide. Some of the red signalling flare compositions are given in Table 9.4.

**TABLE 9.4**
**Red Signalling Flare Compositions**

| Ingredients | Percentage | | | | | |
|---|---|---|---|---|---|---|
| | SR91 [2] | SR97A [2] | SR170A [2] | SR297B [3] | SR307A [3] | SR232 [2] |
| Magnesium Gr. "O" | 35 | 22 | – | – | – | – |
| Magnesium Gr.4 | – | – | 48 | 50 | – | 40 |
| Potassium perchlorate | – | – | 38 | 30 | 28.5 | 29 |
| Strontium nitrate | 41 | 53 | – | – | – | – |
| Strontium oxalate | – | – | 10 | 10 | 26.5 | 12 |
| Chlorinated rubber | 20 | 8 | – | – | – | – |
| Polyvinyl chloride | – | 12 | – | – | – | – |
| Boiled linseed oil | 4 | 5 | 4 | 2 | – | 3 |
| Acaroid resin | – | – | – | 8 | 12 | – |
| Ammonium perchlorate | – | – | – | – | 33 | – |
| Starch | – | – | – | – | – | 16 |

## 9.6   GREEN SIGNALLING FLARE COMPOSITIONS

The intense green signal is obtained from barium monochloride. The chlorine containing substances used are barium chlorate, potassium chlorate, potassium perchlorate and chlorinated organic substances like chlorinated rubber or polyvinylchloride. Magnesium is normally used as fuel.

The chlorine formed during combustion at high temperature also reacts with any barium oxide (formed, if any) to form barium monochloride. Hence these signal compositions are required to be made in negative oxygen balance to avoid formation of barium oxide. Some of the Green signalling flare compositions are enlisted in Table 9.5.

## 9.7   YELLOW SIGNALLING FLARE COMPOSITIONS

Magnesium is used as a fuel in yellow signalling flare compositions. Yellow signal is obtained through atomic spectral radiation of sodium. The sodium compounds get dissociated at high combustion temperature and the electrons in sodium atoms get excited and return back to ground level, thereby emitting radiation in the yellow zone. As it is atomic radiation and not molecular radiation of any compound containing chlorine, no chlorinating agents are required to be infused in the composition. But some compositions still make use of them. Addition of chlorinating agents would produce NaCl whose formation reduces atomic sodium emission, thereby reducing colour quality and light intensity. Some yellow signalling compositions are given in Table 9.6.

## 9.8   WHITE SIGNALLING FLARE COMPOSITIONS

This may be obtained by a combination of two ingredients that produce two different colour signals overlapping each other to form white signal. Table 9.7 gives some white signalling compositions.

## 9.9   ELASTIC SIGNAL FLARE COMPOSITIONS

Cackett [2] has reported elastic signal flare compositions using Thiokol rubber (Table 9.8).

## 9.11   NON-TOXIC SIGNALLING FLARE COMPOSITIONS

Non-toxic signalling flare composition has gained importance over the recent years using environmentally accepted ingredients, mostly with enhanced performance of signal flare compositions, i.e., much better in luminosity, colour purity, and burn rates. The following strategies have been proposed [5]:

   a. Replace $KClO_4$ with nitrate salt
   b. Utilise high nitrogen oxidisers or alternate fuels to increase energy content

**TABLE 9.5**

**Green Signalling Flare Compositions**

| Ingredients | Percentage | | | | | | |
|---|---|---|---|---|---|---|---|
| | SR428 [2] | SR429 [2] | SR486[2] | SR487 [2] | SR703 [2] | SR700 [2] | SR193 [3] |
| Magnesium | – | – | – | – | – | – | 26 |
| Magnesium Gr "O" | – | – | – | 20 | – | – | – |
| Magnesium Gr IV | 43 | – | 35 | – | 42 | 42 | – |
| Magnesium Gr.5 | – | 48 | – | – | – | – | – |
| Potassium perchlorate | 30 | 30 | – | – | 25 | 27 | 16 |
| Barium nitrate | – | – | 47 | 53 | 15 | 17 | 45 |
| Boiled linseed oil | 4 | 4 | 3 | 5 | 6 | 6 | – |
| Barium chlorate | 13 | 12 | – | – | – | – | – |
| Chlorinated rubber | – | – | – | 9 | 12 | – | – |
| Polyvinylchloride | – | – | – | 13 | – | 8 | – |
| Starch | 10 | – | – | – | – | – | – |
| Lactose | – | 6 | – | – | – | – | – |
| Chlorinated diphenyl | – | – | 15 | – | – | – | – |
| Hexachlorobenzene | – | – | – | – | – | – | 7 |
| Copper oxide | – | – | – | – | – | – | 2 |
| Vegetable oil | – | – | – | – | – | – | 2 |
| Gilsonite | – | – | – | – | – | – | 2 |

**TABLE 9.6**
**Yellow Signalling Flare Compositions**

| Ingredients | Percentage | | | |
|---|---|---|---|---|
| | SR581 [3] | SR560 [2] | SR583B [2] | SR590A [2] |
| Magnesium Gr "0" | – | – | – | 35 |
| Magnesium Gr 3 | 40 | 45 | 30 | – |
| Sodium oxalate | – | 5 | – | – |
| Sodium nitrate | 48 | 46 | 40 | 39 |
| Calcium oxalate | – | – | – | – |
| Acaroid resin | 4 | – | 4 | 4 |
| Ivory Nut | 8 | | | – |
| Lithographic varnish | – | 4 | – | – |
| Barium nitrate | – | – | 10 | 10 |
| Mannitol | – | – | 16 | 12 |
| Starch | – | – | – | – |

**TABLE 9.7**
**White Signalling Flare Compositions [4]**

| Ingredient | Percentage | | | | | | | |
|---|---|---|---|---|---|---|---|---|
| Magnesium 30/50 | 58 | 50 | 46 | 48 | 44 | 48 | 48.4 | 36 |
| Sodium nitrate | 37.5 | 44 | 45 | 42 | 44 | 40 | 47.2 | 54 |
| Polyvinyl chloride | – | – | – | 2 | – | – | – | – |
| Laminac | 45 | 6 | 9 | 8 | 12 | 12 | – | – |
| VAAR | – | – | – | – | – | – | 4.4 | 10 |

c. Utilise PVC as chloride source for colour enhancement

d. Replace laminac/lupersol with epoxy (epon/versamid binder)

A good performance has been reported for the below mentioned combination.

a. Standard green illuminating composition (Mk 141) of $Mg/KClO_4/Ba(NO_3)_2/$ PVC/Binder to perchlorate free green illuminating composition $Mg/Ba(NO_3)_2/PVC/Epoxy$

b. Standard red illuminating composition (Mk 124) of $Mg/Sr(NO_3)_2/KClO_4/$ PVC/Asphaltum to perchlorate free red illuminating composition $Mg/Sr(NO_3)_2/PVC/Epoxy$

**TABLE 9.8**
**Typical Elastic Signal Flare Compositions [2]**

| Ingredient | Red signal | Green signal |
|---|---|---|
| Thiokol rubber | 16.5 | 18.0 |
| Magnesium powder Gr "O" | 41.5 | – |
| Magnesium powder Gr 3 | – | 35.0 |
| Potassium perchlorate | 21.0 | 11.0 |
| Strontium nitrate | 21.0 | – |
| Barium nitrate | – | 32.0 |
| Boron | – | 4.0 |
| Performance | | |
| Rate of burning sec/inch | 13.6 | 12.1 |
| Specific intensity candles/sq.inch | 38,400 | 17,200 |
| Specific luminous flux Candles-sec/g | 18,300 | 7,000 |

Readers may refer [6] that gives details about the work on perchlorate-free flare compositions.

## 9.12  SPECIAL REQUIREMENTS OF SIGNALLING FLARE COMPOSITIONS

Apart from general essential requirements of compositions as given in Section 1.6, signalling composition should preferably possess the following special requirements:

a. The composition must provide on ignition such species that are capable of providing desired wavelength.
b. The composition should release sufficient heat to excite the atoms or molecules present in the flame in a vaporised state.
c. The composition should give the desired special effect (flare colour signal) with optimum amount of composition.
d. The flame should have a characteristic colour without any other colour. The colour purity should be minimum 70%–75%.
e. The intensity of the flame colour should be in thousands of candelas so that it is visible properly from a distance.
f. The composition flame temperature should not be excessive so as to dissociate the molecules of the radiating entity.
g. The gaseous reaction products of the composition should preferably not produce radiation in other zones of the spectrum.
h. The composition must burn for the desired time requirement.

Readers may refer [7] for review of green signalling flare pyrotechnics, [8] for theory of coloured flame production, [9] on whether the metal monochloride emitters in pyrotechnic flames are ions or neutrals and [10] on colouring properties of various high-nitrogen compounds in pyrotechnic compositions.

## REFERENCES

[1] *"Military Pyrotechnics"*, War Department Technical Manual, T M 9-1981, Washington 25, DC, 8 December 1943.

[2] J. C. Cackett, *"Monograph on Pyrotechnic Compositions"*, Royal Armament Research and Development Establishment, Fort, Halstead, Sevenoaks, Kent, 1965.

[3] *"Service Textbook of Explosives"*, 1972, Ministry of Defence, UK, Reprinted in India in 1976.

[4] F. L. McIntyre, *"A Compilation of Hazard and Test Data for Pyrotechnic Compositions"*, Contractor Report ARLCD-CR-80047, October 1980, U.S. Army Research and Development Command, Large Caliber Weapon System Laboratory, Dover, New Jersey.

[5] Eric Miklaszewski, *"Removal of Perchlorate Oxidizers from Red- and Green-Emitting Pyrotechnic Flares"*, Naval Surface Warfare Center, Crane Division, IN webinar, July 2016 (sponsored by Strategic Environmental Research and Development Program (SERDP) and Environmental Security Technology Certification Program (ESTCP), Alexandra).

[6] Robert G. Shortridge, Dr. Caroline K. Wilharm, and Ms. Christina M. Yamamoto, *"Elimination of Perchlorate Oxidizers from Pyrotechnic Flare Composition"*, Technical Report WP-1280, March 2007, NSWC Crane Division.

[7] Dr. Jesse J. Sabatini, *"A Review of Illuminating Pyrotechnics"*, Propellants, Explosives, Pyrotechnics, January 2018, Volume 43, Issue 1, Pages 28–47.

[8] B.E. Douda," *Theory of Colored Flame Production"*, RDTN 71, No. 71, 20 March 1964, U. S. Naval Ammunition Depot, Crane, Indiana.

[9] Barry Sturman, *"Metal Monochloride Emitters in Pyrotechnic Flames — Ions or Neutrals? Journal of Pyrotechnics*, Summer 2004, Issue 19, Pages 1–13.

[10] T. Klapötke, J. Hendrik Radies, K. Tarantik, G. Chen, and Amita V. Nagori, *"Coloring Properties of Various High-Nitrogen Compounds in Pyrotechnic Compositions."* Propellants, Explosives, Pyrotechnics, 2010, 35, 213–219.

# 10 Tracer Compositions

## 10.1 ROLE

Tracer compositions are used to:

a. *Trace trajectories of projectiles*: A tracer provides sufficient colour illumination to be seen from the firing point. The burning of tracer compositions,
generally between 3 to 20 seconds gives a trail of smoke or flame, which is
visible to naked eyes to allow the shooter to trail the trajectory relative to the
target for proper aiming of the target. It thus gives an idea of the projectile
trajectory as well as the range and place of hit by the projectile. Projectiles
with long duration of flight and range do not use tracers as it is not practicable
to see tracers in such long ranges. Examples are tracers used in anti-tank
projectiles and small arms bullets to correct the line of fire.

  The small arms bullets enable visual observation during its flight and
produce flash and puff of smoke on hitting the target. These momentary
millisecond flash and smoke, allows firing heavy calibre ammunition with
their own tracers on the target. Burn time generally is 3-4 seconds.

b. *Self-destruction of ammunition*: Tracers also serve the purpose of self-
destruction of anti-aircraft ammunition after elapsed duration when the ammunition fails to act on the target so as to avoid ground casualty and fire/loss.

c. *Alert to shooter*: Tracers are sometimes placed in two or three rounds from
the bottom of magazines of small arms ammunition to alert the shooter that
the weapon will soon be exhausted.

## 10.2 TYPES OF TRACERS

The tracers are of following types:

a. *Integral type tracer*: The tracing composition with priming composition is
pressed in the tracer cavity of the projectile itself, usually a solid shot.

b. *Tracer shell*: The tracer composition with priming composition is pressed in a
separate empty tracer body and then assembled to the base of the shell. This
may be flush with the base of the shell or may protrude out from the base of
the shell.

c. *Tracer cum igniter*: This is filled separately as tracer and serves dual
purpose, apart from tracing the path of the projectile, this also helps in self-
destruction of the shell, if the shell does not function after desired time, by
use of delay element. In other words, self-destruction tracers continue
tracing the path of the projectile and after specified time, (during which the

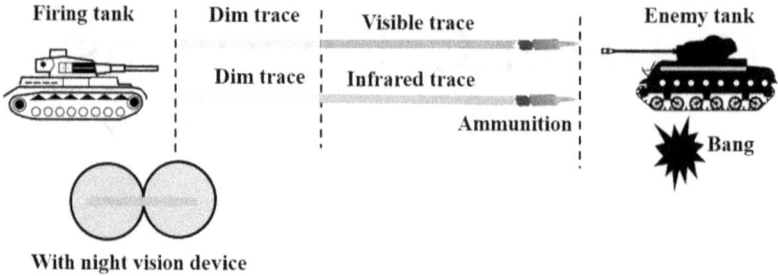

**FIGURE 10.1**  Ammunition with Tracer Firing.

projectile is presumed to hit the target but somehow misses to hit the target), provides heat of combustion to the self-destruction unit for the destruction of the projectile.

The tracers may also be subdivided into three types based on the light energy produced.

a. *Normal tracer/standard tracers*: These tracers burn with bright light. However, bright tracers are found to have two detrimental effects, namely
   i. They would affect the eyes of the firing crew due to glare/bright light of the tracer
   ii. They will disclose the location of the gunner and firing equipment to the enemy
b. *Subdued tracers*: These tracers burn initially without visible bright light for 50–70 m and then burn with full brightness. This helps in keeping the gunner's location hidden as well as the firing equipment position and also prevents the gunner from getting blinded by the flash produced by the tracer. These tracers are filled with a small delay element or subdued igniter mixture, which burns without gas and luminosity, so that the tracers start emitting bright light or infrared light after some distance, thereby avoiding both the above drawbacks (Figure 10.1).
c. *Infrared tracers/dim tracers*: Infrared tracers either do not emit light in visible radiation range or emit very feeble visible light, the main light emitted is in infrared range (non-visible) and may be seen by using night vision infrared devices.

These may also be subdivided into *Day tracers* and *Night tracers*.

## 10.3  TRACER DESIGN

Tracers start burning after firing due to propellant flame and hot combustion products of propellant, thereby allowing tracking of the projectile by smoke trail (during day) or flame trail (during night). The tracers used in the day evolve large quantities of white smoke so that the same is visible in the bright light of the day. Tracers used in the night emit bright light in the dark background of the night. The red colour of tracer is desirable since red has good transmission qualities, having higher wavelength and hence can be seen from long distance. However other colour tracers may also be used if more than one firing equipment is used for the same target.

The main fuel used in tracers is magnesium or sometimes magnesium–aluminium alloy. The oxidisers generally used are barium nitrate (for green) and strontium nitrate (for red) and sodium oxalate for yellow. The binders usually used are shellac, bees wax, phenol formaldehyde, etc.

Tracers are generally made of metallic containers with tracer compositions pressed at high pressure with dwell time. The main composition is pressed in increments (or pre pressed into pellet form) followed by the tracer priming composition. Sometimes booster composition (50:50 of tracer and priming composition) is also used.

The normal principle of pressing tracer is that loading pressure should be 25% more than the setback force. The high pressure with dwell ensures that the tracer composition is not disturbed or thrown out from the tracer due to setback force of firing. The top surface of the tracer is generally sealed with celluloid disc or thin metal foils and held in position by aluminium washer over which the tracer body is turned over so that it avoids ingress of moisture as well as allows picking up flash from propellant quickly. These are fitted at the base of the ammunition.

It is essential to ensure absence of crack in the tracer composition, which otherwise would lead to tracer composition thrown out as a chunk due to penetration of hot gases inside the tracer composition. This is important when working with synthetic binders. The composition using synthetic binders must be pressed within the scheduled time so that the composition does not become hard lump or lose its binding properties. A typical tracer is shown in the Figure 10.2

It shows a tracer body with waxed paper sleeve to avoid heat loss to tracer body. The pressing is done in several layers and topping with priming composition in

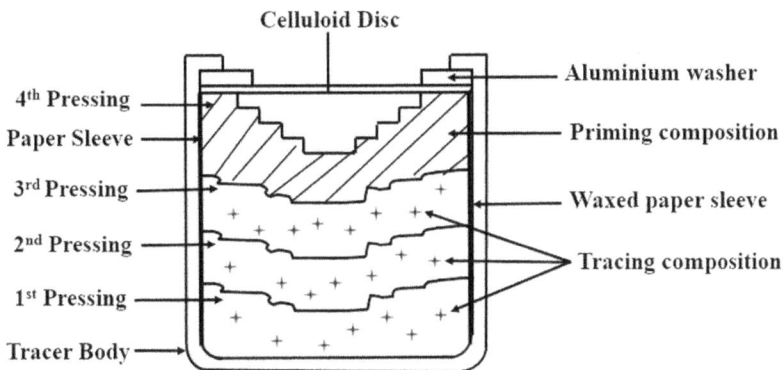

**FIGURE 10.2**  Typical Tracer (Filled).

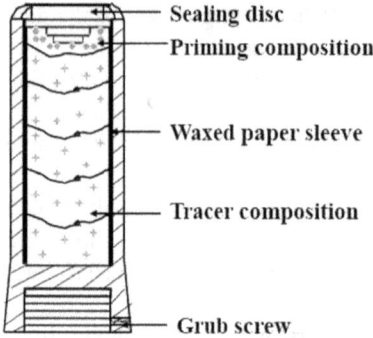

- Sealing disc
- Priming composition
- Waxed paper sleeve
- Tracer composition
- Grub screw

**FIGURE10.3**   Typical Protruded Tracer (Filled).

castellated form for better flash pick-up. It has a celluloid disc to avoid ingress of moisture and closed with aluminium washer and proper sealing.

There are also tracers, which when assembled to the shell, remain protruded outside the projectile (Figure 10.3). They have a long cylindrical tracer body with several layers of tracer composition with priming composition inside paper sleeve to avoid heat loss during tracer burning. It has a grub screw to fix with the shell body.

## 10.4   TRACER LUMINOUS INTENSITY

The following equation is used for finding the minimum luminous intensity required by a tracer to be seen by human eye [1]

$$I = (E_{th} r^2)/\alpha^{r/1,000} \tag{10.1}$$

where $I$ = luminous intensity of the tracer (in Cd), $E_{th}$ = threshold illumination of colour light (in lux), $r$ = distance between the burning tracer and the human eye (metres), $\alpha$ = transparency coefficient of a layer of the atmosphere 1 km thick.

The value of $\alpha$ varies depending upon the ambient conditions prevailing during firing, the values are 0.0004 (medium fog), 0.02 (light fog), 0.67 (light haze), 0.82 (satisfactory visibility) and 0.92 (good visibility).

Further, the luminosity of tracer composition depends upon the diameter of the tracer burning surface. The larger the diameter, the more would be the luminosity of the tracer composition, similar to illuminating composition.

The sensitivity of human eye for all the colours is not the same. It is the highest for red colour. Further sensitivity of human eyes to these colours is considerably less during day than at night. The threshold values of illumination (i.e., minimum values of illumination being perceived by human eye) in lux for human eyes are given in Table 10.1.

For same colour (say yellow), the minimum illumination required during the day for human eyes is $1.0 \times 10^{-3}$ lux. For the same yellow colour, the minimum

**TABLE 10.1**
**Threshold Values of Illumination for Perceiving by Human Eye [1]**

| Colour | Night (lux) | Day (lux) |
|--------|-------------|-----------|
| Red | $0.8 \times 10^{-6}$ | $0.5 \times 10^{-3}$ |
| Green | $1.2 \times 10^{-6}$ | $0.9 \times 10^{-3}$ |
| Yellow | $2.0 \times 10^{-6}$ | $1.0 \times 10^{-3}$ |
| White | $3.0 \times 10^{-6}$ | $1.5 \times 10^{-3}$ |

illumination required during the night to be perceived by human eyes is $2.0 \times 10^{-6}$ lux, which is 500 times less than the minimum requirement during the day.

## 10.5  FACTORS AFFECTING TRACER PERFORMANCE

In addition to the factors affecting performance of pyrotechnic composition (Chapter 4) and Section 9.4, the burning and visibility of a tracer of given aperture size is affected due to the following:

a. The tracer undergoes a very severe drop in pressure as it emerges from the firing barrel since propellant pressure (in several tons per sq. inch) becomes absent and high speed of ammunition causes sudden partial vacuum at the base of the shell
b. Spin imparted to ammunition with tracer (exception being Fin Stabilized Armour Piercing Discarded Sabot tracered (FSAPDST) having low spin)
c. Smoke trail produced by tracer
d. Moving away from the firer as well as external weather conditions (fog, rain, dust, etc.)
e. Velocity of projectile reduces the light output of tracers due to reduced size of flame

The spinning and wind drag increases the tracer burning rate significantly, giving low burn time. Therefore, the dynamic burning time of tracers is always less than those observed in static burn time of the tracer. It is, therefore, necessary to test the burn time of the tracers in an apparatus, which can impart spinning to the tracer at a rate equivalent to that of the ammunition and with air drag equivalent to the one faced by the tracer during actual dynamic firing with ammunition to arrive at more correct tracer burning time. In case it is not possible to test in above equipment, then the tracer static burning time should be kept much higher than the specified dynamic burning time.

The effect of spin on slag and type of cavity shape formed is shown in Table 10.2. The effect of spin on burn time and intensity is shown in Table 10.3.

**TABLE 10.2**

**Effect of Spin on Slag Characteristics of Various Types of Small Arms Tracer Ammunition [2]**

| Tracer | Rpm | Original ch. wt. (g) | Average weight loss (g) | Average loss % | Slag cavity dimensions after burning (64th in.) | Cavity type |
|--------|-----|------|------|------|------|------|
| M20 | 30,000 | 12 | 4.4 | 37.4 | 3 to 4 | Oval & Irregular |
| M20 | 50,000 | 12 | 4.6 | 38.4 | 3 to 4 | Oval & Irregular |
| M20 | 70,000 | 12 | 4.6 | 38.4 | 4 | Oval |
| M20 | 90,000 | 12 | 4.4 | 37.4 | 4 | Oval |
| M1 | 30,000 | 65 | 32.7 | 50.4 | 12 to 13 | Round & Irregular |
| M1 | 50,000 | 65 | 32.4 | 49.8 | 12 to 14 | Round |
| M1 | 70,000 | 65 | 30.8 | 47.5 | 13 | Round |
| M1 | 90,000 | 65 | 30.8 | 47.5 | 14 | Round |
| M10 | 30,000 | 65 | 31.6 | 48.6 | 12 | Irregular |
| M10 | 50,000 | 65 | 29.3 | 45.1 | 12 | Irregular |
| M10 | 70,000 | 65 | 27.8 | 42.8 | 12 | Round & Irregular |
| M10 | 90,000 | 65 | 27.6 | 42.5 | 12 | Round & Irregular |
| M17 | 30,000 | 65 | 30.4 | 46.6 | 10 to 12 | Round & Irregular |
| M17 | 50,000 | 65 | 27.5 | 42.3 | 11 to 13 | Round & Irregular |
| M17 | 70,000 | 65 | 28.5 | 43.8 | 13 to 13.5 | Round & Regular |
| M17 | 90,000 | 65 | 28.4 | 43.7 | 13 to 14 | Round & Regular |

## 10.6   USE OF TRACERS

Tracers are used in small calibre, medium calibre and large calibre ammunition. Examples of few types of small arms ammunition (0.30 caliber, 0.45 caliber, 0.50 caliber, 5.56 mm, 7.62 mm) and medium caliber (20 mm, 25 mm with tracers may be seen in [3]).

Figure 10.4 shows NATO 7.62 mm tracer ball with dim tracer[4].

Figure 10.5 depicts some ammunitions using tracer [5].

A huge list of medium (35 mm, 37 mm, 40 mm) and heavy calibre ammunitions (75 mm, 76 mm, 90 mm, 105 mm, 106 mm, 120 mm and 152 mm) using tracers may be found in reference [6].

**TABLE 10.3**
**Candle Power-Time Data for Standard Small Arms Tracer Ammunition [2]**

| Tracer | Rpm | Duration (sec) | Total intensity (candlepower-sec) |
|---|---|---|---|
| Cal 0.50 M1 | 0 | 13 | 5,500 |
| | 30,000 | 9.5 | 10,600 |
| | 50,000 | 8.2 | 9,820 |
| | 70,000 | 7.2 | 9,650 |
| | 90,000 | 6.4 | 5,300 |
| Cal 0.50 M10 | 0 | 12.5 | 4,560 |
| | 30,000 | 9.5 | 4,670 |
| | 50,000 | 8.2 | 5,740 |
| | 70,000 | 7.2 | 6,320 |
| | 90,000 | 6.3 | 5,600 |
| Cal 0.50 M17 | 0 | 13.7 | 12,000 |
| | 30,000 | 9.7 | 9,000 |
| | 50,000 | 8.3 | 5,200 |
| | 70,000 | 7.4 | 7,840 |
| | 90,000 | 6.5 | 4,000 |
| Cal 0.50 M20 APIT | 0 | 6.6 | 580 |
| | 30,000 | 6.3 | 460 |
| | 50,000 | 6.0 | 600 |
| | 70,000 | 5.7 | 490 |
| | 90,000 | 5.5 | 670 |

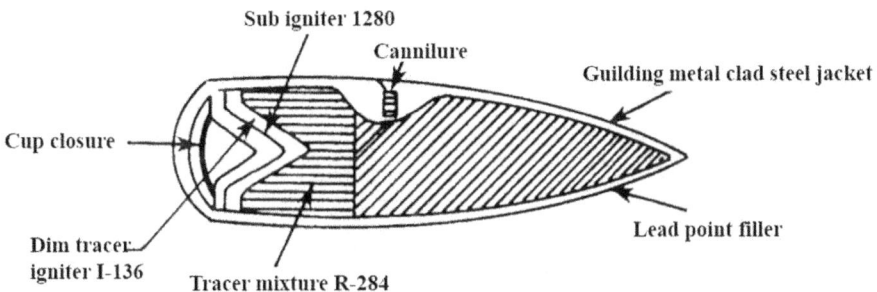

**FIGURE 10.4**   NATO 7. 62 mm Tracer Ball (Reprinted from [4]).

Typical tracer assembly of self-destruction tracers is shown in Figure 10.6. When the primer is struck with firing pin, the flash from cap ignites the primer. The burning of primer ignites the propellant and also imparts velocity to projectile. The propellant also ignites the tracer. The high-explosive bursting charge is detonated by either the fuze functioning or the tracer relay igniting charge, depending upon

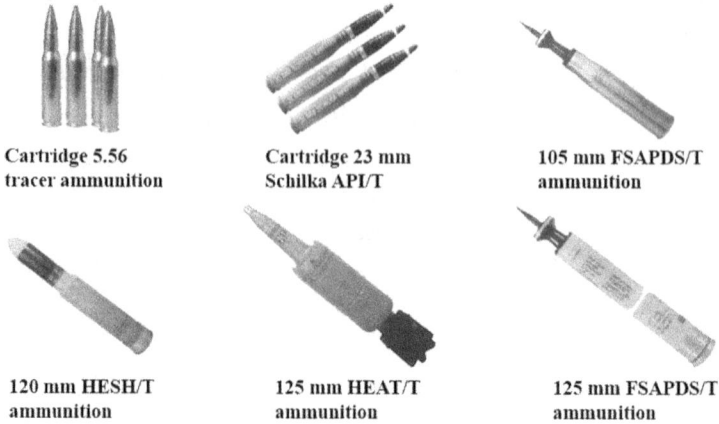

Cartridge 5.56
tracer ammunition

Cartridge 23 mm
Schilka API/T

105 mm FSAPDS/T
ammunition

120 mm HESH/T
ammunition

125 mm HEAT/T
ammunition

125 mm FSAPDS/T
ammunition

**FIGURE 10.5**   Some Ammunitions with Tracer (Reprinted from [5]).

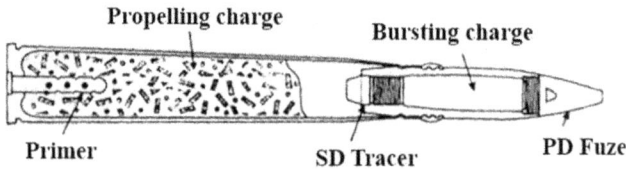

Propelling charge

Bursting charge

Primer

SD Tracer

PD Fuze

**FIGURE 10.6**   Cartridge, 40 Millimetre: HE-T, SD, Mk11, Mk2, MV2870 and SD, M3 Or M3A1, MV2700 (Reprinted from [6]).

whether contact with a target or the burning out of the tracer occurs first. The tracer composition burns with a visible trace for 8 to 10 seconds. Mk11, Mk2, Mk3 and Mk3A1 are nomenclature of tracers used in the projectile.

## 10.7   TYPICAL TRACER COMPOSITIONS

Some of the typical white, red, yellow and green tracing compositions are given in Table 10.4 and Table 10.5.

## 10.8   INFRARED TRACERS

Infrared tracers do not make use of any metallic fuel so that combustion temperature is low, which helps produce infrared output of lower values. A typical infrared tracer composition [8] for medium calibre ammunition is strontium peroxide and barium peroxide, each 34.5%, magnesium carbonate, calcium resinate and barium nitrate, each 10% and silicon (99.9% metal basis) 1%. Another infrared tracer composition is given in Table 10.6.

Another invention [10] relates to non-toxic, boron-containing, IR tracer compositions and to IR tracer projectiles containing such IR tracer compositions for

**TABLE 10.4**
**White and Red Tracer Compositions [1]**

| Ingredient | White | White | White | White | White | Red | Red | Yellow |
|---|---|---|---|---|---|---|---|---|
| Magnesium | 35 | 25 | 36 | 35 | 44 | 30 | - | 33 |
| Al/Mg alloy | - | - | - | - | - | - | 37 | - |
| Barium nitrate | 55 | 65 | 49 | - | 39 | - | - | - |
| Strontium nitrate | - | - | - | 32 | - | 60 | 56 | 40 |
| Barium peroxide | - | - | - | 31 | 3 | - | - | - |
| Sodium oxalate | - | - | - | - | 8 | - | - | 17 |
| Polyvinyl chloride | - | - | - | - | - | - | 7 | - |
| Resin binder | 10 | 10 | 15 | 2 | 6 | 10 | - | 10 |
| Combustion rate (mm.sec$^{-1}$) | 4.7 | 3.1 | 4.0 | - | - | 3.1 | - | - |
| Specific light sum (candle.sec.gm$^{-1}$) | 5,300 | 4,400 | 6,500 | - | - | 4,400 | - | - |

**TABLE 10.5**
**Red, Green, Yellow and White Tracer Compositions [7]**

| Ingredients | Percentage | | | | | |
|---|---|---|---|---|---|---|
| | Red formula 64 | Red formula 65 | Green formula 66 | Yellow formula 67 | Yellow formula 68 | White formula 69 |
| Magnesium | 28 | 28 | 41 | 33 | 43 | 34 |
| Barium nitrate | - | - | 28 | - | 41 | 60 |
| Strontium nitrate | 40 | 55 | - | 40 | - | - |
| Potassium perchlorate | 20 | - | - | - | - | - |
| Barium oxalate | - | - | 16 | - | - | - |
| Strontium oxalate | 8 | - | - | - | - | - |
| Sodium oxalate | - | - | - | 17 | 12 | - |
| Sulphur | - | - | - | - | 2 | - |
| Polyvinyl chloride | - | 17 | - | - | - | - |
| Binder | - | - | 15 | 10 | 2 | 6 |
| Calcium resinate | 4 | - | - | - | - | - |

**TABLE 10.6**

**Infrared Tracer Composition [9]**

| Ingredients | Percentage |
|---|---|
| Potassium nitrate | 70 |
| Shellac | 20 |
| Sodium oxalate | 10 |

**TABLE 10.7**

**Dim Tracer and Main Tracer Composition of a Small Arms Ammunition [12]**

| Ingredients | SR 867 dim trace | SR 390B main trace |
|---|---|---|
| Magnesium | 6.3 | 28.4 |
| Bitumen | 11.0 | - |
| Strontium peroxide | 81.0 | 4.8 |
| Strontium nitrate | - | 50.4 |
| Acaroid | 0.5 | - |
| Linseed oil | - | 11.4 |
| Talc | - | 5.0 |
| Zinc stearate | 1.2 | - |

generating a dim visibility IR trace. It contains boron, potassium perchlorate, sodium salicylate, magnesium carbonate and calcium resinate as binder.

Tracer no. 11 used in 40-mm ammunition [11] has a starter mixture of 65% barium peroxide and 35% of selenium-producing dark ignition, and the tracer does not become visible up to 200 yards from the muzzle end.

Dim tracer and main tracer composition [12] of a small arms ammunition is given in Table 10.7.

## 10.9   SPECIAL REQUIREMENTS OF TRACER COMPOSITIONS

In addition to general essential requirements of compositions as given in Section 1.6, tracer compositions should preferably possess the following special requirements:

a. Burn time should be more than the time taken for the projectile to reach the target.
b. Luminous energy should be very high for good visibility, for both in day and night operations throughout its flight.

c. Performance should withstand the high temperature and pressure of propellant gases.

d. Should have desired compaction strength to withstand stress of firing, i.e., setback, set forward and spin.

e. Performance should not be affected due to low pressure created at the base (base drag) of the projectile during trajectory

f. Must be ignitable by tracer priming composition which should not give bright light so as to expose the equipment, crew and location and be compatible with tracer composition.

g. Should not alter the flight trajectory of the projectile (and hence must leave maximum quantity of slag (35–45% in the tracer shell after combustion). This is important for ammunition where the ratio of tracer weight: projectile weight is comparatively high, especially in small calibre ammunition.

Readers may refer [13] on red flare tracers with superior spectral performance and reference [14] for studies on factors affecting tracer performance.

## REFERENCES

[1] A.A. Shidlovskiy, *"Principles of Pyrotechnics"*, 3rd Edition, Moscow, 1964. (Translated by Foreign Technology Division, Wright-Patterson Air Force Base, Ohio, 1974), American Fireworks News, 1 July 1997. ©1997, Rex E. & S.P., Inc.

[2] R.S. Shulman, *"Factors Affecting Small Arms Tracer Burning"*, Report No. R-1287, Project TSI-46, Pitman-Dunn Laboratories, Frankford Arsenal. Philadelphia, PA, September 1955.

[3] *"Army Ammunition Data Sheets, Small Caliber Ammunition"*, FSC 1305, Technical Manual 43-0001-27, 20 December 1996, Headquarters, Department of the Army, Washington, D.C.

[4] F.L. McIntyre, *"A Compilation of Hazard and Test Data for Pyrotechnic Compositions"*, Contractor Report ARLCD-CR-80047, October 1980. U.S. Army Research and Development Command, Large Caliber Weapon System Laboratory, Dover, New Jersey.

[5] *"Brochure"*, Indian Ordnance Factories (OFB), India.

[6] *"Technical Manual, Army Ammunition Data Sheets, Artillery Ammunition, Guns, howitzers, Mortars, Recoilless Rifles, Grenade Launchers and Artillery Fuzes"* (Federal Supply Class 1310, 1315, 1320, 1390) TM 43-0001-28, Headquarters, Department of The Army, Washington DC, 27 October 2003.

[7] Dr Herbert Ellern, *"Military and Civilian Pyrotechnics"*, ©Chemical Publishing Co., New York, 1968.

[8] Guy Henry III, *"IR Dim Trace for Ammunition"*, 38th Annual Guns & Ammunition Symposium, 26 March 2003, Ordnance and Tactical Systems, General Dynamics.

[9] Steven M. Buc, Gregory Adelman, and Stephen Adelman, *"Development of Alternate 7.62 mm Tracer Formulations"*, Contract DAAA21-93-D-0001, Task Order No. 001, U.S. Army ARDEC Picatinny Arsenal, NJ 07806-5000.

[10] Louise Guindon (Ville De Laval), Carol Jalbert (Lachenaie), Daniel Lepage (Le Gardeur)," *Non-Toxic Metallic Boron Containing IR Tracer Compositions and IR Tracer Projectiles Containing the Same for Generating a Dim Visibility IR Trace"*, Patent Publication number: 20080257194 October 23, 2008.

[11] *"U.S. Explosive Ordnance"*, OP 1664 (Volume 1), 28 May 1947, Navy Department, Bureau of Ordnance, Washington 25. DC.

[12] Dr. Nigel Davies, *"Pyrotechnics, Topics to be covered"*, PowerPoint presentation, Cranfield University, Defence Academy of the United Kingdom, Shrivenham, Swindon, SN6 8LA, UK.

[13] Ramy Sadek, Mohamed Kassem, Mohamed Abdo, and Sherif Elbasuney, *"Spectrally Adapted Red Flare Tracers with Superior Spectral Performance"* Defence Technology, December 2017, Volume 13, Issue 6, Pages 406–412.

[14] T.J. Barton and M.J. Bibby, *"Studies of Factors Affecting Tracer Performance"*. *Seventh International Pyrotechnic Seminar*, 14–18 July 1980, Volume 1, Pages 51–63.

# 11 Photoflash Compositions

## 11.1 ROLE

Photoflash compositions are used in a*erial photography* during night time. Aerial photography involves dropping a photoflash cartridge or a photoflash bomb from an aircraft and recording a film.

## 11.2 CONSTITUENTS OF PHOTOFLASH COMPOSITIONS

Photoflash pyrotechnic compositions contain a fuel and an oxidiser. The fuels used are finely divided metal powders like aluminium, magnesium and magnesium/ aluminium alloy. Aluminium and magnesium have a high combustion enthalpy, high flame temperature, cost less and the combustion products are non-toxic. The rapid scattering and combustion of finely divided magnesium or aluminium forms a cloud of incandescent particles with high temperature and luminosity. Atmospheric oxygen also aids in the combustion.

The oxidants used are mostly potassium perchlorate and barium nitrates.

Photoflash compositions are used as photoflash cartridges and photoflash bombs in loose form (uncompacted). Since the composition is loose, its combustion is very fast, providing high temperature, high gas pressure and high illumination.

Some of the typical combustion reactions that have very high heat output are:

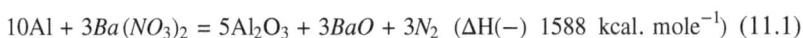

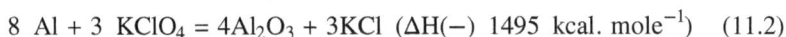

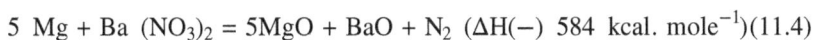

$$10Al + 3Ba(NO_3)_2 = 5Al_2O_3 + 3BaO + 3N_2 \quad (\Delta H(-) \; 1588 \; \text{kcal. mole}^{-1}) \quad (11.1)$$

$$8 \; Al + 3 \; KClO_4 = 4Al_2O_3 + 3KCl \quad (\Delta H(-) \; 1495 \; \text{kcal. mole}^{-1}) \quad (11.2)$$

$$6 \; Al + 3KClO_3 = 3Al_2O_3 + 3KCl \quad (\Delta H(-) \; 1495 \; \text{kcal. mole}^{-1}) \quad (11.3)$$

$$5 \; Mg + Ba \; (NO_3)_2 = 5MgO + BaO + N_2 \quad (\Delta H(-) \; 584 \; \text{kcal. mole}^{-1}) (11.4)$$

$$4 \; Mg + KClO_4 = 4MgO + KCl \quad (\Delta H(-) \; 576 \; \text{kcal. mole}^{-1}) \quad (11.5)$$

## 11.3 MECHANISM OF PHOTOFLASH COMPOSITIONS

The high luminosity of photoflash compositions for short duration is due to:

DOI: 10.1201/9781003093404-11

a. Use of fine metallic powders as ingredients
b. Absence of binders
c. Uncompacted composition (high porosity of composition with surface of ingredients fully exposed)
d. Initiation from strong source (like fuze or explosive charge)
e. Burning of whole mass of composition in short time
f. Clouds of incandescent particles with high temperature and luminosity

## 11.4  DIFFERENCES BETWEEN PHOTOFLASH AND ILLUMINATING COMPOSITIONS

The differences between photoflash compositions and illuminating compositions are shown in Table 11.1.

**TABLE 11.1**
**Differences Between Illuminating and Photoflash Compositions**

| Parameters | Photoflash compositions | Illuminating compositions |
| --- | --- | --- |
| End use | It has very high single flash of high intensity for photography of combat field as photographic films need very less exposure time to capture an image | Reasonable luminosity for considerable time for night vision to navigate and engage targets like enemy troops/equipment like tanks, vehicles and materials, target location for hitting the target, landing of war planes and materials at night and making target for night practice fire |
| Deployment | Dropped as cartridges or bombs from aircraft at very high height to avoid hit by enemy ammunition | Mostly fired as cartridges, bombs and shells from hand held weapons, mortars, howitzers and guns from ground to air |
| Burning type | The burning of whole mass of composition at a time, as flash moves inside the whole mass of composition | The burning is layer by layer as burning front moves from layer to layer of composition |
| Burning time | Burning time is low (in milliseconds) | Burning time is comparatively high (in several seconds) |
| Presence of binder | The compositions do not contain any binder | The compositions contain binder |
| Parachute | The ammunition, generally cartridges and bombs, using photoflash composition do not have parachute | The ammunition using Illuminating candles are assembled with parachute while stars are assembled without parachute |
| Flame size | Flame size is very big | Flame size is comparatively small |
| Sensitivity | Highly sensitive to impact, friction, static electricity due to lose metallic powders | Comparatively less sensitive as compositions are in compacted form |
| Efficiency | Efficiency (candela sec. $g^{-1}$) is less than flares | - |

## 11.5 FACTORS AFFECTING PERFORMANCE OF PHOTOFLASH COMPOSITIONS

The light intensity of photoflash compositions depends on the following:

a. *Chemical constituents of the composition*: This decides the heat of combustion and the flame temperature and presence of solid and fluid hot products exhibiting high emissivity.

b. *Quantity of flash composition and flame size:* Flash intensity and flame size diameter increase with increase in charge, but the specific luminous intensity per gram decreases

c. *The density of flash composition:* The photoflash compositions are mostly used in powder form and are not pressed. This allows it to burn at a very fast rate.

d. *Sieve size of ingredients of flash composition:* The finer the ingredients, faster would be the burn rate due to higher specific surface.

e. *Combustion product's emissivity*

f. *Strength of container containing flash composition:* A metallic container allows composition to burn faster than a weak container made of plastic or cardboard. However, a very thick metallic container would reduce the intensity of flash due to energy loss in breaking the container.

g. *Ambient pressure*: Increases with increase in ambient pressure.

h. *Method of ignition: Ignition by detonator in fuze is essential to have high temperature for ignition*

A typical photoflash composition performance is given in Table 11.2.

Table 11.3 shows the flash intensity and flame size diameter increase with the increase in charge, but the specific luminous intensity per gram decreases.

However, a further increase in composition, the luminous intensity gradually reduces. A graph between luminosity versus time would show steep hike followed by a rapid decline of luminosity.

**TABLE 11.2**
**Flash Duration for Charges of Different Weights [1]**

| Amount of composition (g) | Total flash duration (sec) | Time from start of flash to maximum emission (sec) |
|---|---|---|
| 50 | 0.028 | 0.011 |
| 100 | 0.040 | 0.013 |
| 500 | 0.074 | 0.017 |
| 1,000 | 0.080 | 0.026 |
| 1,400 | 0.120 | 0.036 |

**TABLE 11.3**
**Flash Intensity and Flame Size Diameter [1]**

| Amount of composition in g | Maximum luminous intensity in millions of candles | Specific luminous intensity in thousands of candles per g of composition | Area of projection of flame $m^2$ |
|---|---|---|---|
| 50 | 8.5 | 170 | 0.36 |
| 100 | 15.3 | 153 | 0.75 |
| 200 | 22.6 | 113 | 2.35 |
| 500 | 43.7 | 88 | 3.60 |
| 1,000 | 50.2 | 50 | 6.50 |
| 1,400 | 52.9 | 38 | 7.30 |

## 11.6   FLASH REQUIREMENT FOR PHOTOGRAPHY

The results of night photography depend on [2]:

a. The speed of photographic film
b. Duration of illumination
c. Speed of aircraft
d. The amount of light falling on film which, in turn, depends on
   i.   Reflected light from the target, which depends on illumination from photoflash composition and target reflectance
   ii.  The relative aperture of the lens (f value, which is focal length divided by aperture)
   iii. Transmission characteristics of atmosphere, which allows partial light to reach the camera lens

## 11.7   TYPICAL PHOTOFLASH COMPOSITIONS

Some typical photoflash compositions are at Table 11.4, all of which are aluminium based or a combination of magnesium and aluminium based powders.

## 11.8   DESENSITISATION OF PHOTOFLASH COMPOSITIONS

Photoflash compositions are very sensitive to impact, friction and electrical spark and hence more safety precautions are necessary during mixing and blending. A method of desensitising photoflash compositions has been reported [4] by contact coating from a wet medium employing an electrically conducting carbon black as a coatant. These were used with the Type III Class A photoflash powder consisting of 30% potassium perchlorate, 30% barium nitrate and 40% aluminium. The method is safer than the non-proprietary standard methods previously used, since the material is wet or damp throughout the blending process. Furthermore, the coating, by its ability to leak off voltage, reduces the electrostatic hazards during the blending and

**TABLE 11.4**
**Photoflash Compositions**

| Ingredients | [3] | Type III A [2] | Type III B [2] | Type I [2] | Type II A [2] | Type II B [2] |
|---|---|---|---|---|---|---|
| Aluminium powder 24μ | 4 | 40 | 40 | 26 | – | – |
| Barium nitrate 147μ | 54.5 | 30 | 30 | – | – | 54.5 |
| Potassium perchlorate 24μ | – | 30 | – | 40 | 40 | – |
| Potassium perchlorate coarser 325μ | – | – | 30 | – | – | – |
| Magnesium powder | – | – | – | 34 | – | – |
| Potassium nitrate | – | – | – | – | – | – |
| Strontium nitrate | – | – | – | – | – | – |
| Magnesium aluminium alloy (50/50) | 45.5 | – | – | – | 60 | 45.5 |

further processing of the powder. The product obtained is stable and equals the performance of the uncoated formulations.

Table 11.5 shows the light output of photoflash composition of Type III Class A with a coatant.

## 11.9   SPECIAL REQUIREMENTS OF PHOTOFLASH COMPOSITIONS

In addition to the general essential requirements of compositions as given in Section 1.6, photoflash compositions should preferably possess the following special requirements:

a. Its luminous intensity should preferably be maximum.
b. Its flash duration should preferably be minimum.

**TABLE 11.5**
**Light Output from Coatant Formulations [4]**

| Coatant | Charge wt. (g) | Integral light ($10^5$ candle. sec) | Standard deviation |
|---|---|---|---|
| None | 20 | 1.88 | 0.29 |
| Carbon black Sterling R | 19 | 1.84 | 0.12 |
| Carbon black Royal SR-F-S | 21 | 1.84 | 0.27 |

Coatant at 1% level, loaded in Poppy Cartridges, twelve rounds per set

c. Its radiation of the burning composition should match with that of the spectral sensitivity of the aerial photographic film.

d. It should give high heat of combustion.

e. It should give high flame temperature.

f. Its combustion products should give high emissivity.

## REFERENCES

[1] A.A. Shidlovskiy, "*Principles of Pyrotechnics*", 3rd Edition, Moscow, 1964. (Translated by Foreign Technology Division, Wright-Patterson Air Force Base, Ohio, 1974), American Fireworks News, 1 July 1997. ©1997, Rex E. &S. P., Inc.

[2] "*Theory and Application*", AMCP 706-185, Engineering Design Handbook Military Pyrotechnic series, Part One, Headquarters, US Army Material Command, Washington DC, April 1967.

[3] F.L. McIntyre," *A Compilation of Hazard and Test Data for Pyrotechnic Compositions*", Contractor Report ARLCD-CR-80047, October 1980, U.S. Army Research and Development Command, Large Caliber Weapon System Laboratory, Dover, New Jersey.

[4] S. Dallman, S. Werbel and F.R. Taylor, "*Photoflash Compositions Desensitized by Coatants*", Technical Report ARLCD-TR-80043, December 1980, US Army Armament Research and Development Command, Large Caliber Weapon Systems Laboratory, Dover, New Jersey.

# 12 Screening Smoke Compositions

## 12.1 ROLE

Visibility is related to the distance which a viewer can identify any object relative to the background and illumination. Attenuation (obscuration plus scattering plus reflection etc.) is the phenomenon causing degradation of the visibility. A battlefield requires that the enemy should not be successful in locating the position of the personnel and the firing tanks, equipment, trucks and other battlefield-related inventories. The options available for degrading visibility are:

a. Hiding like concealing behind the tree
b. Blending with nature like covering the truck with nets and tree branches and leaves, etc.
c. Disguising

The natural obscurants are darkness, fog, sandstorm, etc. However, in a battlefield with various terrains and different climatic conditions, the above options are not always feasible. In such cases, a smoke screen has been found to be very good for degrading the enemy's ability to see. It is used for the following purpose.

a. *Screening:* White/grey smoke compositions are used to mask or obscure vision of enemy troops. The white/grey smoke is a nice tool for troops for onward movement or backward retreat in the warfare areas since this will blind the enemy and deprives them of the opportunity to fire on target. These may be considered for following types.
   i. *Offensive Operation:* During offensive operations, smoke is used as a screen for the attacker while an attack is carried out. Some offensive applications include concealing movement of military forces and equipment.
   ii. *Defensive Operation:* During defensive operations, smoke is used effectively to blind enemy observation points to deprive the enemy of the opportunity to adjust fire, to isolate enemy elements to permit concentration of fire and counterattack and to degrade the performance of threat of anti-tank guided missiles. Thermal attenuation screen through the infrared smoke acts as countermeasure to enemies' opto-electronic devices and target acquisition devices operating in intermediate infrared region (3.0–8.0 μm) and far infrared regions (8.0–15.0 μm).

DOI: 10.1201/9781003093404-12

b. *Incendiary*: Incendiary smokes provide flame along with smoke to cause fire.
c. *Training Smokes*: To prepare the troops for firing and working in smoke environments. These mostly produce non-toxic products. In addition, some are used for target practice.

## 12.2 SCREENING SMOKE ARE AEROSOL

The pyrotechnic screening smoke produced is basically an aerosol, a suspension of small solid particles (dispersed phase) in air or gaseous medium (dispersion medium) of size 0.8 μm to 0.2 μm (Shidlovskiy). The small particles of smoke are formed due to the heat of chemical reaction between an oxidiser and a fuel and expel the products from the pyrotechnic reaction zone creating smoke.

These solid aerosols absorb, reflect and scatter the incident light and thus reduce the visibility since light is not able to penetrate the smoke cloud and thus gives a screen effect. The target becomes fully invisible at certain threshold values of the concentration of solid aerosol particles.

The smoke cloud is unable to retain its position in air due to motion of smoke particles, agglomeration of smoke particles, temperature of smoke particles and movement of air. Hence, the smoke screen is lost after some time. The following points should be noted:

a. Size and geometry/shape of aerosol depends on smoke composition
b. Size and geometry/shape of smoke aerosol particles is an important parameter for efficient scattering
c. Efficiency of scattering/absorption varies for different aerosol materials
d. Optimum scattering is given by smoke particles having their size similar to incident radiation wavelength
e. Efficiency of scattering by aerosol varies inversely to incident radiation wavelength $\lambda$ to the power 4, i.e., scattering efficiency is proportional to $\lambda^{-4}$.

### 12.2.1 MECHANISM OF ATTENUATION

Any smoke can minimise visibility. The cause of reduced visibility is due to smoke aerosol particles in air. During such phenomenon, when an incident monochromatic light $I_0$ enters a smoke aerosol dispersed phase, a major part of the incident light energy is absorbed by the aerosol while a part of incident light energy is scattered around different directions. An insignificant amount of reflected light energy moves in the opposite direction of incident light. The transmitted light emerges out in the same direction as of the incident light (Figure 12.1). Absorption, scattering and reflection take place at the atomic and molecular levels. Therefore, the transmitted light $I_t$ would be of substantially reduced intensity. In addition, the smoke also has its own radiation.

The change in intensity is the sum of intensities of absorbance, scattering and reflection.

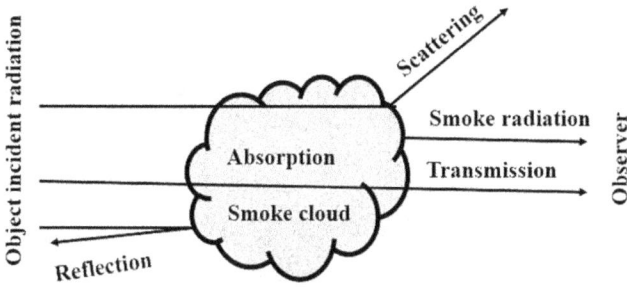

**FIGURE 12.1**  Mechanism of Attenuation.

$$I_o - I_t = I_{absorbance} + I_{scattering} + I_{reflection} \qquad (12.1)$$

Dark or black smoke absorbs more light radiation and scatters or reflects less. White or grey smoke particles possess less absorbance properties but scatter or reflect more.

## 12.2.2 Beer–Lambert Law

Beer–Lambert law is the combination of Beer's (August Beer) law and Lambert's (Johann Heinrich Lambert) law that relate absorbance of a substance to both the concentration and the path length of the sample. It is applicable to only a homogeneous distribution of smoke aerosols at low concentration along the path length of incident monochromatic light. This is used to measure screening efficiency of smoke composition per unit mass of composition.

To derive the equation, consider that an incident radiation $I$ of a particular wavelength passes through smoke aerosols of thickness $dx$ (Figure 12.2).

The Beer–Lambert law states that the decrease in intensity ($dI$) would be proportional to the incident intensity $I$, concentration ($c$) of the absorbing smoke aerosol particles and the length ($dx$) of the smoke cloud where the concentration of smoke particles is taken constant and expressed as

$$dI = (-)\alpha_m I\ c\ dx \qquad (12.2)$$

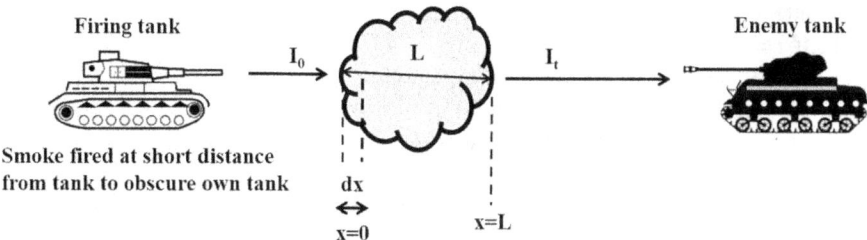

**FIGURE 12.2**  Transmittance of Radiation in Smoke.

$$(-)\ dI/I = \alpha_m\ c\ dx$$

The equation is with (−) ve sign since there is a decrease in intensity. Integrating both sides with limits of integration $I = I_0$ (incident radiation) at $x = 0$ (start) and $I = I_t$ (transmittance radiation) at $x = L$ (end),

$$(-) \int_{Io}^{It} dI/I = \alpha_m c \int_0^L dx$$
$$(-)\ [ln\ I_t - ln\ I_0] = \alpha_m c\ L$$

$$\ln(I_t/I_0) = (-)\alpha_m cL \qquad\qquad (12.3)$$

(12.3) can also be written as

$$I_t/I_o = exp(-\alpha_m cL) \text{ which is an exponential equation} \qquad (12.4)$$

where

$I_o$ = intensity of incident radiation, $I_t$ = intensity of transmitted radiation
$\alpha_m$ = mass extinction coefficient, $(m^2.kg^{-1})$
$c$ = concentration of the smoke aerosol species, $(kg.m^{-3})$
$L$ = length of travel of radiation in smoke cloud (line of sight), (metre)

Extinction coefficient is a measure of effectiveness of aerosol particle as obscurant and is independent of concentration and smoke path length but dependent on wavelength of radiation. Hence its values should be associated with particular wavelength.

## 12.2.3  TRANSMITTANCE FACTOR

$I_t/I_o$ is known as transmittance *factor* $\tau$ where $\tau < 1$ since $I_o > I_t$ (12.5)

It may be seen that transmittance has no units (dimensionless). The value of transmittance factor $\tau$ is between 0 and 1. Attenuation is $(1-\tau) = (I_0-I_t)/I_o$

a. If $\tau = 1$, then $I_t = I_0$, there is no loss of intensity of incident radiation and hence attenuation is zero (i.e., no obscuration).
b. If $\tau = 0$, then $I_t = 0$, there is full loss of intensity of incident radiation and hence attenuation is 1 (i.e., full obscuration).
c. For values of $(\tau)$ between 1 and 0, the visibility depends upon the sensitivity of the eye or the detection instrument.

Since $I_t/I_o = exp(-\alpha_m cL)$, transmittance factor $(\tau = I_t/I_o)$ depends on mass extinction coefficient $(\alpha_m)$, concentration of the absorbing species (c) and the length

($L$) and shall decrease with increasing values of $\alpha_m$, $c$ and $L$. In other words, obscuration shall increase with increasing values of $\alpha_m$, $c$ and $L$.

## 12.2.4 ABSORBANCE

Absorbance ($A$) is defined as

$$A = \ln\ (I_o/I_t) \tag{12.6}$$

or

$$A = \ln(I_0/I_t) = (-)\ln(I_t/I_0) = \alpha_m\ cL \tag{12.7}$$

Hence,

Absorbance = mass extinction coefficient X concentration X path length

$$\tag{12.8}$$

It may be noted that absorbance has no units (dimensionless). A graph between transmittance and length of smoke path (refer Equation 12.4) is an exponential one with continuous decrease in transmittance with increasing length (Figure 12.3) while a graph plotted between absorbance $A$ with smoke path length (or smoke concentration) shows higher absorption with higher smoke path length (or smoke concentration), a straight line passing through the origin (Figure 12.4).

Absorption $A$ may be written as $A = (-)\ln(I_t/I_o) = (-)\ln \tau$ since $\tau = I_t/I_o$

Hence as absorbance increases, transmittance decreases.

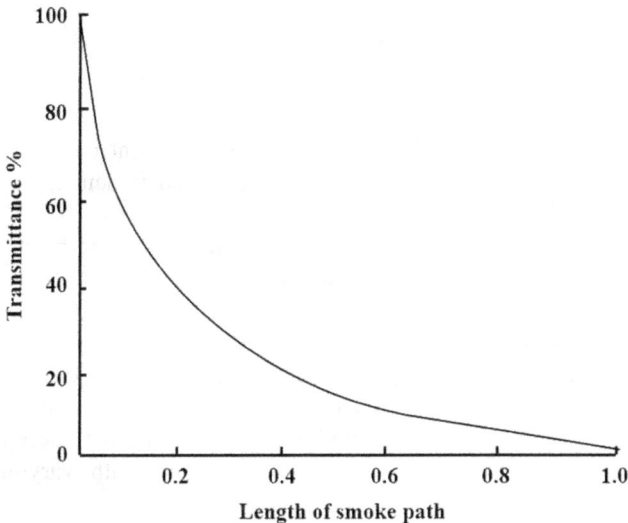

**FIGURE 12.3** Transmittance of Radiation in Smoke.

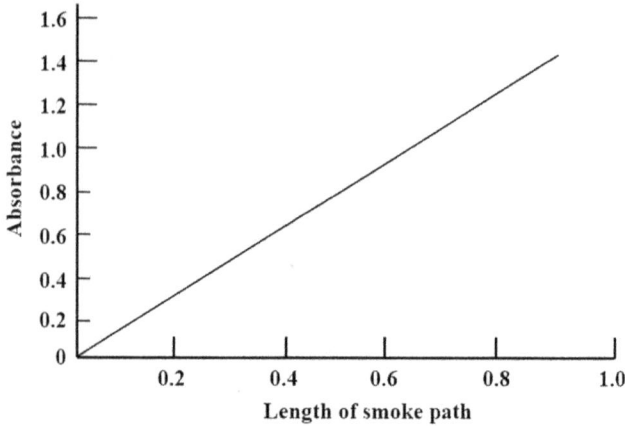

**FIGURE 12.4**  Absorbance of Radiation in Smoke.

## 12.2.5  Obscuration Percent

The percent obscuration shall be $= [(I_o - I_t) \times 100]/I_o$     (12.9)

The mass extinction coefficient ($\alpha_m$) decreases with increase in wavelength of light and is a significant parameter of the smoke particle. The higher the mass extinction coefficient value, the more efficient is the material as attenuator at that wavelength. As attenuation is due to absorption, scattering and reflection, the values of mass extinction coefficient $\alpha_m$ may be considered as the sum of separate coefficients for absorption, scattering and reflection.

$$Mass\ extinction\ coefficient\ (\alpha_m) = \alpha_m(absorption) + \alpha_m(scattering)$$
$$+\ \alpha_m(reflection)\qquad(12.10)$$

It can also be assumed that the [mass extinction coefficient (absorption) + mass extinction coefficient (scattering) + mass extinction coefficient (reflection)] also decreases with increase in wavelength.

The tentative or indicative values of phosphorous smoke and hexachloroethane smoke are shown in Table 12.1. These depend upon the method of manufacture of smoke composition as well as the humidity and are more effective in visible range than in infrared range.

Table 12.2 provides obscuration percentage and extinction coefficient of RP/KNO$_3$ with 5% phenolic resin as binder, where RP was coated with 4-parts binder while KNO$_3$ was coated with 1-part binder and then mixed for safety reasons.

Table 12.3 shows obscuration of RP/KNO$_3$(80/20) with varying humidity conditions.

**TABLE 12.1**

**Typical Values of Mass Extinction Coefficient [1]**

| Material | Visible | Infrared | | | | | mm W |
|---|---|---|---|---|---|---|---|
| | 0.4-0.7 μm | 0.7-1.5 μm | 1.06 μm | 3-5 μm | 8-12 μm | ~ 1 mm |
| Fog oil | ~ 7 | 2.6 | 1.3 | 0.3 | <0.1 | <0.001 |
| Phosphorous smoke | 4 | 1.8 | 0.9 | 0.3 | 0.4 | <0.001 |
| HC smoke | 4 | 2.7 | ~ 2.3 | 0.2 | <0.1 | <0.001 |
| Graphite | 2.4 | 1.5 | 1.5 | 1.5 | 1.5 | <0.001 |
| Brass flake | 0.8 | 1.2 | 1.2 | 1.7 | 1.5 | <0.001 |
| Chaff | 0.04 | – | – | – | – | 0.3 |

**TABLE 12.2**

**Obscuration Percentage and Extinction Coefficients in Visible and Different Infrared Regions [2]**

| Comp* | Obscuration | | | | Extinction coefficient ($m^2/g$) | | | |
|---|---|---|---|---|---|---|---|---|
| RP/KNO3 | 0.4-0.7 μm | 2-2.4 μm | 3-5 μm | 8-14 μm | 0.4-0.7 μm | 2-2.4 μm | 3-5 μm | 8-14 μm |
| 90/10 | 99.7 | 89 | 72 | 62 | 4.36 | 1.66 | 0.95 | 0.73 |
| 80/20 | 99.8 | 90 | 75 | 68 | 4.66 | 1.72 | 1.04 | 0.85 |
| 70/30 | 99.6 | 85 | 66 | 59 | 4.14 | 1.42 | 0.83 | 0.67 |
| 60/40 | 99.5 | 82 | 57 | 51 | 3.97 | 1.28 | 0.64 | 0.54 |

*Notes:*
* with 5% binder phenolic resin, pellet ch. wt. = 2 g, chamber volume = 1.8 $m^3$ RH = 55%

**TABLE 12.3**

**Obscuration of Composition RP/KNO$_3$(80/20) at Different Humidity Conditions [2]**

| RH | Obscuration % | | | |
|---|---|---|---|---|
| | 0.4-0.7 μm | 2-2.4 μm | 3-5 μm | 8-14 μm |
| 15 (low) | 94 | 82 | 70 | 45 |
| 55 | 99.8 | 90 | 75 | 68 |
| 85 (High) | 100 | 100 | 88 | 76 |

*Note*: Charge weight 2 g, chamber volume 1.8 $m^3$, relative humidity 55%

It can be seen from Table 12.2 and Table 12.3 that

a. Obscuration is almost 100% in visual radiation
b. Extinction coefficient decreases as wavelength increases from visual to infrared region
c. Obscuration percent decreases as wavelength increases from visual to infrared region
d. It is easier to obtain obscuration in visual range than in infrared range

## 12.2.6 YIELD

Some smoke compositions based on white phosphorous, red phosphorous and hexachloroethane produce combustion products that absorb moisture from air and increase the dimensions of smoke aerosol particles, enabling better absorbance, scattering and reflection properties, thereby improving screening efficiency and enhancing the yield. As higher temperature will have reduced humidity in atmosphere, a low flame temperature of smoke composition is desirable.

The efficiency of a smoke is determined by its yield factor (mass of smoke particles dispersed per unit mass of payload)

$$\text{Yield factor}(y) = \frac{\text{Mass of aerosol(obscurant) in air by smoke}}{\text{Mass of material before emission of smoke}} \quad (12.11)$$

A yield factor greater than unity shows that the combustion products have absorbed moisture from air. Therefore, yield will be high in higher humidity.

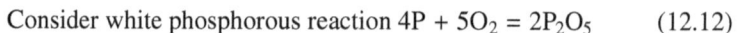

$$\text{Consider white phosphorous reaction } 4P + 5O_2 = 2P_2O_5 \quad (12.12)$$

The 123.90 g of phosphorous (tetrahedral P4 molecule) on combustion in air produces 284 g of phosphorous pentoxide, a yield factor of 2.291 and this increases considerably with increasing humidity as it reacts with humidity as under.

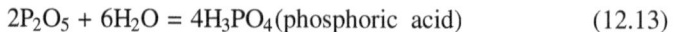

$$2P_2O_5 + 6H_2O = 4H_3PO_4(\text{phosphoric acid}) \quad (12.13)$$

Molecular weight of phosphoric acid is 97.994. Thus, one unit of phosphorous (tetrahedral P4 molecule) gives $4 \times 97.994/123.9 = 3.163$ units of phosphoric acid as aerosol which is high.

The yield factor for hexachloroethane and white phosphorous are humidity dependent and increase substantially with increase in humidity as in Table 12.4.

White phosphorous and red phosphorous have higher total obscuration power, higher yield factor and higher efficiency and hence are better smoke screening agents than the hexachloroethane-based compositions, but are associated with drawback such as pillaring effect (as combustion temperatures are around 850°C) and problem associated with handling and storage.

**TABLE 12.4**
**Yield Factor and Yield** [3]

| Humidity (%) | 100 g hexachloroethane (70% efficient) | | 100 g white phosphorous (100% efficient) | |
|---|---|---|---|---|
| | Yield factor | Yield (g) | Yield factor | Yield (g) |
| 10 | 1.46 | 102 | 3.53 | 353 |
| 20 | 1.52 | 106 | 3.72 | 372 |
| 30 | 1.59 | 111 | 3.91 | 391 |
| 40 | 1.73 | 121 | 4.11 | 411 |
| 50 | 1.89 | 132 | 4.34 | 434 |
| 60 | 2.11 | 148 | 4.65 | 465 |
| 70 | 2.40 | 168 | 5.10 | 510 |
| 80 | 3.25 | 228 | 5.88 | 588 |
| 90 | 5.72 | 400 | 7.85 | 785 |

## 12.2.7 FIGURE OF MERIT

Figure of merit is used to ascertain composition and device efficiency/performance for obscuration. The various figures of merits are as under.

a. *Mass-based Figure of Merit* $(FM_m)$ $(m^2.g^{-1})$ [4]:

Here, the transmittance is related to the initial mass of the composition. Mass-based figure of merit $FM_m$ $(m^2.kg^{-1})$ of smoke is obtained by multiplying extinction coefficient $\alpha_m$ $(m^2.kg^{-1})$ with yield factor.

$$\text{Mass} - \text{based figure of merit } (FM_m) = \text{mass extinction coefficient X} \quad \text{yield factor} \tag{12.14}$$

This is determined by producing smoke aerosols from 1 g to 10 g of smoke composition $(m_c)$ in a smoke chamber of volume $(V)$ 50–90 $m^3$ and observing the transmitted light through a radiometer, after homogenising the chamber atmosphere through a mixing fan. The attenuation of incident light radiation by smoke particles is observed as decreased transmitted light intensity.

A small volume of smoke chamber is allowed to pass through filters and deposited aerosol particle mass $(m_a)$ and dimensions are recorded. The mass extinction coefficient $(\alpha_m)$ is calculated using Beer–Lambert law since the concentration of aerosol particles and the path length of light is known.

From (12.3), $\ln(I_t/I_0) = (-) \alpha_m c\ L$

$$\alpha_m = [-\ln \ (I_t/I_0)]/cL = [-\ln \ (\tau)]/cL \ \text{ but } \ c = m_a/V$$

$$\alpha_m = \frac{[-V \ \ln(\tau)]}{m_a L} \tag{12.15}$$

The yield factor is calculated as $y = m_a/m_c$

where $m_a$ is mass of aerosol and $m_c$ is the mass of the initial composition.

The mass-based figure of merit $FM_m$ (showing relationship with mass of smoke composition $(m_c)$) is calculated as

$$FM_m \ = \ [\alpha_m] \ [y] \ = \ \left[\frac{-V \ \ln (\tau)}{m_a L}\right] \ x \ \frac{ma}{mc} \ = \ \left[\frac{-V \ \ln (\tau)}{m_c L}\right] \tag{12.16}$$

where $V$ = volume of smoke chamber (m$^3$), $\tau$ = transmittance = $I_t/I_0$

$m_c$ = mass of smoke composition (g), $m_a$ = mass of aerosol particles produced inside chamber(g)

$L$ = length of travel of incident light inside smoke chamber (m)

$FM_m$ is used to measure effectiveness of different smoke compositions for obscuration.

b. *Volume-based Figure of Merit (FMv)* (m$^2$.cm$^{-3}$) [4]:

Here the transmittance is related to the initial volume of the composition.

Multiplying $FM_m$(m$^2$/g) by the density of the composition $\rho_c$(g/cm$^3$), one can obtain volume-based figure of merit $(FM_v)$ as under.

$$FM_v = FM_m \rho_c = \frac{[-V \ \ln \ (\tau)] \ \rho_c}{m_c L} = \frac{[-V \ \ln \ (\tau)]}{[m_c/\rho_c]L}$$

$$FM_v \ = \ = \frac{[-V \ \ln \ (\tau)]}{v_c \ L} \tag{12.17}$$

where $v_c$ (cm$^3$) is the volume of the composition.

The ratio $FM_v/FM_m$ is the density (g.cm$^{-3}$) of the composition.

Thus, the overall volume-based effectiveness of a smoke in a payload $FM_v$ (m$^2$.cm$^{-3}$) depends upon the mass extinction coefficient $\alpha_m$ (m$^2$.g$^{-1}$), the yield factor $(y)$ of smoke and the density $\rho$ (g/cc) of the obscurant material (since dense smoke composition would provide more aerosol particles) and may be expressed as

Obscuration effectiveness of payload $FM_v = \alpha_m Y \rho$ (12.18)

Some typical values of figure of merit in visual range are given at Table 12.5.

**TABLE 12.5**
**Typical Values of Figure of Merit in the Visual Range [1]**

| Composition | α (m²/g) | FMv | RH% |
|---|---|---|---|
| Red phosphorous | | 28 | 20 |
| | | 129 | 90 |
| Zinc oxide/aluminium/hexachloroethane | 3.0 | 14 | 50 |
| Terephthalic acid smoke | 4.8 | 3 | 35 |
| Mekon white wax/potassium chloride/sucrose/celite/sodium bicarbonate | 6.0 | 3 | 35 |
| Red phosphorous/manganese dioxide/magnesium/zinc oxide/sodium bicarbonate/linseed oil | 4.4 | 12 | 35 |
| Red phosphorous/manganese dioxide/magnesium/zinc oxide/potassium chloride /linseed oil | 4.6 | 13 | 35 |

c. *Device-based Figure of Merit* [4]:

There are two device-based figures of merit as under.

i. *Device Mass-based Figure of Merit ($FM_{md}$):* Multiply mass-based figure of merit ($FM_m$) by fill fraction ($F_m$) which is the ratio of composition mass to devise mass $m_d$ (g)

$$FM_{md} = FM_m F_m = [\alpha_m] \ [y] \ F_m$$
$$= \frac{[-V \ \ln (\tau)]}{m_d L} \qquad (12.19)$$

ii. *Device Volume-based Figure of Merit ($FM_{vd}$):* Multiply volume-based figure of merit ($FM_v$) by fill fraction ($F_m$) which is the ratio of composition volume to devise volume $v_d$ (cm³)

$$FM_{vd} = FM_v F_v = [\alpha_m] \ [y] \ \rho_c F_v$$
$$= \frac{[-V \ \ln (\tau)]}{v_d L} \qquad (12.20)$$

Readers may refer [4] for more details.

## 12.3 FACTORS AFFECTING SCREENING EFFECT

The overall effect of suspended smoke particles/aerosols in formation of a suitable screen depends on the following factors:

a. The type, aerosol particle size, shape, density, refractive index, concentration and duration of smoke particles, which determine its light absorption, reflection and scattering ability
b. Wavelength of natural light, as it determines smoke cloud reflection towards observer
c. Whether it is a day or night
d. The brightness and colour of the background during the day and its contrast with smoke
e. Obscuration by natural particles in the atmosphere like presence of dust particles due to maneuvering of vehicles and artillery fire
f. The atmospheric conditions like humidity (moisture), rain, fog, snow, etc. and ambient temperature
g. The wind stability and wind speed and direction during observation
h. Whether the smoke canisters fall on dry land or marshy land or jungle or open area
i. The position of the observer with respect to sun and the smoke cloud

Initially, the smoke screen builds up due to aerosols diffusing in air and absorbing moisture from atmosphere, thereby having maximum obscuration. But in due course, because of the decrease in aerosol temperature and further diffusion in air causing lower concentration of aerosol, the obscuration starts reducing. The important features of a good screening smoke are:

a. Good obscuration even at lower concentrations so that its intended purpose is served in visual, infrared and bispectral (both regions)
b. Smoke for long duration (persistence) to provide enough time for obscuration, needing less ammunition for continued obscuration
c. High volume for effect on a large area
d. High density for effective obscuration
e. Non-toxic properties

## 12.4   PRODUCTION OF SCREENING SMOKE

The most commonly used ingredients for white smoke are white phosphorus (WP), red phosphorus (RP), hexachloroethane (HC) and fog oil (standard grade fuel no. 2 also known as fog SGF2). The former three are hygroscopic as they absorb water vapor from the air causing an increase in their diameters making them more efficient reflectors and scatterer of light rays. Fog oils are non-hydroscopic and depend on vaporisation techniques to produce extremely small-diameter droplets to scatter light rays.

The screening smoke may be formed as under.

## 12.4.1 Oil

The mechanism is that when mineral oil is allowed to pass through a hot manifold, it causes oil to vaporise and as it comes into contact with air outside, it condenses, producing aerosols as smoke.

## 12.4.2 Carbon Tetrachloride

Carbon tetrachloride forms zinc chloride through chemical combustion with zinc to give white smoke

$$2Zn + CCl_4 = 2ZnCl_2 + C \qquad (12.21)$$

The heat vaporises zinc chloride. Zinc chloride is hygroscopic and forms smoke particles of zinc chloride droplets as aerosol. However, the smoke is greyish due to presence of carbon. However, carbon tetrachloride is currently not used as newer and better compositions have been discovered. Some typical compositions are given in Table 12.6.

## 12.4.3 White and Red Phosphorous

White phosphorus (consist of tetrahedral P4 molecule) is colourless, transparent and a very good smoke producer. Commercial phosphorous is waxy white and with exposure to light, turns yellowish. Hence both terms white as well as yellow are used. White phosphorous reacts with air to form phosphorus pentoxide and this further reacts with water molecules in air to form phosphoric acid.

**TABLE 12.6**
**Typical Carbon Tetrachloride-Based Smoke Compositions [5]**

| Ingredients | Percentage | | |
|---|---|---|---|
| | PN 325 | PN 327 | PN 436 |
| Carbon tetrachloride | 41 | 46 | 40 |
| Zinc dust | 35 | 37 | 34 |
| Sodium chromate | 2 | 2 | 2 |
| Ammonium chloride | 2 | 2 | – |
| Magnesium carbonate | – | – | – |
| Zinc oxide | 20 | – | 24 |
| Kieselguhr | – | 13 | – |
| Calcium silicide | – | – | – |

$$4P + 5O_2 = 2P_2O_5 \text{(phosphorous pentoxide)} \qquad (12.22)$$

$$P_2O_5 + 3H_2O = 2H_3PO_4 \text{(phosphoric acid)} \qquad (12.23)$$

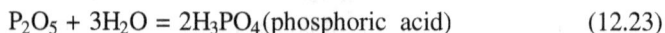

The more the humidity in air, the more the smoke screen would be dense. However, white phosphorous has several drawbacks, which make it difficult to produce and handle for the reasons listed below.

i. It readily reacts with air forming smoke and heat and hence it needs to be produced in absence of air. (It is kept under water as raw material.)
ii. It is dangerous if the composition leaks from the ammunition in storage as it would lead to fire due to large amount of heat produced by its combustion with air. A number of cases of fire have been reported during storage of white phosphorous-based composition.
iii. White phosphorous reacts violently and the heat of combustion is high, resulting in *"pillaring effect"* as smoke screen moves upwards due to convection currents which carry the smoke screen upwards. Thus, the screening effect is lost in a short period.

To avoid such quick combustion at a time, resulting in high temperature and pillaring effect, white phosphorous is used as plasticised white phosphorous by mixing granulated white phosphorous with rubber gel (manufactured by mixing rubber and xylene, etc.). Plasticised white phosphorous has the following advantages:

a. Reduces the heat of combustion
b. Reduces the fragmentation (due to plasticity), thereby reduces the rate of burning
c. Reduces pillaring effect of the smoke

A typical white phosphorous smoke composition is given in Table 12.7.

The best part of white phosphorous is that it does not use any oxidiser and takes oxygen from air and hence large quantities of white phosphorous can be used.

Red phosphorous (made by extended heating of white phosphorous at $220°$–$280°C$) based compositions are used for smoke and fire purposes. Red phosphorous is stabilised for use in pyrotechnic composition. Red phosphorous with

**TABLE 12.7**

**Typical White Phosphorous Smoke Composition [6]**

| Ingredients | Percentage |
|---|---|
| White phosphorous | 65 |
| Plasticiser (neoprene 100 parts, carbon 76 parts xylene 44 parts, litharge 15 parts) | 35 |

impurities of copper and iron in presence of moisture decays fast. Red phosphorous during long storage/ageing reacts with moisture, produces phosphorous acids and phosphine. Red phosphorous has some drawback:

i. Phosphine is toxic and corrodes metals
ii. Phosphorous acid coating on composition affects its ease of ignition

The differences between white or yellow and red phosphorous is given in Table 12.8. Advantages of red phosphorous over white phosphorous are:

i. It produces dense smoke compared to white phosphorous.
ii. It is less toxic than white phosphorous.
iii. It can burn in pots, in shells duly pressed and as pellets.
iv. It acts as incendiary ammunition as an additional feature.

Both white and red phosphorous is hazardous for manufacturing smoke composition due to contact with atmosphere and friction, respectively. Red phosphorous smoke has high mass extinction coefficient and very good absorption of infrared radiation and therefore found useful in attenuation in mid IR (3–5 μm) and far IR (8–14 μm) wavelength bands.

**TABLE 12.8**
**Differences Between White/Yellow and Red Phosphorous**

| Parameters | White/yellow phosphorous | Red phosphorous |
|---|---|---|
| Storage | Under water | Normal |
| Appearance | Translucent soft waxy solid | Hard and dark red crystalline solid |
| Odour | Garlic smell | No odour |
| Toxicity | Toxic | Non toxic |
| Reactivity | Highly reactive | Less reactive |
| Phosphorescence | Shows phosphorescence | No phosphorescence |
| Ignition temperature | Low ($34^0$c), burns easily in air and moisture | High ($240^0$C), does not burn easily in air and moisture |
| Solubility in carbon disulphide | Soluble | Insoluble |
| Melting point | $44.15^0$c | Approx. $587^0$C |
| Reaction with $Cl_2$ | Burns easily with $Cl_2$ forming $PCl_3$ and $PCl_5$ easily | Reacts with $Cl_2$ only on heating |
| Composition filling | Pour filling | As pellets |
| Sensitivity | Highly sensitive to air and moisture | Low sensitive to air and moisture and highly sensitive to friction |
| Special effects | Gives smoke | Gives dense smoke and incendiary effect |

Reviews on red phosphorous by Dr Davies [7] in 1999 and on phosphorous by Ernst Christian Koch [8] in 2008 provide detailed description of the studies on phosphorous-based compositions.

### 12.4.4 HEXACHLOROETHANE

The hexachloroethane-based compositions have advantages over white phosphorous-based composition as *"pillaring effect"* is not pronounced due to low heat of combustion.

The reaction with hexachloroethane and hexachlorobenzene are:

$$C_2Cl_6 + 3Zn = 3ZnCl_2 + 2C + \text{Heat} \qquad (12.24)$$

$$C_6Cl_6 + 3Zn = 3ZnCl_2 + 6C + \text{Heat} \qquad (12.25)$$

Zinc chloride partially gets hydrolysed with moisture in the air and provides grey smoke due to presence of carbon. Addition of zinc oxide provides whiteness to smoke since carbon liberated is consumed by zinc oxide as under.

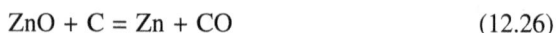

$$ZnO + C = Zn + CO \qquad (12.26)$$

The zinc formed from zinc oxide further reacts with chlorinated carbon compound. There are three types of hexachloroethane smoke.

i. *Hexachloroethane type A:*

Consists of hexachloroethane ($C_2Cl_6$), zinc (Zn) dust, potassium perchlorate ($KClO_4$), ammonium chloride ($NH_4Cl$) as in Table 12.9.

**TABLE 12.9**
**White Smoke type A compositions [9]**

| Ingredients | Percentage | | | |
|---|---|---|---|---|
| | Faster burning upper increment composition | | Main composition | |
| | formula 126 | formula 126′ | formula 127 | formula 127′ |
| Hexachloroethane | 43 | 46.5 | 44 | 46.5 |
| Zinc dust | 36 | 38.5 | 36 | 38.5 |
| Ammonium chloride | 6 | 3 | 10 | 6 |
| Ammonium perchlorate | 15 | – | 10 | – |
| Potassium perchlorate | – | 12 | – | 9 |

**TABLE 12.10**
**White Smoke Type B Composition** [9]

| Ingredients | Formula 128 Percentage |
|---|---|
| Hexachloroethane | 45.5 |
| Zinc oxide | 45.5 |
| Calcium silicide | 9 |

These older mixtures of AN-M8 Grenade consist of a faster burning upper increment (126 or 126') and a main charge of the slower burning mixture (127 or 127'), each giving a choice for the use of oxidiser.

ii. *Hexachloroethane type B:*

Consists of hexachloroethane ($C_2Cl_6$), zinc oxide (ZnO) and calcium silicide ($CaSi_2$) as given in Table 12.10

The hexachloroethane reacts with zinc oxide to form zinc chloride, which being volatile and hygroscopic, reacts with water molecules in the air forming a dense cloud. The main reaction is

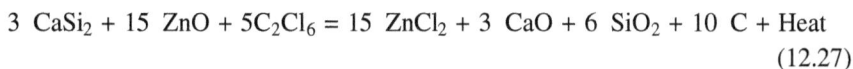

$$3\ CaSi_2 + 15\ ZnO + 5C_2Cl_6 = 15\ ZnCl_2 + 3\ CaO + 6\ SiO_2 + 10\ C + Heat \tag{12.27}$$

iii. *Hexachloroethane type C:*

Consists of hexachloroethane ($C_2Cl_6$), zinc oxide (ZnO) and aluminium (Al) as given in Table 12.11. The reaction proceeds as under.

$$2Al + C_2Cl_6 + 3ZnO = 3ZnCl_2 + Al_2O_3 + 2\ C + Heat \tag{12.28}$$

**TABLE 12.11**
**White (Gray) Smoke Type C Composition** [9]

| Ingredients | Formula 129 Percentage |
|---|---|
| Hexachloroethane MIL-H-235 | 44.5 |
| Zinc oxide MIL-Z-291C Grade A, Class 1 | 46.5 |
| Aluminium powder MIL-A-512A, Type II Grade C, Class 4 | 9 |

**TABLE 12.12**
**Typical White Smoke Compositions**

| Ingredients | Percentage | | | | | |
|---|---|---|---|---|---|---|
| | PN83M | SR269A(M) | SR264 A (M) | PN800 | [6] | SR264A [5] |
| Hexachloroethane | 44 | 44 | 44.5 | 44 | 45.5 | 45 |
| Potassium nitrate | 2 | 2 | 2 | 2 | – | – |
| Zinc oxide | 35 | 40.5 | 44.5 | 47-30 | 47.5 | 45 |
| Calcium silicide | 19 | 13.5 | 9 | 7-24 | – | 10 |
| Aluminium grained | – | – | – | – | 7 | – |
| Zinc stearate | – | – | – | 0.3 | – | – |

A reduction in aluminium percentage leads to increase in burn time with less formation of carbon and thus the smoke becomes brighter.

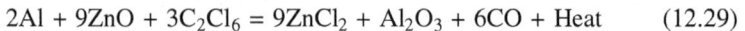

$$2Al + 9ZnO + 3C_2Cl_6 = 9ZnCl_2 + Al_2O_3 + 6CO + Heat \qquad (12.29)$$

There are a number of screening smoke compositions (Table 12.12) containing hexachloroethane and other ingredients.

The burn rate may be increased by an increase in calcium silicide.

### 12.4.5 ANTHRACENE

Anthracene ammonium chloride-based screening smoke composition is used for the purpose of practice, as the composition is less carcinogenic (Table 12.13).

### 12.4.6 TRAINING SMOKE COMPOSITIONS

Some training smoke compositions are given under Table 12.14.

**TABLE 12.13**
**Practice Smoke Compositions [10]**

| Ingredients | Percentage |
|---|---|
| Ammonium chloride | 50 |
| Potassium chlorate | 20-30 |
| Naphthalene or anthracene | 20 |
| Wood charcoal | 0-10 |

**TABLE 12.14**
**Training Smoke Compositions [11]**

| Ingredients | UK Standard PN 907 | US Standard | UK Standard (with TA) |
|---|---|---|---|
| Potassium chlorate | 26 | 23 | 26 |
| Lactose | 26 | – | 26 |
| Sugar | – | 14 | – |
| Cinnamic acid | 33 | – | – |
| Terephthalic acid | – | 56 | 33 |
| Kaolin | 15 | – | 15 |
| Magnesium carbonate | – | 3 | – |
| Stearic acid | – | 3 | – |
| Poly (vinyl alcohol) | – | 1 | – |
| Yield | | | |
| Yield calculated | 0.33 | 0.56 | 0.33 |
| Yield measured | 0.28 | 0.19 | 0.23 |

Both cinnamic acid (m. p. 133°C) and terephthalic acid (sublimates at 300°C) undergoes sublimation and recondensation process exhibiting white smoke but are non-hygroscopic and hence are less effective than HC or Phosphorous smokes.

## 12.4.7 NON-TOXIC SMOKE COMPOSITIONS

The existing effective smoke ingredients have a deleterious effect as displayed in Table 12.15.

Hence a suitable non-toxic material is necessary. A new non-toxic, non-hazardous, benign boron carbide-based smoke composition containing $B_4C/KNO_3$ as a pyrotechnic fuel/oxidiser paired with KCl as a diluent and calcium stearate as a

**TABLE 12.15**
**Deleterious Effect of Smoke Ingredients**

| Ingredients | Deleterious effect |
|---|---|
| White phosphorous | Toxic, incendiary |
| Red phosphorous | Aging and sensitivity issues |
| Hexachloroethane | Toxic |
| Titanium tetrachloride | Corrosive |

**TABLE 12.16**

**Non-Toxic Smoke Compositions (Compiled) [4]**

| Ingredient | Percentage | | | | |
|---|---|---|---|---|---|
| Boron carbide [B$_4$C] AEE 1500 Grit | 13 | 13 | 13 | 13 | 13 |
| Potassium nitrate | 58 | 58 | 58 | 58 | 58 |
| Potassium chloride | 25 | 25 | 25 | 25 | 25 |
| Calcium stearate | 2 | 2 | 2 | 2 | 2 |
| Binder polyacrylate elastomer | 2 | – | – | – | – |
| Binder nitrocellulose | – | 2 | – | – | – |
| Binder polyvinyl alcohol | – | – | 2 | – | – |
| Binder polyvinyl acetate | – | – | – | 2 | – |
| Binder epoxy | – | – | – | – | 2 |
| | Performance | | | | |
| Average burning time (second) | 73.5 | 29.5 | 41.0 | 51.0 | 51.0 |
| *Yield* | 0.53 | 0.74 | 0.63 | 0.75 | 0.69 |
| $\alpha_m$ (m$^2$.g$^{-1}$) | 3.88 | 3.54 | 3.57 | 3.50 | 3.55 |
| $FM_m$ (m$^2$.g$^{-1}$) | 2.07 | 2.62 | 2.24 | 2.64 | 2.44 |

burning rate modifier has been reported [4]. A large number of polymeric binders for granulation have been proposed. Burn rates may be controlled by suitable choice of B$_4$C particle size and the amount of calcium stearate to obtain burning times over a wide range (24–100 seconds) for grenades (Table 12.16).

Readers may refer [12] for non-toxic environmentally acceptable pyrotechnic smoke compositions.

## 12.5 SPECIAL REQUIREMENTS OF SCREENING SMOKE COMPOSITIONS

In addition to the general essential requirements of compositions as given in Section 1.6, screening smoke compositions should preferably possess the following special requirements:

a. The composition should produce a maximum amount of finely dispersed solid or liquid aerosols, with a negligible amount of flame.
b. The composition should also produce maximum gaseous decomposition products to expel aerosols.
c. The temperature of decomposition of the composition should be less so that flaming of composition is avoided.
d. The products of the pyrotechnic composition (aerosols) should be hygroscopic so as to absorb atmospheric moisture and maximise the mass of the aerosols in air.

e. The smoke composition should produce desired volume and density (concentration) of aerosols in all weather conditions.

f. The smoke produced should be stable without much dispersion and rise or fall
   and have good obscuration power even in low concentration and stable in air.

g. The composition should produce negligible slag, which should be porous
   enough, to allow smoke to pass through it to the atmosphere.

## 12.6  INFRARED ATTENUATION BY SMOKE

In a war situation, infrared radiation from tanks and guns is a vital source of detection by enemy through infrared detection devices and subsequent attack of infrared missiles. There are two options to reduce infrared radiation so as to reduce
exposure to enemy as given below.

### 12.6.1  REDUCING TARGETS (LIKE TANK'S, GUN'S)
### INFRARED EMISSION INTENSITY

The intensity of emitted infrared radiation depends upon temperature of the object
and higher the temperature of the tank/gun, higher the intensity of infrared radiation
and this can be reduced by several methods like dissipation of heat by air convection, infrared camouflage paint, heat insulation layer, hot exhaust gas cooling
system, infrared camouflage net and shielding etc.

### 12.6.2  INFRARED ATTENUATING SMOKE

The infrared smoke prevents the infrared radiation from the targets like tank, guns
etc. to reach the detectors of IR sensors by selectively absorbing, reflecting and
scattering these infrared radiations. Thus infrared smoke can conceal the infrared
emission of the target and thus reduce the effectiveness of infrared-guided missiles.

#### 12.6.2.1  Infrared Transmittance

The infrared transmittance may be determined as below [13].

The fundamental equation of smoke transmittance is shown below, according to
the theory of infrared radiation transmission

$$\tau smk = \frac{Lre}{Ltar} \tag{12.30}$$

where $\tau smk$ means smoke transmittance, $Ltar$ refers to the infrared radiance of the
target that the infrared detector receives before the attenuation of smoke and $Lre$
stands for the residue target radiance that infrared detector receives under the attenuation of the smoke screen.

Theoretically, smoke transmittance can be accurately calculated using Equation
(12.30). However, it is impractical to measure $Ltar$ and $Lre$ directly in the field trial
due to environmental factors. As evident, every non-zero temperature object in

nature will emit electromagnetic energy spontaneously. Thus, targets, background and smoke cloud are all electromagnetically radiant. The incident radiation that thermal infrared imager receives is the total amount of radiation in line of sight, and the infrared detector cannot quantitatively distinguish the amount of radiation from the target. To obtain the true value of *Ltar* and *Lre* in Equation (12.30), some other parameters are introduced to represent them as below.

The total radiance that the infrared detector receives before smoke interference in line of sight can be obtained from

$$L(r0) = Ltar + Lsur \tag{12.31}$$

where $L(r0)$ stands for the total radiance that the infrared detector receives before smoke interference in line of sight. This not only includes the radiation from the target *Ltar* but also consists of radiation from the surrounding *Lsur*, such as the background, the sun and the sky.

The total radiance that the infrared imager receives after the release of smoke in line of sight can be obtained from

$$L(r) = L'tar + L'sur + Lsmk \tag{12.32}$$

where $L(r)$ refers to the total radiance that the infrared imager receives after the release of smoke in line of sight, and it contains the target remaining radiance after attenuation by the smoke *L'tar* and surroundings radiance *L'sur*. In addition, the radiation from the smoke *Lsmk* is also included.

Substituting Equations (12.31 and (12.32) into equation (12.30), the smoke transmittance calculation equation is expressed as follows:

$$\tau smk = \frac{L(r) - L'sur - Lsmk}{L(r0) - Lsur} \tag{12.33}$$

For the purpose of calculating smoke transmittance, five radiance variables in Equation (12.33) have to be determined first. With the field trial images provided by the infrared imager, the gray value of each pixel in the image can be extracted by an image processing algorithm. Consequently, a method of converting the radiance to a corresponding gray value in infrared images has been proposed. The detailed procedure to calculate smoke transmittance may be seen in [13].

The schematic diagram of field trial is shown in Figure 12.5.

Effective shielding area is a crucial indicator for the evaluation of the infrared smoke-obscuring effectiveness on the battlefield. Readers may see [14] for details.

A few pyrotechnic smoke compositions with percentage attenuation and emission of IR intensity at two wavelengths 4.9 and 10.6 microns [15] are given in Table 12.17.

Readers may see reference [16] for red phosphorous-based infrared smoke compositions, which can obscure in both the mid IR (3–5 μm) and far IR (8–14 μm) wavelength bands.

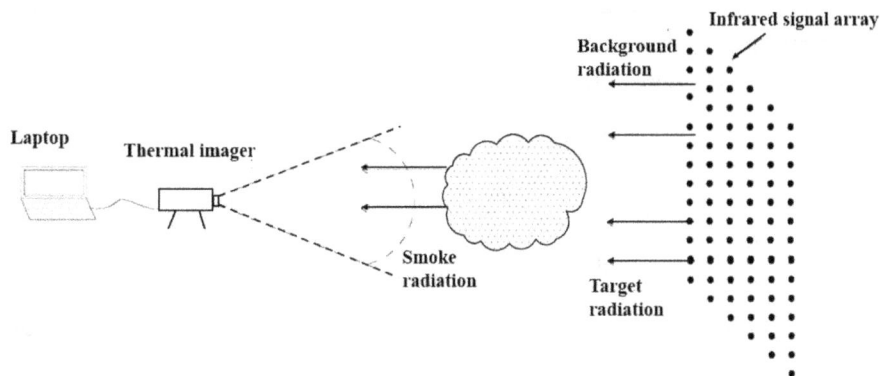

**FIGURE 12.5** Schematic Diagram of Field Trial for Smoke Transmittance (Reprinted from [13]).

**TABLE 12.17**

**Percentage Attenuation and Emission of Infrared Radiation by Pyrotechnic Smoke Compositions [15]**

| Composition | Percentage attenuation | | | Emission at | |
|---|---|---|---|---|---|
| | 2-2.4μm | 3-5μm | 6-13μm | 4.9 μm | 10.6 μm |
| HCE/Mg/Naphthalene 62/15/23 | 95 | 81 | 33 | 751 | 1,452 |
| HCE/Mg/Anthracene 62/23/15 | 94 | 80 | 39 | 517 | 1,208 |
| HCE/Anthracene/KClO$_4$ 30/30/40 | 95 | 81 | 36 | 689 | 1,232 |
| HCB/Mg/Naphthalene 62/15/23 | 95 | 88 | 38 | 821 | 1,602 |
| HCE/Mg//Naphthalene/ZnO 62/15/23/2pts | 92 | 76 | 42 | 728 | 1,476 |
| HCB/Mg-Al/ZnO 73/16/11 | 98 | 96 | 39 | 979 | 1,373 |
| NH$_4$Cl/Anthracene/KClO$_3$ 45/15/40 | 73 | 50 | 7 | 303 | 855 |
| HCE/Al/CaSi$_2$/Si 75/5/10/10 | 52 | 43 | 25 | 307 | 1,151 |
| HCE/Fe$_2$O$_3$/ZnO/Al 40/7/38/15 | 50 | 26 | 6 | 357 | 1,044 |
| HCE/CaSi$_2$/KNO$_3$/ZnO 44/13.5/2/40.5 | 60 | 29 | 6 | 366 | 942 |
| HCE/TiO$_2$/Al 65/28/07 | 21 | 13 | 12 | 306 | 943 |
| HCE/ZnO/Al 45/47/08 | 57 | 25 | 12 | 432 | 999 |

The US patent publication [17] reports slow infrared compositions (Table 12.18) which makes it possible to camouflage a target by preventing the emission of infrared radiance being transmitted by it to make it impossible to detect with a thermal camera. The hexachloroethane and hexachlorobenzene act as oxidants as well as carbon particle producers while naphthalene and anthracene act to produce carbon particles at a temperature above 1,000°C.

**TABLE 12.18**
**Infrared Smoke Compositions** [17]

| Ingredients | Parts | | | | | | |
|---|---|---|---|---|---|---|---|
| Magnesium | 20 | 20 | 20 | 20 | 18.5 | 20 | 20 |
| Hexachloroethane | – | – | – | – | 61.5 | – | – |
| Hexachlorobenzene | 80 | 70 | 70 | 70 | – | 80 | 80 |
| Naphthalene | 10 | 10 | 10 | 10 | 30 | – | – |
| Vinylidene polyfluoride | 10 | – | 5 | 10 | 20 | – | 20 |
| Neoprene | – | 5 | – | – | – | – | – |
| Chlorinated paraffin | – | – | – | – | 20 | – | – |
| Polyvinyl acetate | – | – | – | – | – | 20 | – |

It has been reported [18] that chemical smoke mixture based on hexa-chloroethane consisting of: [20% magnesium powder, 60% hexachloroethane, 15% naphthalene and 5% poly vinyl chloride] could provide relatively high level of thermal attenuation in visual and IR (8–12 μm) range. Readers may see [19] on bursting smoke as an infrared countermeasure.

## 12.7  CASTABLE SCREENING SMOKE COMPOSITION

Castable smoke has several advantages (see Section 13.3) as it can be made in any shape or form, and does not need any press or mould. Readers may refer [20] on manufacture of the castable type of screening smoke composition, reference [21] on environmental screening and ranking for army smoke and obscurants and reference [22] on symposium of smoke/obscurants.

## REFERENCES

[1] Nigel Davies, "*Pyrotechnics Handbook*", Cranfield University, Defence College of Management and Technology, Department of Materials and Applied Science, October 2008. UK. ©Dr N Davies (2008).
[2] G.K. Gautam, A.D. Joshi, S.A. Joshi, P.R. Arya, and M.R. Somayajulu, "*Radiometric Screening of Red Phosphorus Smoke for its Obscuration Characteristics*", Defence Science Journal, July 2006, Volume 56, Issue 3, Pages 377–381. ©2006, DESIDOC.
[3] *http://www.globalsecurity.org/wmd/library/policy/army/fm/3-6/3-6ch2.htm*.
[4] Anthony P. Shaw, Giancarlo Diviacchi, and Ernest L. Black, "*Advanced Boron Carbide-Based Visual Obscurants for Military Smoke Grenades*", 40th International Pyrotechnics Seminar, Colorado Springs, CO, 13–18 July 2014, Pages 170–191.
[5] *www.ammunitionpages.com/download/168/PYROTECHNIC%2520COMPOSITIONS.pdf*.
[6] "*Theory and Application*", AMCP 706-185, Engineering Design Handbook Military Pyrotechnic series, Part One, Headquarters, US Army Material Command, Washington D.C, April 1967.

[7] Dr. Nigel Davies, *"Red Phosphorus for Use in Screening Smoke Compositions"*, Report prepared for CESO (N) Contract Number NBSA5B/2701 Task No. 8, February 1999, Royal Military College of Science, Cranfield University, Shrivenham, Swindon, SN6 8LA, UK.

[8] Ernst-Christian Koch, *"Special Materials in Pyrotechnics: V. Military Applications of Phosphorus and its Compounds"*, Propellants, Explosives, Pyrotechnics, 2008, Volume 33, Issue 3, Pages165–176, ONLINE WILEY.

[9] Dr. Herbert Ellern, *"Military and Civilian Pyrotechnics"*, © Chemical Publishing Co, New York, 1968.

[10] A.A. Shidlovskiy, *"Principles of Pyrotechnics"*, 3rd Edition, Moscow, 1964. (Translated by Foreign Technology Division, Wright-Patterson Air Force Base, Ohio, 1974), American Fireworks News, 1 July 1997. ©1997, Rex E. &S. P., Inc.

[11] Dr. Nigel Davies, *"Pyrotechnics, Topics to be Covered in Course"*, PowerPoint Presentation, Cranfield University, Defence Academy of the United Kingdom, Shrivenham, Swindon, SN6 8LA, UK.

[12] Amarjit Singh, P.J. Kamale, and Haridwar Singh, *"Nontoxic/Environmentally Acceptable Pyrotechnic Smokes"*, Journal of Scientific & Industrial Research, 2000, Volume 59, Pages 455–459.

[13] R. Tang, T. Zhang, X. Wei, and Z. Zhou, *"An Efficient Numerical Approach for Field Infrared Smoke Transmittance Based on Grayscale Images"*, Applied Sciences, 2018, Volume 8, Page 40. © 2018 by the authors. 10.3390/app8010040.

[14] R. Tang, T. Zhang, Y. Chen, H. Liang, B. Li, and Z. Zhou *"Infrared Thermography Approach for Effective Shielding Area of Field Smoke Based on Background Subtraction and Transmittance Interpolation"*. Sensors, 2018, Volume 18, Page 1450. © 2018 by the authors. 10.3390/s18051450.

[15] B. R. Thakur, *"Infrared Intensity Measurement of Different Targets"*, Continued Education Programme on Instrumentation for Testing of Propellants, Pyrotechnics and Allied Devices, High Energy materials Research Laboratory, Pune, India, November 2003.

[16] Thomas M. Klapötke, *"Chemistry of High-Energy Materials"*, 5th Edition, 2019, ©Published by Walter de Gruyter GmbH & Co KG.

[17] Andre Espagnacq and Gerard Sauvestre, *"Pyrotechnical Composition which Generates Smoke that is Opaque to Infrared Radiance and Smoke Ammunition as obtained"*, US Patent No. 4724018 A 09.02.1988.

[18] F.A. Maryam, M.A. Kassem, M. Sh. Fayed, and A.M. Sultan, *"Preparation and Performance Evaluation of a Hexachloroethane-Based IR Chemical Smoke Mixture"*, Paper: ASAT-13-CA-02 13th International Conference on Aerospace Sciences & Aviation Technology, ASAT-13, 26–28 May 2009.

[19] Amarjit Singh, P.J. Kamale, S.A. Joshi, and L.K. Bankar, *"Bursting Smoke as an Infrared Countermeasure"*, ©Defence Science Journal, July 1998, Volume 48, Issue 3, Pages 297–301.

[20] Amarjit Singh, U.K. Patil, and P.K. Mishra, *"Manufacture of the Futuristic Castable Type of Screening Smoke Composition"*, Defence Science Journal, July 1986, Volume 36, Issue 3, Pages 257–264.

[21] Joseph H. Shinn, Stanley A. Martins, Patricia L. Cederwall, and Lawrence B. Gratt, *"Smokes and Obscurants: A Health and Environmental Effects Data Base Assessment, Phase I Report"*, March 1985, Lawrence Livermore National La boratory.

[22] Rush E.Elkins and R. H. Kohl, *"Proceedings of The Smoke/Obscurants"* Symposium V Unclassified Section Volume II, Technical Report DRCPM-SMK-T-OO1-81, April 1981, Pages 423–904.

# 13 Signalling Colour Smoke Compositions

## 13.1 ROLE

Signalling colour smoke ammunitions are used for various purposes like:

a. Communicating in a coded language by sending signal from ground, air or sea
b. Indicating position for tactical landing area
c. Marking and dropping zones
d. Tracking and acquisition (marking targets like recovery of training torpedoes or test flight of high velocity missiles)
e. Distress signal (search and rescue)
f. Wind speed and direction indicators
g. Colourful display in functions

It may be noted that coloured smoke signal is effective in communicating signals between ground to ground, ground to air, ground to sea, sea to ground, sea to air, sea to sea in moderate air current but not from air to ground or air to sea due to high relative motion of coloured smoke cloud due to high air currents at greater altitudes.

The compositions produce smoke of different colours and densities. Communication in war field is most vital as this gives vital clues required for tactical decision-making as well as communication to the battle force. Coloured smoke signals have been found to be very convenient as normal communication like line communication and wireless communication are affected/intercepted in a war field.

Pyrotechnic signal communications using coded languages may be altered and right communication to right person may be possible. It has been observed that smoke colours, namely red, orange, yellow and green, are easily identifiable and can be distinguished from each other. Hence these colours are mostly used in signalling smoke ammunitions. These signals may be sent through canisters, bombs and shells, which make an ideal communication/signalling tool.

These colours are visible on snow-bound areas very well. However, orange is most preferred for naval distress signal ammunitions since orange gives a very good colour contrast with grey/blue air and blue sea. Hence, these are used in a majority of the cases for distress signalling in the sea in all naval signalling ammunitions.

## 13.2 METHODS FOR PRODUCTION OF SIGNALLING COLOUR SMOKE

There are four methods of producing signalling colour smoke as enlisted below:

a. *Bursting*: Bursting of an explosive embedded inside a projectile containing signalling colour smoke composition
b. *Volatilisation and condensation*: Volatilisation and condensation of coloured dye leading to display of colour in air
c. *Dispersion*: Dispersion of finely powdered coloured dyes in air
d. *Chemical reaction*: Chemical reaction leading to formation of coloured smoke emission in air

The dispersion and chemical reaction do not produce the desired signalling colour smoke emission as these lead to dull colour and small quantity of smoke. Hence, the other two methods, namely bursting and volatilisation and condensation are mostly used.

## 13.2.1 MECHANISM OF BURSTING TYPE COLOURED SMOKE COMPOSITIONS

Ammunition (Figure 13.1) with signalling colour smoke compositions comprise a central detonating device containing tetryl (CE) or pentaerythritol tetranitrate (PETN). The percussion fuze initiates the detonating device causing bursting of composition and dispersion of dye in air.

However, the signalling colour smoke produced in this process does not persist for long and has pillaring tendency due to high heat generated during explosion of CE or PETN. They also get easily mixed with the sand or get embedded in the ground, due to percussion fuze, resulting in poor colour of the smoke or even failure. The only benefit is that the ammunition is more accurate and consistent and without dispersion.

Coloured smoke grenades can also be used. The disadvantage is that these emit a puff of colour smoke instantaneously and hence smoke emission time is short unlike canister types which provide smoke for prolonged time.

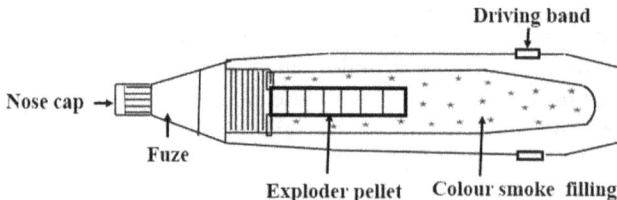

**FIGURE 13.1** Bursting Type Signalling Colour Smoke Shell.

## 13.2.2 MECHANISM OF VOLATILISATION AND CONDENSATION TYPE COMPOSITIONS

The fuel and oxidiser on combustion produce heat to sublime the dye as well as produce copious amount of gases to expel the dye in air. The dye does not take part in the reaction. The volatilisation of dye is an endothermic reaction. Hence the smoke is comparatively cooler than those of phosphorous smoke. Thus, pillaring effect is not exhibited by coloured smoke.

Such compositions (having 40% to 60% of dye), therefore, generate less heat and thus avoid decomposition of dye. The heat produced are just sufficient to volatilise the dye to evaporate in the air emitting colour as well as producing sufficient gaseous products to expel the dye. These gaseous products also help in minimising the agglomeration of dye particles and hence remain dispersed.

An excess quantity of dye is preferred as it acts as a coolant by taking away extra heat of combustion by its own sublimation, giving a dense coloured smoke. The granulation of the dye improves the bonding of the ingredients as well as improves the colour of the smoke and also increases the surface area of the composition for burning, thereby facilitating early smoke formation.

However, some compositions may produce flame at the reaction zone due to combustion of dye vapour, giving black smoke. This may be overcome by the use of flash inhibitors, like sodium bicarbonate ($NaHCO_3$) or potassium bicarbonate ($KHCO_3$), that produce carbon dioxide ($CO_2$) and lower the temperature of the composition as well as dilute the air and avoid direct contact of the dye vapour with air, avoiding flaming of the composition.

To avoid discoloration of the dye being allowed to pass through the composition having hot porous slags, the pellets and the containers are designed such that pellets are made with central hole and the volatised dye is allowed to pass through a central tube into the air, thereby avoiding more contact with hot porous slags.

Lactose, sucrose, dextrose, starch, etc., are used as fuel since these produce copious quantity of gases due to presence of C, H and O atoms in the molecule forming gaseous products CO, $CO_2$ and $H_2O$ and do not produce high heat. Potassium chlorate is chiefly used as an oxidiser.

The use of binders is also very important to ensure that the consolidated composition does not get cracked or become powdered due to shock of discharge, especially in case of gun ammunition.

These compositions need priming for pick-up of ignition. The priming composition generally used is SR214 containing potassium nitrate and lactose, producing very less heat.

As a large amount of pyrotechnic ccomposition are required in a container, the composition may be initially pressed as pre-pellets and then two or three pellets may be further consolidated with higher load.

An alternative is to make a paste of the composition with gum arabic and extrude the pellets in very small diameter and dry and pack them in the container. The molecular weight of potassium chlorate is 122.55 while molecular weight of lactose monohydrate is 360.31. The proportion of the constituents (oxidiser:fuel) may be varied to get different course of reaction with lactose 42.36 percent (eq. 13.1) and 26.87 percent (eq. 13.2) as under.

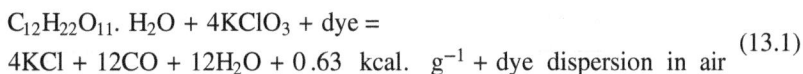

$$C_{12}H_{22}O_{11}. H_2O + 4KClO_3 + dye =$$
$$4KCl + 12CO + 12H_2O + 0.63 \text{ kcal. } g^{-1} + dye \text{ dispersion in air} \tag{13.1}$$

$$C_{12}H_{22}O_{11} \cdot H_2O + 8KClO_3 + dye =$$
$$8KCl + 12CO_2 + 12H_2O + 1.06 \ \text{kcal.} \ g^{-1} + dye \ \text{dispersion in air} \quad (13.2)$$

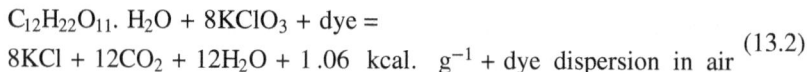

An increase in the ratio of oxidiser:fuel gives higher heat of combustion from $0.63 \ \text{cal.g}^{-1}$ to $1.06 \ \text{cal.g}^{-1}$ as it converts the CO to $CO_2$. Hence ratio of oxidiser: fuel as at (13.1) is preferred which gives lower heat of combustion and thus avoid decomposition of dye.These compositions are used in pots and canisters and assembled in base ejection type signalling colour smoke ammunition shells. These canisters are generally expelled by gunpowder during flight by time fuzes. The canisters pick up ignition on ejection and disperse over a large area and burn layer by layer giving out smoke for a longer duration. Various dyes are used to obtain the desired colour. Often, two dyes are used to obtain better results as single dye may not give desired colour. It may be stated that the colour a human eye perceives is a complimentary colour as dyes absorb a part of the visible spectrum and scatter the rest.

## 13.3  TYPICAL SIGNALLING COLOUR SMOKE COMPOSITIONS

Some typical signal smoke compositions are shown in Table 13.1 to Table 13.7.

A report on castable signalling colour smoke composition [4] shows that the range of concentration of binder should be between 17-20 percent of the composition. The report mentions that castable colour smoke compositions possess a number of inherent advantages over conventional pressed compositions. These are [4]

   a. *Improved mechanical properties and stability*: The susceptibility of such mixtures to shock and deterioration is decreased.
   b. *Improved safety in handling*: Due to wet processing technique, the processing personnel have reduced exposure to dust, and the risks arising from explosive dust are minimised. The compositions are also less sensitive to ignition by impact, friction or electrostatic discharge.

**TABLE 13.1**
**Orange Signalling Colour Smoke Compositions**

| Ingredients | Percentage | |
|---|---|---|
|  | PN 61 type II | PN 61 Type I |
| Dye oil orange | 65 | 58 |
| Potassium chlorate | 17.5 | 21 |
| Lactose | 17.5 | 19 |
| Magnesium carbonate *(Over and above) | 2 | 2 |
| Zinc stearate * (Over and above) | 0.5 | 0.5 |

## TABLE 13.2
### Red Signalling Colour Smoke Compositions [1]

| Ingredient | 1 | 2 | 3 | 4 | 5 | 6 |
|---|---|---|---|---|---|---|
| Dye red | 40 | 47.5 | 40.2 | 47 | 36 | 40 |
| Sodium bicarbonate | 25 | – | 14.3 | – | 1 | 22 |
| Potassium chlorate | 26 | 29.5 | 31.3 | 32 | 35 | 27.4 |
| Sulphur | 9 | – | 12.3 | – | – | 10.6 |
| Magnesium carbonate | – | 5 | – | – | – | – |
| Lactose | – | 18 | – | – | – | – |
| VAAR | – | – | – | 2 | – | – |
| Sugar | – | – | – | 29 | 26.5 | – |
| Asbestos powder | – | – | – | – | 1.5 | – |
| Dextrin | – | – | 1.9 | – | – | – |

## TABLE 13.3
### Green Signalling Colour Smoke Compositions [1]

| Ingredient | 1 | 2 | 3 | 4 | 5 | 6 |
|---|---|---|---|---|---|---|
| Dye yellow | 4 | – | 5.65 | 4.7 | 5 | 15.5 |
| Benzanthrone | 8 | – | – | 9.4 | 10 | – |
| Dye solvent green | 28 | 40 | 39.45 | 32.9 | 33 | 33 |
| Sodium bicarbonate | 22.6 | 24.6 | 14.75 | – | 4 | – |
| Potassium chlorate | 27 | 25.3 | 28.85 | 31.5 | 28 | 31 |
| Sulphur | 10.4 | 10 | 11.3 | – | – | – |
| Lactose | – | – | – | 18 | – | – |
| Magnesium carbonate | – | – | – | 3.5 | – | – |
| Sugar fine | – | – | – | – | 16 | 18.5 |
| Sil-o-cel binder | – | – | – | – | 4 | – |
| VAAR | – | – | – | – | – | 2 |

    c. *Improved performance*: The fact that the pyrotechnic mixture is well dispersed means that it burns at more uniform rate, and with more uniformity.

    d. *Greater flexibility in grain design*: The smoke composition can be very easily cast in a variety of configurations, and is not restricted solely to cylindrical

**TABLE 13.4**
**Yellow Signaling Colour Smoke Compositions [1]**

| Ingredient | 1 | 2 | 3 | 4 | 5 | 6 |
|---|---|---|---|---|---|---|
| Dye yellow | 14 | 18 | 34 | 51 | 41 | 15 |
| Benzanthrone | 24.5 | 32 | 8 | – | – | 32 |
| Sulphur | 8.5 | – | – | – | 9 | – |
| Potassium chlorate | 20 | 23 | 26 | 30 | 23 | 30 |
| Sodium bicarbonate | 33 | – | 3 | – | 27 | 3 |
| Lactose | – | 16 | – | – | – | – |
| Magnesium carbonate | – | 9 | – | – | – | – |
| Sugar | – | – | 15 | 17 | – | 20 |
| Sil-o-cel | – | – | 4 | – | – | – |
| VAAR | – | – | – | 2 | – | – |

**TABLE 13.5**
**Violet Signalling Colour Smoke Compositions [1]:**

| Ingredient | 1 | 2 | 3 | 4 | 5 |
|---|---|---|---|---|---|
| Dye violet | 42 | 42 | 47 | 44 | 47.5 |
| Sodium bicarbonate | 24 | 26 | - | - | 4.5 |
| Potassium chlorate | 25 | 23 | 22 | 30.2 | 28 |
| Sulphur | 9 | 9 | | 11.8 | |
| Lactose | | | 24 | | |
| Magnesium carbonate | | | 7 | | |
| Sugar | | | | | 18 |
| Asbestos | | | | | 2 |
| Potassium bicarbonate | | | | 14 | |

      shapes. This is especially advantageous for the filling of aerodynamically-shaped projectiles.

    e. *Suitability for mass production using an automated production line*: There is no need for incremental filling techniques.

The only drawback is that the pot life of such compositions is low (approximately 3 h to 10 h) and hence cannot be stored for longer time and must be used within its pot life. Another patent [5] refers to mixture as per Table 13.8, which after 15 minutes of mixing under vacuum is casted into tube mold and dried overnight at 160°F.

## TABLE 13.6
### Black Signalling Colour Smoke Compositions [2]

| Ingredient | Percentage | | |
|---|---|---|---|
| Hexachloroethane | 60 | – | – |
| Magnesium | 19 | – | – |
| Naphthalene | 21 | 40 | – |
| Potassium chlorate | – | 45 | 55 |
| Wood charcoal | – | 15 | – |
| Anthracene | – | – | 45 |

## TABLE 13.7
### Plastic Bonded Signalling Colour Smoke Compositions [3]

| Ingredients | Formula 138 parts[a] | Formula 139 Percentage[b] |
|---|---|---|
| Potassium chlorate | 23 | 28.25 |
| Sugar | 18 | 19.5 |
| Green Dye, MIL-D-3277 | 51 | – |
| Red Dye, MIL-D-3284 | – | 35 |
| Potassium bicarbonate | 8 | – |
| Sodium bicarbonate | – | 6.5 |
| Polyvinyl acetate (Plasticised) | 2.2 (mixed in 50 parts of dichloromethane) | 9 (mixed in acetone) |
| Infusorial earth | – | 1.75 |

*Notes*

a  The combined slurry is "cured" by evaporation of the solvent in situ.

b  The plastic binder is added as a solution in acetone. Most of the solvent evaporates during mixing and granulation.

## TABLE 13.8
### Castable Yellow Colour Smoke Compositions [5]

| Ingredients | Quantity | |
|---|---|---|
| Glycil azide polymer triol (GAP) | 20.25 g | 12.13 g |
| Glycil azide polymer azide (GAPA) | 43.3 g | 52.00 g |
| Hexamethylene dibutyl diisocyanate (HMDI) | 1.45 g | 0.87 g |
| Dye Atlasol smoke yellow S | 65 g | 65 g |
| Dibutyl tin dilaurate (DBTL) | 70 µl | 70 µl |
| Burn rate | | |
| Burn rate (in.sec$^{-1}$) | 0.018 | 0.023 |

## 13.4 FACTORS AFFECTING PERCEPTION OF SIGNALLING COLOUR SMOKE

The suspended particle of the smoke cloud scatter, reflect and absorb incident radiation and the overall effect depends on:

a. Duration and rate of evolution of signalling colour smoke
b. Whether it is day or night time (signalling colour smoke clouds cannot be used in nights since it is difficult to identify the colour and it is a serious drawback)
c. The brightness and colour of the background during the day
d. The size and shape of ingredients and quantity of signalling colour smoke
e. The atmospheric condition like humidity (moisture), rain, fog, snow and ambient temperature
f. Wind speed and direction during observation for the persistence of colour smoke in air
g. Whether the smoke containers fall on dry land or marshy land or in a jungle or open area
h. The position of the observer with respect to sun and the smoke cloud

The ideal observation is a calm weather with maximum wind speed of 2–3 m/second and observation angle of 45° to 135° (see Section 4.4). The important feature of a good signalling colour smoke for better visibility is

a. Higher purity of colour of smoke
b. Long duration of signalling colour smoke in air
c. Higher volume of colour smoke, so as to cover more space
d. Higher density of colour smoke

Let us understand how one perceives the signalling colour smoke. The colour dyes get volatilised and condense in air and do absorb some visible spectral range. The light scattered and reflected from these condensed dye particles thus does not contain this specific visible spectral range and gives a complimentary colour. This is in sharp contrast with signalling star flare compositions where emitted wavelengths are perceived as colour.

## 13.5 DIFFERENCES BETWEEN SCREENING SMOKE AND COLOUR SMOKE

Table 13.9 gives difference between screening smoke and colour smoke.

**TABLE 13.9**
**Differences Between Screening Smoke and Colour Smoke**

| Parameters | Screening smoke | Colour smoke |
|---|---|---|
| End use | White/grey smoke for screening, incendiary and training etc. | Coloured smoke for signalling in coded language for communication in combat field, distress signalling for search and rescue, marking and dropping zones etc. |
| Production of smoke | Combustion products are expelled to form screening smoke (aerosol) cloud or infrared smoke cloud | Heat of combustion of smoke composition causes volatilisation of dye which are condensed to form colour smoke cloud |
| Combustion temperature | The combustion is at higher temperature, around 800°c | The combustion is at low temperature, just sufficient to volatilise the dye |
| Pillaring effect | Pillaring effect especially with white phosphorous as the combustion temperatures are high | No pillaring effect as the combustion temperatures are low |
| Humidity effect | Screening smoke aerosols are hygroscopic and it absorbs moisture from air and increase the dimensions of smoke aerosol particles, enabling better absorbance, scattering and reflection properties, thereby improving screening efficiency and enhancing the yield. (Exception cinnamic acid and terephthalic acid based compositions which are non-hygroscopic) | Unaffected by humidity |
| Infrared attenuation | Capable of infrared attenuation with phosphorous based compositions | Not capable of infrared attenuation |

## 13.6 SPECIAL REQUIREMENTS SIGNALLING COLOUR SMOKE COMPOSITIONS

In addition to the general essential requirements of compositions as given in Section 1.6, signalling composition should preferably possess the following special requirements:

a. The composition should have a low ignition temperature.
b. The composition should give the required signalling colour smoke at uniform rate.
c. The colour and smoke density should be such that it is visible from long distance.
d. The composition should give smoke for specified duration so that it may be observed properly, i.e., persist in the air for specified time.

e. The composition should generate just sufficient heat of combustion to sublime the dye but not very high so as to decompose the dye. Thus, the heat of combustion of the composition should be less so that flaming is avoided.

f. The composition should not produce solid hot slag/residue as this would also decompose the dye giving grey/black soot. The slag produced during combustion should be porous to allow volatilised dye to pass through it to the atmosphere.

g. The composition must generate sufficient gas with high pressure to expel the volatilised dye particles in the air.

h. The dye should have properties as mentioned in Section 6.7.2.

Readers may also see reference [6] and [7] on various coloured signalling smoke compositions.

## REFERENCES

[1] F.L. McIntyre, *"A Compilation of Hazard and Test Data for Pyrotechnic Compositions"*, Contractor Report ARLCD-CR-80047, October 1980. U.S. Army Research and Development Command, Large Caliber Weapon System Laboratory, Dover, New Jersey.

[2] A.A. Shidlovskiy, *"Principles of Pyrotechnics"*, 3rd Edition, Moscow, 1964. (Translated by Foreign Technology Division, Wright-Patterson Air Force Base, Ohio, 1974), American Fireworks News, 1 July 1997. ©1997, Rex E. &S. P., Inc.

[3] Dr Herbert Ellern, *"Military and Civilian Pyrotechnics"*, ©Chemical Publishing Co., New York, 1968.

[4] P. Lessard and G. Couture, *"Polymer-Bonded Coloured Smoke Compositions"*, *Propellant, Explosives, Pyrotechnic*, 1988, Volume 13, Pages 58–61.

[5] Edgar Rolan Wilson, Joseph Edward Flangan, Milton Bernard Frankel, and Louis Russel Grant, *"Castable Smoke Producing Pyrotechnic Composition"*, A2, Publication No. 0450147 European Patent Office.

[6] P. Barnes, L. de Yong, J. Domanico, P. Twadawa, and F. Valenta, *"A Comparison between Several Standard Methods used to Characterize the Ignition/Ignition Transfer of Pyrotechnic Compositions - A Collaborative Study"*, Part I data Report MRL-R-1043, Pt 1, February 1987, Department of Defence, Defence Science and Technology Organisation, Materials Research Laboratories, Melbourne, Victoria, Australia.

[7] Harold E.Filter, Erwin M. Jankowiak, George A. Lane, and Don L. Stevens, *"Pour-Castable Compositions for Colored Smoke Signals"*, Proceedings Third International Pyrotechnics Seminar, Antlers Colorado, August 1972.

# 14 Riot Control Compositions

## 14.1 ROLE

These are mainly used to

a. Disperse unruly mobs in a civil disturbance by use of irritants
b. Incapacitate individual rioters by impact
c. Identification of rioters by indelible ink
d. Rescue operations concerning hostages
e. Anti-insurgent operations
f. Industrial security

These compositions are mostly used in cartridges, grenades and to a limited extent in shells. These are non-lethal ammunition as they are incapable of killing a person. However, the user has to be sufficiently trained otherwise it may lead to serious casualties. These are also known as irritants, lacrimators or erroneously called tear gas since these agents are solids dispersed as fine particles or droplets or as a solution.

*Article II(7) of the 1993 Chemical Weapons Convention* defines "*Riot Control Agent*" as: "*Any chemical not listed in a Schedule, which can produce rapidly in human's sensory irritation or disabling physical effects which disappear within a short time following termination of exposure.*"

*Article II(9)(d) of the 1993 Chemical Weapons Convention* provides that riot control agents may not be used as a method of warfare but may be used for certain law enforcement purposes including riot control.

## 14.2 TYPICAL RIOT CONTROL AGENTS (RCA)

The most common chemicals used in law enforcement include O-chlorobenzylidene malononitrile (CS), chloroacetophenone (CN) and dibenz (b,f )-1,4-oxazepine (CR), which are often referred erroneously as "tear gases," and oleoresin capsicum (OC) and pelargonic acid vanillylamide (PAVA) which are often referred to as "pepper sprays." OC is naturally derived from the capsicum species of plant (such as chilli peppers) whereas PAVA is a synthetic formulation of one of the active ingredients in OC. Other less commonly used irritants include N-nonanoyl morpholine (MPK/ MPA) and diphenylaminechlorasine (DM, or Adamsite). Table 14.1 gives some of the properties of riot control agents.

DOI: 10.1201/9781003093404-14

**TABLE 14.1**
**Some Riot Control Agents**

| Class | Name | Formula | Mol. weight | Melting point(°C) | Boiling point (°C) |
|---|---|---|---|---|---|
| Tear agents halogenated | Chloroacetophenone (CN) and o-chlorobenzylidene malononitrile (CS) | $C_8H_7OCl$<br>$C_{10}H_5 Cl N_2$ | 154.6<br>188.6 | 54<br>93 | 247<br>310–315 |
| Tear agents non-halogenated | Dibenz (b,f)-1,4-oxazepine (CR) | $C_{13}H_9ON$ | 195.2 | 72 | 335 |
| Vomiting agents | Diphenylaminechlorarsine (DM) | $C_{12}H_9AsClN$ | 277.5 | 195 | 405–415 |
| Tear agents in solvents | Oleoresin capsicum (OC) and pelargonic acid vanillylamide (PAVA) | $C_{18}H_{27}NO_3$<br>$C_{17}H_{27}NO_3$ | 305.41<br>293.407 | 62–65<br>54 | 210–220<br>– |

## 14.3  TOXICITY OF RIOT CONTROL AGENTS [1]

RCAs are crystalline solids with low vapour pressure. RCAs are typically administered as fine particles, aerosol, sprays or in solutions. Therefore, they are not true gases. The inhalation toxicity of RCAs is often indicated by the expression Ct. This term is defined as the product of concentration (C) in $mg.m^{-3}$ multiplied by exposure time (t) in minutes ($mg \cdot min. \ m^{-3}$). $LCt_{50}$ and $ICt_{50}$ are conventional terms used to describe airborne dosages that are lethal (L) or incapacitating (I) to 50% of the exposed population. The intolerable concentration ($mg.m^{-3}$) $ICt_{50}$ and minimal lethal concentration are provided in Table 14.2 for the most common RCAs.

The ocular irritancy threshold (minimal irritant or minimal effective dose), estimated human LCt50 and safety ratio are provided in Table 14.3 for some RCAs [1]. The modem RCAs are characterised by a high $LCt_{50}$, low effective $Ct_{50}$, low $ICt_{50}$, low minimal irritating concentration and large safety index ratio ($LCt_{50}$/irritancy threshold).

## 14.4  IMMEDIATE SIGNS AND SYMPTOMS OF EXPOSURE TO RIOT CONTROL COMPOSITIONS ON HUMAN BODY

a. The effect of riot control agents depends upon the concentration and the period of exposure. A higher concentration with low period of exposure is more severe than the same amount of riot control agent in low concentration with high period of exposure.

b. It works on the area of contact of a person, like eyes, skin and nose.

c. Confinement/enclosed area lead to more severe effects than in open space with free air.

## TABLE 14.2
## Toxicity Data of Common RCAs [1]

| Agent | Onset | Intolerance concentration ($mg.m^{-3}$) | $ICt_{50}$ ($mg.min.m^{-3}$) | Minimal lethal concentration[a] ($mg.m^{-3}$) |
|---|---|---|---|---|
| CS | Immediate | 5 | 3–10 | 2,500 |
| CN | Immediate | 35 | 20–40 | 850–2,250 |
| DM | Delayed with long recovery period | 5 | 22–150 | 1,100-4,400 |
| CR | Immediate | 1 | 1 | 10,000 |
| Bromobenzyl cyanide (CA) | Immediate | 0.8 | 30 | 1,100 |

*Notes*

a  Estimate for minimal lethal concentration (10 min exposure)

**TABLE 14.3**

**Health Risk Considerations for the Common RCAs [1]**

| Agent | Irritancy threshold[a] (mg.m$^{-3}$) | Estimated human LCt$_{50b}$[b] (mg.min.m$^{-3}$) | Safety ratio[c] | Adverse effects |
|---|---|---|---|---|
| CN | 0.3[a] | 8,500–22,500 | 28,000 | Danger of permanent eye injury, vesiculation, bronchopneumonia, reactive airways, documented fatality cases |
| CS | 0.004[a] | 25,000–150,000 | 60,000 | Same as CN, but fatality cases not authenticated, enhanced persistence compared to CN and CS |
| CR | 0.002[a] | 100,000 | 100,000 | No significant respiratory toxicity |
| OC | 0.003[d] | – | > 60,000 | Eye, skin, respiratory toxicity, significant morbidity in neonates, fatality involving case of in-custody use |
| DM | Approx.1[a] | 11,000–44,000 | 11,000 | No longer used |
| CA | 0.15[a] | 11,000 | 11,000 | Predominantly a lacrimatory agent, no longer used |

*Notes*

a  Ocular irritancy threshold unless indicated otherwise
b  Values obtained from references: Sidell (1997), Maynard (1999), Smith and Stopford (1999), Olajos and Salem (2001)
c  Values derived from estimate of the human LCtso (lower bound)/irritancy threshold (minimal effective dose). Therefore, ranges are not provided for the safety ratios
d  Threshold for respiratory complaints by Capsaicinoids: Stopford and Sidell (2006)

d. The effect is reduced after the person is moved from the area and cleaned properly.
e. The effect remains for a short duration of about 30 minutes maximum.
f. Exposure of high concentration for significant duration may lead to severe injury to the person.
g. Some immediate symptoms of these agents on exposure to eyes, nose, mouth, lung and skin
    i. *Eye*: Tearing, itching, burning, blurred vision, temporary blindness and unintentional closing of eye lids
    ii. *Nose*: Swelling, itching, running nose, burning pain
    iii. *Mouth*: Burning
    iv. Lungs: Difficulty in breathing, coughing, wheezing
    v. *Throat*: Itching, coughing
    vi. *Skin*: Itching, blisters, chemical burns

## 14.6 MECHANISM OF RIOT CONTROL AGENTS

A fuel and an oxidiser are used as evaporating agents to ensure that the riot control agent gets evaporated but not decomposed. Screening property is irrelevant for lachrymatory compositions. This vaporised riot control agent undergoes condensation to form a dispersed phase. Human body has wet surfaces like the eyes, mouth, respiratory tract and perspiration on skin. The riot control agents' aerosol act on these wet surfaces to produce discomfort. The body's internal auto mechanism tries to counter this action by tears and mucus and itching. The physiological pain and the sudden panic incapacitate the rioter or disorient him. Readers may refer [1] for better insights into the mechanism.

The effect depends upon:

a. Particle size of disseminated smoke
b. Volume and duration of smoke

The performance of the composition depends upon several other factors as given below:

h. The efficacy is reduced in heavy rains as the aerosol is washed away.
i. Its effect is reduced by a delayed effect at very low temperature due to rioter wearing more layers of clothes, including warm clothes. However, once the skin gets affected, the rioter has to open clothing at low temperature.
j. Its effect is more in summer due to the action of aerosol on the perspiration from the skin of human body.
k. Its effect is more on sensitive persons.

Sensitivity of human body to RCAs differs substantially amongst individuals and factors inducing individual reactions are mainly their mental state, motivation, physical activity, ambient temperature and humidity. Exposed human body needs fresh air to breath, remove clothes and keep them wrapped in plastic, wash and rinse eyes and affected parts with soap.

## 14.7 TYPICAL RIOT CONTROL COMPOSITIONS IN GRENADES

The agents used in irritant grenades must be of a type that can be easily dispersed into the air. The grenade should be designed so that the maximum concentration of agent per unit weight of grenade is generated and dispersed. The two most effective methods of dispersing the agent are [2]:

a. The agent is mixed with combustible composition that is ignited when the grenade is thrown. The heat of combustion causes the agent to disperse into the air by sublimation. The grenade casing remains intact; the gases are dispersed through small orifices in the casing.
b. The agent, in micro pulverized form, is dispersed instantaneously by the explosion of a detonator. The explosion bursts the grenade casing, and the agent is dispersed as a heavy concentration of small particles.

For a given amount of agent, the burning-type grenade has a longer dispersal time; dispersal times up to one minute are typical. However, the bursting-type produces a higher concentration of the agent. Furthermore, the bursting-type grenade cannot be kicked aside or picked up and thrown back at the thrower, which is possible with the burning type if its dispersal period is too long In addition to the agent, the filler or main charge of a burning-type irritant grenade requires a fuel-oxidiser mixture and a starter mixture. The fuel provides the combustion necessary to disperse the agent. Oxygen for combustion is provided by the oxidiser. The starter mixture ignites the fuel-oxidiser mixture. Generally [2], the agent and the fuel-oxidiser mixture are pressed to the desired shape, which is usually cylindrical, and then coated with the starter mixture. However, in cases where the agent and the fuel-oxidiser mixture used tend to react with one smother, the two must be physically separated. This can be accomplished by encapsulating the agent and embedding it in the fuel-oxidiser mixture before pressing.

In some cases, a single grenade may be required to contain more than one type of agent. If the agents are incompatible, they must be kept physically separated within the grenade. This can be accomplished by loading the agents into separate containers that are ignited simultaneously when the grenade is fired.

Some burning and bursting type of riot compositions as shown in Tables 14.4–14.8.

## 14.8  SPECIAL REQUIREMENTS OF RIOT CONTROL COMPOSITIONS

The Riot Control Compositions require special precautions during manufacture of composition for obvious reasons. In addition to general essential requirements of compositions as mentioned in Section 1.6, riot compositions should preferably possess the following special requirements:

**TABLE 14.4**
**CN Irritant Grenade Composition (Burning Type) [2]**

| CN mixture | Parts by weight | Igniter composition | Parts by weight |
|---|---|---|---|
| Chloroacetophenone | 29 | Potassium nitrate | 70.5 |
| Diatomaceous earth | 5 | Charcoal | 29.5 |
| Sucrose | 17 | Mix with | |
| Potassium chlorate | 24 | | |
| Potassium bicarbonate | 25 | | |
| Press at: | | Nitrocellulose | 4 |
| 5000-7500 lb dead load | | Acetone | 96 |

Agent/igniter proportion: Igniter is poured into grenade as a slurry and then poured out in the manner of a ceramic slip casting. Approximately 5 grams (dry basis) is retained.

## TABLE 14.5
## DM Irritant Grenade Composition (Burning Type) [2]

| DM mixture | Parts by weight | Igniter composition | Parts by weight |
|---|---|---|---|
| Diphenylaminechloroarsine | 52.5 | Potassium nitrate | 70.5 |
| Potassium chlorate | 25.5 | Charcoal | 29.5 |
| Sucrose | 17 | Mix with | |
| Magnesium oxide | 4 | | |
| Press at: | | Nitrocellulose | 4 |
| 5000-7500 lb dead load | | Acetone | 96 |

Agent/igniter proportion: Igniter is poured into grenade as a slurry and then poured out in the manner of a ceramic slip casting. Approximately 5 grams (dry basis) is retained.

## TABLE 14.6
## CS (Encapsulated) Irritant Grenade Composition (Burning Type) [2]

| CS mixture | Parts by weight | Starter mixture | Parts by weight |
|---|---|---|---|
| Orthochlorobenzylidene malononitrile in 92 #00 gelatin capsules | – | Potassium nitrate | 70.5 |
| *Potassium chlorate | 40 | Charcoal | 29.5 |
| *Sucrose | 28 | Mix with | |
| *Magnesium carbonate | 32 | | |
| Press at: | | Nitrocellulose | 8 |
| 5000-7500 lb dead load | | Acetone | 92 |

Agent/igniter proportion: 907/1209 (parts by weight)
*Total sum =165 gms

## TABLE 14.7
## Compositions for Bursting-Type Irritant Grenades [2]

| Mixture | Percentage |
|---|---|
| CS | 95 |
| Aerogel | 5 |
| DM | 95 |
| Aerogel | 5 |
| CN | 92.0 ± 0.5 |
| Magnesium oxide | 8.0 ± 0.5 |

**TABLE 14.8**
**Typical Riot Control Compositions**

| Ingredient | Percentage | | | |
|---|---|---|---|---|
| | [3] | PN770 | [4] | [4] |
| CS(o-chlorobenzylamalononitrile) | 42 | 25 | 40 | 41 |
| Potassium Chlorate | 26 | 30 | 30 | 27 |
| Lactose | 20 | 30 | 18 | – |
| Kaoline | 12 | 15 | – | – |
| Nitrocellulose/Acetone Binder (8/92) *Over and above | 3.6* | – | 3 | – |
| Dextrin/Water ((15/85) Binder | – | – | – | 2 |
| Magnesium Carbonate | – | – | 9 | 12 |
| Sugar | – | – | – | 18 |

a. The composition should instantaneously produce aerosols on combustion.
b. The aerosols produced on combustion should have good irritation power and should be stable in air.

## REFERENCES

[1] Corey J. Hilmas, *"Riot Control Agents"*, *Chapter 11, Handbook of Toxicology of Chemical Warfare Agents*, Second Edition, Editor R.C. Gupta, Academic Press, London, 2009.

[2] *"Engineering Design Handbook Grenades(U)"*, AMCP 706-240, 13 December 1967, Headquarters, U.S. Army Materiel Command, Washington DC.

[3] *"Design of Ammunition for Pyrotechnic Effect"*, AMCP 706-188, March, 1974, Engineering Design Handbook Military Pyrotechnics Series, Part FOUR, Headquarters, US Army Material Command.

[4] F.L. McIntyre, *"A Compilation of Hazard and Test Data for Pyrotechnic Compositions"*, Contractor Report ARLCD-CR-80047, October 1980. U.S. Army Research and Development Command, Large Caliber Weapon System Laboratory, Dover, New Jersey.

# 15 Incendiary Compositions

## 15.1 ROLE

Incendiary compositions are widely used to:

a. Destroy buildings, installations, paddy fields, oil fields, refineries, petrol and diesel dumps and ammunition depots
b. Creating jungle fire to flush out enemy troops, hold out place
c. Destruction of own equipment and documents so as to avoid falling in the hands of enemy

## 15.2 TYPES OF INCENDIARY COMPOSITIONS

Incendiary compositions are of two types, namely solid incendiary compositions and gel incendiary compositions.

### 15.2.1 SOLID INCENDIARY COMPOSITIONS

These compositions contain oxidisers; say a metal oxide or salt oxidiser and metal fuel. The composition in which the main oxidiser is a metal oxide is known as *a thermite*. Thermites have the following properties:

a. It gives a flameless combustion.
b. Its combustion process does not produce any appreciable gas.
c. It has a very high combustion temperature (2,500°C to 3,500°C).
d. It gives molten slags which spread rapidly over the composition for further ignition.

The thermite fuel and metal oxide must satisfy the following parameters:

a. *Thermite Fuel:* The fuel must satisfy following:
   i. Must have a high density
   ii. Evolve maximum amount of heat during combustion
   iii. Form low melting and non-volatile oxides
      The fuels used are aluminium, magnesium, titanium, silicon, zinc boron, zirconium, beryllium, etc., all of which have high heat of combustion. However, aluminium is extensively used due to

DOI: 10.1201/9781003093404-15

    i. Low cost, high density but high reactivity
    ii. Low melting point (660°C) allowing early melting and start-up of combustion in fluid phase with non-volatile oxide products
    iii. High boiling point (2,467°C) allowing achieving higher temperature of combustion

b. *Thermite oxidiser:* The thermite oxidiser must satisfy the following:
    i. Should have high density
    ii. Should have minimum heat of formation
    iii. During combustion it must be reduced to low melting point and high boiling point metal
    iv. It must have oxygen content of more than 25%

The oxidisers used are oxides of boron, silicon, chromium, iron, copper, lead, etc. like boron (III) oxide, silicon (IV) oxide, chromium (III) oxide, manganese (IV) oxide, iron (II) oxide, iron (III) oxide, copper (II) oxide, lead (II) oxide, lead (III) oxide and lead (IV) oxide, etc.

Thermites are difficult to ignite. To improve the performance of thermites, certain ingredients are added to improve ignition, to enhance the heat of combustion, control the burn rate and add strength of the pressing by use of binders.

### 15.2.2 Gel Incendiary Compositions

These compositions do not contain oxidisers and burn in air like combustible carbonaceous liquid petroleum fuel-based incendiary composition (phosphorous and its compounds also burn in air but is not a gel incendiary). These incendiary compositions produce a large flame with high temperature for quick action. Liquid petroleum-based compositions have the following advantage:

a. Form large flame and hence cover more area
b. Give high heat output
c. Are filled by pouring and do not require compaction through pressing machines
d. May be solidified by gelling the liquid petroleum products
e. Are easily available at low cost

However, they are inferior to thermites since:

a. Do not form solid combustion particles like hot slags on combustion
b. Have low combustion temperature around 700°C to 900°C compared to thermites having high combustion temperature of 2,500°C to 3,500°C
c. Have low density (made of petroleum and gel, having low density)
d. Have a tendency to spread the zone of action (to reduce this effect, thickened or gelled composition are used)

e. Have a low shelf life of approximately 6 months compared to approximately 10 years for solid incendiary compositions in ammunitions. They deteriorate and reduce viscosity.
f. Their burning time is less.
g. They cannot be used in projectiles designed to spin

## 15.3 MECHANISM OF INCENDIARY COMPOSITION COMBUSTION

Incendiary involves a metal reacting with a metallic or a non-metallic oxide to form a more stable oxide and a corresponding metal or non-metal of the reactant oxide and can be expressed as

$$M + AO \rightarrow MO + A + \Delta H \tag{15.1}$$

where $M$ is a metal or an alloy and $A$ is either a metal or a non-metal, $MO$ and $AO$ are their corresponding oxides, and $\Delta H$ is the heat generated by the reaction.

These compositions on suitable initiation generate tremendous heat due to breaking of bonds of the reactant and formation of new bonds in the product where the bonds in the product are stronger than the bonds of the reactant, the products are more stable and have a lower energy than the reactants. Consider a thermite reaction of aluminium with iron (III) oxide as under.

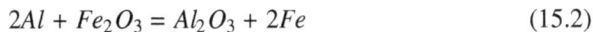

$$2Al + Fe_2O_3 = Al_2O_3 + 2Fe \tag{15.2}$$

As per Chapter 3 on heat of reaction,

*Heat of reaction* $(\Delta H)$

= *Sum of heat of formation of products (from its constituent elements)*
− *Sum of heat of formation of reactants (from its constituent elements)* (15.3)

a. Heat of formation for iron (III) oxide (reactant) is (−) 824.2 kjoules. mole$^{-1}$
b. Heat of formation for aluminium oxide (product) is (−) 1675.7 kjoules. mole$^{-1}$

Hence $\Delta H$ = (−) 1675.7 kjoules mole$^{-1}$ − (−) 824.2 kjoules mole$^{-1}$ = (−) 849.5 kjoules. mole$^{-1}$
Since 1 kcal = 4.184kjoules, hence $\Delta H$ = (−) 202.895 kcal. mole$^{-1}$.
Dividing 202.895 by 213.64 g of starting material, $\Delta H$ = (−) 0.94970 kcal. g$^{-1}$
The reaction can be written as

$$2Al + Fe_2O_3 = Al_2O_3 + 2Fe + 949.70 \text{ cal. g}^{-1} \text{(exothermic reaction)} \tag{15.4}$$

The values of theoretical maximum density, reaction temperature, state of products of combustion and heat of reaction, for some thermite reactions are shown in Table 15.1.

**TABLE 15.1**
**Typical Thermite Reactions [1]**

| Reactant | TMD $(g.cm^{-3})$ | Adiabatic reaction temperature(°K) | | State of products | | (-) Heat of reaction | |
|---|---|---|---|---|---|---|---|
| | | W/o phase change | W/ phase change | Oxide | Metal | cal. $g^{-1}$ | cal.$cm^{-3}$ |
| $2Al + Fe_2O_3 = 2Fe + Al_2O_3$ | 4.175 | 4,382 | 3,135 | Liquid | l-g | 945.4 | 3,947 |
| $8Al + 3Fe_3O_4 = 9Fe + 4Al_2O_3$ | 4.264 | 4,057 | 3,135 | Liquid | l-g | 878.8 | 3,747 |
| $4Al + 3MnO_2 = 3Mn + 2Al_2O_3$ | 4.014 | 4,829 | 2,918 | Liquid | Gas | 1,153 | 4,651 |
| $2Al + MoO_3 = Mo + Al_2O_3$ | 3.808 | 5,574 | 3,253 | l-g | Liquid | 1,124 | 4,279 |
| $3Mg + Fe_2O_3 = 2Fe + 3MgO$ | 3.224 | 4,703 | 3,135 | Liquid | l-g | 1,110 | 3,579 |
| $4Mg + Fe_3O_4 = 3Fe + 4MgO$ | 3.274 | 4,446 | 3,135 | Liquid | l-g | 1,033 | 3,383 |
| $2Mg + MnO_2 = Mn + 2MgO$ | 2.996 | 5,209 | 3,271 | Liquid | Gas | 1,322 | 3,961 |

Note: TMD: Theoretical maximum density

Another Table showing thermite reactions involving iron (II) oxide $Fe_2O_3$ with various fuels is given in Table 15.2.

## 15.4   TYPICAL SOLID INCENDIARY COMPOSITIONS

Some incendiary compositions are at Table 15.3.

**TABLE 15.2**
**Heat of Combustion of Thermite Compositions Containing Iron (II) Oxide [2]**

| Fuel | Thermite composition percentage | | Heat of Combustion $(Kcal.gm^{-1})$ |
|---|---|---|---|
| | Iron (II) oxide [Fe$_2$O$_3$] | Fuel | |
| Aluminium | 75 | 25 | 0.93 |
| Magnesium | 69 | 32 | 1.05 |
| Calcium | 57 | 43 | 0.93 |
| Titanium | 69 | 31 | 0.57 |
| Silicon | 79 | 21 | 0.58 |
| Boron | 88 | 12 | 0.59 |

**TABLE 15.3**
**Incendiary Compositions**

| Ingredients | Percentage | | | | | |
|---|---|---|---|---|---|---|
| | [2] | [2] | [2] | [2] | [2] | SR 801B |
| Aluminium powder | 34 | 26 | 24 | 13 | – | – |
| Barium nitrate | – | – | 26 | 44 | – | – |
| Iron oxide | – | – | – | 21 | – | – |
| Iron scale | – | – | 50 | – | – | – |
| Magnesium | – | – | – | 12 | 50 | 57 |
| Potassium nitrate | – | 65 | – | 6 | – | – |
| Binder | – | – | – | 4 | – | – |
| Boric acid | – | – | – | – | – | – |
| Wood charcoal | – | 9 | – | – | – | – |
| Potassium perchlorate | 56 | – | – | – | 50 | 37 |
| Graphite size 170 or talc powder (French chalk) | – | – | – | – | – | 6 |

**TABLE 15.4**
**Incendiary Compositions for Bullets and Shells [3]**

| Ingredients | SR 365 | SR 379 |
|---|---|---|
| Magnesium aluminium alloy (50/50) | 50 | 47 |
| Barium nitrate | 50 | 50 |
| Paraffin wax | – | 3 |

Some incendiary bullets and shell fillings use SR 365 and SR 379 as given in Table 15.4.

Some typical incendiary composition/Incendiary mixtures (IM) for small arms ammunition are given in Table 15.5 and Table 15.6.

## 15.5   NON-TOXIC INCENDIARY COMPOSITIONS [5]

Potassium perchlorate is an environmental hazard. It can be released into the environment as a result of spillages during manufacture, demilitarization or when ammunition fails to function correctly. The presence of potassium perchlorates in drinking water is a cause for concern as all perchlorates are recognized as a potential hazard to human health. In particular, their ingestion is known to inhibit iodide uptake by the thyroid gland. IM-28 incendiary composition (used in 0.50" caliber ammunition)

**TABLE 15.5**

**Typical Incendiary Compositions for Small Arms [4]**

| Ingredients | Percentage | | | | | | |
|---|---|---|---|---|---|---|---|
| | IM-11 | IM-21A | IM-23 | IM-28 | IM-68 | IM-69 | IM-112 |
| Magnesium aluminium alloy (50/50) | 50 | 48 | 50 | 50 | 50 | 50 | 45 |
| Barium nitrate | 50 | 48 | – | 40 | 24 | 40 | 50 |
| Calcium resinate | – | 3 | – | – | – | – | – |
| Asphaltum | – | 1 | – | – | – | – | – |
| Potassium perchlorate | – | – | 50 | 10 | – | – | – |
| Ammonium nitrate | – | – | – | – | 25 | – | – |
| Zinc stearate | – | – | – | – | 1 | – | – |
| Iron oxide | – | – | – | – | – | 10 | – |
| Tungsten powder | – | – | – | – | – | – | 5 |

**TABLE 15.6**

**Typical Incendiary Compositions for Small Arms [4]**

| Ingredients | Percentage | | | | | | |
|---|---|---|---|---|---|---|---|
| | IM-136 | IM-139 | IM-142 | IM-214 | IM-241 | IM-385 | MOX-2B |
| Magnesium aluminium alloy (50/50) | 49 | 10 | 46 | 25 | 25 | 49 | – |
| Potassium perchlorate | 49 | – | – | 25 | 25 | – | – |
| Calcium resinate | 2 | – | – | – | – | 2 | – |
| Barium nitrate | – | 47 | 48 | – | – | – | – |
| Red phosphorous | – | 40 | – | – | – | – | – |
| Aluminium stearate | – | 3 | – | – | – | – | – |
| Asphaltum | – | – | 5 | – | – | – | – |
| Graphite | – | – | 1 | – | – | – | 1 |
| Zirconium (60/80) | – | – | – | 50 | – | – | – |
| Zirconium (20/65) | – | – | – | – | 50 | – | – |
| Ammonium perchlorate | – | – | – | – | – | 49 | 35 |
| Aluminium powder | – | – | – | – | – | – | 52 |
| RDX/wax (97/3) | – | – | – | – | – | – | 6 |
| TNT | – | – | – | – | – | – | 4 |
| Calcium stearate | – | – | – | – | – | – | 2 |

contains magnesium–aluminium alloy (50%), barium nitrate (40%) and potassium perchlorate (10%). Report [5] recommends magnesium–aluminium alloy 48%, sodium nitrate 48% and calcium resinate 4% as perchlorate-free incendiary composition.

## 15.6 FACTORS AFFECTING EFFICIENCY OF INCENDIARY COMPOSITIONS

The efficiency of incendiary composition depends on:

a. Ingredients of composition and their particle size
b. Spread area of composition/pellets, which would depend upon the quantity of incendiary composition in the ammunition, the assistance of wind and negative impact of rain or moisture or type of land like dry or marshy land, etc.
c. Burning time of incendiary composition, which would depend upon loading density of composition and charge weight
d. Flame temperature of incendiary composition
e. Quantum of heat transfer to combustibles. This quantum of heat transfer would depend on:
   i. The duration of contact of slag of incendiary and combustible material so as to continue heat transfer to combustible material
   ii. The area of contact of slag of incendiary and combustible material
   iii. The temperature difference of slag of incendiary and combustible material, the lower the difference, the earlier the combustible material would pick up fire
   iv. The heat transfer properties of combustion products

It is necessary that the incendiary composition should not produce significant gas, which otherwise would carry away heat from burning surface, which is not desirable. The incendiary compositions are, therefore, designed to have as much hot slag as possible for effective incendiary effect and to be almost gasless, which is unaffected by pressure variation.

## 15.7 TYPICAL GEL INCENDIARY COMPOSITIONS

Sodium palmitate ($C_{16}H_{31}NaO_2$) was used as a thickener for gasoline during World War II and thus the ammunition was named as Napalm (Na-palm). Recent advances have led to the use of polystyrene plastic beads as thickener, which is not very combustible and slows the combustion (Table 15.7).

Another World War II incendiary composition (Table 15.8) is a thickened fuel which on initiation bursts and spreads over the target with high flame and temperature.

These are impact insensitive with high ignition temperature. Igniter compositions required to ignite the incendiary gel should ignite within milliseconds and should produce long and large flame with high temperature for longer duration to ensure proper flame pick-up by gel incendiary.

**TABLE 15.7**

**Incendiary Gel Composition Used in AN 569 Incendiary Bomb [6]**

| Ingredients | Napalm filling (NP Type II) | IM filling | IM filling (Type III) |
|---|---|---|---|
| Napalm thickener | 9 | – | – |
| Gasoline | 91 | – | – |
| Isobutylmethacrylate polymer NR | – | 5 | – |
| Isobutylmethacrylate polymer AE | – | – | 2 |
| Fatty acid (stearic acid) | – | 2.5 | 3 |
| Napthenic acid | – | 2.5 | 3 |
| Aqueous solution of caustic soda (40%) | – | 3.0 | 4.5 |
| Gasoline | – | 87 | 87.5 |

**TABLE 15.8**

**Typical Incendiary Gel Composition (SR425) [3]**

| Ingredients | Percentage |
|---|---|
| Liquid hydrocarbon [SR 400] [benzene 95 rubber 5] | 50 |
| Aluminium powder | 10 |
| Sodium nitrate | 27 |
| Calcium silicate | 10 |
| Boric acid | 3 |

## 15.8 SPECIAL REQUIREMENTS OF SOLID INCENDIARY THERMITE COMPOSITIONS:

Incendiary thermite compositions should possess the following special features

a. Should be ignitable with suitable igniting composition
b. Should have high energy density
c. Should have high combustion temperature preferably above 2,000°C with high heat of combustion
d. Must burn at a rate suitable for the ignition of the material to be ignited like wood, grass, fuel, etc.
e. Their combustion should almost be gasless with formation of slag, which should be of low melting point and non-volatile type and which should spread uniformly
f. Should be difficult to quench the combustion

## 15.9 DIFFERENCES BETWEEN SOLID INCENDIARY AND GEL INCENDIARY COMPOSITIONS

The differences between gel/liquid incendiary and solid incendiary are given in Table 15.9.

**TABLE 15.9**
**Comparison of Different Incendiary Systems [7]**

| Property | Gel/liquid incendiaries | Solid incendiaries |
|---|---|---|
| Heat output (Cal cm$^{-3}$) | 8,000 | 18,100 |
| Temperature ($^0$C) | 670 | 2,000 |
| Duration of burning (sec) | 60 | 150 |
| Stickiness | Sticky | Non-sticky |
| Storage life (year) | 0.5 | 10 |
| Suitability | Not suitable for spinning projectiles | Suitable for spinning and non-spinning projectiles |

Thermite compositions can be used for demining operations [8] of anti-tank and anti-personnel mines by armed forces.

## REFERENCES

[1] S.H. Fisher and Mark C. Grubelish, *"Theoretical Energy Release of Thermites, Intermetallics and Combustible Metals"*, 24th International Pyrotechnics Seminar, Monterey, CA, US, 1 June 1998.

[2] A.A. Shidlovskiy, *"Principles of Pyrotechnics"*, 3rd Edition, Moscow, 1964. (Translated by Foreign Technology Division, Wright-Patterson Air Force Base, Ohio, 1974), American Fireworks News, 1 July 1997. ©1997, Rex E. &S. P., Inc.

[3] *"Service Textbook of Explosives"*, 1972, Ministry of Defence, UK., Reprinted in India in 1976.

[4] *"Design of Ammunition for Pyrotechnic Effect"*, AMCP 706-188, Engineering Design Handbook Military Pyrotechnics Series, Part Four, Headquarters, U S Army Material Command, 5001, Eisenhower Ave, Alexandria, VA 22304, March 1974.

[5] T.T. Griffiths," *Alternative for Perchlorates in Incendiary and Pyrotechnic Formulations for Projectiles"*, Department of Defense Strategic Environmental Research and Development Program (SERDP) Project WP-1424, QinetiQ, August 2009, Fort Halstead, Sevenoaks, Kent, TN14 7BP, UK.

[6] *"Fire Warfare Incendiaries and Flame Throwers"*, Summary Technical Report of Division 11, NDRC Volume 3, National Defense Research Committee Washington, DC, 1946.

[7] J.P. Agrawal, *"High Energy Materials: Propellants, Explosives and Pyrotechnics"*, ©Wiley-VCH Verlag GmbH & Co. KGaA, Weinheim, 2010. ISBN: 978 -3-527-32610-5. Reprinted with permission.

[8] R.A. Walker, D. M. Bergeron, M.P. Braid, B. Jeyakumar, and J.M. L. Mah, *"Experimental Assessment of Two Exothermic Systems to Neutralize Landmines"*. Technical Report DRDC Suffield TR 2006-050, February 2006, Defence Research and Development Canada.

# 16 Simulating Compositions

## 16.1 ROLE

The simulating compositions (also known as flash and sound compositions) mimic the sound of burst, flash and smoke of the regular ammunition. In a conventional war field or anti-terror strikes, the enemy may be deceived by producing sound and flash similar to the sound and flash of actual live rounds. These simulating compositions serve several purposes.

a. Deceiving the enemy into believing that the firing is on (like imitating the effects of ground bursts, gun flashes, grenades, air bursts and boobytraps, etc.)
b. Training of the personnel to acquaint with war-like situations
c. Reduced expenditure of regular service ammunition since these are cheaper for training
d. Minimising the wear and tear of weapon systems
e. Salute in military protocol

## 16.2 TYPICAL SIMULATING FLASH AND SOUND COMPOSITIONS

Some typical simulating flash and sound compositions are given in Table 16.1.

Ellern [2] has given some flash and sound-producing compositions as given in Tables 16.2 to 16.5.

Cracker compositions containing aluminium and potassium chlorate-based compositions are filled in bicat strips. The ignition is through quick match strands. These bicat strips produce flash and bang equivalent to the flash and bang of a battlefield to deceive the enemy.

The whistling composition generally contains fuel and oxidisers as given below.

*Fuel:* Gallic acid ($C_6H_2(OH)_3COOH$), potassium benzoate ($C_6H_5COOK$), sodium benzoate ($C_6H_5COONa$), potassium picrate ($C_6 H_2(NO_2)_3OK$), sodium salicylate ($C_6H_4(OH)COONa$), potassium dinitrophenolate ($C_6H_3(NO_2)_2OK$), potassium hydrogen phthalate ($KC_8H_5O_4$).

*Oxidiser:* Potassium chlorate, potassium perchlorate, potassium nitrate.

Some typical whistling compositions are given in Table 16.6.

Gallic acid and potassium chlorate are highly sensitive to shock and friction and hence their use is restricted.

Formulation efficiency of potassium benzoate: potassium perchlorate has been reported as below [3].

DOI: 10.1201/9781003093404-16

**TABLE 16.1**
**Simulating Flash and Sound Compositions [1]**

| Ingredient | Percentage | | | | |
|---|---|---|---|---|---|
| Potassium perchlorate | – | 64 | 50 | 40 | 73 |
| Sulphur | – | 10 | – | – | – |
| Aluminium | 9- | 22.5 | – | 26 | – |
| Antimony sulphide | – | 3.5 | 33 | – | – |
| Magnesium | – | – | 17 | 34 | – |
| Black powder | 91 | – | – | – | – |
| Gallic acid | – | – | – | – | 24 |
| Red gum | – | – | – | – | 3 |

**TABLE 16.2**
**Simulating Flash and Sound Compositions [2]**

| M100 Gun Flash Simulator | Formula 43 | |
|---|---|---|
| | Ingredient | Percentage |
| | Magnesium | 45 |
| | Potassium perchlorate | 35 |
| | Barium nitrate | 15 |
| | Barium oxalate | 3 |
| | Calcium oxalate | 1 |
| | Graphite | 1 |

**TABLE 16.3**
**Simulating Flash and Sound Compositions [2]**

| M115 Projectile Ground Burst Simulator | Formula 44 | |
|---|---|---|
| | Ingredient | Percentage |
| | Magnesium | 34 |
| | Aluminium | 26 |
| | Potassium perchlorate | 40 |

i. For carbon monoxide only:

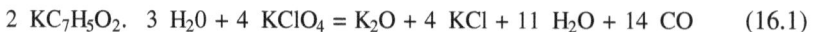

$$2 \ KC_7H_5O_2. \ 3 \ H_2O + 4 \ KClO_4 = K_2O + 4 \ KCl + 11 \ H_2O + 14 \ CO \qquad (16.1)$$

stoichiometric (fuel:oxidiser ratio of 44:56)

**TABLE 16.4**
**Simulating Flash and Sound Compositions [2]**

| Boobytrap Simulator | Formula 45 | |
|---|---|---|
| | **Ingredient** | **Percentage** |
| | Magnesium Gr A Type 1 | 17 |
| | Antimony sulphide Gr I or II, Class C | 33 |
| | Potassium perchlorate | 50 |

**TABLE 16.5**
**Simulating Flash and Sound Compositions [2]**

| Tank Gun Simulator | Formula 46 | |
|---|---|---|
| | **Ingredient** | **Percentage** |
| | Navy SPD propellant powder | 33.33 |
| | Magnesium/aluminium alloy 50:50 | 33.33 |
| | Potassium perchlorate | 33.33 |

**TABLE 16.6**
**Simulating Whistling Compositions [2]**

| Ingredient | Percentage | | | | |
|---|---|---|---|---|---|
| | Formula156 | Formula157 | Formula158 | Formula159 | Formula160 |
| Potassium chlorate | 73 | – | – | – | – |
| Potassium perchlorate | – | – | 70 | – | 72.5 |
| Potassium nitrate | – | 50 | – | 30 | – |
| Red gum | 3 | – | – | – | – |
| Gallic acid | 24 | – | – | – | – |
| Potassium picrate | – | 50 | – | – | – |
| Potassium benzoate | – | – | 30 | – | – |
| Potassium dinitrophenate | – | – | – | 70 | – |
| Sodium salicylate | – | – | – | – | 27.5 |

ii. For carbon dioxide only:

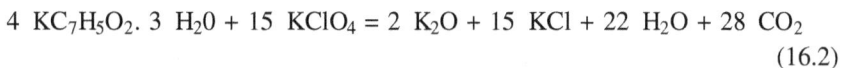

$$4\ KC_7H_5O_2.\ 3\ H_2O + 15\ KClO_4 = 2\ K_2O + 15\ KCl + 22\ H_2O + 28\ CO_2 \tag{16.2}$$

stoichiometric (fuel:oxidiser ratio of 29:71)
Readers can see more information in [3] with a lot of visual presentation.

## 16.3  MECHANISM OF WHISTLING SOUND PRODUCTION FROM SIMULATING COMPOSITIONS

The mechanism of whistling sound production is due to fast generation of gaseous combustion products, creating waves in the air, with same or higher speed of sound. The crystalline nature of the fuel causes intermittent burning, leading to whistling sound.

The cause of intermittent burning has been explained [4] in Figure 16.1. The whistling sound is produced by a combination of oscillatory burning of the composition plus the sound reflecting up and down the length of the tube above the burning surface. When a whistle composition is burned in a tube [4]:

a. A pressure wave rises in the tube.
b. The pressure wave reaches the end of the tube.
c. Part of the pressure wave passes out of the tube and part is reflected back into the tube.
d. When the reflected pressure wave reaches the burning surface, the increased pressure causes the composition to burn faster.
e. This produces a pressure wave that rises in the tube, beginning the process again.

In this way a repeated series of pressure waves is created in the air and is heard as a whistle. The pitch is determined by the length of the tube above the present burning surface.

The frequency of whistle composition depends on the length of the tube and linear burn rate decreases as the frequency increases. Table 16.7 shows the influence of oxidiser/fuel ratios on combustion characteristics of whistling compositions.

Composition M3 was found to be the most suitable due to its high intensity of whistle.

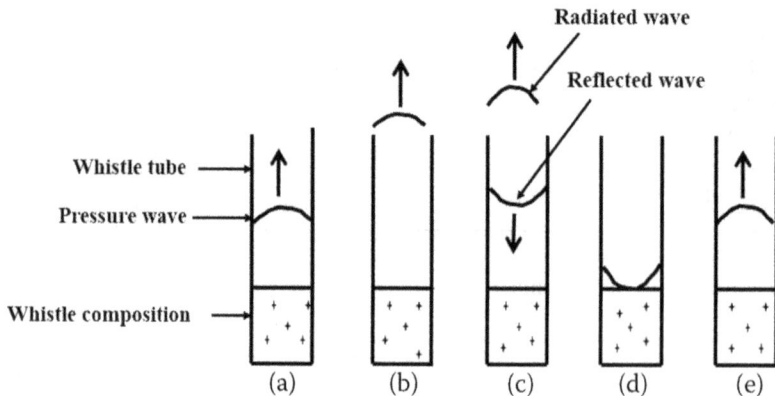

**FIGURE 16.1**  Burning of Whistle Composition (Reprinted from [4]).

**TABLE 16.7**
**The Influence of Oxidiser/Fuel Ratios on Combustion Characteristics of Whistling Compositions [5]**

| Samples | Potassium perchlorate | Sodium benzoate | Kb (%) | Qv (Kcal.kg$^{-1}$) | V0 (l.kg$^{-1}$) | u (mm.s$^{-1}$) | I (dB) | Tb (°C) |
|---------|----------------------|-----------------|--------|---------------------|------------------|-----------------|--------|---------|
| M1 | 60.3 | 34.4 | -42.7 | 576 | 488 | 3.6 | 116 | 465 |
| M2 | 63.3 | 31.4 | -36.3 | 668 | 454 | 4.4 | 119 | 477 |
| M3 | 66.3 | 28.4 | -29.9 | 749 | 407 | 5.1 | 124 | 471 |
| M4 | 69.3 | 25.4 | -23.5 | 837 | 359 | 5.3 | 123 | 450 |
| M5 | 72.3 | 22.4 | -17.1 | 927 | 312 | 5.7 | 122 | 461 |
| M6 | 76.3 | 18.4 | -8.6 | 1068 | 251 | 5.6 | 119 | 474 |
| M7 | 80.3 | 14.4 | -0.1 | 1185 | 203 | 5.2 | 99 | 466 |
| M8 | 84.3 | 10.4 | 8.4 | 969 | 220 | 3.8 | 82 | 463 |

Note: Kb- oxygen balance; Qv- heat of explosion; V0- volume of gaseous products; Tb- autoignition temperature; u- burning rate; I- intensity of whistles

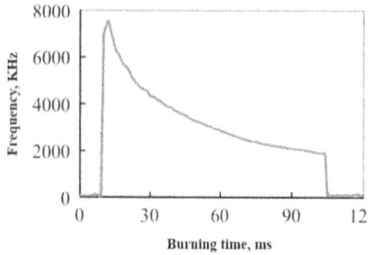

**FIGURE 16.2**   Variation of Frequency of Whistling
Sound with Burning Time (Reprinted from [5]).

## 16.4   MECHANISM OF DECAY OF THE FREQUENCY OF WHISTLING SOUND

The decay of the frequency of whistling sound with burning time is shown in Figure 16.2. This mechanism may be interpreted as follows [5].

As the composition burns with time [5], tube length above the composition is increased. This causes increased time needed for pressure wave propagation from the combustion front to the end of the tube. As is well known, whistling sound is formed by repeat propagation of the pressure wave in the tube and frequency of whistling sound is inversely proportional with the propagation time of pressure wave. Hence, as propagation time increases, the frequency of whistling reduces.

Readers may refer [6] on chemical kinetic theory of pyrotechnic whistles.

## 16.5   SPECIAL REQUIREMENTS OF SIMULATING COMPOSITIONS

In addition to the general essential requirements of compositions as given in Section 1.6, simulating compositions should preferably possess the following special requirements:

a. The luminous energy should be high for good visibility, both in day and night.
b. It must be ignitable by safety fuse or safety match fuzee as well.

## REFERENCES

[1] F.L. McIntyre, "*A Compilation of Hazard and Test Data for Pyrotechnic Compositions*", Contractor Report ARLCD-CR-80047, October 1980, U.S. Army Research and Development Command, Large Caliber Weapon System Laboratory, Dover, New Jersey.
[2] Dr. Herbert Ellern, "*Military and Civilian Pyrotechnics*", ©Chemical Publishing Co., New York, 1968.
[3] Joseph A. Domanico, "*The Secrets of Pyrotechnic Whistles*", Stargate 2000, VP Publications/Webmaster, Crackerjacks, Inc.
[4] K. L. and B.J. Kosanke and Clive Jennings-White, "*Lecture Notes for Pyrotechnic Chemistry, Pyrotechnic Reference Series Number 2*", Section 9, page 16–17@ Journal of Pyrotechnics Inc., 2004, USA Revision 4.0[ISBN 1-889526-16-9].

[5] Dam Quang Sang, Nguyen VanTinh, and Nguyen Van Bo, *"Several Combustion Behaviours of the Whistling Pyrotechnic Based on Potassium Perchlorate and Sodium Benzoate"* Vietnam Journal of Chemistry, 2019, Volume 57, Issue 3, Pages 272–276. © 2019 Vietnam Academy of Science and Technology, Hanoi & Wiley-VCH Verlag GmbH & Co. KGaA, Weinheim.

[6] Gregory Lyons and Richard Raspet, *"Chemical Kinetics Theory of Pyrotechnic Whistles"*, The Journal of the Acoustical Society of America, 2015, Volume 137, Page 2200.

# 17 Delay Compositions

## 17.1 ROLE

Pyrotechnic delays are used to provide a time interval between two successive events as per design requirement of the ammunition. Pyrotechnic delays are used across a wide range of military ammunitions and devices and aerospace platforms.
Some examples are as listed below:

a. Providing delays to ensure sufficient time or distance between personnel and ammunition performance like in 36-mm grenades, where a grenade is thrown through hand (4 seconds) or rifle (7 seconds)
b. Providing delay in fuzes so that the explosive in the ammunition explodes only after a short delay after it has penetrated the target like warhead penetration in runway/bunker
c. Providing delays for self-destruction in fuzes so that the fuze functions if it misses the target after specified delay, like 40-mm L/70 airborne anti-aircraft ammunition
d. Providing delay in fuzes so that fuze functions after elapse of specified time so that the desired range or height of ammunition is achieved, like 81-mm illuminating ammunition
e. Providing delays in illuminating ammunition for performance like opening of main parachute and candle after some delay to ensure that the effect of spin has been reduced like 155-mm illuminating ammunition (Section 17.7.3)
f. Providing delays in rockets, missiles, space vehicles for various synchronised activities

These delay compositions do not fall under the category of main composition, since special effects are produced by the main composition in the ammunition.

## 17.2 DELAY BURNING RATES

The burn rates are required to be consistent for safety and performance of the ammunition. To ensure that the delay timings do not vary much with very minor percentage change in the fuel or oxidiser, various compositions with varied fuel and oxidiser ratios are made and filled in delay columns of standard length. These are tested under identical conditions and the inverse burn rates are determined. These are then plotted on a graph with inverse burn rate on y-axis and fuel percentage on x-axis as shown in Figure 17.1 (based on Table 17.1).

It could be seen that in a delay composition, as the fuel is increased, the inverse burn rate ($sec.cm^{-1}$) decreases, reaching a flat curve. Then, further increase in fuel

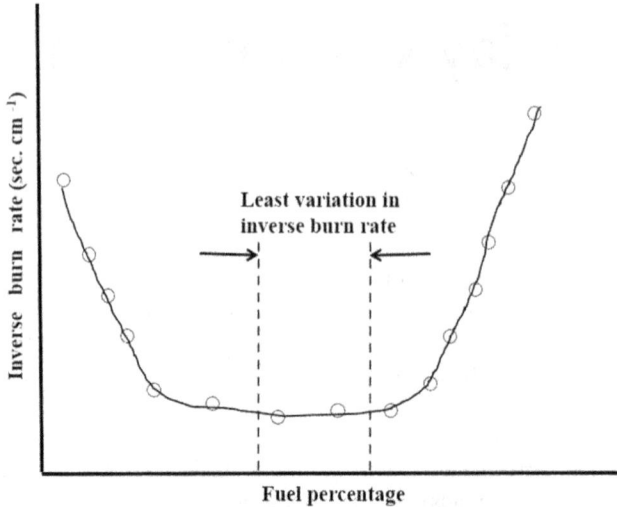

**FIGURE 17.1**   Inverse Burn Rate Variation with Fuel Percentage.

results in increase in the inverse burn rate. This interval of fuel percentage, where inverse burn rate variation is least, is most suitable for delay composition as a minor variation in fuel or oxidiser content would not substantially affect the delay time. The effect of fuel on inverse burn rate and heat of combustion for typical delay composition of Boron–Barium chromate composition is shown in Table 17.1.

The delay compositions may be divided into five types, namely extremely fast, fast, medium, slow and extremely slow. An old method for determination of burning rate of a delay composition is using lead tubes. The lead tube is filled with pyrotechnic delay composition by hand stemming and the ends are closed. The delay tube is allowed to pass through a set of reduction rollers with decreasing size of circular grooves so that the delay tube, being soft, takes the shape of the grooves and the composition is compacted inside the lead tube (Figure 17.2).

These are cut in small lengths and tested for delay timings. If not satisfactory, compositions are modified using new ingredients or with new sieve spectrum for ingredients. Delay burn rates are determined as given in Section 5.1.11.

The size (diameter and length) of the filled delay tube is made as per the design of the ammunition so that the same may be accommodated suitably (Figure 17.2).

## 17.3   DELAY BURNING TRAIN MECHANISM

Delay is filled in a pressed delay column (or semi-circular channel or lead tube). One of its end is filled and pressed with priming composition for receiving ignition from other stimulus and the other end of the delay column is filled with ignition transfer system consisting of relay composition. Delay burning takes place as below:

**TABLE 17.1**

**Typical Effect of Fuel on Inverse Burn Rate and Heat of Combustion of Boron–Barium Chromate Delay Compositions** [1]

| Boron percentage | Charge weight Mg | Average burning time (sec) | Average burning time (sec. $g^{-1}$) | Heat of combustion (cal. $g^{-1}$) |
|---|---|---|---|---|
| 3.0 | 2,130 | 7.56 | 3.549 | - |
| 3.5 | 2,150 | 3.54 | 1.646 | 278 |
| 4.0 | 2,140 | 1.72 | 0.804 | 354 |
| 4.5 | 2,125 | 1.44 | 0.678 | 400 |
| 5.0 | 2,130 | 1.09 | 0.512 | 420 |
| 6.0 | 2,110 | 0.767 | 0.364 | 231 |
| 7.0 | 2,000 | 0.653 | 0.327 | 453 |
| 8.0 | 2,000 | 0.560 | 0.280 | 462 |
| 9.0 | 2,000 | 0.539 | 0.270 | 474 |
| 10.0 | 1,975 | 0.465 | 0.235 | 515 |
| 11.0 | 1,925 | 0.432 | 0.224 | 536 |
| 13.0 | 1,900 | 0.397 | 0.209 | 556 |
| 15.0 | 1,875 | 0.382 | 0.204 | 551 |
| 17.0 | 1,800 | 0.375 | 0.208 | 543 |
| 19.0 | 1,750 | 0.366 | 0.209 | 535 |
| 21.0 | 1,685 | 0.375 | 0.223 | 526 |
| 23.0 | 1,650 | 0.407 | 0.247 | 503 |
| 25.0 | 1,625 | 0.433 | 0.266 | 497 |
| 30.0 | 1,611 | 0.574 | 0.356 | 473 |
| 35.0 | 1,500 | 0.965 | 0.643 | 446 |
| 40.0 | 1,430 | 2.19 | 1.531 | 399 |
| 45.0 | 1,360 | 5.25 | 3.860 | 364 |
| 50.0 | 1,290 | 14.5 | 11.24 | - |

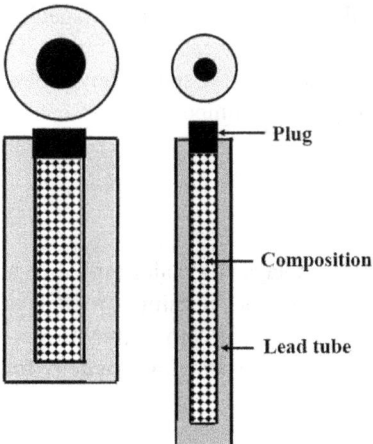

**FIGURE 17.2** Lead Tube Filling.

a. Ignition takes place from a source external to delay .

b. Priming composition (with low energy of activation) picks up flash with formation of hotspots and ignites booster composition (if any) and then ignites the delay composition.

c. The heat generated in delay compositions are mostly low and hence the rate of burning is low, except fast burning delays.

d. Since heat/flash energy of delay composition is mostly weak, a relay composition is used to positively ensure ignition of next explosive in the train of explosives.

e. The direction of combustion, direction of flame, direction of heat transfer and direction of gaseous combustion products takes place as given in Section 3.2.2.

The choice of a fuel and oxidiser depends on the burn rate requirement of delays. A fast-burning delay may be obtained using metallic fuels having higher heat of combustion with an oxidiser having low melting point and low decomposition temperature. A slow burning delay may be obtained using organic fuels with an oxidiser having higher melting point and higher decomposition temperature.

## 17.4   FACTORS AFFECTING DELAY TIME

Performance of the pyrotechnic composition depends on a number of factors (see Chapter 4). The following factors in the design of the delay element are considered to be the most important:

a. Type of ingredients and their characteristics

b. Proportion of ingredients and quantity of composition

c. Particle size of ingredients

d. Uniformity in blending of composition

e. Density of delay composition

f. Moisture content in delay composition

g. Length, diameter and thickness of the delay tube

h. Free space above delay column

i. Delay tube material's thermal properties like conductivity, density, specific heat capacity thermal effusivity and thermal diffusivity

j. Ammunition system working as closed or vented, i.e., free space above delay

k. Type and strength of ignition stimulus and priming composition

l. Spin of ammunition

m. Ambient temperature and pressure

n. Relay charge

However, non-ignition and interruption in burning of delays is a major problem. A higher propagation index gives better propagation of burning in delays (Section 3.15). Section 3.6 may also be referred for factors affecting time to ignition. It is recommended that delay bodies should not have extreme low diameter for delay compositions having low burn rates as lower diameters are at risk of failures or erratic burn rates especially at lower ambient temperatures (Section 4.3.3).

Since delay body materials are better conducting materials than delay pyrotechnic compositions, there is probability of failure in delay combustion, especially since delay compositions have low heat of combustion and hence lower burn rates. This burn rate further reduces due to the heat loss to delay body as well as surrounding atmosphere. The probability of failure becomes very high at extreme low temperatures when the rate of heat dissipation becomes very fast. Refer Table 4.32 for effect of various parameters on rate of combustion.

Hence delay body's material properties, dimension and thickness are very important. As an example, stainless-steel delay body material yield lower linear burn rates than aluminium delay body material due to higher thermal conductivity and thermal effusivity of aluminium (see Table 4.21).

Delay burn rate empirical Equations (4.20) and (4.21) may be referred, given in Section 4.5 and it could be inferred that linear delay burn rate increases with increase in thermal conductivity, increase in reaction temperature, increase of heat transfer, increase of heat of reaction, decrease in temperature difference between initial temperature and maximum temperature of the burning column, increase in thermal diffusivity, decrease in apparent activation energy for diffusion coefficient, decrease in particle size, decrease in specific heat and decrease in density.

## 17.5   ADVANTAGES/DISADVANTAGES OF PYROTECHNIC DELAY COMPOSITIONS

Pyrotechnic delay compositions have an edge over other modes of delays like mechanical delays and electronic delays for the following reasons:

a. It is a very simple chemical formulation, requiring simple machineries or equipment.
b. There are no moving parts/components required for the delays (as in mechanical delays).
c. There is no electrical power required for the delays (as in electronic delays).
d. A large number of delays from short to medium to long may be designed.
e. Electronic delays are much less durable than pyrotechnic delays, being comparatively fragile and, thus, susceptible to damage by the high acceleration or "g" loading experienced when the projectile is fired or the munition is launched or ejected [2].
f. Mechanical delays are less common for timing munitions because of their poor reliability. In particular, pre-wound spring mechanisms fatigue over time, and complex winding or other energising mechanisms are inherently less reliable [2].

The disadvantages of pyrotechnic delay compositions are:

a. The compositions are high energy materials, requiring safety precautions.
b. The compositions are sensitive to friction, impact, electrical energy and flash.

c. The compositions are hygroscopic in nature.

d. There are a large number of factors that affect the performance of the pyr-otechnic compositions (Chapter 4).

e. Pyrotechnic delays do not provide longer delays due to constraint in ac-commodating the delay column in the ammunition [2].

f. Slow burning delays are difficult to ignite and burn with inconsistency in burning rate and thus less reliable [2].

g. There is a risk of failure at very low ambient temperature.

h. The burn time varies sharply (even up to 20 percent of the mean value) at extreme temperatures of −40°c and +60°c.

## 17.6   DELAY-FILLED COMPONENTS

The delay-filled components vary in shape and sizes as per ammunition design. Following are the various types, though the list is not exhaustive.

a. Straight column filled with pyrotechnic composition

b. Semi-circular channel filled with pyrotechnic composition

c. Spiral or helix grooves filled with pyrotechnic composition

d. Lead tube filled with pyrotechnic composition

The most common is a metal tube of aluminium or brass or lead, etc., holding the compressed delay composition. Figure 17.3 depicts a typical filled delay unit in a straight column with incremental filling.

The filling is done in reverse order of burning. Thus, the relay composition is pressed first while priming composition is pressed last. A typical delay composition filling sequence is as presented below:

a. Relay composition (ignition transfer composition for ignition to the next in explosive train)

b. Delay composition 1st increment

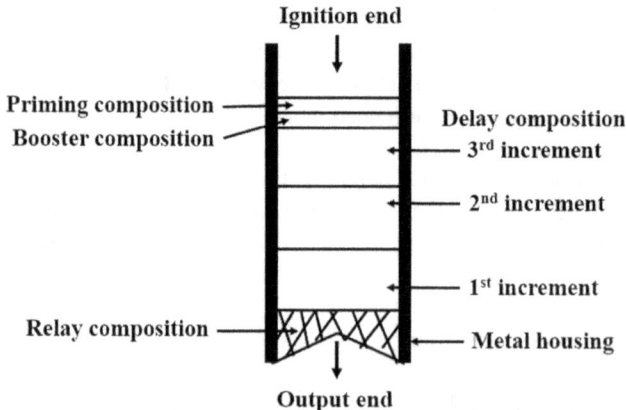

FIGURE 17.3   A Typical Delay Unit.

c. Delay composition 2nd increment

d. Delay composition 3rd increment

e. Booster composition (or intermediate composition or transitional composition)

f. Priming composition (or ignition composition or quick composition or first fire charge or starter mixture)

The delay requires priming (and booster) composition for picking up the ignition stimuli for burning. The top surface of the delay-pressed columns is serrated or made rough or uneven so as to increase the surface area for better pick-up of flash for positive ignition.

The number of increments depends upon the length and diameter of delay tube and bulk density of delay composition. The delays should be filled with increments having heights less than the diameter of the delay unit body. The delays, after their own full combustion, are required to ignite the next composition, outside the delay column. However, since the delay compositions are slow burning, they cannot readily ignite the composition which is outside the delay component. Hence, a flash-producing composition known as relay composition is used at the end, which carries the flash from the delay component to the other composition layers. This relay composition in many cases is shaped conical for better concentration of flash.

## 17.7   CLASSIFICATION OF DELAYS

Various types of delays may be manufactured by selection of suitable ingredients and their particle size and pressing load. The delays may be classified as follows:

a. Based on delay time like short, medium or long delays

b. Based on combustion products like gasless delays (low gas) or gassy delays (slag less)

c. Based on position and functioning in ammunition like projectile flight delays, safety delays, functioning delays, post impact delays, bore safety and muzzle safety delays

d. Based on design of ammunition like vented delays or obturated delays

e. Based on shape of delays like ring-shaped type, cylindrical type or lead fuse type

f. Castable delays

g. Environmentally benign pyrotechnic delays

However, all such classifications are interrelated as short, medium or long delays may be gasless or gassy and they may have another nomenclature due to its position and functioning in the ammunition. Gasless delay is used in obturated ammunition design and gassy delays are used in vented ammunition design.

### 17.7.1   SHORT-RANGE, MEDIUM-RANGE AND LONG-RANGE DELAYS [3]

Typical *short-range, medium-range* and *long-range delay composition* ingredients, mostly using NC as a binder, are as given below [3]]:

a. Short-Range Compositions (milliseconds):
   i. Zirconium, red lead, NC binder
   ii. Ferrosilicon, red lead, NC binder
   iii. Zirconium, barium chromate, NC binder
   iv. Zirconium, lead chromate, NC binder
   v. Zirconium/nickel alloy, potassium perchlorate, NC binder

b. Medium-Range Compositions (1–10 seconds):
   i. Antimony sulphide, barium chromate, NC binder
   ii. Antimony sulphide, barium chromate, NC binder
   iii. Sulphur, barium chromate, potassium perchlorate, NC binder
   iv. Antimony sulphide, lead chromate, potassium perchlorate, NC binder
   v. Mealed gunpowder, sulphur, binder

c. Long-Range Compositions (more than 10 seconds)
   i. Tetra nitro oxanilide, barium nitrate, potassium nitrate, Nc binder
   ii. Tetranitro carbazole, barium nitrate, NC binder

Readers may refer [4] for typical ingredients used in delay compositions for providing milliseconds delay, seconds delay and long delay.

## 17.7.2   Gasless (Low Gas) and Gassy Delays (Slag Less)

Let us now consider the two types of delay compositions, namely gasless delays (low gas) and gassy delays (slag less), based on combustion products.

### 17.7.2.1   Gasless (Low Gas) Delay

These compositions do not produce appreciable gas ($<0.005$ $m^3kg^{-1}$) during burning and are used where no venting of hot gases is possible in the design of the ammunition. Since a gasless delay composition does not produce appreciable gas and thus does not allow the pressure to build inside the ammunition system, the combustion rate of delay composition remains unaffected. Also, these burn faster as the consolidation is increased as consolidation causes the particles of metallic fuel and oxidiser to come closer, paving the way for the heat to be conducted to the next layers more effectively. These are useful in ammunitions under confinement conditions or at high altitudes.

Fuels generally used in gasless (i.e., low gas) delays are zirconium, titanium, tungsten, molybdenum, manganese, silicon, boron, ferrosilicon, nickel–zirconium alloy, selenium while oxidisers used are red lead, barium peroxide, potassium nitrate, barium nitrate and potassium perchlorate (Table 17.2 to 17.5).

Sulphur and organic binders are not used in gasless delay compositions as these produce quite large quantities of gas.

### 17.7.2.2   Gassy Delays

Gassy delays on combustion produce appreciable gases ($>0.02$ $m^3.kg^{-1}$). As the combustion of delay composition continues, the gas pressure continues to build up, causing fast burning of the delay composition. Hence, for unaffected burning of the

## TABLE 17.2
## Gasless Delay Compositions [1]

| Ingredient | Percentage | | | | | |
|---|---|---|---|---|---|---|
| Manganese | 30–45 | – | – | – | – | – |
| Barium chromate | 0–40 | 89–96 | 40–44 | 22 | 70 | – |
| Lead chromate | 26–55 | – | – | – | – | – |
| Diatomaceous earth | – | – | – | – | – | 8 (max) parts |
| Boron | – | 4–11 | 13–15 | – | – | – |
| Chromic oxide | – | – | 41-46 | – | – | – |
| Nickel–zirconium mix | – | – | – | 5/31 | 5/17 | – |
| Potassium perchlorate | – | – | – | 42 | 8 | – |
| Silicon | – | – | – | – | – | 20 |
| Red lead | – | – | – | – | – | 80 |

## TABLE 17.3
## Gasless Delay Compositions [1]

| Ingredient | Percentage | | | | | |
|---|---|---|---|---|---|---|
| Barium chromate | 70–60 | 59–46 | 46–5 | 70–40 | – | 60 |
| Diatomaceous earth | – | 5–12 | 3–10 | – | – | – |
| Nickel–zirconium alloy | – | – | – | – | – | 26 |
| Potassium perchlorate | 10 | 9.6 | 4.8 | 10 | – | 14 |
| Molybdenum | 20–30 | – | – | – | – | – |
| Tungsten | – | 27–29 | 39–87 | 20–50 | – | – |
| Zirconium | – | – | – | – | 28 | – |
| Lead dioxide | – | – | – | – | 72 | – |

## TABLE 17.4
## Gasless Delay Compositions with Inverse Burn Rate [1]

| Ingredient | Percentage | | | | | | | |
|---|---|---|---|---|---|---|---|---|
| Barium chromate | 95 | 90 | 40 | 60 | 32 | 41 | 58 | – |
| Boron amorphous | 5 | 10 | – | – | – | – | – | – |
| Potassium perchlorate | – | – | 10 | 10 | 5 | 5 | 10 | – |
| Tungsten | – | – | 50 | 30 | 58 | 49 | 27 | – |
| Red lead | – | – | – | – | – | – | – | 80 |
| Silicon | – | – | – | – | – | – | – | 20 |
| Celite (added) | – | – | – | – | – | – | – | 3–7 |
| | Performance | | | | | | | |
| Inverse burn rate (sec.in$^{-1}$) | 1.5 | 0.6 | 12.5 | 31 | 1 | 10 | 40 | 4–11 |

**TABLE 17.5**

**Gasless Delay Compositions with Inverse Burn Rate [1]**

| Ingredient | Percentage | | | | | | | |
|---|---|---|---|---|---|---|---|---|
| Barium chromate | 60 | 60 | 30 | 35 | – | – | 42 | 70 |
| Zirconium–nickel alloy (70–30) | 9 | 3 | – | – | – | – | – | – |
| Zirconium–nickel alloy (30–70) | 17 | 23 | – | – | – | – | – | – |
| Potassium perchlorate | 14 | 14 | – | – | – | – | 22 | 8 |
| Lead chromate | – | – | 37 | 33 | – | – | – | – |
| Manganese | – | – | 33 | 42 | – | – | – | – |
| Selenium | – | – | – | – | 16 | – | – | – |
| Barium peroxide | – | – | – | – | 84 | – | – | – |
| Talcum (added) | – | – | – | – | 0.5 | – | – | – |
| Lead peroxide | – | – | – | – | – | 28 | – | – |
| Zirconium | – | – | – | – | – | 72 | 5 | 5 |
| Nickel | – | – | – | – | – | – | 31 | 17 |
| | | | | | Performance | | | |
| Inverse burn rate (sec.in$^{-1}$) | 6 | 11 | 9.45 | 12.5 | 2.3 | <0.5 | 6.5 | 17.5 |

composition due to pressure built up, it is convenient to release the excess pressure by suitably designing the ammunition. In a delay column, which is small, the quantum of gas production would be less and hence the effect of pressure of gases on the combustion of delay composition will also be less. However, if the delay column length is increased to have higher delay time, the effect of gas pressure on combustion will be more pronounced, needing a more free space, thereby necessitating the delay to be bulky. However, it is difficult to accommodate a bulky delay element in ammunitions. Hence, ammunitions using such gassy delays are vented.

Even long delays with very little pressure output are required to be vented as pressure builds up inside in due course, thereby altering the burn rate.

Therefore, if a gassy composition is chosen then it is necessary that the ammunition system is vented so as to allow excess pressure to be vented outside. To avoid the effect of external ambient conditions like moisture, these vents are closed using a seal (aluminium foil or lead foil or wax), which are opened under pressure, thus allowing venting of the gas pressure.

These burn slower as the consolidation pressure is increased due to decrease in the porosity of the consolidated composition, thus not allowing hot gases to penetrate further.

Burning time of a gassy delay composition depends upon the gas pressure and may be represented by the following empirical formula [5].

$$t = \alpha + k/p^n \text{ (sec)} \qquad (17.1)$$

**TABLE 17.6**
**Gassy Delay Compositions**

| Ingredient | Percentage | | |
| --- | --- | --- | --- |
| | | SR227 | |
| Potassium nitrate | 75 | 72 | – |
| Sulphur | 9 | 7 | – |
| Charcoal | 16 | 21 | – |
| Ammonium perchlorate | – | – | – |
| Starch | – | – | – |
| Tetra nitro oxanilide | – | – | 40 |
| Barium nitrate | – | – | 60 |

where $t$ = burning time (sec), $\alpha$ = factor depending upon composition (sec), $k$ = factor depending on composition (dimensionless), $p$ = pressure (lb.in$^{-2}$), $n$ = factor depending upon composition (dimensionless). The numerical values of the factors are $a = 0.26$, $k = 1.0$ and $n = 0.13$ for the 95.4/4.6 barium chromate/boron composition.

The fuels used in gassy delays are organic materials since they produce large quantity of gaseous products, as given below:

a. Starch, sugar, resins, gums, charcoal, chlorinated rubber, polyvinyl chloride, tetra nitro oxanilide, tetra nitro carbazole
b. The oxidisers used in gassy delay are oxy salts like potassium nitrate, barium nitrate and potassium chlorate

Typical gassy delay compositions are given in Table 17.6.

At high temperatures, both types of delay burn faster, thereby reducing the burning time of the delay composition. However, the effect is more predominant with gasless (low gas) delay composition. Gasless delays are better than gassy delay compositions as these provide consistent delay and remain unaffected due to variation in pressure.

Cord type safety fuses with gunpowder filling are also used as delays in many ammunitions (see Chapter 21).

### 17.7.3 Position and Functioning in Ammunition

This classification is based on its functioning in ammunition, as given below:

a. *Projectile Flight Delay*: A delay in functioning of ammunition after desired delay is used in many types of ammunition. For example, one version of 51-mm illuminating ammunition has a delay setting of 3.5 second, 6.0 second,

9.5 second and 11.5 second to enable the user to select desired delay to function the ammunition at corresponding range desired (300 to 900 m). The desired delay element is aligned with the ignition train by setting the ammunition suitably before firing.

b. *Safety Delay*: Delays are provided in hand or rifle grenades which causes the grenades to function after 4 second or 7 seconds. This gives sufficient delay for the user to take safe position.

c. *Functioning Delays*: There are delays provided to ensure that the delay functions after certain activity of the ammunition. An example is 155-mm illuminating ammunition wherein the sub-system canister of the ammunition is initially expelled after the fuze has functioned (i.e., after desired mechanical or electronic delay) and hangs in the air by the auxiliary parachute. The main parachute and candle are then released after some delay, using another delay element, after sufficient reduction in spin of the ammunition, enabling the main parachute to open properly in air with the candle.

d. *Post Impact Delay*: Delays are also provided in some ammunition to ensure that the ammunition does not function immediately on hitting the target but functions after certain delay enabling the ammunition to penetrate well inside the target for more damage. This is especially used in bombs for damaging the runways or bunkers and also in anti-aircraft ammunition like 40-mm L/70. The destruction of the target is less if the ammunition functions immediately on hitting the target as ammunition is still outside the target. The delay enables functioning of the ammunition inside the target and thus ensures that destruction is maximum.

e. *Self-Destruction Delays*: Delays are also used for self-destruction of the ammunition if the ammunition misses the target. This is particularly important for airborne targets as projectiles missing the airborne target are destroyed in the air itself and do not fall over the ground on own troops or civilians. An example is self-destruction delay element in fuze 104Mk12 used in 40-mm L/70 ammunition.

f. *Bore Safety and Muzzle Safety Delays*: Delays are used in some fuzes so that fuze never function inside the barrel or near the barrel. On firing, this delay element burns and takes time such that the projectile is out and away from the barrel and then the fuze is ready to function. An example of bore safety and muzzle safety delay is gunpowder pellet used in fuze 104 Mk12 for 40-mm L/70 ammunition which does not allow the fuze striker to get armed till the delay pellet is burnt completely.

## 17.7.4 Vented or Obturated Delays

A vent in the ammunition system is introduced when the delay system is gassy. This allows the release of gas pressure and thus the gas pressure remains unchanged and does not affect the rate of burning of delay compositions. A vent is not necessary in case of non-gassy delay compositions. Therefore, vented and obturated delays are associated with gassy and gasless delays, respectively.

## 17.7.5 Shape of Delays Like Ring-Shaped, Cylindrical or Lead Fuse Type

Lead fuse type and cylindrical shapes are common and are explained in Figures 17.2 and 17.3. The ring-shaped are generally semi-circular in shape as shown in Figure 4.11. Safety fuses are long cord type with gun powder filling. Safety fuses are used in making detonators to function by suitably crimping with detonators. It also provides means for ignition to pyrotechnics like Igniter safety fuse, electric as well as for mining purposes.

## 17.7.6 Castable Delays [6]

Castable delay compositions were formulated by the Thiokol Chemical Corporation. These compositions contained potassium dichromate with iron and boron powders dispersed in a polysulphide binder. The best results were obtained with a polysulphide liquid polymer, LP-3, in which the powders were dispersed. The mixture was injected into the delay bodies and the polymer was converted to an elastic rubber by heating to 170°F. A small quantity of sodium tetraborate dec-ahydrate was used as the curing agent. It releases its water of hydration at about 170°F to initiate the cure. This injection system for filling the bodies with delay mixtures can be readily adapted to quantity production. It also may make it possible to design delay compositions in unusual shapes that could not be loaded with dry powder. The burning time of T6E3 units loaded with one of these compositions is shown in Table 17.7.

The burning time of these compositions was shown to be almost independent of pressure from 30 to 2,100 psia. This characteristic may allow the delay composition to be ignited directly from burning propellant in a mortar fuze or rocket motor, since the high propellant gas pressure would not affect the burning time of the delay.

## 17.7.7 Environmentally Benign Pyrotechnic Delay Compositions

Many delay compositions [7] like $W/BaCrO_4/KClO_4/VAAR$; Zr-Ni alloy/$BaClO_4$/ $KClO_4$; $W/BaCrO_4/KCLO_4$/Diatomaceous earth; $Mn/BaCrO_4/PbCrO_4$; Boron (amorphous)/$BaCrO_4$ contain ingredients like barium chromate, potassium per-chlorate, potassium chlorate and lead chromate. These chromates, perchlorates and heavy metals like lead and barium are environmentally hazardous compounds due

**TABLE 17.7**
**Burning Time of Castable Delay Composition in T6E3 Delay Unit [6]**

| Temperature (°C) | Average(sec) | Minimum(sec) | Maximum(sec) |
|---|---|---|---|
| −70 | 15.65 | 13.78 | 16.43 |
| 80 | 13.10 | 12.59 | 15.75 |
| 170 | 13.03 | 12.11 | 14.01 |

**TABLE 17.8**

**Standard Deviation and Limit Values of Fe-BaO$_2$ Composition Burn Rate [10]**

| Iron content (%) | Minimum burn rate (mm.s$^{-1}$) | Maximum burn rate (mm.s$^{-1}$) | Standard deviation (mm.s$^{-1}$) |
|---|---|---|---|
| 15 | 18.29 | 22.94 | 1.22 |
| 20 | 18.92 | 23.33 | 1.01 |
| 25 | 21.99 | 23.73 | 0.48 |
| 30 | 17.95 | 20.47 | 0.59 |
| 35 | 12.27 | 14.38 | 0.50 |
| 40 | 8.67 | 9.63 | 0.28 |
| 45 | 5.42 | 6.60 | 0.25 |
| 50 | 3.84 | 4.35 | 0.14 |

to toxicity issues. They pose health hazard especially when they percolate into soil and water. However, due their reliability, the compositions are still in use. Some environmentally benign pyrotechnic time delay compositions developed include compositions based on combinations of Mn/MnO$_2$, Ti/C-3Ni/Al, W/KIO$_4$/Sb$_2$O$_3$, B4C/NaIO$_4$/PTFE and W/MnO$_2$.

Readers may see references [7], [8] and [9] on development of environmentally benign time delay compositions.

### 17.7.8  Burn Rates of Some Delay Compositions

Table 17.8 shows standard deviation and limit values of Fe-BaO$_2$ delay composition burn rate.

The paper [10] mentions that susceptibility to ageing necessitates subjecting gasless mixtures to a *stabilisation process*. The stabilisation process typically consists of conditioning the composition at an elevated temperature, for an extended period, such as a week. The stabilisation process is meant to accelerate the ageing processes to an extent that the solid-state products of the pre-ignition reactions create a barrier between components of the composition, stopping further ageing. By making use of this process, one can ensure that the burning rates of a particular composition will be consistent and stable over time.

## 17.8  SPECIAL REQUIREMENTS OF DELAY COMPOSITIONS

In addition to the general essential requirements of compositions as given in Section 1.6, delay compositions should preferably possess following special requirements:

a. It should have a well-defined burning rate with a given composition.
b. It should have consistency in delay time to achieve close tolerance of delay time.

c. It should have a wide range of burn rates with variation of ingredients.

d. Its burn rate should not change with minor variation of ingredients.

e. It should have high functional reliability to ensure quick ignition.

f. It could be pressed in suitable shape and size of delay bodies to meet end requirement of assembly in the ammunition.

g. It must match the ammunition requirement by having specified burn time and combustion products.

Readers may refer [11] for development of delay composition and its consistency of delay time, [12] regarding development and testing of precision pyrotechnic delay with incorporation of a Single Bridgewire Apollo Standard Initiator (SBASI) for initiation in assembly, [13] for development and performance studies of pyrotechnic compositions for pressure-generated and delay cartridges and [14] on review of gasless pyrotechnic time delays.

## REFERENCES

[1] *"Theory and Application"*, AMCP 706-185, Engineering Design Handbook Military Pyrotechnic series, Part One, Headquarters, US Army Material Command, Washington DC, April 1967.

[2] Gregory D. Knowlton, Bruce B. Anderson, and Theodore B. Gortemoller, *"Heat Transfer Delay"*, Patent WO 2001039586A2.

[3] K. R. K. Rao, "Pyrotechnics", *Proceedings of Seminar on Technological Development of Pyrotechnics"*, Ordnance Factory Dehuroad, India, 30–31 December 1983.

[4] A. Bailey and S.G. Murray," *Explosives, Propellants and Pyrotechnics"*, ©Brassey's World Military Technology, 1989, ISBN 1857532554, Redwood Books, Trowbridge, Wiltshire, UK.

[5] *"Design of Ammunition for Pyrotechnic Effect"*, AMCP 706-188, Engineering Design Handbook Military Pyrotechnics Series Part FOUR, Headquarters U S Army Material Command, 5001, Eisenhower Ave, Alexandria, VA 22304, March 1974.

[6] Raymond H. Comyn, *"Pyrotechnic Research at DOFL Part II. Pyrotechnic Delays"*, TR-1015, 15 February 1962, Armed Forces Technical Information Agency, Arlington Hall Station, Arlington 12, Virginia (Diamond Ordnance Fuze Laboratories, Ordnance Corps, Department of The Army, Washington 2S, D.C.

[7] Jay C. Poret, Anthony P. Shaw, Eric J. Miklaszewski, Lori J. Groven, Christopher M. Csernica, and Gary Chen, *"Environmentally Benign Energetic Time Delay Compositions: Alternatives for the [5US Army Hand -Held Signal"*, 40th International Pyrotechnic Seminar, Colorado, Springs, CO, 13–18 July 2014, Pages 305–314.

[8] Eric J. Miklaszewski, Anthony P. Shaw, Jay C. Poret, Steven F. Son, and Lori J. Groven, *"Performance and Aging of Mn/MnO2 Environmentally Friendly Energetic Time Delay Composition"*, Sustainable Chemistry and Engineering, 2014, Volume 2, Pages 1312–1317, American Chemical Society.

[9] Eric J. Miklaszewski, Anthony P. Shaw, Jay C. Poret, Steven F. Son, and Lori J. Groven *"Ti/C – 3Ni/Al as a Replacement Time Delay Composition"*, Propellants, Explosives, Pyrotechnics, 2014, Volume 39, Pages 138–147.

[10] M.M. Gerlich and A.T. Wojewodka, *"Study of Gasless Compositions used in Time-Delay elements Fe/BaO2 Composition"*, Journal of Thermal Analysis and Calorimetry, 2019, Volume 139, Pages 3473–3479. 10.1007/s10973-019-08721-8

[11] Azizullah Khan, Zulfiqar H. Lodhi, and Abdul Qadeer Malik, *"Development and Experimental Investigation on Delay Time Consistency of Modified Si/PbO/Pb$_3$O$_4$/ FG Pyrotechnic Delay Composition"*, Engineering, Technology & Applied Science Research, 2017, Volume 7, Issue 6, Pages 2167–2170.

[12] S.J. Salter, R.E. Lundberg, and G.L., McDoZlgdl, *"SBASI-Actuated Pyrotechnic Time Delay Initiator"*, NASA Contractor Report CR-2357, June 1975, National Aeronautics and Space Administration, Washington, DC 20546.

[13] Azizullah Khan, *"Development and Performance Studies of Pyrotechnic Compositions for Pressure-generated and Delay Cartridges"*, Ph.D. Thesis, December 2018, School of Chemical and Materials Engineering (SCME) National University of Sciences and Technology (NUST) H-12, Islamabad, Pakistan.

[14] Walter W. Focke, Shepherd M. Tichapondwa, Yolandi C. Montgomery, Johannes M. Groblerand, and Michel L. Kalombo, *"Review of Gasless Pyrotechnic Time Delays"*, Propellants, Explosives, Pyrotechnics, 30 May 2018, Volume 44, Issue 1.

# 18 Infrared Flare Compositions

## 18.1 ROLE

It is well known that some pyrotechnic compositions give energy output in the form of visible light radiation and heat. These are in the visible range of spectral band. This visible range of light consists of wavelengths 0.380 microns to 0.780 microns, approximately. By changing some ingredients of the pyrotechnic composition, it is possible to obtain infrared radiation. The special features of infrared radiation are:

a. Invisible to human eyes as it gives radiation in infrared zone and are used mostly for night warfare as infrared illuminating flares and as infrared decoys for infrared seeking missiles.
b. The temperature generated during infrared radiation is less than that generated by visible radiation.
c. The infrared radiation has lower frequency than visible radiation and hence has lower energy than visible radiation (Equation 9.1).
d. It has better capability of penetrating through smoke, fog and more useful in night warfare using night vision devices.

The infrared radiation from infrared compositions are used for:

a. *Covert Illumination:* This is done by deploying infrared flares attached to parachutes in the night over the battlefield and tracking the enemy movement and its equipment through night vision devices.
b. *Covert Photography:* Night photography using infrared films.
c. *Decoys:* Decoys save the aircraft by release of burning infrared composition pellets that produce just more infrared radiation than the aircraft itself, attracting the infrared heat seeking missile as decoy target and thus saving the aircraft.
d. *Tracking the path of wire-guided missile.*
e. *Training the aircrew* in accurate firing of combat missiles.

## 18.2 ELECTROMAGNETIC SPECTRUM

It is considered appropriate to discuss the electromagnetic spectrum for better understanding of infrared flares. There are no sharp defined boundaries in these regions. Light and infrared are the only parts of the electromagnetic spectrum that humans are able to directly sense [1]. The short wavelength side edge of infrared begins where our eyes' response ends, which is approximately 0.7 μm (700 nm).

DOI: 10.1201/9781003093404-18

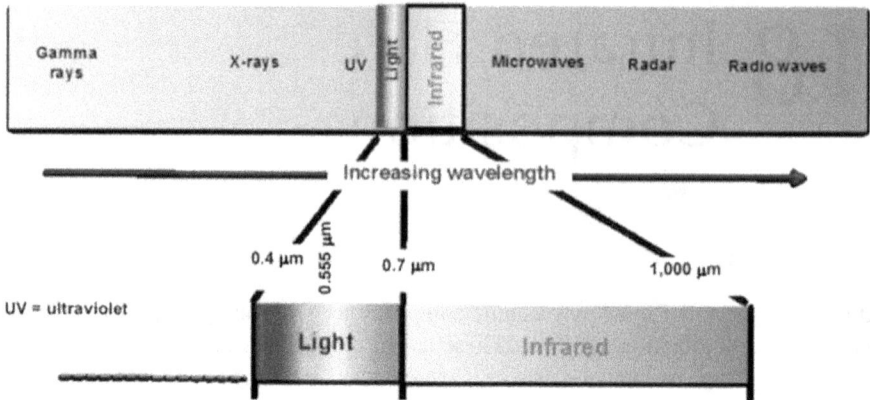

**FIGURE 18.1** Electromagnetic Spectrum (Reprinted from [1]).

The long wavelength limit is less sharply defined but is usually specified as about 1,000 μm. The practical long wave limit with today's sensor technology goes only to about 14 μm. Figure 18.1 shows the location of infrared on the electromagnetic spectrum. Properties vary greatly across the IR, with several sub-bands of particular interest to aircraft.

Remember GXULIMR as the first letters of each electromagnetic radiation (shown in the Figure 18.1) where wavelength increases towards right while frequency decreases towards right. Infrared radiations have higher wavelengths than light (visible radiation) but less than microwave radiation.

## 18.3    INFRARED RADIATION FROM TARGET

The range adjacent to the visible spectrum is called the "near infrared" and the longer wavelength part is called "far infrared." Infrared radiation is capable of greater penetration in fog and smoke,. The values of the infrared wavelengths are ; near infrared 0.78 to 2.5 μm, middle infrared 2.5 to 50 μm and far infrared 50 to 1,000 μm.

All tanks on the ground, war ships in the sea and fighter aircrafts and helicopters in the air emit infrared radiations in various wavelengths. This again depends upon the shape and size of the tanks, war ships, fighter aircrafts and helicopters. These infrared radiations are not fully allowed to pass through the atmosphere due to presence of molecules of water vapour, carbon monoxide, carbon dioxide, ozone, methane and dust particles. The infrared radiations are selectively absorbed and scattered by these molecules and dust in the air. Atmospheric transmittance plays a significant role in infrared signature levels of aircraft. The attenuation of infrared radiation of the aircraft in the atmosphere depends on the wavelength of infrared radiation, temperature and mixture of gases in the air. Typical atmospheric transmittance for tactical purposes is shown in Table 18.1.

The distribution and details of all infrared signatures of fighter aircraft are shown in Figure 18.2 [3].

**TABLE 18.1**

**Atmospheric Transmittances for Tactical Purpose** [2]

| Range (km) | Transmittance | | | |
|---|---|---|---|---|
| | 0.25–0.28 µm | 2.0–3.5 µm | 3.5–5.0 µm | 8.0–12.0 µm |
| 0.5 | 0.366 | 0.608 | 0.725 | 0.921 |
| 1.0 | 0.139 | 0.546 | 0.667 | 0.877 |
| 2.0 | 0.022 | 0.477 | 0.593 | 0.808 |
| 5.0 | 0.001 | 0.379 | 0.467 | 0.657 |
| 10.0 | – | – | 0.354 | 0.482 |
| 20 | – | – | 0.233 | 0.288 |

Radiation from each component in the aircraft has a different spectral distribution [1] and, consequently, propagates through the atmosphere with different degrees of attenuation. The total infrared signature of an aircraft is the sum of its components, but each component does not make an equal contribution in all aspects. As Figure 18.3 shows, a component's contribution to the total infrared signature of an aircraft depends upon the aspect angle. For a typical aircraft or helicopter, the dominant mid-wave signature component(s) in each region are as follows [1]:

1. *Tail:* Engine hot parts
2. *Rear Quarter:* Hot parts and exhaust plume
3. *Beam to Forward Quarter:* Airframe and exhaust plume
4. *Nose:* Airframe and intakes

Table 18.2 shows the likely range of aircraft signature.

Similarly, a typical infrared radiation emitted by a war ship is given below.

a. Exhaust 3–5 microns intensity 3–5 kw
b. Hull and deck 8–14 microns intensity 0.5–2 kw

Three sub-bands (near IR 0.7-1.5 µm, Mid Wavelength IR 1.5-6.0 µm and long wavelength IR 7.0–14 µm) are of particular military interest. Of these, the MWIR from approximately 1.5 to 6.0 µm is of the greatest concern to aircraft because that is where most of the antiaircraft missiles operate [1].

An infrared seeking missile detects the combined radiation and identifies the aircraft as a probable target. The improved versions of missiles can differentiate the radiations from various sources. Therefore, these targets may be easily detected through infrared detection equipment and destroyed by infrared-guided surface-to-air missiles (SAMs) or infrared-guided air-to-air missiles (AAMs) by sensing the infrared radiation emitted by the target.

(a)

Plume

Aerodynamically
heated surfaces

Hot engine parts

Rear fuselage

Sunshine / Skyshine

Earthshine

(b)

aircraft hot
parts emission

engine hot parts, exhaust
nozzle, tailpipe

aircraft plume
emission

radiations from hot $CO_2$,
water vapour ($H_2O$)

reflectd earthshine

radiation from earth reflected
off the airframe

TOTAL IR
SIGNATURE

skin emission

aerodynamically heated
skin, plume heated skin

reflected sunshine

reflected solar radiation
off the airframe, canopy

reflected skyshine

sky radiations reflected
off the airframe

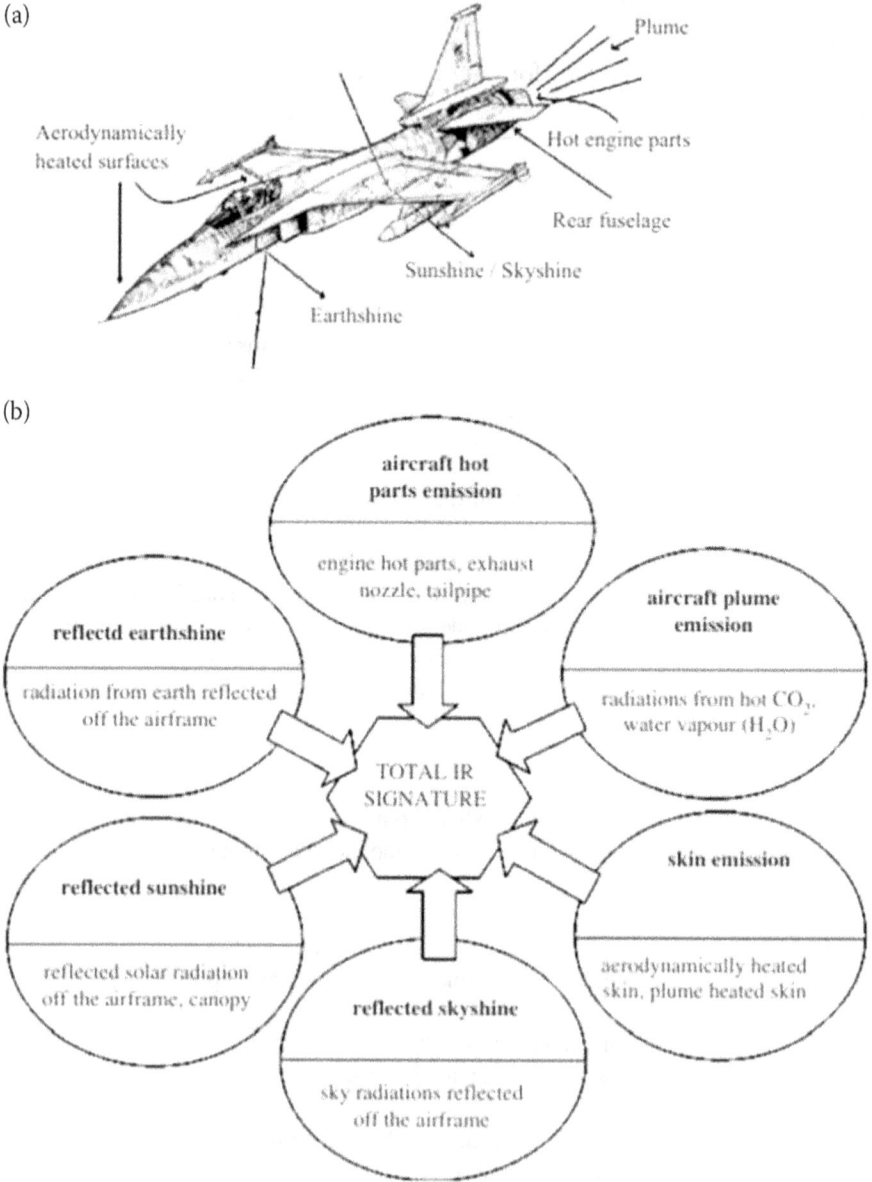

FIGURE 18.2   Sources and Details of IR Signature of Aircraft: (a) Sources of IR Radiance
from Typical Fighter Plane and (b) Distribution and Details of IR Signature of Aircraft
(Reprinted from [3]).

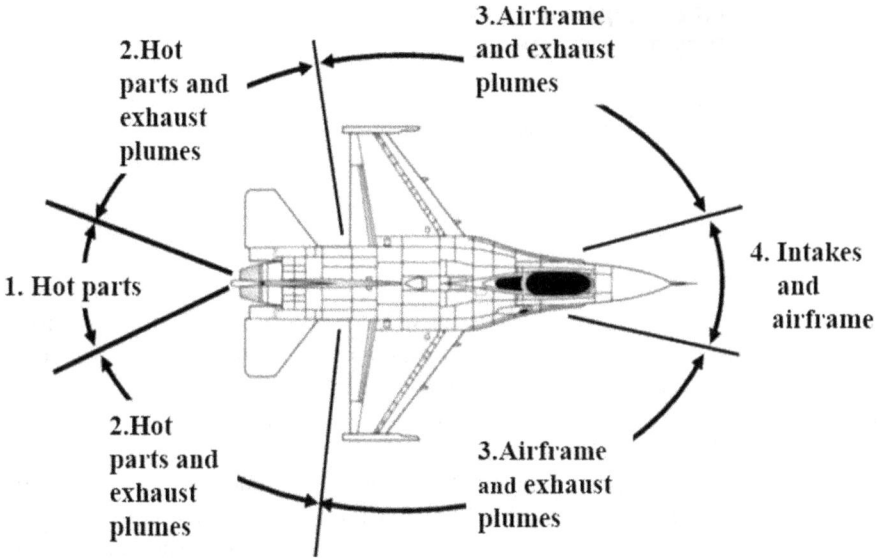

**FIGURE 18.3**  Aircraft Component Signature Dominance with Aspect Angle (Reprinted from [1]).

**TABLE 18.2**
**Typical Aircraft Signature Levels [2]**

| Aircraft type | Intensity ($W.sr^{-1}$) | |
|---|---|---|
| | 2–3 μm | 3–5 mμ |
| Rotary wing | 10–100 | 10–300 |
| Fixed wing (propeller) | 20–200 | 200–500 |
| Jet fighter | 50–1,000 | 100–10,000 |
| Jet transport | 100–1,000 | 100–5,000 |

Aircraft infrared defence has three main elements [1]:

a. The suppression of aircraft emissions to reduce the range at which the aircraft can be acquired and tracked.
b. A warning receiver to detect missile launch and cue a countermeasure response.
c. Countermeasure devices and systems, of which there are two types: off-board (decoys) and onboard (jammers).

## 18.4 MANUFACTURE OF INFRARED FLARE COMPOSITIONS FOR DECOYS

Infrared flare decoy compositions are saviours of fighter aircrafts, helicopters and ships against infrared seeking missiles.

There are two methods for manufacture of composition for decoy flares.

    a. Compositions for pressing
    b. Compositions for extrusion

Both pressed and extruded flares have their merits and performance differences, often in their burn characteristics. The pressed flare has a more homogeneous form giving a more consistent burn characteristic, whereas the extruded flare is much cheaper to produce but is more liable to break on ejection.

Carbon black is considered a black body with maximum emissivity (near 1) and hence infrared flare composition should be designed to produce carbon with heat of combustion so as to release infrared radiation through the carbon particles. The infrared compositions for decoys are manufactured mostly using magnesium powder, PTFE (polytetrafluoroethylene) and viton as binder cum plasticiser. The viton coating on magnesium prevents aerial oxidation of magnesium and improves the homogeneity of composition and facilitates manufacture of pellets. These infrared compositions contain higher magnesium content (55%–65%).

The major reaction products are magnesium fluoride, magnesium and solid carbon. The proportion of carbon as a combustion product remains relatively constant over a large range of fuel or oxidant concentrations.

The combustion products [4] are:

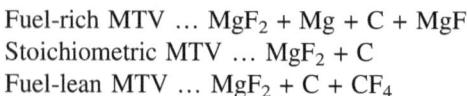

Fuel-rich MTV ... $MgF_2 + Mg + C + MgF$
Stoichiometric MTV ... $MgF_2 + C$
Fuel-lean MTV ... $MgF_2 + C + CF_4$

These equations list only the major reaction products (each with at least 10% of the reaction product mass). The other products that form are mainly $CF_2$ (g) and HF (g). The HF (g) is formed from the decomposition of Viton, which contains hydrogen. It does not come from the Teflon.

A simplified combustion scheme for stoichiometric (all Mg, H and F converted to $MgF_2$ and HF) and fuel-rich MTV compositions is given by the reaction [5]:

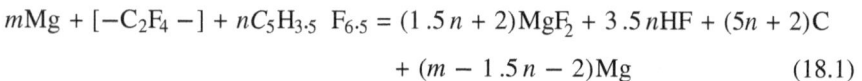

$$m\text{Mg} + [-C_2F_4 -] + nC_5H_{3.5} F_{6.5} = (1.5n + 2)\text{MgF}_2 + 3.5n\text{HF} + (5n + 2)\text{C}$$
$$+ (m - 1.5n - 2)\text{Mg} \qquad (18.1)$$

For stoichiometry, a value of $m$ corresponds to every $n$. For $n = 0$ (no Viton), the stoichiometric ($m = 2$) Mg mass fraction corresponds to 32.7%.

Ernst–Christian Koch [6] has presented Equations (18.2) and (18.3) as the idealised reaction steps of the combustion of Mg-polytetrafluoroethylene-based

(Mg - PTFE) pyrotechnic composition with an excess of magnesium in air, Mg and PTFE react within a primary anaerobic zone close to the pyrolant surface yielding condensed magnesium fluoride, carbon (soot) and an excess of vaporised magnesium. Upon mixing with air in the outer parts of the flame, carbon and magnesium are oxidised producing CO, $CO_2$ and MgO. The aerobic afterburn reaction ((18.3)) is more exothermal than the primary anaerobic reaction (Equation(18.2)) and largely dominates the radiant performance of a black body infrared flare.

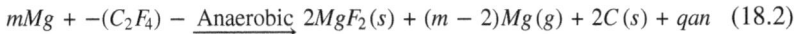

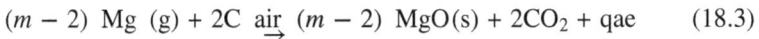

$$mMg + -(C_2F_4) - \underline{\text{Anaerobic}} \; 2MgF_2(s) + (m-2)Mg(g) + 2C(s) + qan \quad (18.2)$$

$$(m-2) \; Mg \; (g) + 2C \; \underline{\text{air}} \; (m-2) \; MgO(s) + 2CO_2 + qae \quad (18.3)$$

Like all pyrotechnic compositions, it is possible to modify proportion of ingredients to have fast, medium and slow inverse burn rate composition by using various percentages of magnesium as cut and blown and oxidiser PTFE as coarse and fine. The fast burn rate MTV composition is applied to the grooves of the pellets as priming mix for facilitating flash pick-up, improving rise time and uniform burn rate.

## 18.5   INFRARED DECOY FLARE DESIGN [4]

In the design of decoy flares, the amount of Viton in the formulation is largely governed by the manufacturing method. For example, if the flare is pressed, low percentages of Viton may be used; if the flare is extruded, larger quantities are required. This leaves the amount of magnesium and Teflon to be varied to meet the specification for the end use of the system. It would seem logical that the magnesium to Teflon ratio is based on three major requirements:

a. Ignitability
b. Burn rate
c. Performance (decoy effect)

First, in the design of pyrotechnic systems, it is usual to have a fuel-rich system to enhance the system ignitability.

Second, the burn rate for MTV formulations increases with decreasing reaction temperature. Results have shown that decreasing reaction temperature occurs for increased magnesium percentage. For decoy applications, the MTV formulation must have a very high bum rate, which requires a high magnesium percentage.

Finally, the infrared output of MTV flares are largely due to the grey body radiation from the hot carbon particles. Results have shown that the mass fraction of C(s) is not very dependent on the percentage of magnesium for a wide range of temperatures and pressures. Therefore, high magnesium percentages do not significantly reduce the decoy effect of the MTV formulation. These facts indicate that although the magnesium percentage in standard MTV compositions is far from ideal

in terms of heat of combustion and reaction temperature, it has probably been opti-
mised for burn rate and ignitability, while retaining acceptable decoy performance.

Thus, although the maximum reaction temperature and the maximum heat of
reaction occur near the stoichiometric MTV formulation, i.e., approximately 33%
magnesium, standard extruded MTV decoy flares use a formulation based on
54% magnesium. The ratio for extruded infrared flare reported is 54:30:16, since
higher magnesium content improves ignitability and lowers combustion temperature, thereby increasing the combustion rate. However, other compositions in
the ratio of 55:40:5 have also been used.

## 18.6 SPECTRALLY MATCHED INFRARED PAYLOADS

The aircraft emits spectral radiations in two or more bands. MTV compositions
were adequate for heat-seeking missiles in a single infrared band. The new missile
seeker heads are tuned to respond to actual signature of the aircraft. These missile
seeker heads are intelligent in distinguishing between radiation emanated from
aircraft at low temperatures to that of an infrared induced fake hot cloud. These
missiles with intelligent seeker heads are also equipped to analyse the two band
intensities and their ratios and compare with the one from the fake cloud produced
by infrared flares and reject/ignore the latter for the mismatch.

Spectrally matched infrared payloads provide a selective emission spectrum to
match the actual signature of the aircraft (Table 18.3).

## 18.7 PYROPHORIC FLARES [8]

The MTV flares are quite effective against older type missiles that seek heat in a
single infrared band. However, modern missiles employ counter–counter measures

**TABLE 18.3**
**General Composition of Spectrally Matched Infrared Payloads [7]**

| | | | | | | |
|---|---|---|---|---|---|---|
| Potassium perchlorate | X | X | X | | | |
| Ammonium perchlorate | | | | X | | |
| HTPB | | | | X | | |
| Kraton | | X | | | | |
| Viton | X | | X | | | |
| Benzene tetracarboxylic anhydride | | X | | | | |
| Potassium benzoate | X | | | | | |
| Boron | | | X | | | |
| Teflon | | | X | | | |
| Red phosphorous | | | | | X | |
| Polyvinyl acetate | | | | | X | |
| Diethylene glycol dinitrate | | | | | | X |
| Nitrocellulose | | | | | | X |
| Nitroglycerine | | | | | | X |

(CCM). Their more refined seeker heads use two or more spectral bands in an attempt to distinguish between the flare and the aircraft. Both infrared and ultra-violet (UV) bands may be used. Trajectory discrimination may also be used by some seeker heads and the physical size of the heat source will become more important in the future as imaging seekers are developed. Alternatives to MTV flares have therefore been considered in recent years and in particular flares that use the combustion of pyrophoric liquids to generate an intense heat source have been shown to be particularly effective.

Pyrophoric flares have the following principal advantages [8]:

a. The infrared emission from the flames produced by some pyrophoric liquids is similar to that produced from burning aviation kerosene which is largely a molecular emission of carbon dioxide and water. A continuum component from radiating hot particles can be added in a controlled manner by varying the pyrophoric fuel composition. Thus, the infrared spectral emission profile can be made to closely match that of a jet aircraft exhaust plume and hot engine metal.

b. The ultraviolet radiant intensity from pyrophoric flames is much less than that from MTV flares so that a much closer spectral match is achieved with a jet aircraft exhaust plume.

c. The flame from a pyrophoric flare can be several metres in length and it is therefore much closer in physical size to a jet engine plume than is the MTV flare which is typically less than a metre in length.

d. The trajectory of a launched pyrophoric flare can be varied by altering the aerodynamic properties of the container, whereas the trajectory of an MTV flare is fixed by the properties of the burning surface of the pellet used in the flare.

e. Since pyrophoric fuels use air as the oxidant, the fuel may be stored sepa-rately from the oxidant. MTV flares on the other hand, comprise an intimate mixture of oxidant and fuel so that when they are ignited, they are very hard to extinguish.

f. Under normal conditions, pyrophoric liquids ignite spontaneously when sprayed into air and thus no ignition mechanism is required. To effectively protect high-performance jet aircrafts from modern missiles, a pyrophoric infrared decoy flare must function effectively under extreme conditions of high air speed, high altitude, and low temperature.

The radiant intensity/time profile of the pyrophoric flare depends upon the fuel mass, the oxygen pressure and the fuel and oxygen exit aperture or orifice dia-meters. These parameters are easily adjustable to obtain the desired profile.

Pyrophoric flares are either liquids or solid-based materials and ignite instantly upon contact with air. The drawback is that manufacture of pyrophoric infrared flare material is cumbersome.

## 18.8   FACTORS AFFECTING INFRARED FLARE OUTPUT

Infrared output of a decoy depends upon following factors:

a. Use of good radiating material, i.e., type and proportion of ingredients for infrared output and burn time
b. Rate of heating up of the radiating material for quick release of radiation, i.e., suitable priming of pellet
c. Mass burn rate of pellet, i.e., density of pressed pellet
d. Size and geometry of pellet
e. Flame temperature of burning pellet as emissivity is dependent on temperature of the radiating material) and this temperature should be less than illuminating compositions to ensure that the emission does not shifts towards visible radiation (Section 4.2.5).
f. Nature, concentration and emissivity of combustion products for radiation in specified wavelengths (in tune with infrared detection device sensitivity in specified wavelengths)

The infrared pellet output and burn time are varied to suit the type of fighter aircraft or helicopter for its infrared signature.

The shape of the pellet, i.e., geometry of the pellet can be made different (as circular, rectangular or with central hole or having serrations over the body giving different shapes) with various sizes, each having different burning characteristics (Section 4.2.16). The UK and the US flares are similar in shape (mostly rectangular), size, form and performance in infrared output and burn time. Similarly, the Russian and French flares are similar in shape (mostly circular), size, form and performance in infrared output and burn time.

Density is important as it prevents breaking of pellet and thereby gives consistent performance. It is generally pressed in double acting press (simultaneous pressing from top and bottom). However, pellets are also made by extrusion method where paste of composition is allowed to pass through a die under hydraulic pressure at high temperature to form infrared cakes which are further cut to proper size and grooved as required for enhancing surface area and coated with priming mix composition for quick flash pick up.

## 18.9   SPECIAL REQUIREMENTS OF INFRARED DECOY
## FLARE COMPOSITIONS

In addition to the general essential requirements of compositions as given in Section 1.6, infrared flare compositions should meet the following special requirements:

a. The composition should give the desired infrared radiation effect with optimum amount of composition to make it attracting target equivalent to the aircraft for a short duration, away from the aircraft.
b. The radiation rise time should be the least for instantaneously attracting the heat-seeking missiles.

c. The burn time should be more than the missile engagement time.

d. It should withstand extreme fluctuations of temperature, pressure and wind speed at high altitudes.

There are two major drawbacks of MTV flares [8].

a. The MTV flares upon release quickly decelerate and travel in ballistic trajectory away from the aircraft.

b. The MTV flare's spectral distribution does not match fully with that of a typical aerial target.

## 18.10   ILLUMINATING INFRARED FLARE COMPOSITIONS

For covert operations, it is desirable that near infrared (NIR) illuminant flare produces both infrared as well as visible light output in a favourable manner. The NIR illuminant illuminates the war field/target with the high intensity NIR. The highly illuminated war field/target may be observed through infrared sensitive night vision goggle/device only since the low visible light radiation makes it impossible to be seen through the naked eye. This helps in night vision for observation, recognition and action.

The NIR pyrotechnics are found useful in perimeter protection, emergency aircraft landing and enemy and tanks movement. The ammunition used are handheld flares for signal purpose or parachute flares for observing a large war field through mortars, artillery, rocket or bomb. NIR illuminants also provide a larger illuminated area after ignition compared to visible flares.

Infrared illuminating flares are required to have a high infrared intensity, faster burn rate and higher *"concealment index"* (measured as the ratio of infrared radiation to visible radiation). A higher visible radiation would give a lower concealment index. Fuels preferred are those with lower burning temperature like silicon. The oxidisers are generally caesium nitrate and potassium nitrate. Some compositions do contain hexamine to improve infrared illumination performance.

A very high value of concealment index is preferred. A few typical illuminating infrared compositions with silicon, caesium nitrate, hexamine and epoxy resin are given in Table 18.4.

Another set of infrared illuminating composition has been given in Table 18.5.

Infrared intensity is measured using infrared radiometer through suitable filters and is expressed as watts/steradian, written as $kw.sr^{-1}$. Infrared efficiency is the product of the infrared intensity and burn time per gram of composition.

## 18.11   SPECIAL REQUIREMENTS OF INFRARED ILLUMINATING FLARE COMPOSITIONS

In addition to the general essential requirements of compositions as given in Section 1.6, illuminant flares composition should meet following special requirements:

**TABLE 18.4**

**Infrared Illuminating Compositions [9]**

| Ingredient | Percentage | | | |
|---|---|---|---|---|
| Silicon | 9 | 9 | 9 | 9 |
| Potassium nitrate | – | 30 | – | 30 |
| Caesium nitrate | 70 | 40 | 70 | 40 |
| Epoxy resin Epon 828 | 4 | 4 | 4 | 4 |
| Hexamine | 16 | 16 | 16 | 16 |
| Nitrocellulose | 1 | 1 | – | 1 |
| Black powder | – | – | 1 | – |
| | **Performance** | | | |
| Burn time (sec) | 17.67 | 30.41 | 19.61 | 17.49 |
| Burn rate ($sec^{-1}$) | 0.042 | 0.0289 | 0.44 | 0.050 |
| NIR intensity (w) | 44.8 | 36.0 | 41.3 | 50.3 |
| NIR efficiency (w-s. $g^{-1}$) | 22.6 | 31.0 | 23.1 | 25.1 |
| Visible intensity (cd) | 156.4 | 134.6 | 143 | 174 |
| Efficiency (2-cd. $g^{-1}$) | 78.9 | 116.1 | 80.1 | 87.1 |
| Infrared/visible light intensity | 0.286 | 0.267 | 0.289 | 0.2 |

**TABLE 18.5**

**Infrared Illuminating Compositions [10]**

| Ingredient | Percentage | | |
|---|---|---|---|
| Silicon | 10 | 10 | 16.3 |
| Potassium nitrate | 70 | – | – |
| Caesium nitrate | – | – | 78.7 |
| Rubidium nitrate | – | 60.8 | – |
| Hexamethylenetetramine | 16 | 23.2 | – |
| Epoxy resin | 4 | 6 | 5 |
| | **Performance** | | |
| Wavelength emitted (micrometre) | 0.76 | 0.79 | 0.8–0.9 |

a. High near infrared radiation (>25 w) and low visible radiation (<350 cd)
b. Fast burn rate
c. Favourable concealment index INIR/IVIS (near infrared light intensity/visible light intensity ratio)
d. No chunking out of composition (causing less infrared illumination, low burn time and risk of fire due to burning chunks on ground)

## 18.12   TRACKING FLARE COMPOSITIONS

*Tracking flares* use fuel as magnesium, silicon, zirconium or alloys of iron–silicon or magnesium–aluminium with oxidisers as barium peroxide or iron (iii) oxide and charcoal or graphite as additive and polymeric binders. These compositions provide both infrared radiation and visible radiation for tracking.

A patent [11] refers to tracking infrared flares to be attached to rockets for tracking purposes. The patent mentions that in the generation of infrared radiation for tracking purposes, four considerations are of importance.

    a. The amount of heat that is liberated during the burning of the flare, since emissivity depends upon the temperature of the radiating material.
    b. The rate of release of the heat since this rate determines how fast the radiating material is heated up.
    c. The particular material which is radiating, as some materials are better emitters than others.
    d. The particular wavelength of the radiation emitted, since the detecting apparatus is sensitive to only certain wavelengths.

Table 18.6 gives a comparison of several flares as to total infrared radiation emitted at ambient and at 65,000 feet. Data on typical metal-nitrate flares are

**TABLE 18.6**
**Comparison of Infrared Tracking Flares [11]**

| Name of Flare | Composition | Watts/Steradian per sq. in. burning surface (0.8 to 3.5 microns) | |
|---|---|---|---|
| | | Ambient | 65,000 feet |
| 1. BuOrd Mk 21 Mk O | 54% Mg<br>34% sodium nitrate<br>12% Laminac | 677 | 500 |
| 2. Applicants' flare | 54% Mg<br>23% Teflon<br>23% Kel-F | 2,283 | 1,070 |
| 3. Army "Rita" flare | 66.7% Mg<br>28.5% sodium nitrate<br>4.8% binder | 1,000 | – |
| 4. Optimum aluminium-<br>    Teflon | 48% Aluminium<br>52% Teflon | 1,700 | – |
| 5. Optimum boron-<br>    Teflon | 56% Boron<br>44% Teflon | 445 | – |
| 6. Optimum zirconium-<br>    Teflon | 54% $ZrH_2$<br>46% Teflon | 428 | – |

Note: Teflon = polytetrafluoroethylene and Kel-F = polytrifluorochloroethylene

presented, as well as data on flares using other metals than magnesium with teflon.

The reaction of magnesium with Teflon and Kel-F (polychlorotrifluoroethylene) produces sufficient heat, and produces it fast enough to be quite useful as an infrared source in the desired wavelength. A further advantage is the fact that carbon, which is produced during the burning of the flare, is a very good emitter in the desired wavelength range. Ignition of the flare is reliable and the burning is steady and even.

Readers may see reference [12] for comparison between MT and MTV combustion and effect of viton on combustion characteristics.

## REFERENCES

[1] Jack R. White, *"Aircraft Infrared Principles, Signatures, Threats, and Countermeasures"*, NAWCWD TP 8773, Naval Air Warfare Center Weapons Division, Point Mugu, CA, September 2012.

[2] Neal Brune, *"Expendable Decoys"*, Chapter 4, Volume 7, Countermeasure Systems, The Infrared and Electro Optical Systems Handbook, Editor David H. Pollock, Executive Editor, Joseph S Accetta and David L. Schumaker, Copublished by Infrared Analysis Information Center, Environmental Research Institute of Michigan, Ann Arbor, Michigan, USA and SPIE Optical Engineering Press, Bellingham, Washington USA. ©1993 The Society of Photo Optical Instrumentation Engineers.

[3] Shripad P. Mahulikar, Hemant R. Sonawane, and G. Arvind Rao, *"Infrared Signature Studies of Aerospace Vehicles"*, October–November 2007, Progress in Aerospace Sciences, Volume 43, Issues 7-8, Page 218–245, ©Elsevier Ltd.

[4] L.V. de Yong and K.J. Smit, *"A Theoretical Study of the Combustion of Magnesium /Teflon/Viton Pyrotechnic Compositions"*, MRL Technical Report MRL-TR-91-25, December 6 1991, Material Research laboratory, Cordite Avenue, Maribyrnong Victoria, Australia.

[5] Arie Peretz, *"Investigation of Pyrotechnic MTV Compositions for Rocket Motor Igniters"*, Journal of Spacecraft and Rockets, 1984, Volume 21, Issue 2, Pages 222–224 , @American Institute of Aeronautics and Astronautics, Inc. 1982.

[6] Ernst-Christian Koch, Arno Hahma, Volker Weiser, Evelin Roth, and Sebastian Knapp, *"Metal-Fluorocarbon Pyrolants. XIII: High Performance Infrared Decoy Flare Compositions Based on $MgB_2$and $Mg_2Si$ and Polytetrafluoroethylene/ Viton"*, Propellants, Explosives, Pyrotechnics, July 2012, Volume 37, Issue 4, Page 2, Copyright Wiley-VCH GmbH. Reproduced with permission.

[7] Ernst-Christian Koch, *"On the Sensitivity of Pyrotechnic Countermeasure Ammunitions"*, 35th International Pyrotechnics Seminar at Fort Colling, Colorado, USA, July 2008, Page 533, Copyright Wiley-VCH GmbH. Reproduced with permission.

[8] John L. Halpin, Maurice Verreault, and Simon A. Barton, *"Flame-~Stabilized Pyrophoric IR Decoy Flare"*, United States Patent Number 5,136,950, 11 August 1992.

[9] Patricia M. Farnell, Russell Broad, and Stuart Nemiroff, *"Munitions Using Infrared Flare Weapon Systems"*, Patent Publication Number WO 1998002712 A1, 22 January 1998.

[10] *"Military Explosives"*, Department of The Army Technical Manual No. TM 9-1300-214, Headquarters, Department of Army, Washington DC, September 1984.

[11] George T. Hahn, Paul G. Rivette, and Rodney G. Weldon, *"Infrared Tracking Flare"*, United States Patent No. 567921, 21 October 1997.

[12] Farid C. Christo, *"Thermochemistry and Kinetics Models for Magnesium /Teflon/ Viton Pyrotechnic Compositions"*, DSTO-TR-0938, December 1999, Weapons Systems Division, Aeronautical and Maritime Research Laboratory, Melbourne.

# 19 Priming and Booster Compositions

## 19.1 ROLE

All pyrotechnic compositions need a steady heat from external stimuli (like flame or hot particles from caps, squibs, impulse cartridges safety fuses, fuzes and lasers, etc.) to reach their ignition temperature. It is the lowest temperature to which a composition must be heated so as to cause a spontaneous combustion and is described as *"ignition temperature"* (see details in Section 3.5). It is easy to ignite a loose pyrotechnic composition but difficult to ignite a pressed pyrotechnic composition. Therefore, most of the pressed pyrotechnic compositions require *"igniting compositions"* for their ignition. These igniting compositions are of two types.

a. *Priming composition* (also known as ignition composition or quick composition or first fire charge or starter mixture) which receives the first ignition stimulus in the composition and ignites the main composition.
b. *Booster composition* (also known as intermediate composition or transitional composition or transfer mix) consisting of a mix of priming and main composition in definite proportion and receives ignition from priming composition and ignites the main composition.

These priming and booster compositions do not fall under the category of main composition. These are different from initiatory compositions which are used in caps/detonators (sometimes referred as *"priming mix"* or *"primer mixes"* compositions).

Priming compositions aid in ignition of many pyrotechnic compositions by readily igniting themselves from external stimuli either alone or in combination with booster compositions to ignite the main composition.

## 19.2 MECHANISM OF IGNITION BY PRIMING AND BOOSTER COMPOSITIONS

The composition with low energy of activation (low ignition temperature) may be ignited easily by an external stimuli. However, there are compositions with moderate ignition temperature and activation energy, which require priming compositions having low ignition temperature to ignite the composition See para 3.14.5 and Figure 3.22 as to how priming composition aids in ignition of composition. But there are compositions which have considerable high energy of activation and high ignition temperature. Such pyrotechnic compositions need priming cum booster compositions. In the sequence of pressed pyrotechnic compositions, priming compositions have low ignition temperature while main composition has high

DOI: 10.1201/9781003093404-19

ignition temperature while booster compositions have intermediate ignition temperature. Hence, it is easy for priming compositions to ignite the booster composition, which, in turn, can easily ignite the main composition.

As ignition depends upon the amount of heat transferred to the main composition from the combustion of priming composition or priming cum booster composition, a priming composition or booster composition with higher combustion temperature with hot slags will provide better ignition to the main composition. It has been observed that a priming composition or in combination with booster composition, which allows sufficient time for transfer of heat to the main composition, is better for ignition.

Priming and booster compositions in small quantities are pressed at lower pressing load than the main composition to ensure quick pick-up of ignition from external stimuli and hot slags as combustion products ensure sufficient time for transfer of heat to the main composition. They are not filled loose as this would lead to very fast combustion of priming composition and very low duration of heat supply from priming or priming cum booster combination (Figure 19.1).

Pyrotechnicians, over the years, have devised suitable priming and booster compositions for ignition as given below.

a. The priming compositions are pressed or pasted over the main composition and are the first to receive the external stimuli and thereby ignite the main composition.
b. The booster compositions are incorporated between priming and main composition for step-by-step ignition of main composition having very high ignition temperature.

During combustion of a priming composition, some quantity of gaseous products of combustion moves away from the composition layer to be ignited and thus the ignition depends upon only heat transmitted by the flame and hot residue of priming composition. Hence, a solid hot residue on combustion of priming composition with negligible gas output is desirable to continue heating the layer to raise the

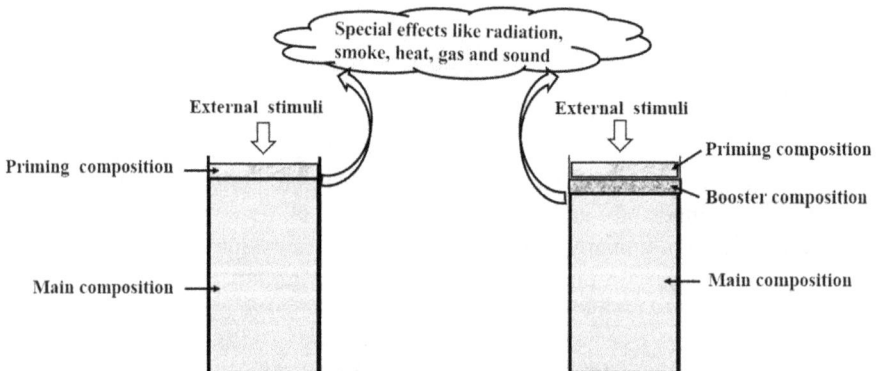

**FIGURE 19.1** Priming and Booster Composition Over Main Composition.

temperature of the layer of the most of the main pyrotechnic compositions to ignition temperature.

*Priming alone or in combination with booster compositions are pasted or pressed with the pyrotechnic composition to be ignited, thus become one unit. This differs from igniter compositions for ignition of propellants where the igniter compositions are kept away from the composition to be ignited and compositions are ignited through hot combustion products of igniter composition.*

All the main compositions like illuminating composition, smoke composition, signal composition, incendiary composition, tracer composition, delay composition, infrared flare composition, simulating composition, etc. need these igniting compositions for ignition.

The use of a priming composition depends upon the ignition and performance characteristics of the main composition to be ignited. Thus, low combustion temperature priming compositions are used in smoke and delay compositions while high combustion temperature priming compositions with hot slag are used in illuminating, tracer and incendiary compositions. Hence, a suitable choice of these compositions is necessary for each type of composition.

## 19.3   TYPES OF PRIMING COMPOSITIONS

There are several types of priming compositions used for igniting a pyrotechnic composition:

### 19.3.1   COMPOSITION BASED ON GUN POWDER

Sulphur-free gunpowder (SFG or PN 623) (70.5 parts of potassium nitrate and 29.5 parts of charcoal) is used for igniting some compositions having low ignition temperature, since gunpowder containing sulphur is not preferred as sulphur reacts during long storage and also is not compatible with chlorates and metal powders.

Priming is done over the star pellet. In such cases, there are three options:

i. Applying a fine layer of binder solution over the bare star pellet and sprinkling priming composition dust over the wet surface and allowing it to dry slowly.
ii. Applying a thick paste of priming composition with binder over the bare star pellet and allowing it to dry slowly. It may be noted that very quick drying of pellet may form a dry coat of binder which may not allow further evaporation of solvent from the star.
iii. Wrap the star pellet in primed cambric. Primed cambric is made by dipping the cloth in slurry of SMP (sulphur-free mealed gunpowder containing potassium nitrate 70.5% and charcoal type A 29.5%) with binder like gum arabic. It is further sprinkled with sulphur-free mealed gunpowder and dried.

Primed cambric is strong and flexible. This allows star to burn from all sides. However, it has a disadvantage that the primed composition adhering to the cambric can crack when twisted.

Activated carbon fabric has been found to be better than cloth cambric. It is coated with a concentrated solution of potassium nitrate. It can be easily deposited on a charcoal cloth by dipping the cloth into aqueous solution and allowing the cloth to dry.

Fibrous activated carbon impregnated with potassium nitrate burns as under.

$$5C + 4KNO_3 = 2K_2CO_3 + 3CO_2 + 2N_2 \qquad (19.1)$$

Quick match strands were used in the earlier ammunitions but these have now become obsolete.

## 19.3.2 Compositions producing Hot Slags

These compositions contain inorganic fuel like magnesium, boron, silicon, ferrosilicon, zirconium, aluminium, zirconium–aluminium alloy and inorganic oxidisers like potassium nitrate, potassium perchlorate, barium nitrate, oxides of lead, oxides of iron, etc. Readers may refer [1] on slag-producing priming compositions and their use.

Priming compositions containing titanium and zirconium are highly sensitive; the priming compositions containing magnesium and boron are moderately sensitive while priming compositions with silicon, calcium silicide and aluminium are least sensitive.

Boron and zirconium-based compositions are found to be extremely good priming composition for compositions with high heat of combustion like illuminating, signalling and tracer compositions. The fact that boron has a very high heat of combustion as well as low melting point of its combustion product $B_2O_3$, allows molten hot slag to move over a greater area for quick ignition of composition, compared to other fuels. However, these need more safety measures during manufacture. The boron potassium nitrate priming composition reaction is as under.

$$10B + 6\ KNO_3 = 5B_2O_3 + 3K_2O + 3N_2$$

The infrared flares use boron, magnesium, potassium perchlorate, barium chromate and viton as priming composition or use dip coat of fast burning MTV composition on the infrared flare pellet.

The pyrotechnic compositions which are difficult to ignite often use composition SR252 (Table 19.1).

SR 252 contains SMP with ferrosilicon and potassium nitrate. Here, the ferrosilicon acts as hot particles for retention of heat after the SMP burns out. This composition is readily ignitable and forms fusible silicates which remains in hot molten stage to provide further heat to the main composition. The presence of metals increases the temperature of the flame. Composition SR 251 containing 50:50 potassium nitrate and fine silicon also act as a good priming composition.

Priming composition for some illuminating compositions is given in Table 19.2.

Some typical tracer priming compositions are given in Table 19.3.

**TABLE 19.1**
**Priming Composition SR 252**

| Ingredient | Percentage |
|---|---|
| Potassium nitrate | 40 |
| ferrosilicon | 40 |
| Sulphur-less mealed gunpowder (SMP) | 20 |

**TABLE 19.2**
**Priming Composition for Illuminating Composition [2]**

| Ingredient | Percentage |
|---|---|
| Potassium nitrate | 75 |
| Magnesium | 15 |
| Iditol | 10 |

**TABLE 19.3**
**Tracer Priming Compositions**

| Ingredient | Percentage | | | | |
|---|---|---|---|---|---|
|  | SR 399 | [2] | [2] | [3] | SR 867 |
| Magnesium powder | 12 | 18 | 13 | 12 | 6.3 |
| Barium nitrate | – | – | 48 | – | – |
| Barium peroxide | 86 | 80 | 30 | 88 | – |
| Phenol formaldehyde | – | – | – | – | – |
| Binder | – | 2 | – | – | – |
| Iditol | – | – | 9 | – | – |
| Strontium peroxide | – | – | – | – | 81 |
| Acaroid resin | 2 | – | – | – | 0.05 |
| Bitumen | – | – | – | – | 11 |
| Zinc stearate | – | – | – | – | 1.2 |

**TABLE 19.4**
**Priming Compositions for Gasless Delay Compositions [4]**

| Ingredient | Percentage | | | | |
|---|---|---|---|---|---|
| Lead peroxide | 70 | – | – | – | – |
| Boron | 30 | 10 | – | – | – |
| Barium chromate | – | 90 | – | – | – |
| Zirconium | – | – | 51 | 65 | 33 |
| Ferric oxide | – | – | 39 | 25 | 50 |
| Diatomaceous earth | – | – | 10 | 10 | – |
| Titanium | – | – | – | – | 17 |

Some priming compositions for gasless delay compositions are given in Table 19.4.

In some pyrotechnic composition, the ingredient percentages and sieve spectrum of main compositions are modified and used as priming compositions (also known as "*fast burning*" or "*quick burning*" composition) and are found to be very effective. As an example, a slurry of MTV priming composition is pasted over the pellet, as in the case of infrared flares for ease of picking up of flash and burning of whole surface of pellet to achieve quick ignition.

### 19.3.3 Containing Organic Fuels with Oxidisers like Potassium Nitrate or IDITOL

These priming compositions (Table 19.5) do not generate high heat and are useful in signalling colour smoke compositions as they avoid burning of dyes, which otherwise would give out black soot.

### 19.3.4 Elastic Priming Compositions

Some elastic priming compositions are given in Table 19.6.

**TABLE 19.5**
**Priming Compositions for Signalling Colour Smoke**

| Ingredient | Percentage | |
|---|---|---|
| | [2] | SR 214 |
| Charcoal | 15 | – |
| Potassium nitrate | 75 | 60 |
| Iditol | 10 | – |
| Lactose | – | 40 |

**TABLE 19.6**
**Elastic Priming Compositions [5]**

| | Percentage | |
|---|---|---|
| | SRE794 | SRE 796 |
| Boron | 20 | 20 |
| Potassium nitrate | 55 | – |
| Potassium perchlorate | – | 55 |
| Polysulphide rubber | 22.5 | 22.5 |
| S101 Surfactant | 0.3 | 0.3 |
| Magnesium oxide | 0.2 | 0.2 |
| Curing agent | 2 | 2 |

**TABLE 19.7**
**Small Arms Ammunition Tracer Priming Composition SR 867 [5]**

| Ingredient | SR 867 Percentage |
|---|---|
| Strontium peroxide | 81 |
| Bitumen | 11 |
| Magnesium | 6.3 |
| Acaroid resin | 0.5 |
| Zinc stearate | 1.2 |

A typical priming composition for small arms ammunition is shown in Table 19.7.

### 19.3.5 SPECIAL REQUIREMENTS OF PRIMING COMPOSITIONS

In additon to the general essential characteristics of compositions given in Section 1.6, priming compositions should meet the following special requirements:

a. They must be readily ignitable (i.e., low energy requirement/having low ignition temperature, says 250°C or less) from the external stimuli, since the duration of functioning of external stimuli is very less.
b. Combustion temperature should be sufficiently high and it should generate sufficient heat during combustion and should provide heat energy for sufficient duration for the composition to pick-up ignition (i.e., raise the temperature of booster and or main composition to its ignition temperature).

c. Should preferably also give hot slag, which should remain red hot for some time though the initial flame is over. Fluid hot slags are preferred than solid hot slags since the former can move over a large area and enter into voids/pores over the composition and transfer heat. However, hot slags are not desirable for colour smoke compositions so as to avoid burning of dye.

d. Should not give more light so as to avoid revealing the location of user or firing position (for ammunitions which are fired from ground).

e. In gasless (low gas) delay compositions, they should not produce more gas.

f. Should not interfere with the special effects (heat, sound, radiation, smoke, incendiary, etc.) of the main pyrotechnic composition.

g. Their combustion should be accomplished in a controlled manner. A very fast and vigorous combustion may break the combustion front leading to extinction of further combustion.

h. Should be compatible with the main composition to be ignited.

i. Should be in direct contact with the main composition to be ignited or in contact with the booster composition which should be in contact with the main pyrotechnic composition.

## 19.4   BOOSTER COMPOSITIONS

Booster composition generally contains 50:50 mixture of pyrotechnic priming composition and the pyrotechnic main composition. It can be said to be an intermediary phase from priming to main composition. It has properties intermediate between the priming composition and the main composition. Booster composition allows smooth transition of burning from priming composition to the main composition.

However, the single booster composition percentage may be replaced by three booster composition (say 70:30, 50:50 and 30:70 of priming:main composition) interposed between priming and main composition for smooth ignition of the main composition.

It is the main pyrotechnic composition, on which special effect performance of the ammunition depends. However, any malfunctioning of priming composition and/or booster composition in the chain of compositions would result in non-ignition as well as non-performance of the main composition.

## REFERENCES

[1] Prof. A. Bailey and Dr. S.G. Murray, "*Explosives, Propellants and Pyrotechnics*", 1989, ©Brassey's World Military Technology, UK.

[2] A. A. Shidlovskiy, "*Principles of Pyrotechnics*", 3rd Edition, Moscow, 1964. (Translated by Foreign Technology Division, Wright-Patterson Air Force Base, Ohio, 1974.) Publisher: American Fireworks News (1 July 1997) ©1997, Rex E. & S.P., Inc.

[3] F.L. McIntyre A., "*Compilation of Hazard and Test Data for Pyrotechnic Compositions*", Contractor Report ARLCD-CR-80047, October 1980, U.S. Army Research and Development Command, Large Caliber Weapon System Laboratory, Dover, New Jersey.

[4] *"Theory and Application"*, AMCP 706-185, (April, 1967), Engineering Design Handbook Military Pyrotechnic series, Part ONE, Headquarters, US Army Material Command, Washington D.C.

[5] Dr. Nigel Davies, *"Pyrotechnics, Topics to be Covered"*, PowerPoint presentation, Cranfield University, Defence Academy of the United Kingdom, Shrivenham, Swindon, SN6 8LA, UK.

# 20 Flame and Smoke Compositions

## 20.1 ROLE

Flame and smoke compositions are single composition capable of producing flame and smoke as distress signalling for use in day-night marker. The normal flame producing compositions (chapter 9) require high combustion temperature for exhibiting flame colour while coloured smoke compositions (chapter 13) need low combustion temperature to avoid burning of dyes. Hence single composition with different kind of materials have been used for flame and smoke composition.

Details of some of the compositions are given further.

## 20.2 FLAME AND SMOKE COMPOSITIONS

For giving distress signals in the navy, red phosphorous-based compositions are used, which provide both yellow flame and white smoke, thereby making the signal visible in both day and night time. Also, some compositions may emit red or yellow smoke. Some typical compositions are given below in Tables 20.1 to 20.4

a. A typical combined smoke and light composition for drift and float smoke signal composition is given in Table 20.1 [1]where phosphorous evaporates and ignites at the orifice.
b. This composition Table 20.2 [2] is rubbery and has better performance due to higher phosphorous content.
c. This is a solvent-free RP composition produced in the vertical mixer and vacuum cast into the actual MK25 cardboard tubes as in Table 20.3 [3].
d. A combined coloured smoke/flame signal [4] involves the volatilisation of metals and metal derivatives to provide both colour-emitting species for the flame and, by way of condensation, suitably coloured reflecting inorganic pigments for the smoke. The smoke pigments can be either vaporised directly from the composition or can be formed in the flame zone and eventually condensed to submicron-sized particles suitable for the formation of a smoke cloud. The most successful clouds were obtained from materials which were comparatively volatile since non-volatile pigments tended to form larger particles which rapidly fell out of the combustion zone producing an ash around the flare grain and a low-density smoke cloud. It was observed that the metal halides and oxides were the best candidates for the coloured smokes. The most vivid colours were produced by metallic iodides consisting of bismuth and lead followed by molybdenum and misch metal. To a lesser extent, other materials also gave coloured smoke in small laboratory mixes

DOI: 10.1201/9781003093404-20

**TABLE 20.1**
**Typical Smoke and Light Composition [1]**

| Ingredient | Percentage |
|---|---|
| Magnesium | 8 |
| Red phosphorous | 51 |
| Manganese dioxide | 35 |
| Zinc oxide | 3 |
| Linseed oil | 3 |

**TABLE 20.2**
**Typical Smoke and Flame Composition [2]**

| Ingredient | Percentage |
|---|---|
| Red phosphorous | 71 |
| Sodium nitrate | 15 |
| Aluminium flakes | 8 |
| Calcium carbonate | 1 |
| Titanium isoproxide | 1 |
| Polybutadiene (Taktene) binder | 4 |

**TABLE 20.3**
**The Baseline RP Composition in the MK25[3]**

| Ingredient | Weight % |
|---|---|
| Polyglycerol nitrate (PGN) polymer | 21.11 |
| Triethylene glycol dinitrate (TEGDN) | 11.55 |
| Desmodur N-100 curative | 2.34 |
| Red phosphorus (RP) | 62.5 |
| Gas generant (GG) | 2.5 |

**TABLE 20.4**

**Typical Smoke and Flame Composition [4]**

| Composition | DL-Q131 | DL-Q134 | DL-Q135 | DL-Q136 |
|---|---|---|---|---|
| Smoke colour | White | White | Orange | Orange |
| Flame colour | Red | Red | Yellow | Yellow |
| Binder epoxy/anhydride | 7.0 | 7.0 | 8 | 15.69 |
| Aluminium | 39.0 | 30.0 | – | – |
| Strontium nitrate | 37.0 | 44.5 | – | – |
| Strontium chloride | 15.0 | 14.5 | – | – |
| Aluminium sulphate | 2.0 | – | – | – |
| Ammonium chloride | – | 4.0 | – | 3.92 |
| Bismuth | – | – | 17.8 | 18.30 |
| Ammonium iodate | – | – | 17.8 | 15.03 |
| Lead iodate | – | – | 20.8 | 24.18 |
| Ammonium iodide | – | – | 32.6 | 17.65 |
| Magnesium | – | – | 3.0 | 5.23 |
| Burn rate (inch.sec$^{-1}$) | | | | |
| | 0.046 | 0.022 | 0.058 | 0.016 |

but generally when a binder was added the colour and density were markedly degraded. Four such marker compositions are given in Table 20.4.

The compositions DLQ-131, DLQ-134 and DIQ-135, were cured at 135'F for 18 to 36 hours while DLQ-136 candles were given an 18-to-24-hour ambient pressure precure to allow dissipation of a cure exotherm and then placed in an oven at 100 to 130'F for 48 to 72 hours.

e. Typical smoke and flame compositions [5] are SR 414 and SR 622. The former contains 60% red phosphorous coated with small quantity of mineral oil, 28% manganese dioxide, magnesium powder and calcium silicide while the latter contains 50 parts of oil coated with red phosphorous, calcium sulphate and oxalic acid. The fillings are consolidated by hand-stemming or by light mechanically applied pressure. When SR 414 is ignited, the magnesium, calcium silicide and some of the phosphorous reacts with manganese dioxide and the heat generated vaporises the excess phosphorous which burns spontaneously in atmosphere. The oxides of phosphorous so produced react with moisture to produce minute droplets of phosphoric acid in the form of dense white aerosol.

SR 622 when ignited results in a reaction between calcium sulphate and phosphorous that leads to formation of phosphorous sulphide which is emitted as a vapour with excess phosphorous and burns to produce phosphorous oxides and sulphur dioxide. SR 622 burns less vigorously and for producing large amount of gas, 3% oxalic acid is included.

## REFERENCES

[1] Dr. Herbert Ellern, *"Military and Civilian Pyrotechnics"*, Chemical Publishing Co. New York, 1968.

[2] Dr. Bernard E. Douda, *"Survey of Military pyrotechnics"*, 24 May 1991, Naval Weapons Support Center Ordnance Engineering Department Crane, Indiana, USA.

[3] *"Castable, Solvent-Free, Red Phosphorus Smokes for Target Markers"*, SERDP Program PP-1180 Final Technical Report Data Item Number A006, 30 June 2003.

[4] Graham C. Shaw and Philip S. Shadlosky, *"Smoke/Flame Pyrotechnic Marker Compositions"*, Technical Report AFATL-TR-73-199, September 1973, Air Force Armament Laboratory, Airforce Systems Command, USA Airforce, Eglin Air Force Base, Florida.

[5] J. C. Cackett, *"Monograph on Pyrotechnic Compositions"*, Royal Armament Research and Development Establishment, Fort, Halstead, Sevenoaks, Kent 1965.

# 21 Gunpowder Compositions

## 21.1 ROLE

Gunpowder is one of the oldest pyrotechnic compositions and has been used extensively over the years. It is used in a variety of ways in ammunitions as:

a. Expelling charges for ejection of canisters or candles by shearing pins in bombs and shells
b. Igniters for propellants as flash carriers in primers where the holes in the primer allows hot combustion debris to ignite the whole propellant
c. Priming composition for pyrotechnic compositions
d. Delay composition in ammunitions and some fuzes
e. Safety fuse
f. Quick match composition
g. Blank charges for saluting purpose
h. Can be used as an explosive

## 21.2 GUNPOWDER COMPOSITION

Gunpowder consists of three ingredients as below:

a. *Potassium nitrate* (approx.75%), a solid hygroscopic substance, acts as an oxidiser as it provides oxygen for combustion. Thus, three-fourth of a gunpowder comprises this oxidiser.
b. *Sulphur* (approx.15%), a solid, used as a fuel as well as acts as an agent that lowers ignition temperature of the composition and also cements potassium nitrate and charcoal due to ease of flow under pressure. It also increases the number of gaseous products on combustion.
c. *Charcoal* (approx.10%), a solid, used as a fuel. Charcoal does not consist of pure carbon; rather, it consists of partially pyrolysed cellulose, in which the wood is not completely decomposed. Simplified and represented as carbon C. It can be best summed up by its empirical formula: $C_7H_4O$.

To reduce the ignition of gunpowder by static electricity, the gunpowder granules are coated with graphite. The quality of gunpowder depends upon the purity of sulphur and potassium nitrate but depends mainly on quality of charcoal, which is obtained through carbonisation of wood in absence of air. The quality of charcoal depends upon the temperature of carbonisation (varies from 260°C to 1,500°C) of wood and also the quality of wood for parameters like the type of soft wood tree, the

DOI: 10.1201/9781003093404-21

soil condition, age of the tree, the climatic condition under which tree grew and the part of the tree used like trunk or branch.

## 21.3  GUNPOWDER MANUFACTURE

The process of manufacture of gunpowder is different from processing of other pyrotechnic compositions. The process involves the following sequence.

   a. Grinding
   b. Weighing
   c. Preliminary mixing
   d. Incorporation
   e. Pressing
   f. Granulation
   g. Glazing
   h. Drying
   i. Finishing and blending
   j. Packing

## 21.4  GRADES OF GUNPOWDER

There are various grades of gunpowder. The British grades are given in Table 21.1.

## 21.5  PROPERTIES OF GUNPOWDER

The gunpowder properties are given at Table (Table 21.2).

## 21.6  ADVANTAGES AND DISADVANTAGES OF GUNPOWDER

The gunpowder has the following advantages:

---

**TABLE 21.1**
**Grades of Gunpowder**

| Size | Passing through | Retained by |
|---|---|---|
| P3 | 0.75" | 0.375" |
| G3 | 0.375" | 0.187" |
| G7 | 0.25" | No. 8 sieve (.081") |
| G12 | No. 8 sieve (.081") | No. 16 sieve (.0395") |
| G20 | No. 16 sieve (.0395") | No. 25 sieve (.0236") |
| G40 | No. 25 sieve (.0236") | No. 52 sieve (.0118") |
| Mealed | No. 150 sieve (.0041") | No. 240 sieve (.0026") |
| Fuze powder | 0.0336" | No. 25 sieve (.0236") |

---

**TABLE 21.2**
**Some Properties of Gunpowder**

| Parameters | Values |
|---|---|
| Heat 0utput | 665–740 Cal.g$^{-1}$ |
| Figure of insensitivity | 90 (picric acid 100 as standard) |
| Gas output | 260 cc.g$^{-1}$ |
| Ignition temperature | 300–370°C |
| Burning property | Burns vigorously |
| Moisture effect | Substantial (may even fail to ignite) |

a. The ingredients are readily available and are less costly.
b. The ingredients are not highly reactive and environmentally safe.
c. The composition can be formed into pellets easily.
d. The composition is safe in handling and comparatively insensitive to shock, having figure of insensitivity 90.
e. The hot debris formed during combustion are useful in burning propellants.
f. The composition provides heat as well as gaseous products for both ignition as well as propelling the star or candle or canister from the ammunitions.
g. The composition can be ignited with ease.
h. The composition is stable at moderately high temperature.
i. The composition can be stored for long duration as dry composition.
j. Acts as three-in-one high energy material (useful as propellant, explosive and pyrotechnic)

Despite the above advantages, gunpowder has been superseded by a variety of propellants due to the following deficiencies/disadvantages:

a. *Non-homogeneous mixture*: It is a mixture made by mechanical mixing and thus non-homogeneous.
b. *Variation in burning* time: A large variation is observed in burning due to variation in source of charcoal, purity of ingredients and moisture content.
c. *Irregular burning*: The porous structure of composition leads to hot gases penetrating deep inside and breaking of grains during burning, leading to very high rise of pressure.
d. *Large quantity of smoke*: Combustion converts 42.98% of the mass of black powder to gas. In military applications, gunpowder thus produces thick smoke as a by-product, which may give soldier's location away to an enemy observer. The smoke may also impair aiming for further rounds of ammunition.
e. *Hygroscopic nature*: As composition contains three-fourth of the ingredient as sodium nitrate, which is hygroscopic, this poses restriction in use as moisture affects its performance. Hence, its use is restricted in the navy and in

moist weather. However, it is still extensively used in pyrotechnic ammunitions maintaining the production building humidity 55% to 65% with proper
sealing of ammunition joints.

f. *Large quantity required as propellant*: The modern-day propellants are highly
energetic and hence to achieve high energy loadings, the gunpowder quantity
required would be much higher than with modern propellants with high calibre heavy projectiles. Hence, modern propellants have replaced gunpowder
as a propellant.

g. *Large quantity of solid residue*: 55.91% combustion products are solid. These
form a thick coat of soot inside the barrel. The thick soot fouls and causes
problem in subsequent firing. This requires frequent cleaning of the gun
barrel.

h. Further, the residue is hygroscopic and alkaline and forms caustic substance
in presence of moisture from air (the potassium oxide converts to potassium
hydroxide). This causes corrosion of the barrel.

## 21.7   GUNPOWDER COMBUSTION PRODUCTS, LOADING DENSITY AND HUMIDITY EFFECT

The burning of gunpowder does not take place as a single reaction, and the by-
products are not easily predicted, especially since charcoal formula varies. The
sulphur reduces the ignition temperature of the gunpowder. On applying a stimulus,
potassium nitrate decomposes to produce oxygen. This oxygen causes sulphur to
burn and heat thus produced results in burning of charcoal with production of heat
as under.

$$\text{Gunpowder combustion} = \text{Gaseous products} + \text{Solid products} + \text{Heat}$$
(21.1)

A simple equation for combustion of gunpowder is

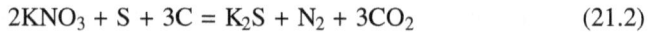

$$2KNO_3 + S + 3C = K_2S + N_2 + 3CO_2 \tag{21.2}$$

A balanced simplified equation is

$$10KNO_3 + 8C + 3S = 2K_2CO_3 + 3K_2SO_4 + 6CO_2 + 5N_2 \tag{21.3}$$

However, since charcoal is not carbon but pyrolysed cellulose with empirical formula $C_7H_4O$, the equation may be written as

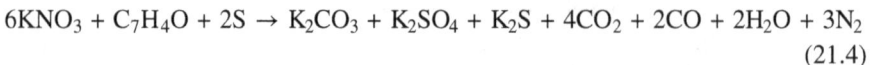

$$6KNO_3 + C_7H_4O + 2S \rightarrow K_2CO_3 + K_2SO_4 + K_2S + 4CO_2 + 2CO + 2H_2O + 3N_2 \tag{21.4}$$

An approximate equation for the burning of black powder [1] is:

**TABLE 21.3**

**Gunpowder Loading Density Variation with Loading Pressure [2]**

| Pressing load (1,000psi) | 25 | 50 | 60 | 65 | 70 | 75 |
|---|---|---|---|---|---|---|
| Density (g.cc$^{-1}$) | 1.74 | 1.84 | 1.86 | 1.87 | 1.88 | 1.89 |

**TABLE 21.4**

**Gunpowder Moisture Gain [2]**

| Temperature °C | R.H.% | Gain % |
|---|---|---|
| 26 | 75 | 0.75 |
| 25 | 90 | 1.91 |
| 30 | 90 | 2.51 |

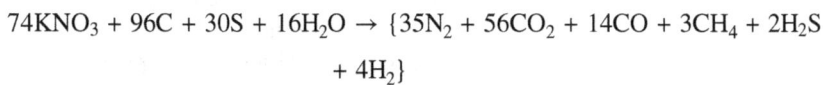

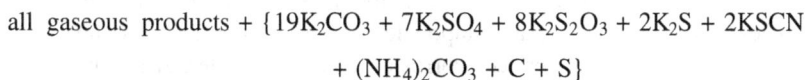

$$74KNO_3 + 96C + 30S + 16H_2O \rightarrow \{35N_2 + 56CO_2 + 14CO + 3CH_4 + 2H_2S$$
$$+ 4H_2\}$$

all gaseous products + $\{19K_2CO_3 + 7K_2SO_4 + 8K_2S_2O_3 + 2K_2S + 2KSCN$

$$+ (NH_4)_2CO_3 + C + S\}$$

all solid products + 665 Cal g$^{-1}$

$$(21.5)$$

The above reaction corresponds to a composition containing saltpetre (75.7%), charcoal (11.7%), sulphur (9.7%) and moisture (2.9%). Gunpowder is used as loose as well as pressed composition. The loading density variation is given in Table 21.3.

Gunpowder is hygroscopic and absorbs moisture. Table 21.4 shows gain in moisture content at various humidity.

The early application of gunpowder is given in Table 21.5.

## 21.8   SULPHUR-FREE GUNPOWDER

Sulphur has strong effect on propellants and metal powders. Also, sulphur forms sensitive products with potassium chlorate. It is useful in many pyrotechnic ammunitions as absence of sulphur avoids formation of unstable and sensitive mixture with other ingredients. It has higher ignition temperature of 500°C compared with gunpowder having ignition temperature 300°C–370°C and higher calorific value of 870 cal. g$^{-1}$ compared to gunpowder of 665–740 cal. g$^{-1}$. Due to increase in potassium nitrate content, it is more hygroscopic in nature.

**TABLE 21.5**

**Early Applications of Gunpowder [1]**

| Application | Composition | | |
|---|---|---|---|
| | KNO$_3$ | Charcoal | Sulphur |
| Lift charge or burst charge | 75 | 15 | 10 |
| Priming powder | 70 | 30 | 0 |
| Blasting powder | 68 | 18 | 14 |
| Rocket propellant | 62 | 28 | 10 |
| Delay fire | 62 | 18 | 20 |
| Sparking composition | 60 | 12 | 28 |
| White smoke | 50 | 0 | 50 |
| Fire extinguishing smoke | 85 | 15 | 0 |

Sulphur-free gunpowder contains potassium nitrate (70–80) and charcoal (20–30) by parts. The process of manufacture is same as normal gunpowder. It is also known as *sulphurless fine grit or PN 623*. It is also used as priming composition. However, it is difficult to manufacture since it takes more time to grind the charcoal which is almost double in the gunpowder. Absence of sulphur (a binding material) also leads to problem in consolidation in absence of a binding material. It is also less easy to ignite than gunpowder, since sulphur reduces the ignition temperature. However, it gives less corrosion than gunpowder. A simple reaction for sulphur-free gunpowder would be

$$6KNO_3 + C_7H_4O \rightarrow 3K_2CO_3 + 4CO_2 + 2H_2O + 3N_2 \qquad (21.6)$$

Another equation is

$$10KNO_3 + 2C_7H_4O \rightarrow 5K_2CO_3 + 4CO_2 + 5CO + 4H_2O + 5N_2 \qquad (21.7)$$

DEFSTAN 13-166, revision 12, January 18, 2013 on gunpowder, sulphurless mentions SFG 12, SFG 20, SFG 40, SFG 90, all in granulated and glazed form and SMP in fine powder form (Table 21.6).

Sulphur-free gunpowder (with stoichiometric ratio of potassium nitrate and charcoal) combustion is as follows.

$$4KNO_3 + 5C = 2K_2CO_3 + 2N_2 + 3CO_2 \qquad (21.8)$$

The above reaction corresponds to a composition containing saltpetre (87.1%) and charcoal (12.9%). Another equation considering carbon as pyrolysed cellulose with chemical formula $C_7H_4O$ is

**TABLE 21.6**
**Sulphur-Free Gunpowder Sieving Requirements**

| Type | Sieving as per BS 410 | Requirement |
|------|----------------------|-------------|
| SFG 12 | Retained on a 2.00 mm sieve | Nil |
| | Retained on a 1.00 mm sieve, % m/m | 75 min |
| | Retained on a 850 μm sieve, % m/m | 95 min |
| SFG 20 | Retained on a 1.00 mm sieve | Nil |
| | Retained on a 600 μm sieve, % m/m | 75 min |
| SFG 40 | Retained on a 600 μm sieve | Nil |
| | Retained on a 300 μm sieve, % m/m | 75 min |
| SFG 90 | Retained on a 300 μm sieve | Nil |
| | Retained on a 125 μm sieve, % m/m | 75 min |
| SMP | Retained on a 106 μm sieve | Nil |
| | Retained on a 63 μm sieve, % m/m | 25 max. |

$$6KNO_3 + C_7H_4O \rightarrow 3K_2CO_3 + CO_2 + 6CO + 2H_2O + 3N_2 \qquad (21.9)$$

## 21.9 MATCHES

A number of products are made from gunpowder which are used for ignition of pyrotechnic composition.

A *black match* is a type of fuse consisting of a set of cotton strings coated with dried slurry of sulphurless gunpowder and glue. Slurry is made using 1:4 isopropyl alcohol and water, i.e., 20 ml of isopropyl alcohol and 80 ml of water. Alcohol reduces the surface tension of water and makes it wetter and helps in its absorption by cotton threads. It is dried at ambient or hot temperature inside a room. It is dusted with sulphurless mealed gunpowder when nearly dry. It burns at the rate of approximately 3 metre per 30 seconds.

## 21.10 FUSES

*Safety fuse* is a medium for conveying flame at a slow and uniform rate to initiate an explosion in either a charge of gunpowder or a detonator. It provides the time delay required for the person to reach the cover of safety before the actual explosion takes place.

Safety fuse consists of a gunpowder core in a tube formed by jute yarn, which is covered with layers of bitumen (asphaltum) and has an outer wrapper of tough yarn or polymer. It is made in a standard diameter designed to be crimped in to blasting cap (detonator). Safety fuse are manufactured with a specified burn rate per metre.

For example, 115 seconds per metre means a length of fuse 100 cm long will take around 115 seconds to burn. Details on two typical safety fuses are given below.

a. *Safety Fuse (Polycoated)* [3]

   Polycoated fuse is a highly water-resistant fuse in which gunpowder is wound in jute yarn and coated with a layer of bitumen. This base fuse is then wrapped in a paper tape tube and is provided with a continuous coating of low-density polyethylene (LDPE) or polyvinyl chloride (PVC). The LDPE/PVC coating makes the safety fuse water-resistant and thus suitable for use in wet conditions. The fuse is given a red or yellow colour by default.

b. *Safety Fuse (Blue WP)* [3]

Blue WP fuse contains gunpowder wound in jute yarn and coated with a layer of bitumen. It is given an additional covering of polyethylene tape and again coated with a layer of bitumen. For identification purposes and to cover the stickiness of the bitumen layer, the fuse is coated with a layer of blue colour. This product is less water-resistant and is recommended for use only in dry areas. Figure 21.1 shows safety fuse polycoated and safety fuse blue.

The comparative parameters of the above safety fuses are shown in Table 21.7.

c. Micro Cord/Visco Fuse [4]

**FIGURE 21.1**   Safety Fuse Polycoated and Safety Fuse Blue (Reprinted from [3]).

**TABLE 21.7**

**Comparative Data on Typical Safety Fuses (compiled from [3])**

| Parameters | Safety fuse polycoated | Safety fuse blue WP |
|---|---|---|
| Diameter (mm) | $5.0 \pm 0.1$ | $4.7 \pm 0.1$ |
| Burn rate (sec.m$^{-1}$) | $115 \pm 10\%$ | $115 \pm 10\%$ |
| Standard core load (net explosive qty.) (gm.m$^{-1}$) | $5.0 \pm 10\%$ | $4.5 \pm 10\%$ |
| Gap sensitivity (mm) | 30 | 30 |
| Standard length (m) | 10, 250(reel), 1000(reel) | 10, 7.32 |

**TABLE 21.8**
**Comparative Data on Typical Visco Fuses [4]**

| Parameters | Particulars | | |
|---|---|---|---|
| Diameter (mm) | 1.8 ± 0.1 | 2.0 ± 0.1 | 2.5 ± 0.1 |
| Burn rate | 105 ± 10% | 105 ± 10% | 105 ± 10% |
| Colour | Green | Pink | Pink |

Micro cord is made with four external layers, the first is yarn wrapped around the inner gunpowder core, then a second layer of yarn is wrapped around in the opposite direction to prevent unwrapping, then, another layer of yarn is added in the opposite direction to further strengthen the micro-cord. The fourth and final layer comprises a nitrocellulose lacquer, which helps to waterproof the fuse and stops the black powder core from degrading.

It is distinguished from safety fuse in that micro cord burns externally, produces a substantial amount of external sparking which provides lateral transmission of flame and is rather fully consumed as it burns instead of leaving a distinguished carcass. It can also be used to create delays in the rate of firing multiple fireworks linked together.

It provides a much safer and reliable alternative to the quick match fuse which was traditionally used in fireworks for creating delays. A comparative data on typical micro cord/visco fuses is shown in Table 21.8.

Military safety fuse requirements are of various higher dimensions and burning rates. The typical burning speeds may vary from 80 seconds to 170 seconds per metre having diameters between 5.0 mm to 5.5 mm and are used as delays for detonators.

Safety fuses must meet the following requirements:

a. Must be able to ignite from external stimuli
b. Must give a uniform specified burn rate when ignited
c. Combustion should be slow and non-violent
d. Propagation of burning should not stop till end
e. Must not ignite by mechanical shock/force

Readers may refer [5] for comprehensive review of gunpowder.

## REFERENCES

[1] Michael. S. Russell, "*The Chemistry of Fire Works*", RSC, Cambridge, 2002. Published by The Royal Society of Chemistry, Thomas Graham House, Science Park, Milton Road, Cambridge CB4 0WF, UK. Copyright Michael S. Russell, 2009. Reproduced by permission of The Royal Society of Chemistry.

[2] "*Properties of Materials used in Pyrotechnic Compositions*", AMCP 706-187, Engineering Design Handbook-Military Pyrotechnic Series, Part Three, Headquarters, US Army Material Command Washington DC, October 1963.

[3] *https://safetyfuse.in/* > of M/S Commercial Explosives (India) Private Limited, Nagpur, India.

[4] *https://safetyfuse.in/micro-cord-visco-fuse/* of M/S Commercial Explosives (India) Private Limited, Nagpur, India.

[5] Ronald A, Sasse, "*A comprehensive Review of Black Powder*", Technical Report BRL-TR-2630, 1985, US Army Ballistic Research Laboratory, Aberdeen Proving Ground, MD.

# 22 Initiation of Ignition in Ammunition

## 22.1 GENERAL

Ammunitions mostly have more than one high energy compositions, each having different sensitivities and combustion output. These high energy materials are generally arranged in a sequence from higher sensitivity material (with lower quantity and lower yield) to lower sensitivity material (with higher quantity and higher yield). The initiation of ignition in a pyrotechnic ammunition is done through the *"Ignition train." Ignition train is a sequence of ignition in the ammunition system where each high energy compositions are initiated in sequence from higher sensitivity to lower sensitivity of composition.* The quantity of composition of higher sensitivity (like cap compositions) is kept small while the quantity of lower sensitivity of composition (like smoke or illuminating composition) is kept higher for special effects.

Almost all the ammunitions fired through a weapon system are assembled with caps (some refer it to primers) containing initiatory composition for igniting the chain of events leading to functioning of the ammunition. However, some ammunitions do not require weapon system to fire but are assembled with cap-like hand flare red and para flare red. However, some ammunitions do not require weapon system and do not have caps as well. Examples are generator smoke no. 5, marker smoke white and ground flare indicating MK-1 yellow, which are required to be ignited through port fire friction.

Let us consider 105-mm BE screening smoke cartridge and shell. The cartridge case contains the cap, primer and the propellant. The projectile (Figure 32.5) consists of a fuze (after removing transit plug and assembling fuze), burster bag and four canisters. The initiation of combustion of 105-mm BE screening oke ammunition takes place in two phases as below.

*Phase I Explosive train during firing (inside weapon system)*: The cap, on initiation, gives flash to primer, filled with gunpowder, which starts burning and flash, hot gases and solid hot products are emitted out of the holes in the primer and get distributed inside the cartridge case, filled with propellant. This causes uniform burning of the propellant, raising the temperature with high gas pressure. This causes the projectile to move inside the barrel and eject out with high velocity and spin. The explosive train thus starts with cap inside the primer to the gunpowder in the primer and finally to the propellant for ejection of the projectile.

*Phase II Explosive train in the projectile (outside weapon system)*: The setback caused during ejection causes the timer of the fuze to start. The fuze functions after set delay in the air. The gunpowder in the fuze magazine gives flash to gunpowder in burster bag which, in turn, gives flash to four canisters. In a canister, priming

DOI: 10.1201/9781003093404-22

composition picks up flash and burns the main smoke composition in the canister, emitting smoke. The burster bag also generates sufficient pressure for the separation of base plate of the projectile and allows four canisters to eject and fall in the air, giving a streak of smoke. The canisters continue to burn on ground providing dense white screening smoke for about 45 seconds (Figure 32.6). The explosive train thus starts with fuze to burster bag and finally to canister composition.

## 22.2   EXTERNAL ENERGY FOR INITIATION OF IGNITION

The pyrotechnic compositions are metastable and sensitive to certain external stimulus. Hence, it is necessary to subject these compositions to such external stimulus for combustion. The application of an external energy stimuli leads to initiation of a chain of events culminating into functioning of the ammunition. Some conventional external stimuli are:

a. Mechanical/impact energy (stab or percussion caps, both filled with initiatory compositions)
b. Friction energy (pull wire igniter, striking card, safety match, igniter safety fuse, striking using friction composition)
c. Electrical energy (caps using conducting compositions, electric squib/impulse cartridge composition)
d. Thermal energy (primer/ igniter, laser beam, fuze)

All these produce heat/thermal energy like flame/flash/spark and hot gases, hot fluids and hot particles.

The type of initiatory composition used for caps depends upon the impact or electrical sensitivity of the composition and its method of firing. For example, initiatory compositions having impact sensitivity are suitable as stab or percussion cap (primer), while initiatory compositions having electrical sensitivity are suitable as electrical caps.

Let us discuss all the applications used for ignition of pyrotechnic compositions.

## 22.3   INITIATORY COMPOSITIONS

Primary explosives are not used directly as initiating substance. Their performance is modified by addition of certain substances as listed below:

a. Improving oxygen balance by addition of oxidisers such as potassium chlorate or potassium perchlorate, etc.
b. Increasing sensitivity by addition of substances like tetrazene or ground glass
c. Increasing output by addition of fuels such as antimony sulphide or secondary explosives such as TNT or PETN
d. Use of binders for homogeneity of composition

Therefore, the initiatory compositions consist of initiatory explosive compound with oxidising agent, fuel and sensitiser, occasionally with a suitable binder and or

**TABLE 22.1**

**Ingredients Used for Initiatory Compositions**

| Type | Ingredient |
|---|---|
| Primary explosives (initiators) | Mercury fulminate, lead azide, lead styphnate |
| Fuel | Antimony sulphide, calcium silicide, boron, zirconium |
| Oxidiser | Potassium chlorate, potassium perchlorate, barium nitrate, lead peroxide, manganese dioxide |
| Sensitiser | Sulphur, lead thiocyanate, ground glass, tetrazene |
| Binder | Nitrocellulose, gum arabic, shellac |
| Secondary explosive | PETN, TNT |

secondary explosive. The quantum of initiatory primary explosive is kept low to ensure that no detonation takes place and acts as sensitiser to make the initiatory composition sensitive to be ignited. Its function is to convert mechanical energy from external energy stimulus to chemical energy in the form of deflagrating pyrotechnic reaction resulting into large volume of hot gases and solid hot particles. Table 22.1 gives the various ingredients generally used in initiatory composition.

*The primary explosive lead azide is mostly used in detonators while mercury fulminate and lead styphnate are used for ignition, the latter is also sensitive to electrical initiation.* Tetrazene acts as a sensitiser. The presence of tetrazene considerably lowers the stab sensitivity. For example, 5% addition of tetrazene in compositions based on red lead oxide and boron with stab sensitivity of about 200 mJ, the sensitivity falls between 10 mJ and 20 mJ. However, there are some initiatory compositions that do not contain any explosive ingredients. These initiatory compositions are highly sensitive to above external stimuli and provide hot flame, hot gases and incandescent solid particles for ignition. Lead azide is incompatible with copper, copper alloys, zinc and cadmium and, hence, lead azide-based initiatory compositions are pressed in metal cups made of aluminium or aluminium alloys while mercury fulminate is not compatible with aluminium and hence mercury fulminate-based initiatory compositions are pressed in metallic cups of copper duly plated with compatible metal.

### 22.3.1 FACTORS AFFECTING EFFICIENCY OF INITIATORY COMPOSITIONS

The efficiency of initiatory compositions depends on many factors:

a. *Factors related to composition:*
   i. Composition ingredient characteristics including purity
   ii. Particle size and its distribution
   iii. Particle shape and its distribution
   iv. Moisture content

     v. Ingredient percentage
    vi. Blending
   vii. Quantity of composition
  viii. Loading density (extreme high pressings may result in dead pressed for some compositions)

b. *Factors related to design of cap:*
     i. Type of confinement
    ii. Material of cup and material, hardness and thickness of disc for stab caps as higher energy for penetration would be required for materials having higher hardness and thickness
   iii. Material, hardness and thickness of cup for percussion caps as higher energy for indentation would be required for materials having higher hardness and thickness
    iv. Type of secondary explosive, if any
     v. Contact area of initiatory composition and secondary explosives, if any

c. *Factors related to strength of cap rigidity while assembling* (i.e., tightness or looseness of cap on assembly):
A loose cap is not desired as it may lead to failure since impact force is diminished.

d. *Factors related to mode of ignition:*
     i. Firing pin material strength
    ii. Firing pin contour/design (i.e., radius of tip)
   iii. Degree of penetration-Firing pin energy (sufficiently high to cause puncture in stab caps or indentation in percussion caps)
    iv. Loose anvil causing anvil movement shall decrease the sensitivity (for percussion caps only)
     v. Eccentric hit by firing pin reduces sensitivity

Caps are filled with initiatory compositions and assembled at the base of the primer and provides flash, hot fluid, hot gases and hot particles as output on getting external stimuli.

Caps are of three types viz. stab caps, percussion caps and rim fire caps.

## 22.3.2 Stab Caps

Stab caps receive mechanical impulses to initiate a chain of explosive initiation events. The stab composition consists of fuel, oxidiser and a primary explosive (initiator) as under.

a. *Fuels:* Antimony sulphide, lead thiocyanate, calcium silicide, carborundum
b. *Oxidisers:* Potassium chlorate, barium nitrate and lead oxide
c. *Initiators:* Lead styphnate, lead azide and tetrazene

In case of stab caps, the striker pierces through the disc and crushes the composition, causing ignition of the cap composition leading to hot flame, high pressure, hot gases and hot slag.

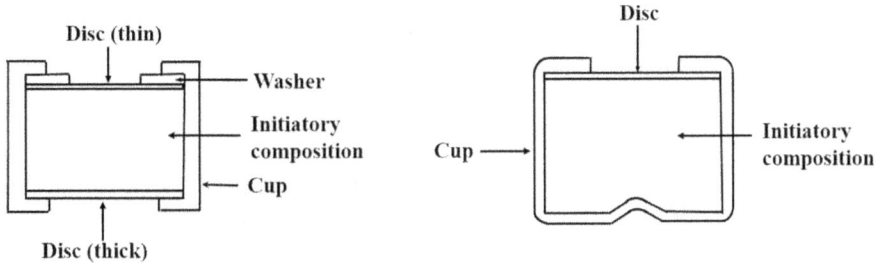

**FIGURE 22.1**   Typical Stab Caps.

The initiatory composition is pressed in a small container, i.e., cup empty body (Figure 22.1) and suitably sealed by metal foil and sealant/varnish to avoid moisture. The quantities in the caps are very meagre and the composition sensitivity depends upon the primary explosive (initiator).

Striker dimensions are important and several dimensions are in use. As an example, Naval Ordnance Laboratory prefers the following [1]:

Firing pin flat 0.010 inch ±0.003 inch, Angle of cone: 28 degree ±2 degree

Radius of corner: Sharp preferred, 0.002 inch maximum

In another typical case, the striker pin, made as truncated cone of steel or aluminium alloys, with flat diameter varying from 0.08 mm to 0.20 mm with included angle 30° or flat diameter 0.203 to 0.381 with included angle 26°–4° is allowed to hit the stab primer.

A standard firing pin for stab initiation has been given in Figure 22.2 [2].

The kinetic energy required for the firing pin shall be more for higher thickness and hardness of the foil which is required to be penetrated. The kinetic energy of the striker is concentrated and dissipated over a small area of the cap composition, thereby causing the ignition of the cap. This is also termed as "*hotspot*" for initiation of ignition. A "hotspot" (localised higher than average temperature of surrounding composition) can be formed in several different ways including:

a. Adiabatic heating due to compression of gases entrapped in porous composition
b. Intercrystalline friction between the pyrotechnic composition ingredients or breaking of granular structure of the ingredients of the composition (mostly for initiatory compositions)

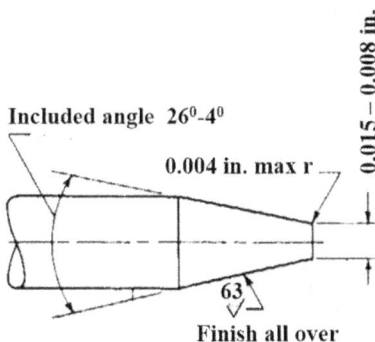

**FIGURE 22.2**   Standard Firing Pin for Stab Initiation (Reprinted from [2]).

c. Viscous heating of the flowing ingredients as these escapes from between the impacting surfaces

d. Localised adiabatic shear of ingredient

e. Plastic deformation at a sharp point

Any or all of the above lead to initiation of ignition of composition. Existence of "*hotspot*" is confirmed as there is ignition delay in the composition, which is sufficient to raise the temperature to its ignition temperature. It is likely that instead of one "*hotspot*," a number of "*hotspots*" in proximity results in ignition. Increase in diameter of "*hotspot*" or a greater number of "*hotspots*" leads to better ignitability. The typical hot spot dimensions are 0.0001mm to 0.01 mm and duration of .01μs -1μs and temperatures more than 500°C. Very small *hotspots* with lower duration and temperatures than above values shall not be effective in ignition. Apart from physical, chemical and thermal properties of the pyrotechnic composition, the *hotspot* plays a significant role in achieving ignition temperature of the pyrotechnic composition due to size of "*hotspot*" and duration of "*hotspot*".

It may be noted that the truncated cone of striker pin allows the striker to penetrate the disc as well as the composition in the cap. Hence, it is necessary that the composition is well-consolidated so that the striking pin faces resistance to enter inside the composition, thereby converting the kinetic energy into heat energy at a short distance as "*hotspot*," thereby raising the temperature much higher than ignition temperature of the composition. Thus, higher consolidation gives higher sensitivity of the composition (Table 22.2).

The striker pin's flat diameter size is determined based upon the sensitivity of the composition and lower the sensitivity, higher would be the flat diameter.

The work conducted at the US Naval Ordnance Laboratory showed a definite relationship existing between the energy input requirements and firing pin velocity for the initiation of primer Mk 102. There appear to be two possible conditions under which initiation will not occur [1].

a. Where the total transfer of energy is large but rate of transfer is low

b. Where the rate of transfer is high but total energy is low

**TABLE 22.2**

**Sensitivity Variation of Stab Caps with Loading Pressure [2]**

| Loading pressure (1,000 psi) | Drop test height (in.) |
|---|---|
| 15 | 1.31 |
| 25 | 0.91 |
| 40 | 0.77 |
| 60 | 0.68 |
| 80 | 0.57 |

In other words, there is a minimum energy requirement as well as minimum velocity requirement for firing stab primers. Between these two extreme conditions, there seems to be a hyperperbolic relationship between impact velocity and total energy requirements (Figure 22.3).

Some typical stab cap compositions are given in Table 22.3.

Stab primers are generally used for initiating detonators due to higher brisance but are also used for ignition and mechanical action. These are mostly used in medium calibre ammunitions but some are also used in high calibre ammunitions.

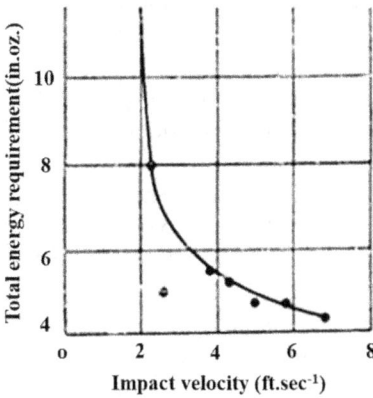

**FIGURE 22.3** Plot of Impact Velocity Versus Total Energy Required for 100% Firing (Reprinted from [1]).

**TABLE 22.3**
**Stab Cap Compositions**

| Ingredient | Percentage | | | | | | |
|---|---|---|---|---|---|---|---|
| | 08 891 mixture [1] | FA 70 [1] | [3] | [3] | [3] | PA 100 [1] | NOL 130 [3] |
| Lead azide | – | – | 29 | – | – | 5 | 20 |
| Potassium chlorate | 45 | 53 | 33 | 45 | – | 53 | – |
| Antimony sulphide | 22 | 17 | 33 | 22 | 5 | 17 | 15 |
| Carborundum | – | – | 5 | – | – | – | – |
| Lead thiocyanate | 33 | 25 | – | 33 | – | 25 | – |
| Basic lead styphnate | – | – | – | – | 38 | – | 40 |
| Tetrazene | – | – | – | – | 2 | – | 5 |
| Barium nitrate | – | – | – | – | 39 | – | 20 |
| Lead dioxide | – | – | – | – | 5 | – | – |
| Calcium silicide | – | – | – | – | 11 | – | – |
| TNT | – | 5 | – | – | – | – | – |

### 22.3.3 PERCUSSION CAP

Percussion caps receive mechanical impulse to initiate a chain of explosive initiation events. The Percussion composition also consists of fuel, oxidiser and a
primary explosive (initiator). The common ingredients are:

    a. *Fuels:* Antimony sulphide, lead thiocyanate, powdered aluminium, calcium
       silicide, zirconium, boron
    b. *Oxidisers:* Potassium chlorate, barium nitrate, lead dioxide, leads peroxide
       and lead oxide
    c. *Initiators:* Mercury fulminate, lead thiocyanate and lead styphnate
    d. *High explosives:* TNT and PETN

Secondary high energy materials are used to improve the brisance power. The
quantity of compositions in the caps is very meagre and the composition sensitivity
depends upon the primary explosive (initiator).

    In a percussion primer, the explosive is backed by a metal anvil. The cap
composition is crushed between the anvil and the bent surface of cap (due to striker
hit), causing ignition of the cap composition leading to flash, hot gases and hot
particles with low brisance. (Figure 22.4).

    The striker (flat on the point or hemispherical tip) is allowed to hit the cap
assembled with anvil. The hemispherical design is preferred since the striker tip
does not puncture the base of the cap and allows the composition to be crushed
between the cap and the anvil.

    It is necessary that indentation should not lead to puncture of the cap causing
back venting. Similarly, smaller indentation may lead to misfire as the composition
is not crushed to desired level. An indentation of 0.508 mm to 0.762 mm has been
found satisfactory for functioning; the lower calibre caps need lower values of
indentation. As energy required for functioning varies for each initiatory composition, some trials are necessary to arrive at the striker dimension.

    Such instantaneous crushing of the composition causes ignition of the cap
composition due to momentum of the striker causing breaking of granular structure

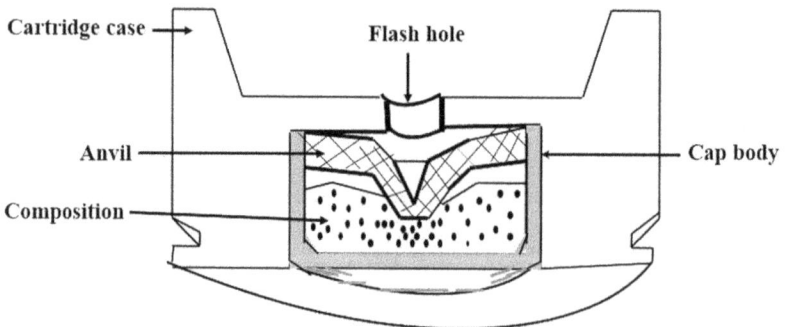

**FIGURE 22.4**  Percussion Cap Assembled to Cartridge Case.

**TABLE 22.4**

**Cap Compositions with Mercury Fulminate [4]**

| Ingredient | Percentage | | | | |
|---|---|---|---|---|---|
| | A | B | C | D | E |
| Mercury fulminate | 37.5 | 11.0 | 32 | 25 | 19.0 |
| Potassium chlorate | 37.5 | 52.5 | 45 | 40 | 33.4 |
| Antimony sulphide | 25.0 | 36.5 | 23 | 35 | 42.8 |
| Sulphur | – | – | – | – | 2.4 |
| Mealed gunpowder | – | – | – | – | 2.4 |

of the ingredients of the composition. A higher momentum is required for the firing pin for higher composition thickness between anvil and cup.

A study of the effect of firing pin alignment on primer sensitivity indicates that there is little effect if the eccentricity is less than 0.02 in. Above this eccentricity, sensitivity decreases rapidly.

A number of percussion cap compositions with mercury fulminate are given in Table 22.4.

The finer particle sizes of ingredients were designated as composition A1, B1 and C1.

Mercury fulminate has a short life in humid climate despite varnish protection as moisture diffuses inside. Further, the combustion products mercury from mercury fulminate, potassium chloride from potassium chlorate and sulphur dioxide from antimony sulphide has a strong corrosive action on the barrel.

Hence, to avoid this problem, lead styphnate has been incorporated. However, since lead styphnate is less sensitive to impact, tetrazene is added to sensitise the composition and potassium chlorate replaced by barium nitrate. Table 22.5 shows composition developed over A and B by Research Department, Woolwich.

Cap compositions without mercury fulminate are listed in Table 22.6.

Some more percussion caps are given in Table 22.7 and Table 22.8.

These caps are assembled to the primer directly by ringing or turn over and then the primer is threaded to the cartridge case.

**TABLE 22.5**

**Composition RD1651(L Mix) [4]**

| Ingredient | Percentage |
|---|---|
| (2:4 LDNR, RD1337) | 50 |
| Barium nitrate | 45 |
| Tetrazene | 5 |

**TABLE 22.6**

**Cap Compositions without Mercury Fulminate [4]**

| Ingredient | Percentage | | | |
|---|---|---|---|---|
| | QF | Q3 | Friction Tube | RD1610 |
| Potassium chlorate | 36.5 | 54.0 | 44.6 | 54.0 |
| Antimony sulphide | 54.5 | 40.0 | 44.6 | 13.0 |
| Mealed gunpowder | 3.0 | – | 3.6 | – |
| Sulphur | 3.0 | 6.0 | 3.6 | – |
| Lead thiocyanate | – | – | – | 23.0 |
| Powdered glass | 3.0 | – | 3.6 | – |
| PETN | – | – | – | 10.0 |

**TABLE 22.7**

**Percussion Cap Compositions [3]**

| Ingredient | Percentage | | | | |
|---|---|---|---|---|---|
| | M-39 | PA101 | FA70 | NOL60 | FA959 |
| Lead styphnate | – | – | – | – | 35 |
| Basic lead styphnate | – | 53 | – | 60 | – |
| Tetrazene | – | 5 | – | 5 | 3.1 |
| Potassium chlorate | 37.05 | – | 53 | – | – |
| Barium nitrate | 8.68 | 22 | – | 25 | 31.0 |
| Antimony sulphide | – | 10 | 17 | 10 | 10.3 |
| Lead thiocyanate | 38.13 | – | 25 | – | – |
| Powdered glass | 10.45 | – | – | – | – |
| Powdered aluminium | – | 10 | – | – | – |
| Powdered zirconium | – | – | – | – | 10.3 |
| Lead dioxide | – | – | – | – | 10.3 |
| TNT | 5.69 | – | 5 | – | – |

### 23.3.4 Rim Fire Caps:

Rim fire caps are filled with pyrotechnic composition at the annular ring of the cap. A typical rim fire cap composition is given in Table 22.9.

### 22.3.4 DIFFERENCES BETWEEN STAB AND PERCUSSION CHARACTERISTICS

The differences between stab and percussion characteristics are enlisted in Table 22.10.

**TABLE 22.8**

**Percussion Cap Compositions [3]**

| Ingredient | Percentage | | | | | | | |
|---|---|---|---|---|---|---|---|---|
| | 1 | 2 | 3 | 4 | 5 | 6 | 7 | 8 |
| Potassium chlorate | 35 | – | – | – | 50 | 53 | – | – |
| Potassium Perchlorate | – | – | – | 50 | – | – | – | – |
| Antimony sulphide | 30 | 7 | 15 | – | 20 | 17 | 37.05 | 10.3 |
| TNT | 3 | – | – | – | 5 | 5 | 5.69 | – |
| Tetrazene | – | 12 | 4 | – | – | – | – | 3.1 |
| Barium nitrate | – | 22 | 32 | – | – | – | 8.68 | 31 |
| Aluminium | – | – | 7 | – | – | – | – | – |
| Lead thiocyanate | 17 | – | – | – | – | 25 | 38.18 | – |
| Lead styphnate (normal) | – | 36 | 37 | – | – | – | – | 35 |
| Zirconium | – | 9 | – | 50 | – | – | – | 10.3 |
| Lead dioxide | – | 9 | – | – | 25 | – | – | 10.3 |
| Calcium silicide | 15 | – | – | – | – | – | – | – |
| PETN | – | 5 | 5 | – | – | – | – | – |
| Ground glass | – | – | – | – | – | – | – | 10.45 |
| Boron | – | – | – | – | – | – | – | – |

**TABLE 22.9**

**$VH_2$ (Von Herz) Composition [4]**

| Ingredient | Percentage |
|---|---|
| Lead styphnate | 38 |
| Barium nitrate | 39 |
| Calcium silicide | 11 |
| Antimony sulphide | 5 |
| Tetrazene | 2 |
| Lead dioxide | 5 |

For gasless cap composition containing red lead oxide 85.5%, boron 9.5% and tetrazene 5.0%, a study was carried out [5] on pressure versus sensitivity to percussion and mass of composition versus sensitivity as given below.

a. *Pressure versus Sensitivity to Percussion* [5]:

The relationship between sensitivity to percussion and applied pressure was investigated over a range of pressures from 2.8 MPa to 22.4 MPa in steps of 2.8 MPa.

## TABLE 22.10
## Differences between Stab and Percussion Characteristics (compiled from [1])

| Stab | Percussion |
|---|---|
| Sharp firing pin (truncated conical) comes in contact with the cap composition as firing pin penetrates through the cap. | Blunt firing pin (hemispherical) does not come in contact with the cap composition as firing pin only hits the cup and makes a dent. |
| It has disadvantage that it allows back venting through the perforated cap and hence unsuitable for closed system. | It has advantage that it provides gas seal as the cup is not punctured. |
| Ignition is obtained by crushing/fracture/ friction between sharp pointed firing pin and cap composition. | Ignition is obtained by crushing/fracture /friction of composition between cup and anvil due to indentation of firing pin. |
| Sensitivity of stab mixtures is more than the percussion mixtures hence input energy of firing pin (kinetic energy) required is lower than percussion. Most stab caps function with high reliability between 3.5 mJ to 35 mJ of energy. | Sensitivity of percussion mixtures is less than the stab mixtures hence percussion caps require more mechanical energy for functioning. Hence higher input energy (momentum) of firing pin is required. Percussion caps require higher firing energies, in the range of 125–420 mJ. |
| It is mostly used in initiating detonators. In fuzes; stab caps are used when instant functioning and high sensitivity are required. | It is used for initiating explosives, propellant igniters, delays etc. |
| This gives flame, hot gases but with higher brisance and hence likely to break pressed pyrotechnic composition. | It gives flame, hot gases with low brisance and hence does not break the pressed pyrotechnic composition |
| Kinetic energy appears to be the determining magnitude for stab initiation. | Momentum appears to be the determining magnitude for percussion initiation. |

The filling weight was kept constant at 20 mg. 15 caps at each pressure were subjected to the 25 g striker and the same number of caps to the 56 g striker. The results are shown in Table 22.11.

The Table 22.11 indicates that higher pressing load leads to increase in the sensitivity since the composition offers higher resistance at higher pressing load. It is also seen that compositions are having higher sensitivity for lighter mass of striker compared to heavier striker for which no suitable explanation could be offered except that it an expected trend.

b. *Mass of Composition versus Sensitivity* [5]:

Compositions with masses ranging from 10 mg to 30 mg were pressed into caps using a pressure of 17 MPa. The sensitivity to percussion was determined using a 56 g striker and the results are shown in Table 22.12.

The Table 22.12 reveals that sensitivity of the composition reduces with increase in charge mass in the caps. This is due to the fact that higher mass of composition

**TABLE 22.11**

**Pressure versus Sensitivity of Percussion Caps [5]**

| Pressure load (MPa) | Height of composition (mm) | Sensitivity, Mean (mJ) | |
|---|---|---|---|
| | | 25 g | 56 g |
| 2.80 | 1.30 | – | 138 |
| 5.61 | 1.27 | 111 | 130 |
| 8.41 | 1.14 | 112 | 124 |
| 11.22 | 1.14 | 109 | 121 |
| 14.02 | 1.17 | 107 | 120 |
| 16.83 | 1.14 | 101 | 115 |
| 19.60 | 1.14 | 103 | 103 |
| 22.40 | 1.09 | 106 | 107 |

**TABLE 22.12**

**Mass of Composition versus Sensitivity of Percussion Caps [5]**

| Weight of composition (mg) | Average height of composition(mm) | Sensitivity, Mean (mJ) |
|---|---|---|
| 10 | 0.86 | 108 |
| 15 | 0.99 | 111 |
| 20 | 1.14 | 117 |
| 25 | 1.27 | 121 |
| 30 | 1.42 | 126 |

acts as cushioning material and hence does not provide the resistance required to the striker for composition to initiate. A low charge mass gets higher resistance due to the resistance offered by the cap material.

## 22.3.5 Special Requirements of Stab, Percussion and Rim Fire Cap Compositions

In addition to the general essential requirements of compositions as given in Section 1.6, cap compositions should meet the following special requirements.

a. *Sensitivity to external stimuli*: It should be sensitive to the external stimuli within desirable limits to ensure regular ignition.
b. *High initiating efficiency*: It should have sufficient ignition power, i.e., it should be able to initiate ignition even with small quantity.
c. *Resistance to dead-pressing*: It could be pressed at high pressures for stab and percussion initiatory compositions without being dead pressed (i.e., difficult

to ignite). Mercury fulminate and diazo dinitrophenol (DDNP) may be dead pressed at high pressing load.

d. *Combustion products*: It should produce adequate quantum of incandescent combustion products and adequate quantity of hot gases and combustion products to ensure proper ignition.

## 22.4 FRICTION ENERGY ON FRICTION COMPOSITIONS

Relative sliding motion of two pyrotechnic compositions against each other causes kinetic friction (or sliding friction or dynamic friction). This kinetic friction resists relative lateral motion of solid pyrotechnic surfaces in contact and converts kinetic energy into thermal energy enough to ignite the pyrotechnic composition. Friction compositions are a combination of two complimentary compositions, namely friction pyrotechnic composition (flame producing) and striker pyrotechnic composition.

The basic features are as below:

a. The ignition is through friction stimulus and not impact.
b. This friction stimulus is by rubbing through striker pyrotechnic composition over friction pyrotechnic composition.
c. The outputs are flame, glow and gases. They do not give brisance.
d. These differ from stab or percussion caps in that they do not contain any initiatory explosives.

There are several ways for friction to take place.

### 22.4.1 PULL WIRE/CORD IGNITER

Pull wire/cord igniter produce flame by friction of a roughened steel wire (or thread coated with striker composition) through another friction pyrotechnic composition.

The working principle of friction composition is that the cup contains the match tip friction composition through which passes a cotton thread (or roughened metal wire) coated with striker composition (red phosphorous and shellac). On pulling the thread or wire, the flame is generated due to friction.

### 22.4.2 STRIKER CARD

In this method, the friction composition is pasted over main composition in dome shape or flat shape, suitably covered by tape or other means and which is removed during ignition by striker card having striker composition.

Typical striker and friction pyrotechnic compositions are given in Table 22.13 and Table 22.14.

McLain [7] has reported a friction composition consisting of striker composition (lacquer 61, pumice 2.2, red phosphorous 26 and butyl acetate 10.8) and friction composition (shellac 40, strontium nitrate 3, quartz 6, charcoal 2, potassium perchlorate 14, potassium chlorate 28, wood flour 5 and marble dust 2).

## TABLE 22.13
### Striker and Friction Compositions [6]

| Striker composition | | | Friction composition | | |
|---|---|---|---|---|---|
| $KClO_3$ | Oxidant | 51 | Red phosphorous | Fuel | 37.2 |
| Ground glass | Frictional sensitizer | 15 | $Sb_2S_3$ | Fuel | 33.5 |
| Animal glue | Binder | 11 | Animal glue | Binder | 9.3 |
| ZnO | Filler | 7 | Iron ochre | Filler | 7.0 |
| $Fe_3O_4$ | Filler | 6 | Dextrin | Binder | 7.0 |
| S | Fuel | 5 | $MnO_2$ | Catalyst | 3.4 |
| $MnO_2$ | Catalyst | 4 | $CaCO_3$ | Burn rate modifier | 2.0 |
| $K_2Cr_2O_7$ | Oxidiser | 1 | Ground glass | Frictional sensitiser | 0.6 |

## TABLE 22.14
### Striker and Friction Compositions

| Striker composition PN 445 | | Friction composition PN 196 | |
|---|---|---|---|
| Red phosphorous | 80 | Potassium chlorate Gr II | 60 |
| Shellac varnish APC 224 (Shellac 20%) | 20 | Charcoal type SG | 10 |
| | | Shellac varnish APC 224 (Shellac 20%) | 50 ml |

### 22.4.3 FRICTION MATCH

These are of two types:

a. *Strike Anywhere* (SAW), which are highly sensitive to friction and can be ignited on rubbing against an inert rough surface.
b. *Safety Matches*, which are of lower sensitivity to friction and, therefore, need another reactive friction composition to be ignited.

*Strike Anywhere Match* composition is given in Table 22.15.

The striker composition formula 25 can ignite on any hard surface while formula 26 is its base.

*Strike Anywhere Match* chiefly contain potassium chlorate powdered glass, zinc oxide and phosphorous (III) sulphide. When the striker is struck against a rough surface, the friction leads to heat generation causing combustion reaction of potassium chlorate and phosphorous (III) sulphide leading to burning of the match stick. Strike anywhere matches are more hazardous than safety matches since it can be ignited by rubbing against any rough surface.

**TABLE 22.15**

**Strike Anywhere Match Composition [8]**

| Ingredient | Role | Striker composition formula 25 | Base composition formula 26 |
|---|---|---|---|
| Animal glue | Binder and fuel | 11 | 12 |
| Starch | Extender for glue | 4 | 5 |
| Paraffin wax | Combustible | – | 2 |
| Potassium chlorate | Oxidant | 32 | 37 |
| Phosphorous sesquisulphide | Primary fuel | 10 | 3 |
| Sulphur | Fuel | – | 6 |
| Rosin | Binder and fuel | 4 | 6 |
| Dammar gum | Binder and fuel | – | 3 |
| Infusorial earth | Burn rate modifier | – | 3 |
| Powdered glass | Frictional material | 33 | 21.5 |
| Potassium dichromate | Burn rate catalyst | – | 0.5 |
| Zinc oxide | Stabiliser | 6 | 1 |

A *Safety Match consists* of a match box whose sides are pasted with the normal friction composition. The wooden sticks are treated with chemicals to ensure stick are non-inflammable. The wooden sticks are provided with a friction composition in the form of a large bulb at its one end. The bulb on scratching the sides of the match box provides intense heat with a dull red glow for 10 seconds minimum, which is sufficient to raise the ignition temperature of the safety fuze composition. The burnt particles of the bulb composition remain adhered to the fine wooden sticks and do not cause any flame, burst, spark or fumes. A typical safety match composition is given in Table 22.16.

Safety matches are safe because they do not spontaneously lead to combustion. It needs to strike against a special coated friction surface to get ignited. Some match heads contain sulphur (sometimes antimony (III) sulphide or phosphorous (III) sulphide) and

**TABLE 22.16**

**Safety Match Composition [9]**

| Match head | | Friction surface | |
|---|---|---|---|
| Ingredient | Percentage | Ingredient | Percentage |
| Potassium chlorate | 42 | Red phosphorous | 56 |
| Lead (II, IV) sulphide | 42 | Antimony (III) sulphide | 38 |
| Gum arabic | 16 | Gum arabic | 6 |

oxidizing agents (generally potassium chlorate), with powdered glass, colorants, fillers and a binder made of glue, starch, gum arabic, etc. The friction surface generally consists of powdered glass or silica (sand), red phosphorus, binder and filler.

Let us understand how the safety match works. When one strikes a safety match, the friction generates heat, converting a small amount of red phosphorus to white phosphorus vapor. White phosphorus spontaneously ignites, decomposing potassium chlorate and liberating oxygen. At this point, the sulphur starts to burn. Safety matches ignite due to the extreme reactivity of phosphorus with the potassium chlorate in the match head. The reaction is

$$6P + 5KClO_3 = 3P_2O_5 + 5KCl \qquad (22.1)$$

In case of phosphorous (III) sulphide, the reaction is

$$3P_2S_3 + 11KClO_3 = 3P_2O_5 + 11KCl + 9SO_2 \qquad (22.2)$$

The friction priming compositions have low autoignition temperature as shown in the Table 22.17.

## 22.4.5 Special Requirements of Friction Compositions

In addition to the general essential requirements of compositions as given in Section 1.6, friction compositions should meet the following special requirements:

**TABLE 22.17**
**Autoignition Temperature of Some Friction Compositions [10]**

| Friction priming composition | Percentage | Autoignition temperature |
|---|---|---|
| Strike Anywhere Match mix | 11.0 animal glue + 4.0 starch + 32.0 potassium chlorate + 6.0 zinc oxide + 10.0 phosphorous sesquisulphide ($P_4S_3$) + 33.0 powdered glass + 4.0 rosin | 120–150 |
| Safety Match mix | 11.0 animal glue + 5.0 sulphur +51.0 potassium chlorate + 7.0 zinc oxide + 4.0 manganese dioxide + 15.0 powdered glass + 1.0 potassium dichromate + 6.0 black iron oxide ($Fe_3O_4$) | 180-200 |
| Friction Primer mix | 42.0 potassium chlorate + 42.0 antimony sulphide + 3.0 sulphur +2.0 calcium carbonate + 3.0 meal powder + 3.0 ground glass + 5.0 gum arabic | 139 |
| Friction Primer mix | 53.0 potassium chlorate + 22.0 antimony sulphide + 9.0 sulphur +1.0 calcium carbonate +10.0 ground glass + 5.0 gum arabic | 137 |
| Friction Primer mix | 63.0 potassium chlorate + 32.0 antimony sulphide + 5.0 gum arabic | 152 |

a. Should provide reliable ignition over a range of friction stimuli
b. Combustion should be slow and non-violent
c. Combustion should not allow hot dross or spit to fall
d. Must provide adequate duration of burning
e. Should be sufficiently moisture proof
f. Should be unlikely to be ignited by mechanical shock/force

## 22.5 ELECTRICAL ENERGY ON ELECTRICAL SENSITIVE COMPOSITIONS

Some electrical cap conducting compositions work on electric impulse *by use of graphite* in the composition which imparts electrical *conductivity*.

Also, electrical resistance wires are capable of producing heat when suitable current is passed. This has been used to devise ignition systems like electric squib, impulse cartridges, igniter safety fuze, electric where heat generated by resistance wire are sufficient to cause ignition. Let us discuss these compositions.

### 22.5.1 ELECTRICAL CAP CONDUCTING

In case of electrical cap conducting composition, the firing current passes through the electrically sensitive cap composition (containing a mix of lead styphnate (RD1303) and graphite, the latter acts as sensitiser) causing the cap composition to ignite (Table 22.18).

It further ignites SR 227 (gunpowder variant consisting of 72% of potassium nitrate, 21% of charcoal and 7% of sulphur) which in turn produces flash for primers filled with gunpowder to burn. These electrically conductive caps are assembled to the primer by ringing or turn over and the primer is threaded to the cartridge case. Cap conducting compositions are given in Table 22.19 and Table 22.20.

Figure 22.5 depicts M52 DEFA 30 mm cap and Figure 22.6 depicts M52A3B1 20 mm cap.

Time to ignition and sensitivity of conducting compositions depends upon:

a. Type and origin of conducting ingredients (like finely milled graphite, natural or synthetic or conductive carbon black)
b. Particle size of conducting composition (lead styphnate and finely milled graphite or conductive carbon black)

---

**TABLE 22.18**
**Electric Conductive Composition [4]**

| Ingredient | Percentage | |
|---|---|---|
| | RD 1653 | RD 1654 |
| Graphite (Acheson grade 615) or DOHM's air floated) | 9–10 | 6–7 |
| Lead styphnate (RD1303) | To 100 | To 100 |

**TABLE 22.19**
**Electric Conducting Compositions [8]**

| Ingredient | Formula 16 | Formula 17 | Formula 18 |
|---|---|---|---|
| Zirconium fine <5μ | 7.5 | 6–9 | 15 |
| Zirconium coarse >10μ | 32.5 | 30–35 | – |
| Zirconium hydride | – | – | 30 |
| Lead dioxide | 25 | 18–22 | 20 |
| Barium nitrate | 35 | 15–25 | 15 |
| PETN | – | 15–23 | 20 |

**TABLE 22.20**
**Pyrotechnic Priming CC Mixtures Used in Explosive Ordnance [11]**

| Ingredient | M52DEFA[a] | | N8 Igniter[b] | N43 Primer[c] | M52A3B1[d] |
|---|---|---|---|---|---|
| | Conducting mix | Priming mix | | | |
| Lead styphnate | 95.0–95.5 | 48 | 97 | 98.5 | 40 ±2.5 |
| Graphite (finely milled) | 4.5–5.0 | 2 | 3 | 4.5 | – |
| Carbon black, conducting | – | – | – | – | 0.75 ± 0.25 |
| Barium nitrate | – | 12 | – | – | 44.25 ± 2.5 |
| Potassium perchlorate | – | 28 | – | – | – |
| Calcium silicide | – | 10 | – | – | 13.0 ± 2.5 |
| Titanium | – | – | – | – | – |
| Arabic gum | – | – | – | – | 1.0 ±0.25 |
| Styphnic acid | – | – | – | – | 1.0 ± 0.25 |

*Notes*
a Double base primer for 30-mm Aircraft Cannon Ammunition
b An igniter for an electrical fuze
c Part of a primer for 4.5-inch naval ammunition (Australian Ordnance)
d Primer for 20 mm F/A-18 ammunition (Australian Ordnance)

There is a large variation in resistance of the composition that varies due to humidity, pressing load and porosity. Any shock or jolt alters the resistance significantly. The reason attributed is that the graphite acting as conductive particle in the composition is not uniformly distributed and its position changes even with a small shock or jolt. The resistance changes even during storage. The minimum initiation energy requirement varies drastically as some may function even at

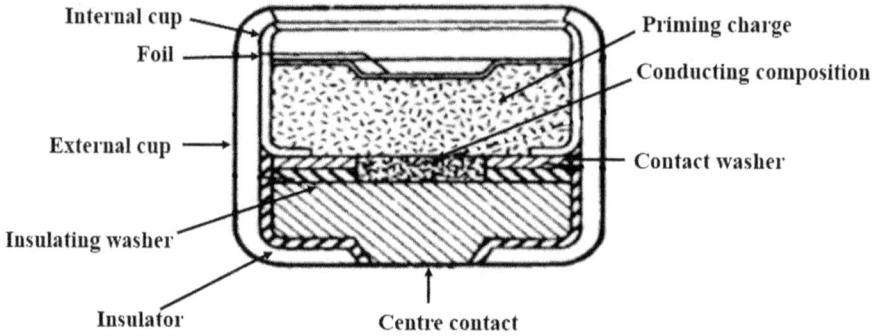

**FIGURE 22.5**  Construction of M52 DEFA 30 mm Cap (Reprinted from [11]).

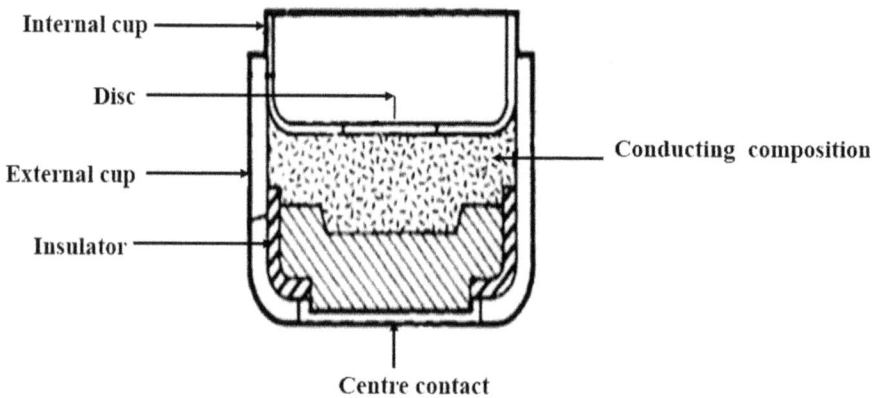

**FIGURE 22.6**  Construction of M52A3B1 20 mm Cap (Reprinted from [11]).

low electrical initiation energy while the other may not function even at little higher electrical initiation energy. However, all generally function at substantially high electrical initiation energy as current path are short circuited by micro arcs between conductive graphite particles.

## 22.5.2  ELECTRICAL SQUIB

Squibs are also known as electro-explosive devices, initiated by electrical energy. Mostly, lead ferrocyanide, lead thiocyanate, lead styphnate, lead azide, LMNR (lead mono nitro resorcinate) are used as fuel. Potassium chlorate and perchlorate are used as oxidisers. These have low ignition temperature and hence a resistance wire is sufficient to ignite when energised with current. The ignition temperatures of a few squib compositions are given in Table 22.21.

Brass foils are insulated by cardboard and have a bridge wire over which ignition mixture is formed as a bulb. The squib composition is made into slurry with binder, generally nitrocellulose binder, and coated over the resistance wire. The charge

**TABLE 22.21**

**Ignition Temperatures of Squib Compositions [12]**

| Composition | Ignition temperature °C |
|---|---|
| Lead styphnate | 282 |
| Lead azide | 390 |
| Lead ferrocyanide + potassium perchlorate+ calcium silicide + nitrocellulose | 320 |
| Lead thiocyanate + potassium chlorate + calcium silicide + nitrocellulose | 195 |

mass required to be used in a squib is achieved through several coatings of the composition over the bridge wire. The squib composition is then finally coated with a NC varnish to ensure that it is resistant to moisture effect.

Calcium silicide acts as a fuel that produces hot incandescent particles which retain the heat energy and aid in ignition of the squib composition.

Some squibs (not containing calcium silicide) are coated with 50:50 mixes of aluminium powder and NC varnish so that hot slag retains heat for some time after the flash. A typical squib is shown in Figure 22.7.

Some typical squib compositions are given in Table 22.22.

Some more electrical cap formulations are given in Table 22.23.

It may be mentioned that squib compositions based upon lead styphnate are of disruptive type and produces shock pressure or detonation wave while squib compositions based on lead thiocyanate and lead ferrocyanide are igniferous type for ignition.

The hardware components used are lead wire or contact pins, plug and the resistance wire.

Heat is generated, on providing electric impulse to wire, which is more than the ignition temperature of the squib composition. The resistance wire generally used is nichrome, platinum -iridium, platinum- rhodium, tungsten, cupro-nickel etc. which have high and uniform specific resistance; without kinks/twist; with higher melting

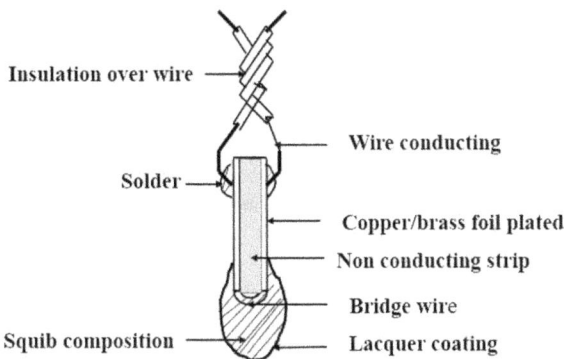

**FIGURE 22.7** Typical Squib.

## TABLE 22.22
## Squib Compositions [13]

| Ingredient | Percentage | | | | |
|---|---|---|---|---|---|
| Potassium chlorate | 50 | 65 | 70 | 47 | – |
| Lead ferrocyanide | 40 | – | – | 29.4 | 40 |
| Calcium silicide | 10 | 5 | – | 17.6 | 10 |
| Lead thiocyanate | – | 30 | 30 | 6 | – |
| Binder* over and above | 2 | 2 | 2 | 1 | 2 |
| Potassium perchlorate | – | – | – | – | 50 |

## TABLE 22.23
## Electric Cap Formulations [3]

| Ingredient | Percentage | | | | | |
|---|---|---|---|---|---|---|
| Potassium chlorate | 8.5 | 55 | 25 | 60 | – | – |
| Lead mononitrate resorcinate | 76.5 | – | – | – | – | – |
| Nitrocellulose | 15 | – | – | – | – | – |
| Lead thiocyanate | – | 45 | – | – | – | – |
| Diazo dinitro phenol | – | – | 75 | 20 | – | – |
| Charcoal | – | – | – | 15 | – | – |
| Nitro-starch | – | – | – | 5 | – | – |
| Potassium perchlorate | – | – | – | – | 66.6 | 66.6 |
| Titanium | – | – | – | – | 33.3 | – |
| Aluminium | – | – | – | – | – | 33.3 |

point than ignition temperature of squib composition and low specific heat for ignition of squib composition even at lower electrical energy. The energy (E) generated by resistance wire is given by

$$E = i^2Rt \tag{22.3}$$

where $i$ = current in amperes, $R$ = resistance in ohms/metre, $t$ = time in seconds.

Thus, for a wire with resistance of 2 ohms with 1 ampere current for 0.01 seconds, the energy (E) generated in the resistance wire would be

$$E = (1)^2 x2x. 01 = .02 joules \ or \ 20mJ \tag{22.4}$$

The maximum "*No fire current*" (NFC) is the maximum current below which the squib would not function (99.9% reliability), while the minimum "*All fire current*" (AFC) is the minimum current above which the squib would definitely function (99.9% reliability). These NFC and AFC are determined by the *Bruceton staircase method*.

The performance of a squib depends upon the properties of the squib plug and the resistance wire and the squib composition. Squib functioning mechanism is through following sequence.

a. *"All Fire"* electrical energy is given to squib
b. The resistance wire gets heated, transmitting heat to squib composition
c. *"Hotspots"* are generated in squib composition
d. Combustion starts in whole squib composition in microseconds
e. The thermal energy (through heat, flame and hot combustion products) is provided to next high energy material in assembly of the ammunition or device

There is a delay in functioning of the squib, i.e., the time interval between application of current and functioning of squib, mostly in milli or microseconds. This delay time (measured by photocell and counter) depends on the following parameters (see Section 5.1.10).

a. *Bridge wire material specific resistance* which depends on the bridge wire length and diameter (an increase in bridge wire resistance reduces the delay in functioning)
b. *Magnitude of firing current* (an increase in current reduces the delay time)
c. *Ignition temperature of squib composition* (lower the ignition temperature, lower is the delay time)
d. Ambient temperature (higher the ambient temperature, lower the delay time)

Closed vessel testing gives *pressure–time curve*. Time taken to reach 10% of $P_{max}$ is specified as maximum delay. The major specification requirements for squibs are its NFC, AFC and delay.

### 22.5.2.2 Use of Squibs

Electrical squib composition is used in a variety of electro-explosive devices as given below:

a. *Igniter Safety Fuze, Electric*: It is also abbreviated as ISFE (Figure 22.8). It consists of a copper sleeve to which is assembled a squib, i.e., fuzehead (brass foils insulated by cardboard and having a bridge wire over which ignition mixture is formed as a bulb. It also specifies no afire current for safety and all fire current for functioning.

The copper sleeve is punctured so that a tip of the copper sleeve is projected inside. The safety fuze is inserted at the bottom of the sleeve, by pressing it up to the pierced position. This intrusion of pierced copper does not allow any further movement of the safety fuze and thus ensures no damage to the bulb. A rubber tape is used over the pierced joint. On supply of electrical energy, the ignition of bulb

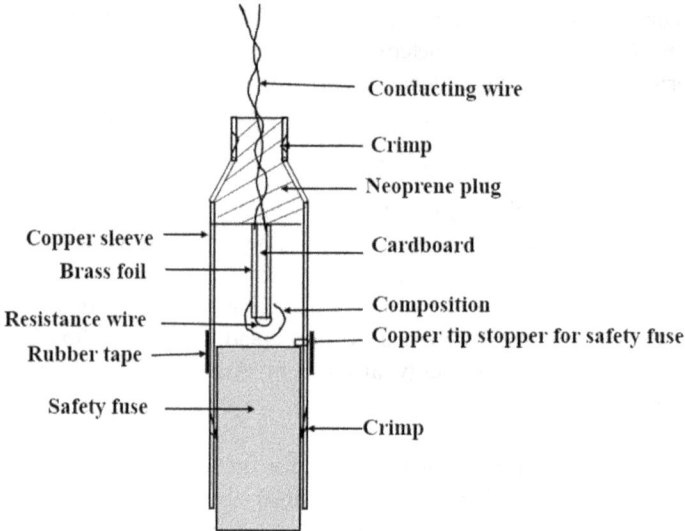

**FIGURE 22.8**   Igniter Safety Fuze, Electric.

takes place, which, in turn, ignites the safety fuze. The length of the conducting wire is kept long for initiation of squib from far distance and twisted to avoid any accidental initiation due to interference of radio frequency (RF).

Another method is to use conventional electrical squib assembled with a pyrotechnic composition or propellant for propelling Grenade Hand 80PWP MK1 through AFVs.

Another use of squibs is made for detonating composition on similar lines instead of propellant or gunpowder.

b. *Impulse Cartridges*: Infrared flares and chaff cartridges use impulse cartridges. They conventionally contain pyrotechnic igniting composition (first fire charge) with boron: potassium nitrate in 30:70 proportion pressed over resistance wire (0.85 to 1.15 ohm) of platinum tungsten. The insulation resistance is kept at 10 mega ohms (minimum). The impulse cartridge contains booster charge and smokeless propellant which is ignited to push the contents (flare or chaff) in the cartridge case with a velocity of 10–50 m/second. The pyrotechnic train for impulse cartridges is

*pyrotechnic igniting composition => booster pyrotechnic composition => propellant*

Boron/$KNO_3$ is a gas-generating material with high combustion temperature and hot particles as combustion product and suitable for rocket motor igniter and gas generation.

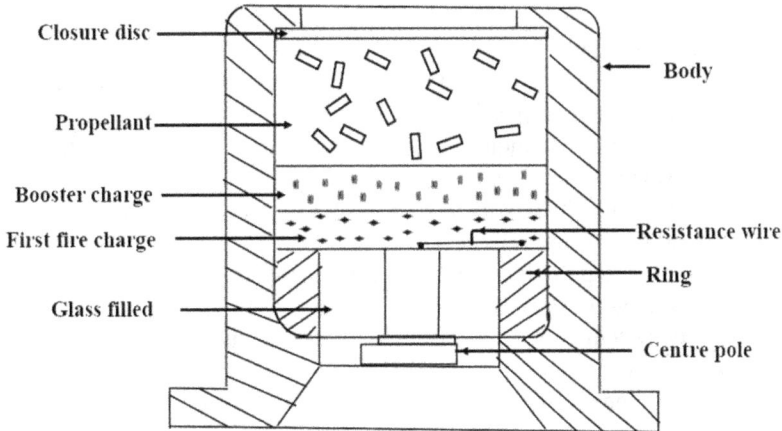

**FIGURE 22.9**  Impulse Cartridge BBU-35-B.

As a safety measure, these have been specified with "*maximum no fire current*" at 1 A/1 W current for 5 minutes and "*minimum all fire current*" at 4.25 A at 15–28 volts for 15 milliseconds. Some squibs are provided with thread for thread assembly or made plane for push-fit assembly. A typical squib is shown in Figure 22.9.

Recent advances [14] in impulse cartridges have shown that the best igniter and booster compositions in terms of impulse cartridge functionality based on the peak pressure (bar) and time to peak pressure (ms) are the mixtures 60% Zr, 40% KClO$_4$ (blended with NC dipping grade lacquer 2:1 and pasted over bridge wire) and booster composition as 30% B, 10% Mg, 60% KClO$_4$, with 4% additional binder fish glue. The bridge wire material was 80/20 (nickel/chromium) of diameter 0.046 mm with wire length 4 mm (length to diameter ratio of bridge wire 87, resistance of wire per mm 0.097 ohms.

The above compositions passes all of the requisite safety tests of MIL-DTL23659E, 2007 which requires that the igniter composition must qualify the required safety tests including maximum no fire current, static discharge and stray voltage, to ensure safe handling, transportation and storage, along with a reduction of hazard during its life cycle as follows [14]:

a. *Maximum no fire current*: Subjected to 1 W/1 A current for 5 minutes, no fire
b. *Static discharge test*: Subjected to 25,000 V simulated human electrostatic discharge, no fire
c. *Stray voltage test*: Subjected to 2,000 pulses of direct current. Each pulse was of 300 millisecond duration and pulse rate was 2 pulses per second. Each pulse had a minimum amplitude of 100 ± 5 mA, no fire

### 22.5.2.3  Advantages and Disadvantages of Squibs/Impulse Cartridges

These have following advantages:

a. There are no moving parts like striker or piston or spring part.
b. Its design is simple and inexpensive.
c. It is easier to hermetically seal through glass to metal seal, ceramic to metal seal or thermosetting plastic.
d. It requires a limited amount of input energy (operates on low current).
e. It has a small/compact size.
f. It provides immediate delivery of energy to the desired location.
g. It may be fired from a distance.
h. It has high energy output (high power/weight ratio).
i. It possesses good safety record.
j. Its safety system may be checked by non-destructive testing of continuity at "No fire current."
k. The electric energy supply may be made fail proof by having two or more circuits (redundant system).
l. It has a high reliability of functioning.
m. It has long-term storage capability.

However, it has the following disadvantages as well.

a. It requires an electrical source.
b. It has safety hazard in case of accidental current discharge.
c. There are possibilities of inadvertent firing due to electrical faults.
d. It has delicate soldering of fine wires.
e. It is a single shot unit with no possibility to check functioning before use.

Because of the crucial nature of many aerospace pyrotechnic applications, superfluous repetition/redundancy, i.e., squibs with parallel paths to perform a single function are used to avoid any single point failure.

### 22.5.4 Special Requirements of Electrical Squib Compositions

In addition to the general essential requirements of compositions as given in Section 1.6, squib compositions should meet the following special requirements.

a. should have low ignition temperature
b. Should meet the requirement of *"maximum no fire current"*(NFC) and *"minimum all fire current"*(AFC)
c. Should meet delay (of functioning) requirement and closed vessel testing

## 22.6  PRIMERS, IGNITERS AND POWER CARTRIDGES

Ignition of pyrotechnic compositions and propellants through thermal energy is accomplished using primers, igniters and mechanical works by power cartridges (initiated through caps or impulse cartridges).

These are capable of initiating ignition of

a. Propellants
b. Pyrotechnic solid compositions
c. Inflammable gel incendiary compositions

For igniting propellants, there are three methods.

## 22.6.1 IGNITER FOR IGNITION OF PROPELLANTS IN CARTRIDGE CASES OR BAGGED CHARGES

*Primer* is used for igniting the propellant in small, medium and high calibre ammunitions. Gunpowder is used in primers for conventional ammunition. Primer is a mechanical tubular component with a large number of flash holes and is assembled with cap at the base and contains gunpowder. It receives flash from the cap and assists in uniform burning of the propellant. Primer propellant combination is important as it provides desired rate of burning of propellant. It is assembled at the base of the cartridge case.

The primer along with cap and cartridge case (Figure 22.10) is responsible for obturation of the ammunition during firing. These primers are generally provided with obturating device, i.e., preventing backward escape of propellant gases by copper ball or copper cone and thus prevents gas escape between cap and primer body.

Primers may be classified as under.

a. *Mechanical Primers*:
   Examples are
   Short primers: Primer 1,5,18,30,33, *Long primers*: Primer 18A and Primer 19B
   If the propellant is required to be ignited at the base, then short primers are used. If the whole propellant is required to be burnt, long primers are used, which can be inserted full length inside the propellant.
b. *Electrical Primers*:
   Examples are primer L1A4, Primer 1A/L
c. *Electro Mechanical Primers*:

These electro-mechanical primers have a combination of both mechanical as well as electrical firing system. Example is Primer KB 5Y or GUV-7 (as primer for 125 mm HE and HEAT ammunition)

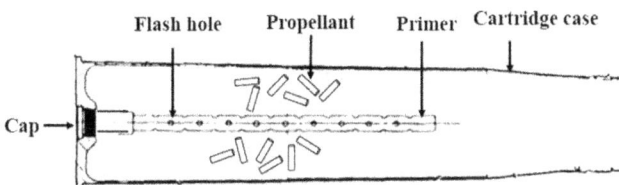

**FIGURE 22.10** Cap, Primer and Cartridge Assembly.

## 22.6.2 IGNITER FOR IGNITION OF BASE BLEED PROPELLANTS IN PROJECTILE

A projectile during its trajectory encounters drags of three types, namely

a. *Base drags*
b. *Skin drags*
c. *Nose drags*

The most influential drag is the base drag as it does not allow the projectile to reach its desired destination due to the vacuum created by it at the base. As the projectile moves in the air, it creates a rarefied zone or vacuum just behind the base of the projectile. This causes the surrounding air to move towards the created vacuum so as to equalise the pressure. This movement of air at the base of the projectile causes the shell to be unsteady as well as to have increased air resistance at the nose of the projectile since rarefied zone or vacuum is created at the base side of the projectile. This slight pull effect felt at the base of the projectile is termed as base drag (Figure 22.11).

Projectile performance can be improved by reducing air resistance as under.

a. Low resistance projectiles having longer and slimmer bodies
b. Afterbody boat tailing
c. Projectiles with own source of power (rocket-assisted projectile)
d. Reducing base drag by ejecting stream of gases to reduce the vacuum

To increase the range of projectiles, base bleed propellants (annularly shaped propellant) are used at the base (rear) side of projectile within a tubular base-bleed housing comprising a combustion chamber. The gaseous combustion products of base bleed propellant assist in reducing the base drag.

These propellants are slow burning propellants. Since sudden pressure drop when the projectile leaves the muzzle may cause failure of burning of base bleed

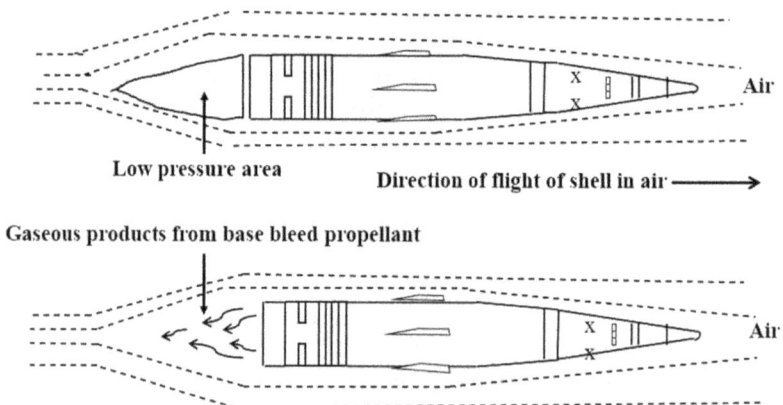

FIGURE 22.11 Reduction of Base Drag by Base Bleed Propellant.

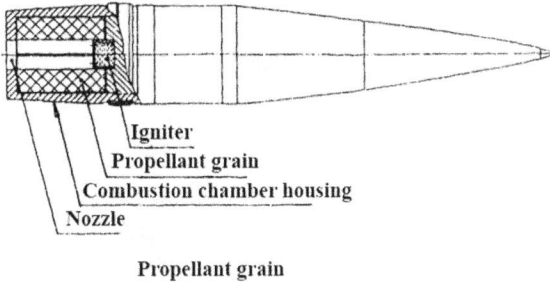

Propellant grain

**FIGURE 22.12** Low Drag Experimental Projectile with Base Bleed Unit (Reprinted from [15]).

propellant, an igniter is assembled along with base bleed propellant, which burns for 2–3 seconds and thus ensures burning of base bleed propellant during flight.

Challenges in developing base bleed charges are the large set back forces (up to 18,000 g) of the projectile, the high rate of spinning of the projectile (200–300 s$^{-1}$), as well as the pressure drop upon firing the projectile, which may cause quenching of the combustion.

Base bleed igniter cum base bleed propellant can increase the range by almost 30%. As an example, shell 155 mm (Naschem) has a range of 24,000 m while with base bleed igniter it has a range of 32,000 m.

The igniter composition (consisting of a pyrotechnic composition which is substantially insensitive to pressure variations) is compacted in a metallic body with perforations, covered by aluminium foil and assembled at the base of the ammunition [15] along with base bleed propellant (Figure 22.12).

The base bleed igniter composition ensures uninterrupted burning of base bleed propellant during flight. The base bleed propellant produces gases, thereby reducing the vacuum at the base on its ongoing trajectory and thus reduces the base drag, enabling the projectile to have higher range.

Thus, base bleed igniter ensures that the base bleed propellant burns during flight and remains unaffected due to drop in pressure as the projectile comes out of the muzzle, thereby reducing the base drag of the projectile and enhances the range. The drawback with base bleed unit consisting of base bleed igniter and base bleed propellant is that it has a high dead weight and a failure in base bleed unit shall result in short range.

The base bleed igniter must have following properties.

a. It must withstand the high pressure of the firing

b. It must get ignited with propellant gases and flame

c. It must be capable of reigniting the extinguished base bleed propellant

d. It must not quench and remain ignited during crucial pressure drop (i.e., low sensitivity to pressure drops) at the shell bottom

e. It must be hermetically sealed and unaffected by extreme conditions of environment

f. Its part should not fall out after ignition so as to affect the flight stability of the projectile

### 22.6.3 IGNITER FOR IGNITION OF PROPELLANTS IN MISSILES

Igniter induces and acts as a first fire charge for the controlled combustion of the solid propellant fuel at high temperatures and pressures. The igniter composition is granulated and made in the form of pellets so that they are capable of withstanding the shock and other external forces to retain their identity. The size of pellets and their quantity is determined with a lot of trials. These pellets are filled in metallic containers assembled with two squibs (for redundancy). The container used for igniter has several flash holes covered with aluminium foil or varnished cloth disc and the total area of flash holes depends upon the propellant configuration. The container is sufficiently strong and it does not rupture into the fragments of the container to avoid damage to the propellant or choke the nozzle.

The igniters developed for all systems by HEMRL Pune India [16] are shown in Figure 22.13.

On ignition through electrical squibs, the composition burns the cloth disc or ruptures the aluminium foil and the flash passes through the perforations of metallic container and ignites the propellants of the missile. The energy output from ignition depends on the following.

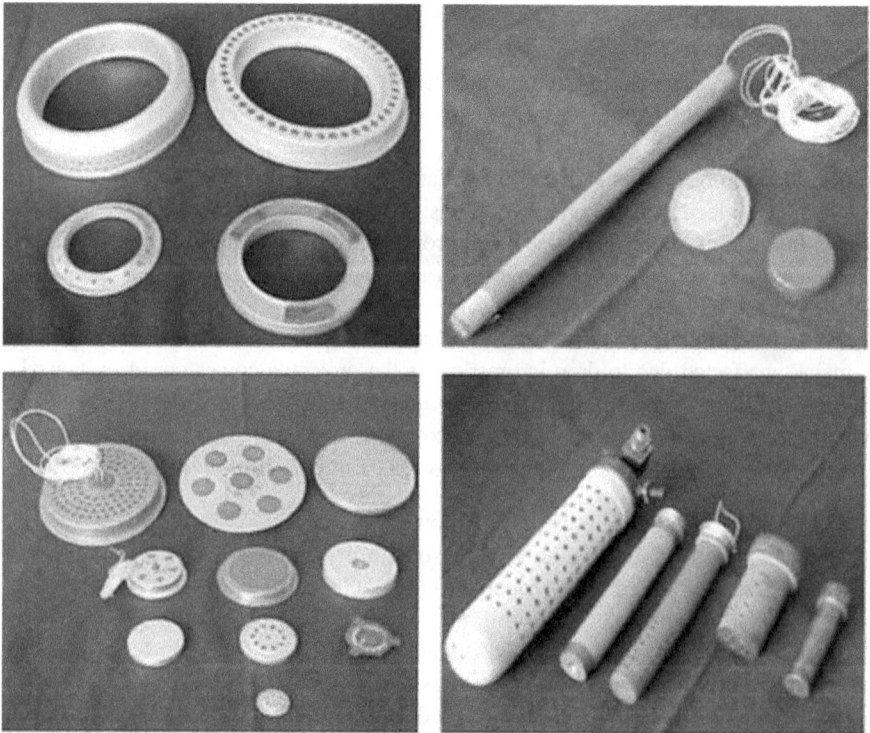

**FIGURE 22.13**  Igniters Developed for all Systems by HEMRL, Pune, India (Reprinted from [16]).

a. Form of the pellets
b. Composition of the pellets
c. Particle size of ingredients in the composition
d. Quantity of the pellets which determines generated pressure
e. Density of the pellets
f. Position of igniter in the system
g. Design of the system

The combustion of the igniter composition serves two important purposes:

a. Generate a heat flux in the form of hot, dense gases which rapidly ignite the propellant grain on all exposed (non-inhibited or bonded) surfaces
b. Pressurise the chamber to a level such that the burn rate of the propellant is sufficient to maintain this pressure

Some typical formulations are recommended for pelleted igniters [17].

a. Boron–Potassium nitrate composition: Boron 23.7%, potassium nitrate 70.7%, binder laminac 5.6%. Characterised by ease of ignition at very low pressures (high altitudes), high gas content and low sensitivity of burning rate to pressure.
b. Aluminium–Potassium perchlorate composition: Aluminum 35.0%, potassium perchlorate 64.0%, vegetable oil 1.0%. High energy content, but difficult to ignite at low pressures. Burning rate strongly pressure-dependent.
c. Mg/Teflon-Basic composition prepared in variety of formulations and configurations; three examples listed in Table 22.24. Generally characterised by very low pressure burning rate exponents and a low percentage of permanent gas content; efficient igniters, having energy output strong in the infrared region. Energy content approximately equivalent to that of boron–potassium nitrate.

MTV compositions having high energy content, low cost with ease of manufacture of igniter pellets is a preferred material for igniters.

**TABLE 22.24**
**Solid Motor Rocket Igniters [17]**

| Ingredient | | Percentage | |
|---|---|---|---|
| Magnesium | 60 | 32.5 | 54 |
| Teflon | 40 | 67.5 | 30 |
| Graphite | 1 | – | – |
| Additive | – | 2 | – |
| Viton | – | – | 16 |

**TABLE 22.25**

**Thermal Properties of Different Pyrotechnic Grains [18]**

| Performance | B/KNO₃ | Al/KClO₄ | Mg/Teflon |
|---|---|---|---|
| Burning rate, mm.sec⁻¹ | 43.2 | 9.9 | 10.2 |
| Pressure exponent | 0.32 | 1 approx. | 0.22 |
| Heating value, cal. g⁻¹ | 1,550 | 2,490 | 2,200 |

Another report [18] provides thermal properties of pyrotechnic grains of B/KNO$_3$, Al/KClO$_4$ and Mg/Teflon as given in Table 22.25.

### 22.6.4 Power Cartridges for Initiation of Mechanical Works

Power cartridges (also known as propellant actuated devices) consist of an initiating device like cap or electric squib, suitably pressed pyrotechnic composition (consisting of priming composition, booster composition and main composition) and a single/double base propellant, all encased in a metallic container. Caps are mechanically fired while squibs are electrically fired. The propellant on burning produce intense heat and pressure, which are utilised for its mechanical performance, generally through gas pressure. Examples are:

 i. *Piston and Bellow Devices* (actuators, thrusters, cable and hose cutters, valves and switches)
 ii. *Explosive Bolts and Explosive Nuts*
 iii. *Aircraft Systems* (seat ejection system, release of fuel tank and storage equipment, emergency and rescue system)
 iv. *Spacecraft Systems* (launch and control system, emergency, stage separation system, fairing release system, recovery and landing system)
 v. *Missile* (safety and arming system, ignition system, control system, stage separation system, destruction system, gyro system)

There are large varieties of power cartridges to suit end requirement. However, some of the power cartridges are shown in Figures 22.14, 22.15 and 22.16.

 i. Cartridge, Impulse, Mark 3 MOD 1[19]: Pilot ejection navy light ejection seat (Douglas Integrated Harness System)
 ii. Cartridge, Impulse, Mark 10 MOD 0[19]: Release and ejection of stores
 iii. Cartridge, Impulse, Mark 64 MOD 0[19]: LAU-25/A aircraft flare launcher
 iv. Cartridge, Delay, Mark 84 MOD 0[19]: Missile stage release mechanism
 v. Generator, Gas, Gyro Spin-Up [19]: S-3A aircraft crew escape system
 vi. Cartridge, Actuator, Mark 28 MOD 0[19]: Thruster, P6M aircraft

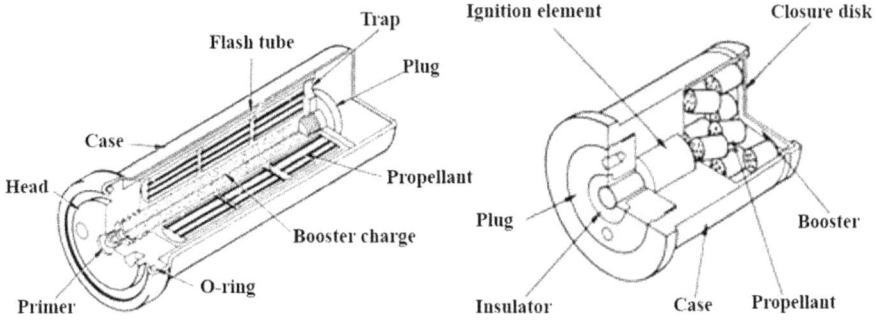

FIGURE 22.14   Cartridge, Impulse, Mark 3 Mod 1 and Cartridge, Impulse, Mark 10 Mod 0 (Reprinted from [19]).

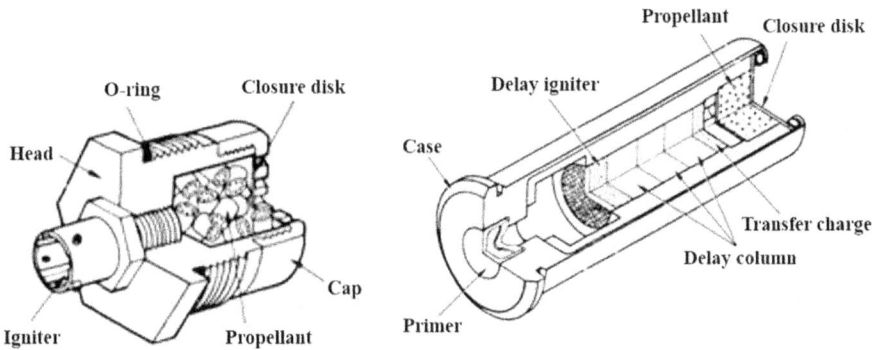

FIGURE 22.15   Cartridge, Impulse, Mark 64 Mod 0 and Cartridge, Delay, Mark 84 Mod 0 (Reprinted from [19]).

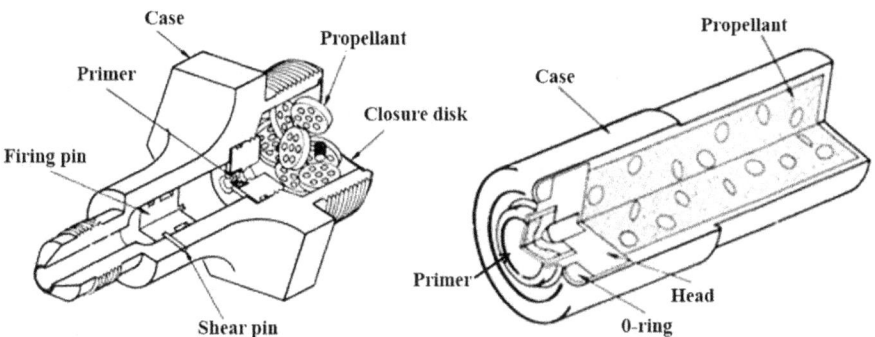

FIGURE 22.16   Generator, Gas, Gyro Spin-Up and Cartridge, Actuator, Mark 28 Mod 0 (Reprinted from [19]).

PROJECT MERCURY
Pyrotechnic Devices

REEFING
LINE
CUTTER
1 REQD

3 EXPLOSIVE
BOLTS AND A
3 SEGMENT
CLAMP RING

BALLISTIC
MORTAR
1 REQD

REEFING
LINE
CUTTERS
4 REQD

DEPLOYMENT
GUNS
2 REQD

SOFAR
BOMBS
2 REQD

PROPELLANT
ACTUATED
DISCONNECT
2 REQD

EJECTOR
BAGS &
GAS
GENERATORS
2 REQD

LINEAR
EXPLOSIVE/
BOLT
POPPING

3 EXPLOSIVE
BOLTS AND A
3 SEGMENT
CLAMP RING

EXPLOSIVE
BOLT
1 REQD

EXPLOSIVE
ACTUATED
DISCONNECT

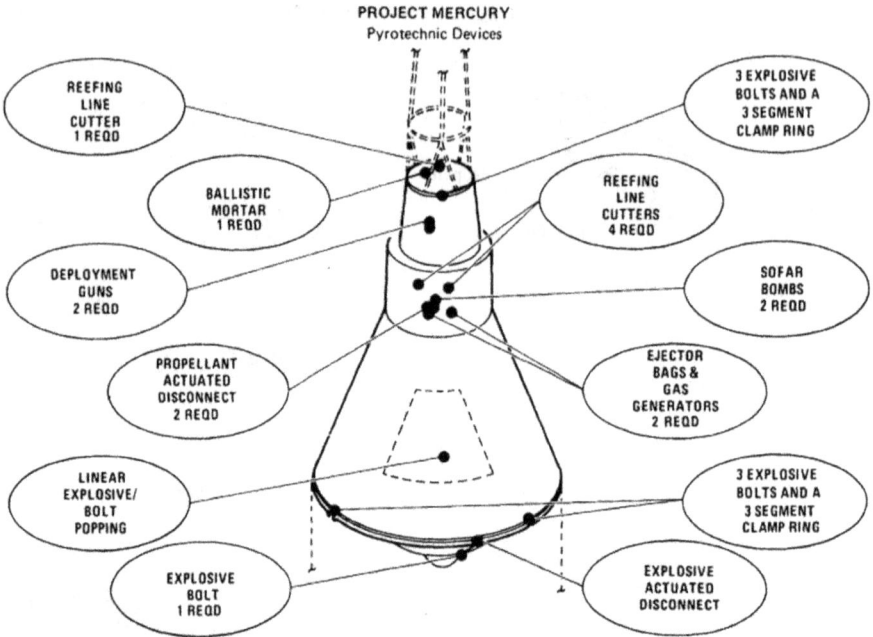

**FIGURE 22.17**   Pyrotechnic Devices in Mercury Space Vehicle (Reprinted from [20]).

The pyrotechnic devices used in project mercury space vehicle [20] is shown in Figure 22.17.

These power cartridges are so designed to function in space vehicles on a sequence of events as under [20].

a. *Igniter:* Initiation of the squib system (a heat sensitive pyrotechnic composition) having ability to ignite the propellant next in the train of events by heat, shock or hot particles.

b. *Chemical Conversion:* The chemical conversion involves the release of controlled amounts of gas and heat from a propellant of moderate burning such that various rates of energy application can be applied.

c. *Work Output:* The next function is the change from chemical to kinetic energy. The most general form is the acceleration of a mechanical piece, such as a piston/cylinder arrangement.

d. *Mechanical Function:* The work output or actual work performed are in the form of cutting, breaking, pressurising, pushing and pulling, among others.

The pyrotechnic system, which can be initiated by five external stimuli, is shown in Figure 22.18, consisting of four essential elements, namely the igniter, a chemical conversion source, the work output mechanism and the mechanical function required.

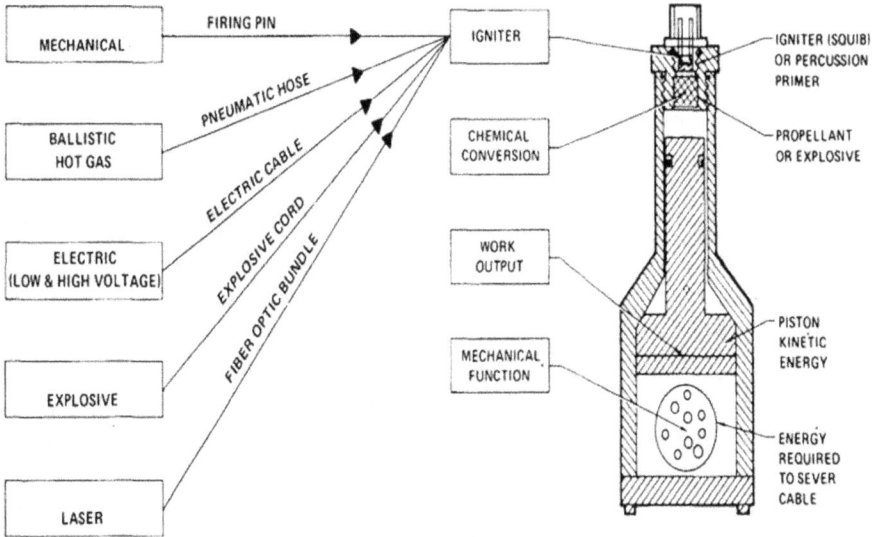

**FIGURE 22.18** Pyrotechnic System for Aerospace Vehicles (Reprinted from [20]).

Readers may also refer to NASA memorandum 106810 Revision A [21] for the catalogue of pyrotechnically actuated devices and systems.

### 22.6.4.1 Special Requirements of Igniter for Rocket Motor/Aerospace Vehicles

In addition to the general essential requirements of compositions as given in Section 1.6, igniter compositions should meet the following special requirements [17].

a. The igniter composition should be readily ignitable over the environmental range required by the application.
b. The exhaust products of the igniter composition should produce the pressurisation required to meet the rocket motor ignition objectives.
c. The igniter composition must be non-hygroscopic or adequately protected against moisture absorption.
d. The igniter composition should have the burning-rate properties required to achieve the necessary energy output rate without over pressuring the igniter hardware.
e. The igniter composition should have sufficient available energy to produce the energy output rate required for motor ignition.
f. The energy output of the igniter composition should be such that efficient transmittal of its energy to the propellant surface is achieved.
g. The igniter must produce propellant ignition and sustained combustion without shock or other adverse effects on the motor.
h. The igniter should perform with ignition delay, ignition time interval and transient characteristics controlled in accordance with end use requirements.

i. The igniter should perform its required function under the specified operational conditions of temperature, altitude, launch loads, acceleration force and vibration levels.

j. The energy release rate of the igniter should not produce excessive pressure peaks in the motor or strains in the propellant.

Readers may refer [22] on ignition system for solid rocket motors by HEMRL Pune, India.

### 22.6.5 IGNITER FOR IGNITION OF SOLID AND INFLAMMABLE GEL PYROTECHNIC COMPOSITIONS

Port Fire Friction (Figure 22.19) is used for igniting solid pyrotechnic compositions like generator smoke no. 5, generator smoke orange 3A and flare ground indicating MK-1 yellow. High intensity of the flame from port fire friction causes ignition to pyrotechnic compositions.

It burns three to four minutes and drops hot burning dross. It is filled in a paper container with maximum three to four folds/layers so that this may also burn with the composition. The composition may be filled and pressed by manual tapping and pressing using a mandrel and then filled with clay or wood piece (approx. 10 cm) so that Port Fire Friction may be held in hand while burning. The top end is primed with friction composition PN 196 and another end portion is assembled with striker cap pasted with composition PN 445.

Some typical compositions are given in Table 22.26.

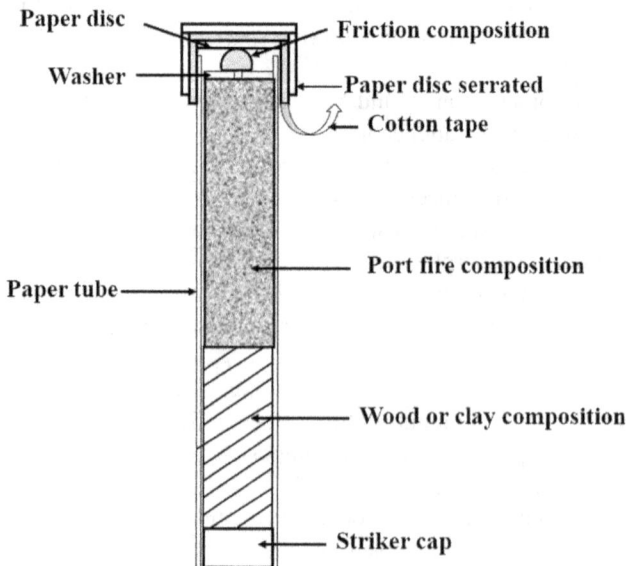

**FIGURE 22.19** Port Fire Friction.

**TABLE 22.26**
**Port Fire Friction Compositions**

| Ingredient | Percentage | | |
|---|---|---|---|
| Potassium nitrate | 67 | 63 | 75 |
| Sulphur | 22 | 16 | 23 |
| Gunpowder | 11 | 11 | – |
| Antimony sulphide | – | 10 | – |
| Charcoal | – | – | 2 |

For igniting inflammable incendiary gel materials, flame-producing composition may be preferably tapped in a cardboard (or metallic container) with priming composition at the top. This tamping or manual ramming of composition ensures that the flame length of the composition is high. It should be of sufficient length with higher burning time for several seconds to ensure proper picking up of flame and burning of the inflammable composition. These are generally ignited through a fuze. The priming composition has a special feature that it retains the heat in the form of hot red slags and hence ensures positive pick-up of ignition by igniter composition, which provides long flame for several seconds to ignite gel composition.

## 22.7 LASER IGNITION

Laser is an acronym for *light amplification by stimulated emission of monochromatic radiation*. Laser is used to heat a large number of materials like melting, welding, drilling or igniting materials by sending a high-intensity, coherent beam of radiation. Laser ignition is not a photochemical phenomenon but purely an optical ignition system. Laser ignition of explosive material system can be subdivided into three parts, namely

1. A laser firing unit (typically laser sources are Ruby, Nd: YAG, CO2, Nd: glass, argon ion, laser diodes, etc.)
2. Fibre optics cable made from a glass or plastic core that carries light surrounded by glass cladding that (due to its lower refractive index) reflects "escaping" light back into the core, resulting in the light being guided along the fibre for transmission of energy and a sealed optic window
3. An explosive (pyrotechnic, propellant or high explosive) material to receive laser energy to be ignited

### 22.7.1 MECHANISM OF LASER IGNITION [23]

The process of laser ignition involves absorption of laser energy by pyrotechnic composition. Laser beam provides energy to the pyrotechnic composition, thereby raising its energy till it exceeds the activation energy of the composition resulting in "*hotspots*" formation (Section 22.3.2). As explained in Chapter 3, some heat is dissipated to atmosphere as heat loss while a part is used up in exhibiting special effects and a part is used up for sustaining burning. The absorption of laser energy by pyrotechnic composition follows the Beer–Lambert equation (Section 12.2.2)

$$I_t = I_0 e^{-\alpha L} \tag{22.5}$$

where $I_0$ is the intensity of the incident laser light on the composition, $I_t$ is the intensity of the transmitted laser light through the composition, $\alpha$ is the absorption coefficient of the composition for the laser light and $L$ is the optical path length in the composition. The absorbed energy induces a temperature rise within the material.

Ignition sensitivity can be enhanced by use of carbon black (approx. 3%). The laser energy absorbed ($A$) by the composition is calculated as percentage of the incident laser energy, by

$$A = 100\% - \text{Reflection}\% - \text{Transmission}\%$$

Experimental arrangement for laser ignition of the sensitised explosive samples is given in Figure 22.20.

According to reference [24] lasers have the following advantages:

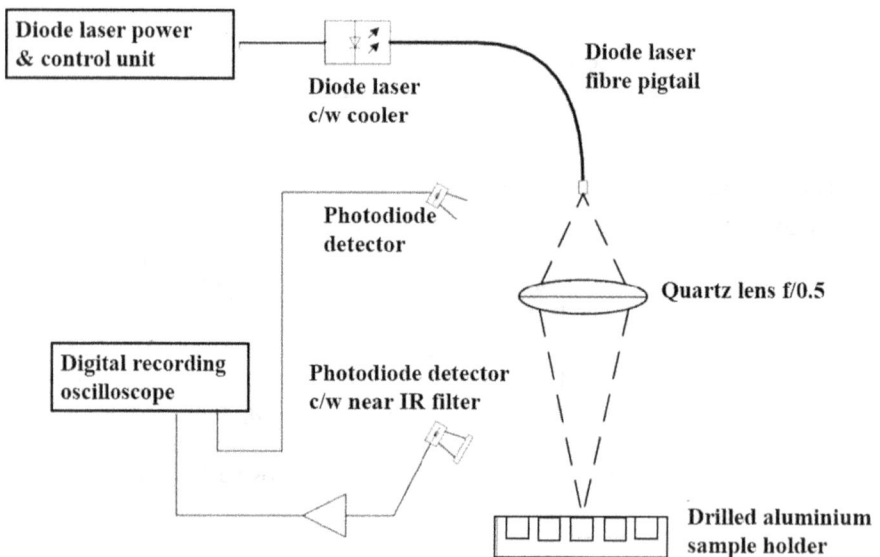

**FIGURE 22.20** Experimental Arrangement for Laser Ignition of the Sensitised Explosive Samples (Reprinted from [23]).

a. Able to deliver a precise, programmed and repeatable quantity of radiant energy to the energetic material.
b. Its emission spectrum is monochromatic and essentially independent of environmental parameters (pressure, temperature) which make the laser ideal for studying ignition behaviour.
c. Ignition systems based on lasers have the advantage of being relatively immune to induced electromagnetic radiations.

However, real ignition systems operate via convective, conductive and radiative energy transfer and are non-monochromatic; therefore, the laser does not model real ignition systems.

## 22.7.2 FACTORS AFFECTING LASER IGNITION

Laser ignition of pyrotechnic composition depends upon:

a. Laser power density and wavelength
b. Beam area focused on the pyrotechnic composition
c. Pyrotechnic composition characteristics like ignition temperature, particle size, compaction, reflectivity, thermal diffusivity, specific heat, etc.
d. Confinement conditions of pyrotechnic composition

Therefore, a correct combination of above factors should lead to pyrotechnic ignition. Ignition delay of laser ignition is given under Section 3.6. Table 22.27 shows the average energy density required to initiate pyrotechnic compositions by laser.

**TABLE 22.27**
**Sensitivity of Compositions to Laser [25]**

| Composition | Uncorrected for sample reflectivity, average energy density to initiate * $(J/in^2)$ | Based on sample reflectivity, adjusted energy density to initiate ** $(J/in^2)$ |
|---|---|---|
| Zr -KClO$_4$ | 8.2 | 7.1 |
| Delay mix 176 | 14.1 | 9.5 |
| Boron pellets (B/KNO$_3$) | 19.1 | 18.6 |
| Delay mix 177 | 21.0 | 12.8 |
| Mg/Teflon | 73 | 12.7 |

*Notes*
  * Energy density value is the mean value of all the conditions tested
 ** Adjusted energy density = energy density − (reflectance x energy density)

### 22.7.3 ADVANTAGES OF LASER IGNITION OVER CONVENTIONAL BRIDGE WIRE INITIATION [26]

Some of the more obvious ones are:

a. They are immune to accidental firings from electromagnetic fields, electrostatic discharge or stray electrical energy.
b. Built in self-test/checking of the integrity of the entire initiating system may be carried out without affecting the safety or the reliability of the system.
c. Most pyrotechnics may be initiated with low quantities of energy (mJ) and so low energy diodes may be used.
d. Safety and arming (S&A) systems may be made wholly electronic.
e. Since the output energy and power of the laser can be chosen to be higher than bridge wire devices, less sensitive explosives or pyrotechnics may be used.
f. Since the bridge wire is eliminated, corrosion of the bridge wire is removed as a failure mechanism thus the safe/service life is lengthened and conductance of the initiator after firing is eliminated.
g. The laser initiator cost can be lower than corresponding 1A/1W devices.
h. The laser initiator can be smaller and lighter than corresponding bridge wire devices.
i. Laser systems may be reused several times (multiple shot capability) compared to the single shot bridge wire systems.
j. The initiation system may be constructed in-line, removing the need for complex mechanical safety and arming systems.
k. Multi-point simultaneity is easily achieved.
l. Distributed initiation systems are cheaper to manufacture.

### 22.7.4 LASER MATCH HEAD [27]

The laser match head is a convenient means for ignition of pyrotechnic devices less prone to unintentional ignition from electromagnetic fields, radiofrequency absorption, electromagnetic pulses, electrical discharge or stray electrical energy than conventional bridgewire match heads connected to electrical cables. The laser match head can also be coated with layers of different pyrotechnic slurry to make the match head less friction and impact sensitive, and to give the match head different thermal ignition properties.

It is made by placing pyrotechnic composition on the tip of the optical fibre, rather than making it part of an assembly separated from the optical fibre by a window or lens. The laser match head reported here uses a slurry of boron/potassium nitrate/Viton (SR 44) as the pyrotechnic ignition composition. Sometimes a coat of SR 252 (consisting of silicon, potassium nitrate and sulphurless mealed gunpowder) is applied over the SR 44 composition. SR 44 acts as ignition composition while SR 252 acts as flash composition as at Figure 22.21. A 1-W pulsed diode laser is used to ignite the match head at pulse energies of about 2 mJ.

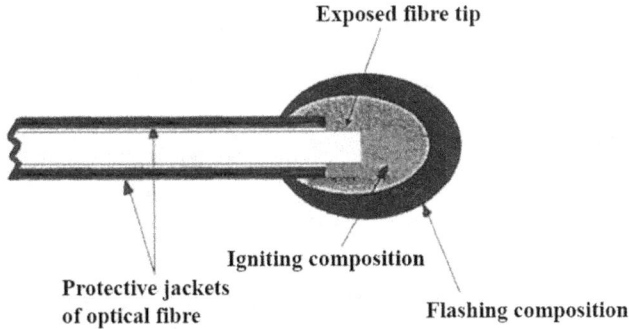

FIGURE 22.21 A Laser Bi-Composition Match Head Formed with Two Different Pyrotechnic Compositions (Reprinted from [27]).

Readers may refer to book under reference [28] on laser ignition of energetic materials as well as reference [29] on NASA laser systems on pyrotechnically actuated systems.

Readers may also like to see reference on military laser technology systems in references [30] and [31] for "*hotspot* "formation, which is vital for ignition pick-up.

## 22.8 FUZES

In the chain of events of initiation, fuzes play an active role typically for projectiles and bombs as they function after a set time delay to enable pyrotechnic composition to exhibit its special effects. Fuzes are used in illuminating, screening and colour smoke mortar bombs and shells, photoflash and incendiary bombs.

## REFERENCES

[1] "*Ordnance Explosive Train Designers' Handbook*", NOLR 1111, April 1952, US Naval Ordnance Laboratory, USA.

[2] "*Explosive Trains*", AMCP 706-179, January 1974, Engineering Design Handbook Explosive Series, Headquarters US army Material Command, USA.

[3] F.L. McIntyre, "*A Compilation of Hazard and Test Data for Pyrotechnic Compositions*", Contractor Report ARLCD-CR-80047, October 1980U.S. Army Research and Development Command, Large Caliber Weapon System Laboratory, Dover, NJ.

[4] "*Service Text Book of Explosives*", 1972, JSP 333, Ministry of Defence UK, Reprinted in India in 1976.

[5] John R. Bentley and Paul P. Elischer, "*Development of a Gasless Pyrotechnic Cap*", REPORT MRL-R-776 May, 1980 Department of Defence, Defence Science and technology Organisation, Materials Research Laboratories, Melbourne, Victoria, Australia.

[6] A. A. Shidlovskiy, "*Principles of Pyrotechnics*", 3rd Ed., Moscow, 1964. (Translated by Foreign Technology Division, Wright-Patterson Air Force Base, Ohio, 1974.) Publisher: American Fireworks News (July 1, 1997) ©1997, Rex E. & S. P., Inc.

[7] J.H. McLain, *"Pyrotechnics from the View Point of Solid-State Chemistry"*, The Franklin Institute Press, Philadelphia, PA, 1980.

[8] Dr.Herbert Ellern, *"Military and Civilian Pyrotechnics"* Chemical publishing Co., New York, 1968.

[9] Dr.Nigel Davies, *"Pyrotechnics, Topics to be Covered in Course"*, PowerPoint Presentation, Cranfield University, Defence Academy of the United Kingdom, Shrivenham, Swindon, SN6 8LA, UK.

[10] Gregory D Knowiton, Bruce B Anderson and Theodore B Gortemoller, *"Heat Transfer Delay"*, Patent WO 2001039586 A2June 7, 2001.

[11] Vojtech Pelican, *"Application of Conducting Composition Fuseheads in Pyrotechnic Devices"*, Journal of Pyrotechnics, Winter 2006, Issue 24, Pages 3–10.

[12] S. S. Jagtap, *"Electro-Explosive Devices"*, Continued Education Programme Course on Pyrotechnics", High Energy Materials Research Laboratory, Pune, India, 4-8 August, 2003.

[13] G. C. Gandhi, S. K. Sahu, S. M. Garud and M.P Thakar, *"Development of Pyrotechnic Compositions for Electro Explosive Devices"*,Proceedings of National Workshop on Power Cartridges, Armament Research and Development Establishment, December 05, 2001.

[14] Azizullah Khan, Abdul Qadeer Malik, Zulfiqar H. Lodhi, *"Development and Study of High Energy Igniter/Booster Pyrotechnic Compositions for Impulse Cartridges"*, Cent. Eur. J. Energ. Mater. 2017, 14(4): 933–951. Copyright © 2017 Institute of Industrial Organic Chemistry, Poland.

[15] Nils-Erik Gunners, Kurt Andersson, Yngve Nilsson *"Testing of Parts and Complete Units of The Swedish Base-Bleed System* in *"Proceedings of the First international Symposium on Special Topics in Chemical propulsion: Base Bleed"*; Athens Hilton, Greece, 23–25 November 1988.

[16] *"Igniters for Solid Rocket Propellants"*, Defence Research and Development Organisation, India.

[17] *"Solid Rocket Motor Igniters"*, NASA SP-8051, 1971, National Aeronautics and Space Administration, Lewis Research Center (Design Criteria Office), Cleveland, Ohio.

[18] Paisan Apinhapat and Narupon Pittayaprasertkul, *"Experimental Investigation on Pyrotechnic Igniter for Solid Rocket Motor"*, 5th International Conference (at Taipei, Taiwan) on Chemical Engineering and Applications IPCBEE vol.74 (2014) © (2014) IACSIT Press, Singapore.

[19] *"Power Cartridge Handbook"*, NAVAIR Report No. 7836, Fourth Edition, 15 Aug, 1973, Department of the Navy, Naval Air Systems Command, Washington DC.

[20] E. R. Lake, S. J. Thompson, and V. W. Drexelius, " A *Study of the Role of Pyrotechnic Systems on the Space Shuttle Program"*, Report no, NASA CR-2292, September, 1973, National Aeronautics and Space Administration, Washington, DC.

[21] Mario Castro-Cedeno, Thomas L. Seeholzer, Floyd Z. Smith, Micahel A. Politi and Paul R. Steffess, *"Application catalog of Pyrotechnically Actuated devices/systems"*, NASA Technical memorandum 106810 Revision A, August 1996.

[22] *"Ignition System for Solid Rocket Motors"*, Technology Focus, Volume 28, Issue 1, February 2020 ISSN No. 0971–4413.

[23] Xiao Fang, Sheikh R. Ahmad, *"Laser Ignition of an Optically Sensitised Secondary Explosive by a Diode Laser"* Central European Journal of Energetic Materials, 2016, 13(1), 103–115.

[24] L. de Yong, B. Park and F. Valenta, " A *Study of the Radiant Ignition of a Range of Pyrotechnic Materials using a $CO_2$ Laser"*, MRL Technical Report MRL-TR-90-20,

DSTO Materials Research Laboratory, Cordite Avenue, Maribyrnong, Victoria, 3032 Australia ©, Commonwealth of Australia 1990 AR No. 006- a.15.

[25] Vincent J. Menichelli, and L. C. Yang, *"Sensitivity of Explosives to Laser Energy"*, Technical Report 32-1474, April 30, 1970, Jet Propulsion Laboratory, California Institute of Technology, Pasadena, California.

[26] Leo de Yong, Tam Ngyuen, and John Waschl, *"Laser ignition of Explosives, Pyrotechnics and Propellants: A Review"*, DSTO- TR-0068, May 1995, Department of Defence, Defence Science and Technology Organisation, Weapons System Division, Published by DSTO Aeronautical and Maritime Research Laboratory, Melbourne, Victoria.

[27] Augustine Lee, Michael Stringer and Kenneth Smit, *"Laser Match Head for Pyrotechnic Ignition"*, Weapons Systems Division Systems Sciences Laboratory DSTO-TR-1448 DSTO Systems Sciences Laboratory Edinburgh South Australia 5111 © Commonwealth of Australia 2003 AR-012-800, May 2003.

[28] S. Rafi Ahmad, Michael Cartwright, *"Laser Ignition of Energetic Materials"*, 2015, Publisher ©John Wiley and Sons.

[29] William W. St. Cyr, *"Second NASA Aerospace Pyrotechnic Systems Workshop"*, Proceedings of a workshop sponsored by Pyrotechnically Actuated Systems Programme, NASA Conference Proceedings - 3258, Office of Safety and Mission Quality, NASA Washington DC, 8–9 February 1994.

[30] David H. Titterton, *"Military Laser Technology Systems"*, © 2015Artech House Canton street, Norwood, MA.

[31] A.M., Mellor, D.A., Wiegand, K.B., Isom, *"Hot Spot Histories In Energetic Materials"* Materials Research Society Symposium Proceedings on "Structure and Properties of Energetic Materials" Volume 296, Symposium held November 30-December 2, 1992, Boston, Massachusetts, U.S.A.Copyright 1993 by Materials Research Society. Pittsburgh. Pennsylvania U.S.A.

# Section 2

Pyrotechnic Ammunitions
and Devices

# 23 Pyrotechnic Ammunition and Device Manufacture

## 23.1 GENERAL

Pyrotechnic ammunitions and devices are required to meet the following requirements:

a. Should be safe to fill, assemble, handle, transport and use.
b. Should be reliable and produce desired effect at the desired place, time or target.
c. Should be consistent in functioning and not malfunction.
d. Should be robust and withstand the stresses of firing set back, set forward, spinning etc. as applicable.
e. Should remain unaffected at all weathers and altitudes.
f. Should have compatible compositions and components.
g. Should have adequate shelf life.
h. Ability to be produced on mass scale.
i. Should preferably have low cost with ready availability of materials.
j. Should preferably produce non-toxic products on functioning.
k. Should have proper sealing, painting/anodising and marking for identification.
l. Should be packed and issued in sturdy packages with nomenclature, name of consigner, name of consignee, gross weight, net weight and stock number as applicable.

The pyrotechnic ammunition differs from other high explosive ammunitions (Table 23.1) in many aspects.

Figure 23.1 shows various military pyrotechnic ammunitions.

Pyrotechnic ammunitions are mostly any one from the following types:

a. Small arms
b. Mortar bombs
c. Artillery shells
d. Cartridges
e. Grenades (hand or rifle)
f. Bombs (aircraft)
g. Container pots
h. Miscellaneous ammunitions, etc.

## TABLE 23.1
## Differences between Pyrotechnic Ammunitions and High Explosive Ammunitions

| Parameter | Pyrotechnic ammunitions | High explosive ammunitions |
|---|---|---|
| Initiation | Generally initiated by initiatory compositions, friction compositions, electrical compositions, port fire friction composition, etc. through flash. | Initiated by detonators or detonation of high explosive pellets. |
| Rate of combustion | The rates of combustion of pyrotechnic composition in the ammunition are slower (seconds.cm$^{-1}$). Thus, release of energy is much slower. | High explosive ammunition explodes in milliseconds. Thus, release of energy is very fast. |
| Blasting, hollow charge effect and scab effect | Due to slow release of energy, pyrotechnic ammunitions are not capable of exhibiting. blasting effects, hollow charge effect on armour plate and scab effect on armour plate. | Many high explosive ammunitions are capable of exhibiting. blasting effects, hollow charge effect on armour plate and scab effect on armour plate. |
| Special effects | Special effects by pyrotechnic ammunitions are heat, smoke, radiation (visible and infrared), gas and sound. | High explosives special effects are detonation, shattering, breaking, shock waves, etc. |
| Place of functioning and special effect | Many pyrotechnic ammunitions function in air but their effects are visible at a large distance on ground like illuminating ammunitions, infrared illuminating ammunitions, smoke ammunitions and photoflash ammunitions, etc. | High explosive ammunitions largely function on ground (barring a few functioning in air) and their special effects are near their functioning point. |
| Fuzing | Pyrotechnic projectiles which are fired dynamically or bomb dropped from air require fuze assembled at the nose of the shell or bomb. In no ammunition, any fuze is assembled at the base. These fuzes mostly produce flash for ignition except a few cases of detonation in photoflash and incendiary ammunitions. | High explosive projectiles which are fired dynamically or bomb dropped require fuze assembled at the nose, but in a few ammunitions, it is assembled at the base of the projectile like those of anti-tank projectiles like HEAT for hollow charge effect and HESH for scab effect. These ammunition produce detonation due to presence of high explosive like CE or TNT or RDX/wax. |
| Safety of aircrafts, helicopters ships, tanks etc. | Military pyrotechnic ammunitions like infrared flare decoys are useful in safety of the fighter aircraft and helicopters against infrared seeking | No high explosive ammunition has this feature. |

**TABLE 23.1 (Continued)**

**Differences between Pyrotechnic Ammunitions and High Explosive Ammunitions**

| Parameter | Pyrotechnic ammunitions | High explosive ammunitions |
|---|---|---|
| | missiles Similarly, infrared smoke act against infrared guided anti-tank missiles by attenuating infrared radiation of the tank. | |
| Application | The pyrotechnic ammunitions are useful not only in military applications but also for civil administration, mining, merchant navy, space and colourful functions. | High explosive ammunitions are largely restricted to military applications and a few for civil mining requirements. |
| Environmental effects | Many of the pyrotechnic ammunition performance is severally affected by environment such as wind direction, wind velocity and nature's background light, humidity, moisture, dust, smoke, heavy fog, snow, and rain. | High explosive ammunition performance is mostly unaffected by environmental effects |
| Shelf life | The shelf life of pyrotechnic ammunitions is much less than high explosive ammunitions. | The shelf life of high explosive ammunitions is considerably much higher. |

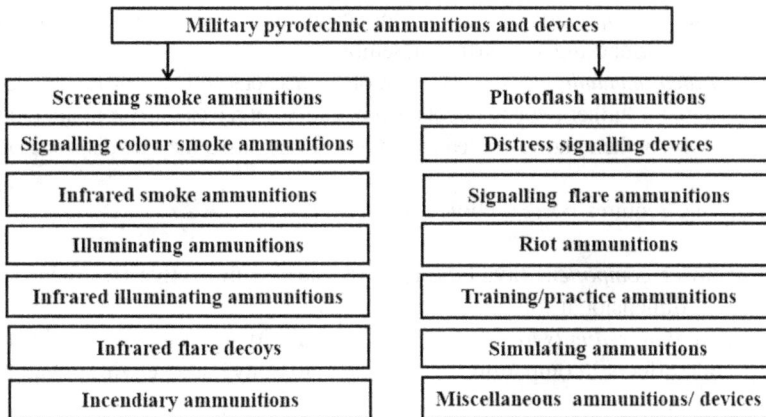

FIGURE 23.1　Military Pyrotechnic Ammunitions and Devices.

The various classifications of pyrotechnic ammunitions based on special effects with type of ammunitions are given below.

a. Illuminating ammunitions and infrared illuminating ammunitions for night warfare (cartridges, bombs, shells containers, etc.)
b. Smoke ammunitions and infrared smoke ammunitions for screening/masking or infrared attenuation (bombs, grenades, shells, etc.)
c. Day signalling ammunitions and night signalling ammunitions (cartridges, grenades, bombs, shells, etc.)
d. Riot control ammunitions for civil administration for mob control
e. Incendiary ammunitions (bullets, bombs, shells, etc.)
f. Training and practice ammunitions (cartridges, grenades, bombs, shells, etc.)
g. Simulating ammunitions (simulating sound or flash or both)
h. Distress signalling devices (signalling, location and rescue operation for day and night)
i. Infrared decoys creating false target for the infrared-guided missiles

## 23.2   BASIC REQUIREMENTS IN PYROTECHNIC AMMUNITION MANUFACTURE

Manufacture of ammunition/component lots mostly involves three major stages:

a. *Input:* Ammunition technology documents, process buildings, process machinery and equipment, process tools, gauges, storage buildings, utilities (electrical power, water, steam, etc.), manpower, ingredients, empty components, etc.
b. *Process:* Filling, pressing, assembly, sealing, painting, stenciling, packing and storage
c. *Output:*
    i. *Composition batch*: Testing of composition and performance
    ii. *Filled component lot:* Testing for acceptance in laboratory or proof range and earmarking for further assembly
    iii. *Filled ammunition lot*: Testing/proof for acceptance in proof range and storage for further issue to depots. Ammunition lot rejected in proof are required to be broken and reworked In extreme cases, these may be disposed off.

Manufacturing requires correct choice of the following:

a. *Hardware component* (availability, cost, surface protective treatment as required, dimensional tolerance, etc.)
b. *Chemical ingredients for composition* (availability, purity, cost, suitability for special effects, compatibility, stability, toxicity and environmental effects, etc.)
c. *All ancillary materials* (availability like paper, sealant, plastic, mill board, felt, celluloid, etc.)
d. *Manufacturing process and proof* (process schedule, tool schedule, gauge schedule, inspection schedule, test schedule, proof schedule, safety, etc.)

The steps of pyrotechnic ammunition manufacture involve:

a. Receipt of ingredients, components
b. Storage of ingredients, components and their testing as and when received
c. Grinding and sieving of ingredients, drying as required
d. Composition manufacture (weighing, mixing of ingredients, granulation as required and drying as required)
e. Blending of composition and tests as required
f. Filling/pressing of composition
g. Filled component sealing and tests, as required
h. Assembly of ammunition
i. Sealing of ammunition
j. Painting, stenciling, packing of ammunition
k. Proof/testing of ammunition
l. Dispatch of ammunition lot to depots
m. Disposal of waste explosives, defective filled components and ammunition

The problems in pyrotechnic ammunition manufacture are:

a. These use high energy materials that are prone to safety hazards.
b. There are restrictions on quantity and safety distance for manufacture and storage and require a robust infrastructure.
c. Design of military pyrotechnic ammunition involves various technologies like high energy material technology, metals technology, clothing technology, rubber technology, plastic technology, sealant technology, ballistics and explosive safety.
d. Several tests and proofs are required for the military pyrotechnic composition, components and ammunition.
e. Difficulty in assessing the behaviour of a new military pyrotechnic composition, component and ammunition.
f. Difficult and time consuming to measure the effects of modification in composition or component.
g. Development of a new ammunition requires long time.
h. Long periods are required for shelf-life assessment of military pyrotechnic ammunition.
i. Occurrence of failures are often encountered during development as well as during bulk production.
j. Disposal problems of defective composition, components and ammunition.
k. Ensuring safety, reliability and consistency of pyrotechnic ammunition requires a robust infrastructure and management committed to quality, safety and customer focus.

A pyrotechnician must understand the following:

a. Properties of various ingredients
b. Factors affecting performance of pyrotechnic compositions (Chapter 4)

c. Sensitivity to impact, percussion, friction, thermal, electrostatic charge (spark) (Chapter 2)

d. Compatibility with other ingredients and hardware components (Chapter 27)

e. Evaluation of pyrotechnic composition performance (Chapter 5)

f. Evaluation of pyrotechnic ammunition performance (Chapter 29)

g. Toxicity of composition and its combustion products

h. Waste disposal methods

i. Storage restrictions (chapter 26)

j. Properties and role of various metallic and non-metallic hardware components like cartridges, bombs, shells, container pots, tail units, parachute, springs, rubber components, plastic components, paper components, celluloid components and surface protection treatment of metallic components, if required, etc.

k. Testing and proof requirements. A good proof range with latest equipment enables the manufacturer to observe the performance of the pyrotechnic ammunition properly.

l. Safety aspect in manufacture and handling. This is considered very important. As pyrotechnic materials are high energy materials, these are required to be handled with utmost care. All precautions should be taken in various stages of storage of ingredients, grinding and sieving of ingredients, mixing of ingredients as composition, storage of composition, filling/pressing of composition, sub-assembly of components, assembly of components in ammunition, painting, stencilling, packing, transportation and storage of ammunition.

m. Safe quantity of composition or filled components or pyrotechnic ammunition to be handled at a time and exposure to limited number of personnel has to be judiciously arrived at and suitable general safety directives (GSD) or work instructions (WI) or standing instructions (SI) to be made.

n. Composition mixing is most hazardous compared to pressing, assembly and packing for a particular composition. Again, amongst compositions, the most hazardous compositions are photoflash compositions, whistle compositions, chlorate-based compositions, zirconium-based compositions and red phosphorous based compositions.

o. Proficiency and skill in manufacture to ensure defect-free ammunition

## 23.3 PYROTECHNIC AMMUNITION DESIGN AND DEVELOPMENT

There are a large number of ammunitions and devices with varying designs, shape, place of firing, mode of firing and special effects. Some filled components are also required during manufacture of military pyrotechnic ammunition like stab caps, percussion caps, primers, igniters, squibs, power cartridges, fuzes, etc. as many of these components are required to be assembled in the ammunition.

An assorted range of pyrotechnic ammunitions and devices with various shapes and sizes and special effects is shown in Figure 23.2.

**FIGURE 23.2** An Assorted Range of Pyrotechnic Ammunitions and Devices (Reprinted from [1]).

Ammunition designer is required to consider following major factors for design and development of ammunition or device.

a. Functioning reliability, consistency and safety of ammunition. A design is considered complete only if all the three parameters are satisfactory.
b. Shape, size and mass of payload like pots, grenades, canisters, cartridges, bombs, shells, etc. suitable for ammunition/weapon system, wherever required
c. Functioning characteristics of ammunition or device like bursting, emission, ejection cum emission, ejection from dispenser of aircraft, penetration cum emission (see Section 29.3),
d. Firing device/equipment/weapon system like hand, pistols, rifles, mortars, howitzers, guns, dispensers in aircrafts, submarines, torpedoes and proof equipment, if any
e. Hardware material drawings, specifications and tolerances like, cartridge body, grenade body, bomb body, shell body, tail units, fuzes, other components etc.
f. Other filled components (caps, primers, delays, tracers, bursters, etc.) or non-filled components like springs, discs, parachute, etc. as per design
g. Pyrotechnic main composition of desired special effects and output, keeping in view compatibility, toxicity and combustion products as well as suitable priming or priming cum booster composition for ignition
h. Quality control check points, suitable tools and gauges, testing instruments
i. Quantity and mode of filling/pressing of composition like in situ, pre-pelleting, cast filling or extrusion

j. Sealing of components and ammunition to avoid deterioration of composition due to atmospheric moisture as well as other environmental effects

k. Type of ignition mechanism (mechanical energy like impact, percussion and stab energy, frictional energy, electrical energy like electrostatic discharge, i.e., spark energy and thermal energy like flame, hot gases and from fuzes).

l. Reliable ignition train to ensure continuity of functioning of explosives by suitable heat transfer at all stages. For example, in 105-mm smoke ammunition, cap on external stimulus must provide flash to ignite the propellant. The propellant, in turn, must produce sufficient pressure to eject the shell. The fuze must function after specified time and provide flash to burster composition so as to ignite and eject the canisters providing smoke.

m. Ejection mechanism, wherever required

n. Ammunition to function at desired time with desired output and duration in all environmental conditions

o. Place of actual firing and use of developed ammunition by users as some are fired on ground, while many are fired in air, dropped from air, over the sea and under the sea

p. Range of functioning (static, short range, medium range or long range) for deciding a suitable propellant where propellant chemical energy is converted into kinetic energy of the projectile, say a mortar or shell or bullet

q. Mode of evaluation for assessing performance of developed composition/component/ammunition through its specification in laboratory, static proof, simulated dynamic proof and dynamic proof

r. Forces acting on ammunition which are fired through weapons
   i. Internal (setback, set forward, acceleration, etc.)
   ii. External during flight (deceleration, drag effect, stability of projectile in flight, consistency and range achieved)
   iii. Terminal performance (fuze ignition, ejection of pellets, canisters and flares, bursting, emission, penetration, para deployment, etc.)

s. Mechanical and environmental effect on ammunition to ensure sturdiness and shelf-life assessment

t. Relevant package design including internal fitments of packages

u. The design should ensure ease of handling, compatibility, safety in handling and suitable for mass production with economy in cost pertaining to input material, manpower and area required for manufacture (preferably automatic operations)

v. Mode of smooth transfer of design data to production line after successful developmental trials

As almost all ammunitions contain many filled and empty components, it is necessary that all such components and the high energy material must work in harmony as desired to meet the ultimate requirement. A sustained effort is required to arrive at a preliminary design and many tests/proofs are required to finalise the design of ammunition. A developed ammunition to be fired in weapon system is considered satisfactory if it meets the requirement of proper ignition, proper ejection, stability of projectile during flight and proper functioning. For projected ammunitions, most cases, range, consistency and functioning reliability of meeting the desired specification parameters like achieving specified burn time, illumination, smoke obscuration, infrared illumination, simulating effects, infrared flare decoys parameters, etc. are specified.

The design and development of ammunition generally involves the following sequence:

a. Conceptualisation
b. Design
c. Manufacture of prototype
d. Validation of design
e. Acceptance by user
f. Mass production

There are various routes for ammunition development as under.

a. *As per GSQR by users*:
   Ammunition development is mostly based upon the requirements of the users who formulate the General Staff Qualitative Requirement (GSQR) of the ammunition. This is akin to preliminary specification of the desired ammunition. These qualitative requirements form the basis of boundaries for design.

   The designer (Research & Development organisation) or the manufacturing agency (if capable of taking such developments) undertakes the design work. The designer designs the ammunition, taking into account the various materials, components, their drawings and specifications and proof requirements. The ammunition is produced from mere design on drawing board to actual ammunition.

   Initially, small developmental batches are produced and subjected to environmental and mechanical handling tests and are proved in proof ranges to confirm the process and material for the ammunition. A large number of trials/proofs with modifications are done (in association with quality assurance organisation if required). When the ammunition design is successful, a pilot lot of the ammunition is offered to users for acceptance. If the ammunition is accepted by users after user's own exhaustive trials, the design documents are sealed (if required by quality assurance organisations responsible for process audit and proof and approval of a bulk lot). This needs formulation of complete drawings, specifications, tests, proof, packages and user instructions by the designer.

   During entire development stage, the designer and user are required to interact frequently and arrive at the desired design. If the design is not satisfactory, the designer may find an optimum design of the ammunition with a modified QR, as permissible by the user.

   The design documents including proof schedule after sealing are forwarded to manufacturing organisation. The designer has to make the documents for the process, tool schedule, gauge schedule, inspection schedule, test schedule, etc.

Subsequently bulk lots are manufactured by manufacturing organisation and after acceptance of ammunition lots in proof, forward the lot to depots.

However, to ensure proper functioning, occasionally some of the mechanical/ environmental tests (see Section 23.5 & 23.6) are also carried out during regular bulk manufacture of lots (say one in 10 lots) especially at high temperature and low temperature to ensure that the ammunition is giving satisfactory results.

In some countries, the manufacturing agency is assigned for production, proof and sentence.

Despite taking all precautions by manufacturer in manufacturing and proof of ammunitions meeting specification requirements, periodic proof or check proof of ammunition is done at depots. This involves visual examination to ensure that ammunition is unaffected by corrosion or other defects. The ammunition is then tested/proved to ensure that ammunition continues to meet the specified qualitative requirements of safety, serviceability (functioning with desired output) and consistency during its full life. Ammunition lots not meeting the above are segregated and disposed so that only serviceable ammunitions are held in stock in depots. Sometimes, these are also used for training/practice ammunition, if it poses no safety hazard.

b. *As per Reverse Engineering of Recovered Ammunition*:

The user may forward a few samples of ammunition and request the designer (R&D organisation or the manufacturing agency if capable of taking such developments) to develop ammunition based on reverse engineering technology. In this case, the samples of ammunition are broken down to individual components and their designs are made. Both hardware material and composition are tested and indigenous equivalents as suitable alternatives, from market sources, are made. Manufacture of lots is done as above.

c. *As per Transfer of Technology from Another Source/Country*:

It is also possible to obtain the Transfer of Technology (TOT) of ammunition from some source/country. In such case, full manufacturing, proof and packing documents from the source/country is made available to the manufacturing organisation. This requires arriving at indigenous material equivalents by the manufacturer for all hardware as well as chemicals and plant and equipment, including proof equipment.

In some transfer of technology, CKDs (components knocked down) are also supplied along with the documents, which are very useful. Special plants or equipment may also be procured from the source/country.

In such transfer of technology, only manufacture, inspection and proof documents are transferred. The design documents (i.e., intricacies of ammunition design) are mostly not forwarded which forms the basis of such design. This is a serious drawback of transfer of technology.

d. *As per Design Agency*:

It is also feasible by the designer (Research & Development organisation) to develop ammunition on designer's own specification of the ammunition and then offer to the users.

e. *As Design Change for Product Improvement*:

The designer of the ammunition may not always provide the ultimate ammunition design. This is due to the fact that all types of problems that ammunitions may face during bulk manufacture cannot be anticipated during design and development stage by the designer. Hence, these *marginal design ammunitions* may likely to exhibit some failures or the packages may exhibit defects. Therefore, product improvement is taken as an ongoing exercise and must be given a thrust by the manufacturer. It is the responsibility of the manufacturer to carry out these minor design developments and make the ammunition even better in consultation with inspection agency, if required.

This design change or product improvement is also necessitated if some material used in manufacture suddenly becomes scarce. Then alternative materials are required to be used ensuring that the material is compatible and the ammunition meets all design criteria.

Further, as the technology base expands, major improvements in existing design should be done to an advanced and improved version of the ammunition (with new drawings and specifications) with respect to

a. Improving special effects
b. Increasing the range of the ammunition
c. Reducing toxicity by use of ingredients which are not toxic and do not produce toxic products.
d. Introducing better package and fitments
e. Improving shelf life

Figure 23.3 shows the five routes of designing an ammunition.

## 23.4  FEATURES OF A PYROTECHNIC AMMUNITION AND DEVICE

The pyrotechnic ammunitions and devices have a large number of designs like pots, grenades, cartridges, bombs, shells and miscellaneous shapes. The details of these designs is as under.

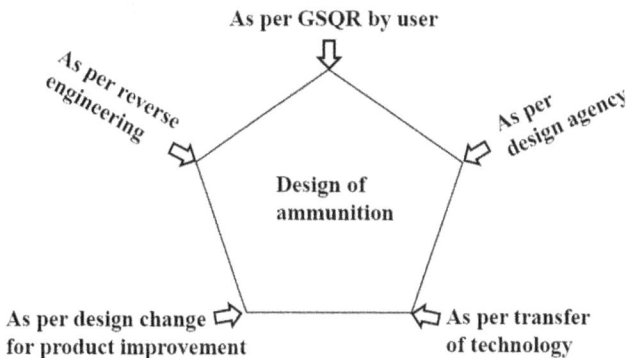

**FIGURE 23.3**  Five Routes for Design of Ammunition.

a. *Pots*: These are generally cylindrical in shape containing the pyrotechnics. Some have their own ignition systems while many are required to be initiated by external means through port fire friction. Pots are generally meant for smoke emission, both screening smoke as well as colour smoke.

b. *Grenades*: These contain the pyrotechnic composition, meant for short ranges/distances. They have their ignition system along with a delay component for safety. Grenades are thrown either by hand or weapon. Size and weight affect the range. Examples are illuminating grenades, colour smoke grenades and riot ammunitions.

c. *Cartridges*: Cartridges hold the payload and contain the ignition system. It makes it easier to load the ammunition in firing weapons. It has a strong base and sufficient strength to withstand the stresses of handling, transportation and firing, it allows the payload to move out of the cartridge system. Examples are signal cartridges, illuminating cartridges, small arms incendiary, photoflash cartridges, infrared flares etc.

d. *Bombs*: These hold the payload and contains its own ignition system. It has fins to stabilise in its trajectory. Examples are smoke bombs, photoflash bombs, incendiary bombs.

e. Shell: A shell holds the payload. Its contour is aerodynamic which ensures that it moves along the designated projectile route and function. It has a strong base to absorb the propellant thrust. It has driving bands to
   i. Imparts spin to the shell
   ii. Acts as forward obturation
   iii. Engages with rifling when loaded to avoid slip back
       Examples are smoke shells, infrared illuminating shells, incendiary shells, etc.

f. *Miscellaneous Shapes*: This consists of miscellaneous devices of varied shapes and sizes like distress signaling ammunitions, simulating ammunitions, signal fuzee, fog signals, flare trip wire, etc.

## 23.5 ENVIRONMENTAL TESTS

Ammunitions are likely to encounter some of the following environmental conditions.

a. High humidity (moisture)
b. Tropical climate
c. Desert climate (high temperature during day and low night temperature)
d. High altitude (pressure variation and temperature variation)
e. On sea (salt and humidity)
f. Under sea (water pressure under sea)
g. Rain
h. Dust

The objective of these environmental tests is to establish functioning aspects of delicate filled components/empty components/safety devices in the ammunition, which are likely to be adversely affected due to varied environmental conditions. These are necessary to ensure that the ammunition would function satisfactorily in all environmental conditions. These environmental tests may include, amongst others:

a. High temperature test
b. Low temperature test
c. Temperature cycling
d. Thermal shock
e. High humidity test
f. Wind test
g. Salt spray test
h. Snow/ice test
i. Vacuum test
j. Water immersion/rain test
k. Shelf-life test

## 23.6 MECHANICAL HANDLING TESTS

The objective of these mechanical tests is to establish the safety and strength of delicate components/safety devices in the ammunition, which are likely to be adversely affected due to shocks during transportation, storage and firing. These are necessary to ensure that the ammunition would function satisfactorily under all stresses of handling during manufacture, transport, loading and firing during entire shelf life of ammunition. These mechanical tests include:

a. Shock
b. Vibration
c. Bump
d. Jolt
e. Jumble
f. Air drop (or para drop)
g. Rough handling tests

All the above tests are conducted in a laboratory or static proof range or dynamic proof range or as a combination thereof.

These environmental and mechanical test methods are not the exact representation of the actual conditions encountered by the ammunition. Rather, these are deliberate exaggeration of severe conditions to which the ammunition might be exposed. These environmental and mechanical tests provide following information.

a. Ingress of moisture due to humidity or rain
b. Deterioration of composition due to moisture and temperature

c. Corrosion of components due to moisture effect

d. Deterioration of rubber seals and gaskets and plastic components

e. Cracks in composition

f. Vaporisation and sublimation of some ingredients in vacuum or high temperature

g. Effect on burn rate and compaction of composition

h. Melting of any ingredients/component

i. Malfunctioning of insulation protection

j. Alteration in electrical and electronic components

k. Cracking on surface coating

l. Growth of bacteria, insects, fungi in tropical atmosphere due to moisture/ heat/radiation

m. Loosening/warping of components due to high and low temperatures

n. Loss of moisture in wood (warping), paper and textiles components due to heat

A satisfactory environmental tests and mechanical handling tests reveal that

a. The ammunition is safe and serviceable, meeting all parameters of functioning.

b. There is no damage to ammunition and the package.

c. The packages are suitable for survival in service condition.

d. The ammunition design and package are suitable for service requirements.

The ammunition during design and development stage is therefore subjected to some of the environmental (climatic) and mechanical handling tests/proof to ensure that the ammunition is capable of delivering the desired effects with safety, consistency and reliability, in all rough conditions of handling, transportation, storage and in the midst of different climatic conditions.

Table 23.2 shows in brief the work involved in design and manufacture of ammunition.

## 23.7  LIFE CYCLE OF AMMUNITION LOTS

The life cycle of ammunition i.e., the stages of ammunition development and manufacture from conceptual stage onwards is shown in Figure 23.4. The figure explains various stages through which an ammunition development and manufacture passes in its life cycle. It starts with conceptual stage of ammunition on drawing board and ends with disposal of rejected ammunition or gets expended by field units.

After confirmation of GSQR, the conceptual design is made on paper and after design and development, a prototype is made and subjected to mechanical and environmental tests. If satisfactory, manufacture of pilot lot in small numbers is undertaken and subjected to proof. If satisfactory, design and specification and proof is sealed by the competent authority. Rejection at any stage necessitates going

**TABLE 23.2**
**Brief of Work Involved in Manufacture of Ammunition**

| Design and development | Manufacture cum -shop floor inspection | Proof |
|---|---|---|
| a. Special effects required | a. Material schedule | a. Prooof schedule |
| b. Shape and mass of ammunition | b. Drawing schedule | b. Proof range |
| c. Matching weapon system | c. Process schedule | c. Proof observation equipment |
| d. Hardware specification | d. Tool schedule | d. Firing devices or equipment |
| e. Chemical ingredients specification | e. Inspection schedule | e. Proof of filled components |
| f. Ancillary materials specification | f. Gauge schedule | f. Proof of bulk lots |
| g. Compatibility of materials | g. Technology schedule | g. Proof reports |
| h. Toxicity of ingredients and combustion products | h. Test schedule | |
| i Ignition mechanism | i SI/GSD/WI | |
| j. Ejection mechanism | j. Mixing of composition | |
| k. Package design including fitments | k. Lab testing of composition | |
| l. Prototype sample | l. Filling/Pressing of composition | |
| m. Pilot lot manufacture | m. Assembly of bulk lot | |
| n. Mechanical handling tests | n. Sealing of ammunition | |
| o. Environmental tests | o. Painting and stencilling of bulk lot | |
| p. Proof of pilot lot | p. Packing of bulk lot | |
| q. Safety, reliability and consistency of ammunition | q. Painting and stencilling of package | |
| r. Shelf-life assessment | r. Dispatch of proof samples for proof | |
| s. Transfer of design documents to manufacturer | s. Dispatch of bulk ammunition lot to depots after satisfactory proof | |

back to design stage. Based on these documents, regular manufacture of ammunition lots are undertaken. The passed/accepted ammunition lots are sent to depot for storage where further periodic proof or check proof of ammunition is done, as required. Rejection of the bulk lot is required to be reworked by the manufacturer of the bulk lot. However, continuous failure of ammunition or device shall again necessitate going back to the designer.

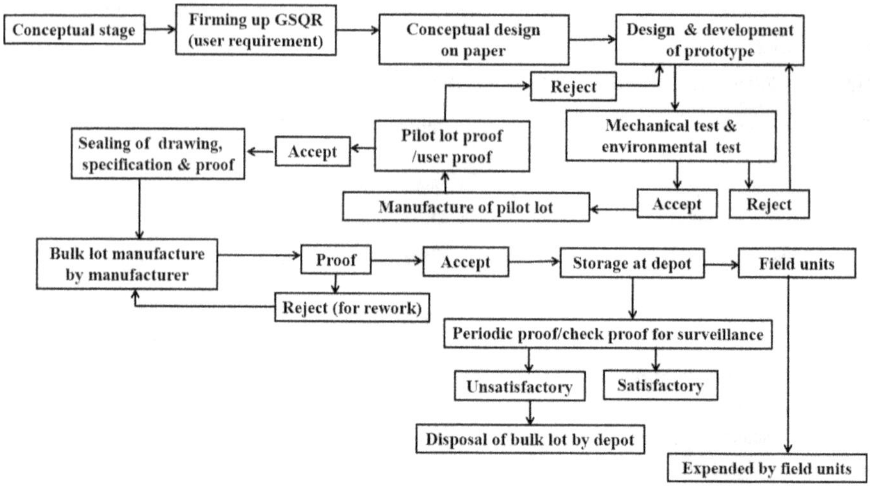

**FIGURE 23.4**   Life Cycle of Ammunition.

## 23.8 DOCUMENTS, SPECIFICATIONS AND DRAWINGS

All ammunitions have a large number of technological documents, specifications and drawings. These are given in Table 23.4.

Readers may refer to references [2]–[4] for details on designing pyrotechnic ammunitions, types of tests, life cycle management of ammunition, environmental tests of ammunition and tests on packaging.

---

**TABLE 23.4**
**Documents, Specifications and Drawings**

1. *Drawing and Specification of Empty Hardware Components:*

   Caps, primers, tracers, fuzes, tail units, bombs, shells, pots, cartridges, etc. with drawing dimension, material, surface finish, surface coating and proof etc.

   *Drawing and Specification of Empty Hardware Component Sub-Assembly*:

   Drawing, proof and test requirement (proof schedule) of empty hardware components, sub- assembly including proof components required:

2. *Specification of Chemical Ingredients and Test Requirements*:

   These give specification of various ingredients used in pyrotechnic composition, its testing methods, special test equipment details for testing. These are standard specifications for chemicals like British standard specifications (BSS) or Joint service specifications (JSS) or MIL standard specifications.

3. *Specification of Composition, Test Methods and Equipment*:

   These give specification of composition, its test methods and special test equipment or instruments, if any.

4. *Safety Certificate of Compositions:*

## TABLE 23.4 (Continued)
## Documents, Specifications and Drawings

The safety certificate gives sensitivity of compositions to external stimuli like impact, friction, heat, spark, etc. It also provides details if the composition is not compatible with certain materials.

5. *Drawing and Specification of Filled Hardware Component/Sub-Assembly* : These give sub-assembly filling details like composition used, quantity used and pressing load and dwell time, height of filling, etc.

6. *Drawing and Specification of Filled Ammunitions:*

These give assembly of various components in the ammunition, filling details of ammunitions like composition used, quantity used and pressing load and dwell time, including the painting and marking details for ease of identification.

7. *Package Drawing of Filled Component, Sub-Assembly and Filled Ammunition:*

The pyrotechnic components, sub-assemblies and the filled ammunition are suitably packed for transportation and storage. These give method of packing in such a way that the material is snug packed as well as withstand the environmental and mechanical effects.

8. *Proof and Test Requirement of Filled Hardware Components, Sub Assembly and Filled Ammunition Including Proof Components Required for Proof:*

This provides the following.

a. Filled component /filled ammunition nomenclature

b. Type of proof like filled component test /proof (like delay proof or luminosity proof, etc.) or filled ammunition proof like static proof or dynamic proof

c. Sample size (as per the sampling plan, may be single or double sampling plan with sample number on samples of ammunition like 1P, 2P, 3P, etc.)

d. Proof range (place of proof)

e. Proof components required (as appendix to proof schedule)

f. Method of proof

g. Observations required

h. Documentation (record of observation on specified format)

i. Defects and defect classification (like critical defect, major defect and minor defect)

j. Sentencing criteria (gives acceptance and rejection criteria)

k. Safety precaution during proof

l. Sentencing authority

9. *Quality Plan of Filled Ammunition:*

The quality plan provides complete details about manufacturing the ammunition. All ammunition should have a separate quality plan. These include:

a. Flow chart (symbolic presentation with brief)

b. Material schedule (materials like ingredients and empty hardware with specification and drawing number)

c. Process schedule (depiction of all the operations with materials, machines, equipment/fixtures, tools and gauges used)

d. Inspection schedule (details of inspection, visual or dimensional or gauging, sample size and inspection equipment and likely defects and classification of defects).

e. Tool schedule with tool drawings (details of tools for each operation with drawing number, tool material and important dimensions and tool life in terms of number of outputs).

f. Inspection schedule with gauge drawings (gauges used in each operation with gauge number, drawing, material, and important dimension and gauge life).

*(Continued)*

## TABLE 23.4 (Continued)
## Documents, Specifications and Drawings

g. Technology schedule (a consolidated schedule of process and inspection schedule in brief for each operation and includes drawing number and specification number of input material and output product during the operation)

h. Test schedule (this includes tests at laboratory, static proof range and dynamic proof range with batch/lot size, sample size, observations required, acceptance criteria, etc. (Chapter 29)

10. *Machinery and Equipment Specification*:

There are a large number of machinery and equipment required for pyrotechnic ammunition manufacture. Each machinery and equipment have its own specification.

11. *Proof Stock Component Specification*:

Many filled components and filled ammunition need proof stock components. These are proof tested materials. These components are assembled with the component under proof and the performance of the filled component or ammunition is recorded. For example,

a. Fuze proof requires filled bomb or shells and propellant as proof stock component

b. Cartridge proof requires shells and propellant as proof stock component

c. Primer proof requires filled cartridge and inert shell as proof stock component

d. Tracer proof requires inert ammunition without tracer as proof stock component

e. Tail unit proof requires inert bomb without tail unit as proof stock component

Similarly, sometimes standard ammunition is required as proof component during proof of small arms ammunitions.

12. *Proof Equipment Specification*:

The ammunition after manufacture is required to be proved as per the stipulated proof schedule which requires various proof equipment, both in static proof and dynamic proof.

13. *Document on Shelf Life of Ammunition*:

All pyrotechnic ammunitions have a shelf life during which the performance remains satisfactory. This is important as it provides a basis for arriving at the quantum of storage in depots vis a vis rate of use and projecting future ammunition requirement.

14. *Safety Instructions*:

It is essential to lay down the safety instructions/directives for plant, process and personnel including personnel protective appliances and measures. Fire safety cover and electrical safety cover are also required.

15. *List of Direct and Indirect Materials*:

The list provides direct and indirect materials. In case of Transfer of Technology (TOT) from one country to another, certain raw materials may not be available indigenously. Hence, it is essential to arrive at proper indigenous equivalent raw material in lieu of specified material in technical documents.

16. *Cut Models of Ammunition and CKDs*:

These provide visual picture of the ammunition and is easy to understand the working of the ammunition. In case of transfer of technology, it is better to have initially some cut models as well as some CKDs (components knocked down) from the original manufacturer for understanding the finer points.

Sometimes, animation film on performance of ammunition at various stages also gives a good understanding about the ammunition.

Readers may also refer to engineering design handbooks mentioned in [5]–[11] providing a compilation of principles and fundamental data on design, manufacture and safety of pyrotechnic ammunitions while references [12] and [13] are about the general ammunition.

## REFERENCES

[1] *"Product Brochure"*, Indian Ordnance Factories, Ordnance Factory Dehuroad, Pune-412113, Maharashtra, India.

[2] Jovana Carapace, Eric J. Deschambault, Paul Haltom, and Benjamin King, *"A Practical Guide to Life-Cycle Management of Ammunition"*, Published in Switzerland by the Small Arms Survey © Small Arms Survey, Graduate Institute of International and Development Studies, Geneva 2018. First published in April 2018.

[3] *"Weapons and Ammunition Safety Manual (H VAS-E)"*, M7762-000242, edition 2000, Swedish Defence Materiel Administration.

[4] *"Test Method Standard Design and Test Requirements for Level A Ammunition Packaging"*, MIL-STD-1904B (AR), 9 March 2016, superseding MIL-STD-1904A (AR), 1 April 1992.

[5] *"Theory and Application"*, AMCP 706-185, April 1967, Engineering Design Handbook Military Pyrotechnic series, Part ONE, Headquarters, US Army Material Command, Washington DC.

[6] *"Safety, Procedures and Glossary"*, AMCP 706-186, October 1963, Engineering Design Handbook Military Pyrotechnic series, Part TWO, Headquarters, US Army Material Command, Washington DC.

[7] *"Properties of Materials Used in Pyrotechnic Compositions"*, AMCP 706-187, October 1963, Engineering Design Handbook-Military Pyrotechnic Series, Part THREE Headquarters, US Army Material Command, Washington DC.

[8] *"Design of Ammunition for Pyrotechnic Effect"*, AMCP 706-188, March 1974, Engineering Design Handbook Military Pyrotechnics Series Part FOUR, Headquarters US Army Material Command, 5001, Eisenhower Ave, Alexandria, VA 22304.

[9] *"Bibliography"*, AMCP 706-189, October 1966, Engineering Design Handbook Military Pyrotechnics Series Part FIVE, Headquarters US Army Material Command, Washington DC.

[10] *"Explosive Train"*, AMCP 706-179, January 1974, Engineering Design handbook Explosive Series, Explosive Trains, Headquarters, US Army Material Command, 5001, Eisenhower Ave, Alexandria, VA 22304.

[11] *"Design Guidance for Producibility"*, Engineering Design Handbook, AMCP 706-100, August 1971, Headquarters, US Army Material Command.

[12] *"Ammunition, General"*, TM-9-1300-200, 30 September, 1993, Department of the Army Technical Manual, Headquarters, Department of the Army, Washington DC.

[13] Major Theodore C. O'Hart, *"Elements of Ammunition"*, © John Wiley And Sons, Inc., New York, Copyright 1946.

# 24 Filling, Pressing and Assembly

## 24.1  GENERAL

Filling and pressing of composition for each type of ammunition is different as each one has its own drawing, design, specification, proof and special effects. Hence, an overall view is presented below.

## 24.2  FILLING OF PYROTECHNIC COMPOSITIONS

Pyrotechnic compositions are used in ammunitions in two forms:

a. *Pour-filled composition* (as powder composition or gel/liquid composition which burn very fast due to more exposed surface area. Their burn rates cannot be controlled.)
b. *Pressed composition including extrusion* (as star, pellet or candle which burn layer by layer and hence their burn rates are not fast and can be controlled)

Gel/liquid compositions are specific for incendiary gel compositions. The details hereunder are for filling/pressing of powder compositions. Filling of powder composition may be done by two methods.

a. *Filling of composition by weighing*:
   Filling by weighing is resorted to where compositions are sticky and cannot be filled by volumetric method. Weighing is generally done on
   i.  Electronic weighing balance (for mass in milligrams and grams like those in certain delays where accuracy of mass of composition is most important)
   ii.  Direct reading single pan balance
   iii.  Two pan balance

b. *Filling of composition by volumetric method*:
   Volumetric method is used when the composition is free flowing. It is useful where the mass to be filled is more or a large number of fillings are required to be done. This is done by *two methods*
   i.  *Filling by scoop:*

Scoops are used to take out a small fixed volume of the composition. As each composition have different bulk densities as well as the quantity required for filling are different, various scoops are made for the composition. Minor variations in

DOI: 10.1201/9781003093404-24

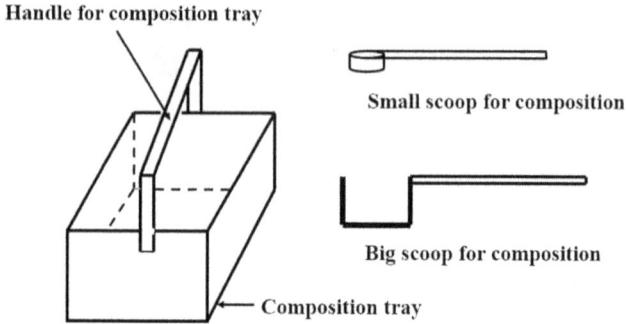

**FIGURE 24.1**   Composition Tray and Scoop.

weight are adjusted by use of wax or aluminium discs pasted on the inside bottom of scoop to obtain the volume and hence the weight of the composition.

For a composition to be taken out through a scoop with desired quantity of composition, the scooped composition is levelled with the handle attached to the composition tray (Figure 24.1).

After measure of composition by scooping, the composition is poured through funnel. This is suitable for dry composition, which should not stick to the funnel. Examples are

a. Solid incendiary composition
b. Photoflash powder composition
c. Smoke composition used in some practice bombs for air force
d. Gunpowder used as expelling charge in signalling cartridges
e. Gunpowders used as expelling charge (as burster bags) for candles in illuminating ammunitions and smoke canisters in smoke ammunition

It has the advantage that it does not require heavy machinery/equipment for filling. This method is used mostly for filling and pressing of star, pellet or candle in mould as explained in Section 24.5.

ii. *Multiple filling of small components*:

A multiple filling (Figure 24.2) followed by multiple pressing is used for small components like caps and small delays by using a charge plate system comprising of

a. Hopper containing composition
b. Top plate for transfer of composition
c. Dosing plate for volumetric filling
d. Bottom guide plate for filling composition inside the component
e. Component holding tray

The hopper and the top plate are always in alignment while dosing plate is kept out of alignment while the bottom plate for filling and the component holding tray are kept in alignment with each other.

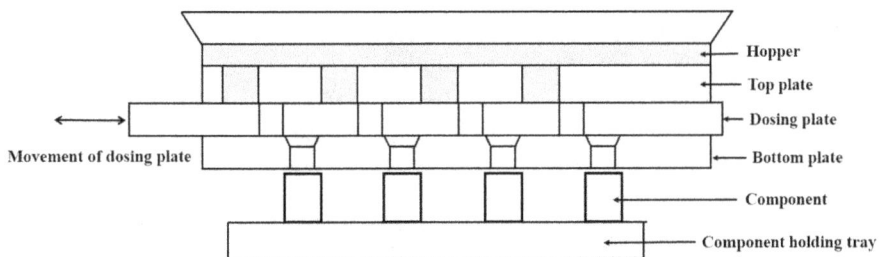

**FIGURE 24.2**  Multiple Filling of Components.

Composition is poured over the hopper and distributed uniformly so that each hole in the top plate is filled with composition, supported by the dosing plate. Now, the dosing plate is moved to a fixed distance, allowing the composition to fall in dosing plate holes for volumetric filling. The hole size of dosing plate is made as required for volumetric filling of each composition.

The dosing plate is then moved, allowing the dosing plate out of alignment with top plate but in alignment with bottom plate with guide holes, which are always in alignment with the component kept on component tray. Thus, the volumetric mass of composition gets filled in the component. A vibrator is also sometimes used so that full charge is transferred to the components.

It is possible to have many layers of charge filling in small components by taking out the filled component holding tray to another set of filling equipment (Figure 24.2).

The component holding tray with composition filled in multi components is taken to a multiple pressing machine which consists of a single row of pressing punches in alignment with the pressing plate components. The machine has such arrangement that once first row of filled components in pressing plate are pressed, the pressing plate moves forward for another pressing of components of the second row of components and so on. The filled components are then taken out.

Another method is of wet filling for typical caps where pressing is not done individually but by bigger load distributed upon small caps.

## 24.3   TABLET MAKING MACHINES

Small bare pellets may be made in tablet making machines where the composition is fed through a funnel/hopper and filling is done volumetrically in the mould through the moving part of the machine. The pellets are pressed continuously through the same mould and pressed pellets are allowed to pass through a chute and collected. It is essential that the composition should be free flowing so that volumetric filling may be done. Tablet-making machines are very useful if the pellet requirement is high.

## 24.4   PRESSING OF COMPOSITION IN MOULD

The pressing of pyrotechnic compositions may be done by

a. *Stop loading* method where pressing punch or drift is stopped at a certain desired height of pressed composition or

b. *Pressure loading* where the pressing punch or drift is given a desired load over the composition without stop

The former stop method is more useful in mass production as it gives the same height of the pellet, which is useful in proper assembly, while in the latter case, the height may vary. The density variation in pellets is more in the stop loading method while the density variation is less in pressure loading. Barring few pyrotechnic compositions like photoflash compositions and solid incendiary compositions for bomb fillings, mostly all compositions are suitably compacted in a press (hand press, hydraulic press or pneumatic press). Flow of composition is affected by several factors of particles of the ingredient in pyrotechnic composition and mould wall surface during filling and pressing as under.

a. *Wall surface:* Smoothness of the wall surface leads to free flow of composition

b. *Particle size and shape:* Fine particles adhere to walls and cause more friction; angular particles cause more friction compared to particles of smooth surface

c. *Moisture content:* Higher moisture content increases frictional resistance

Most compositions are used in pressed form using uniaxial pressing in suitable moulds (Figure 24.3) by hand press or pneumatic or hydraulic press with proper dwell time. A pellet may have varied types of shapes. These shapes are possible by suitably designing of punch and mould. These shapes give varied surface area of the pellet and improve flash pick up and modify the mass burn rate of pellet.

Bare pellets of composition are mostly used as stars in small cartridges though some higher calibre ammunition (like 155-mm red phosphorous smoke ammunition)

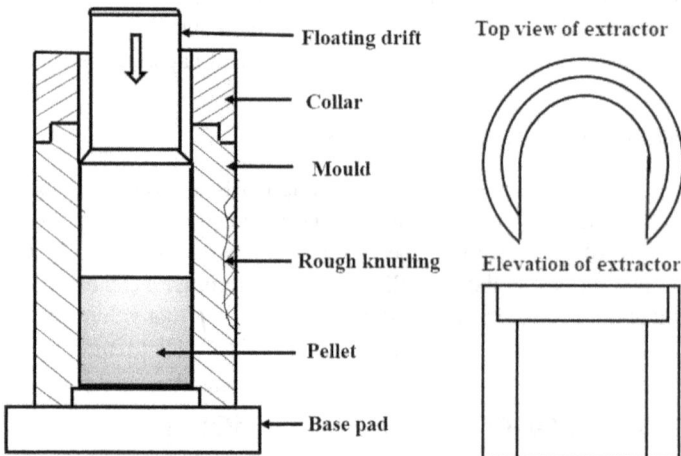

**FIGURE 24.3** Uniaxial Pressing in Moulds.

also uses bare pellets. The pellets may need two or three increments of composition to accommodate required amount of composition. To have quick interchangeability, interchangeable tools are used which consists of the following:

a. Mould
b. Base pad
c. Floating punch or drift
d. Collar/distance piece
e. Extractor

Generally steel EN-31 is used and hardened to 52–55 Rockwell hardness. The inside surface of the mould and outside surface of the drift and the top of the base pad are given a smooth finish to avoid friction during pressing. The working side of the tools are thereafter chrome plated and polished to have a mirror finish with plating thickness 0.04 mm. The gap between the punch/drift and mould may be kept 0.025 to 0.04 mm. A part of the external surface of the mould is made rough for ease of handling during use. However, in many cases, to improve flash pick-up by the pressed pellet, the floating punch/drift head is made serrated or protruded or castellated (step ladder shape). This gives extra surface area of the pellet on the flash pick-up side and aids in better flash pick-up (see Section 4.2.16).

The exactly weighed or scooped charge mass is poured into the mould fitted with the base pad. The drift is assembled over the composition. A distance piece is kept over the mould so that the pellet height remains constant. After pressing is over, the base pad is removed and the mould set is kept over the extractor and pressed. The bare pellet is ejected out of the mould and falls in the extractor. A suitable cloth/felt pad is kept below the extractor to avoid any damage to pellet.

*The dimensions of the mould set are governed by the pellet dimension, the volume and mass of the composition, compaction density and density gradient in the pellet.*

The mould set for grooved pellet is shown in Figure 24.4.

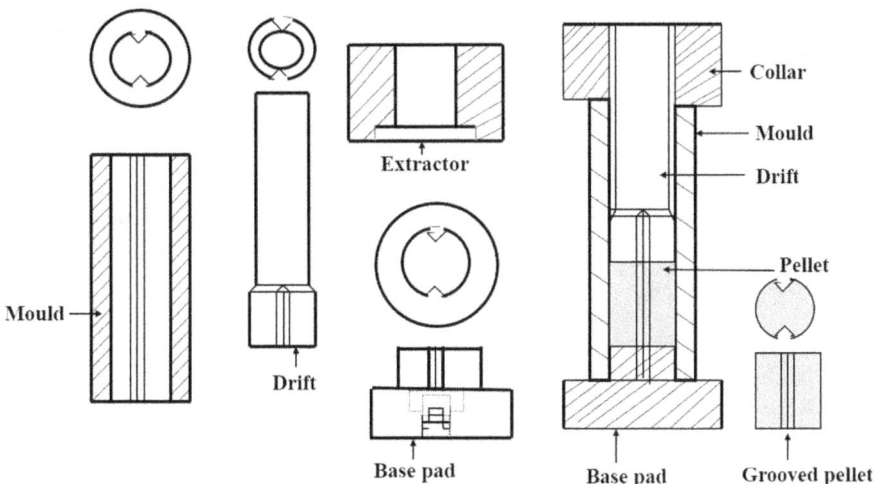

**FIGURE 24.4**   Mould Set for Grooved Pellet.

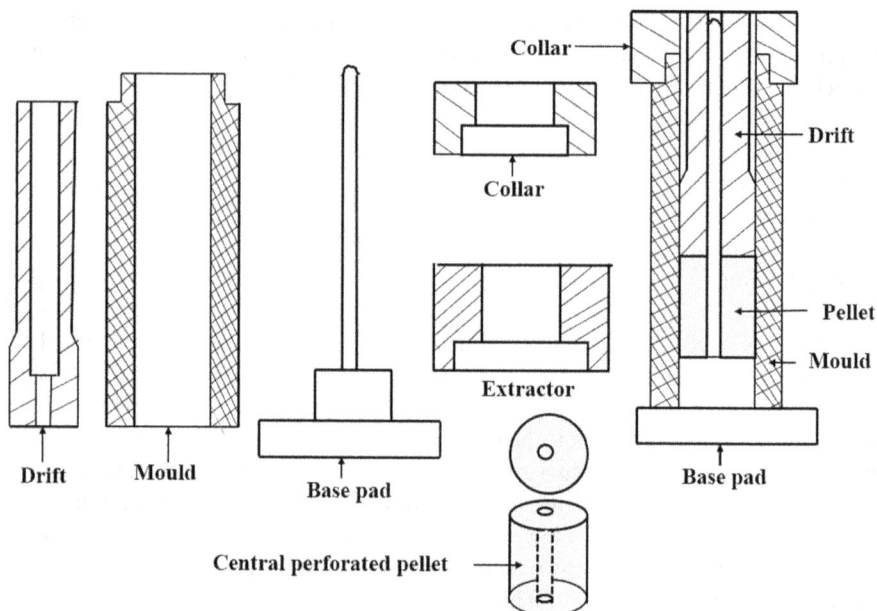

**FIGURE 24.5** Mould Set for Central Perforated Pellet.

The mould set for central perforated pellet is shown in Figure 24.5.

Sometimes, a dual operation is carried out in a press by pressing the composition in one mould as well as extracting the pellet from the already pressed composition in another mould. This improves the productivity.

It should be ensured that the pressing machine is inspected regularly for the friction in the moving parts of the machine. The surface finish of the punch/drift and mould and alignment of punch/drift and mould and cleanliness of press machine should be ensured as burrs and explosive dusts are hazardous. The dwell time is also to be fixed and checked regularly. All pressings are required to be done behind shield (for small filled components) or in a cubicle (for bigger and more hazardous filled components) by remote operation, ensuring that the room doors are interlocked when the press is on so that no person can enter when the pressing is on.

The pressure of the press is required to be checked through load cells daily. Limit switches provided for cubicles should work satisfactorily. The pressing should be stopped if any abnormal sound is heard or any part of the press gets heated up, and to be investigated properly after cool down period. The cause of ignition during pressing could be due to

a. Pyrotechnic composition sensitivity
b. Inadequate clearance between pressing punch/drift and mould
c. Rate of pressing
d. Foreign metallic matter causing friction
e. Adiabatic compression of air pockets/voids in the composition
f. Electrostatic discharge

The risk of hazards is not pronounced in case of pressing in paper container but it increases prominently with metallic containers, wherein particles of composition trapped between the metal surfaces of the container and the floating punch/drift may generate sufficient heat by friction to ignite the bulk of the composition. Presence of grit materials like sand also enhances the risk of hazard. To avoid friction during pressing, lubrication is done using any of the following materials:

a. Graphite  for mould and drift
b. Zinc stearate blended with the composition in small proportion, say 0.5% maximum for lubrication (aids in free flow of composition for automatic pellet pressing)

Some pyrotechnic composition containing epoxy resin-based binders get dried very fast. These compositions are required to be pressed immediately after maturation in approximately 45–60 minutes. The problems associated with such compositions are as follows:

a. If the composition is pressed immediately, it may result in cracks on the pellet as gaseous products come out of the pressed pellet. Such cracked pellets give undesired results.
b. If the composition is pressed after considerable time, the composition would not give required strength and may show low compaction strength.

Compositions, which are having lower bulk density, are required to be made as pre-pellets and two or more pre-pellets are then consolidated in the container so as to accommodate these compositions in more quantities.

Interlocking of ingredients is necessary when two pellets are required to be further consolidated and hence pressure for consolidation of pellets are kept much higher than the pre-pellet pressing. In many compositions, pellet along with loose composition is also used in consolidation for proper interlocking of the pellets.

In some cases, like illuminating composition, the composition is pressed in metallic container (with longer inner paper liner to accommodate composition) to avoid heat loss during burning by heat transfer to metallic container as well as avoid friction during pressing. The extra paper inner liner is then trimmed to size.

Pellets of tracers are pushed in tracer bodies (internally affixed with paper) and then pressed with priming and booster compositions, with top of the composition closed with celluloid/thin metallic disc to avoid ingress of moisture. However, in screening as well as colour smoke compositions, where heat generated during combustion is low and heat loss through container body is not significant, the pre-pellets are consolidated in the metallic canisters directly without any paper liner.

## 24.5  MECHANISM OF PRESSING IN MOULDS

Production of compacted composition requires suitable compaction force. This force should be optimum to ensure that neither the press is damaged due to excessive pressing force nor the composition remain in semi-compacted condition. Compaction of a pyrotechnic composition may be defined as *"rearrangement of ingredients leading to increase in density and decrease in porosity and volume by an external compaction force"* It is thus a combined process of compression (displacement of gaseous phase from composition thereby increasing density and decreasing bulk volume) and consolidation (consolidation of particles to increase mechanical strength) of a two phase (solid and gas) system by application of mechanical force.

The duration of compression involves *consolidation time, dwell time, contact time, ejection time* and *residence time*. These are shown in Figure 24.6 [1].

a. *Contact time:* Time duration for compression and decompression, excluding ejection time.
b. *Consolidation time:* Time duration to maximum compaction force.
c. *Dwell time:* Time duration at maximum compaction force. It is the amount of time the punch/drift head is in contact with the compressed composition and counted when the compression force applied is above 90% of its peak value.
d. *Residence time:* Time duration during which the pressed composition is within the mould.
e. *Ejection time:* Time duration during which ejection occurs.

The compaction of a pyrotechnic composition is completed in five steps:

a. *Particle rearrangement and release of entrapped air resulting in volume reduction*: When compressed initially, the ingredient particles are rearranged under low compaction pressures to form a closer packing structure. The finer particles enter the voids between the larger ones and give a closer packing. This results in the

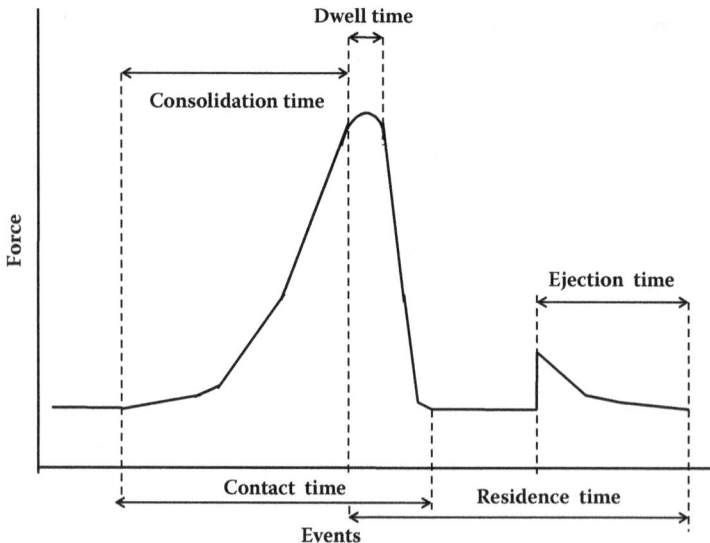

**FIGURE 24.6**  Events During Compaction Process (Reprinted from [1]).

removal of entrapped air, through the sides of the punch/drift and the mould and slight reduction in porosity of the composition and slight increase in density. Thus density increases due to removal of voids.

   b. *Bonding and interlocking*: Further pressing of the ingredient causes the particles of the composition to have more intimate contact and further reduction (70%–80%) in porosity of the composition.

   c. *Deforming*: A further pressing causes further compressing of the compacted composition, resulting in the particles of the composition deformed at their point of contact and more decrease of porosity of the pressed composition. The pores are almost sealed off and the material behaves like a solid with isolated pores.

The process includes composition particle rearrangement, the deformation (brittle to plastic) of particle contacts and fragmentation of the particles. This is due to fact that some ingredients (like aluminium, titanium) behave like plastic material while some ingredients (like glass, silicone) behave like brittle material. We may understand deformation as under.

Every material resists deformation and this resistance increases with increase in compaction load. They have also a critical value of force for deformation. If the compaction force is released prior to the critical force for deformation of the composition, the material is deformed elastically and thus may spring back to its original position. However, if this critical value of deformation stress is exceeded, the composition material would deform plastically and remain deformed and better compacted. This is the reason that a *dwell time* during compaction is necessary for proper compaction.

For a given compaction pressure, soft granules lead to higher compact densities than hard granules. The Figure 24.7 shows loose composition and pressed composition in a press.

   d. *Decompression*: Removal of mechanical force
   e. *Ejection*: Ejection of pellet from the mould

The pressing of pyrotechnic composition allows the composition to

   a. Accommodate in a small space in the ammunition

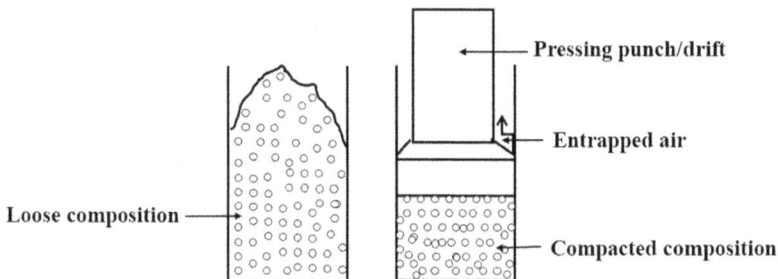

**FIGURE 24.7**  Loose and compacted Composition.

b. Burn layer by layer and burn more uniformly and thus giving special effects for a longer duration, which is required for almost all types of composition like illuminating, signalling, smoke composition, etc.

c. Sustain the various external forces like transportation forces causing severe jolts, bumps and vibrations and firing forces like set back force during discharge from the gun/barrel, rotational/spinning forces during spinning, deceleration in the trajectory and impact. This is due to increases mechanical strength of the pressed composition due to compaction of ingredient particles. The pressed composition is thus able to sustain these forces and does not break, crack or become powdered to render it unusable for producing any special effect.

d. Decrease the permeability and further compressibility of the composition.

## 24.6   OTHER METHODS OF FILLING AND PRESSING

In addition to loose filling or pressing methods, some other methods are as following:

a. *Cast filling*:

Cast filling, in paste form, is done in cases of some phosphorous-based smoke compositions, delay compositions and illuminating compositions, signal flare compositions. These have several advantages over conventional pressed compositions (see Section 13.3).

b. *Extrusion*:

Some pellets like infrared pellets are made through extrusion press by some manufacturers. However, extrusion is required to be done at high temperatures (around 150 $^{0}$C) in a hydraulic press through a die. It is hazardous and needs special precautions since sensitivity of the composition increases with temperature.

c. *Manual ramming*:

Some compositions are not required to be pressed but to be filled by manually ramming to accommodate the mass of the composition. The composition is thus neither loosely filled nor is highly pressed. The density of composition may vary, and performance depends upon the skill of the personnel filling and ramming the composition. This method is also known as *"tamping"* and is safe for less-sensitive compositions but is unsafe for very sensitive compositions (Figure 24.8).

d. *Screw feeding*:

This is a method wherein composition is screw fed in the shell. This requires the composition to have plasticity and least sensitive. This method can be used only with less-sensitive plastic compositions.

e. *Lead tube rolling filling*:

Many delay compositions are pressed by filling them in lead tubes and rolling the same in rollers of successive smaller diameters until the desired size is obtained. This method also helps in establishing delay composition (Section 17.2).

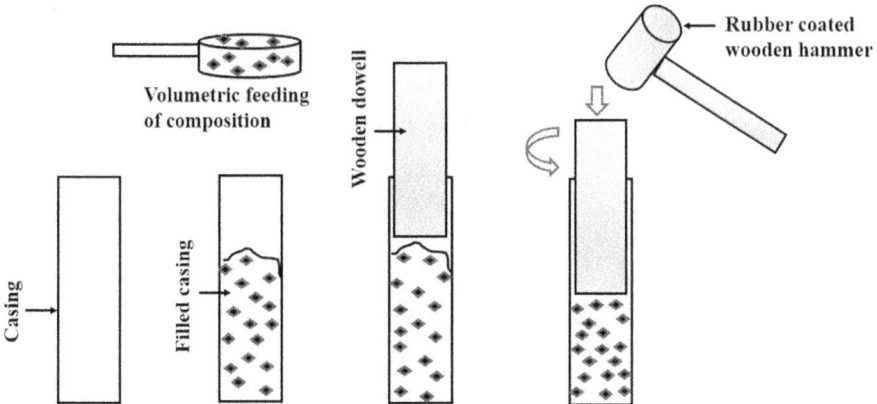

**FIGURE 24.8** Manual Ramming of Composition.

## 24.7 COMPACTION/CRUSHING STRENGTH OF PELLETS

The compaction/crushing strength of bare pellets is checked in compaction strength equipment. Load is applied on the pellet and computer-generated data is observed when the pellet starts crumbling. The compaction strength (crush load) for bare pellets for some heavy calibre ammunitions are specified and sample pellets are checked from the lot, before assembly, to ensure intactness of the pellet at such crush loads. This is essential since heavy calibre ammunitions are subjected to very high stresses during firing.

## 24.8 FACTORS AFFECTING COMPACTION STRENGTH OF THE PELLET

The compactness of the pellet depends upon the following:

a. The properties of the ingredients of the composition (ingredients of composition with higher hardness are less prone to compaction)
b. The proportion of various ingredients of the composition (the higher proportion of ingredients having higher density/hardness will give less compaction strength)
c. The particle size and particle size distribution of the ingredients (the finer the ingredients, better is the compaction)
d. The process of manufacturing the composition like mode of addition of ingredients, mixing time, drying time etc.
e. Dry powder or granulated composition
f. The amount of the binder and its adhesive character and its process of mixing into the composition
g. The slower rate of compaction and rate of decompression and larger dwell time gives higher density and compaction strength
h. Entrapped air (higher entrapped air gives lower density)

**TABLE 24.1**

**Data on Crushing Strength after Exposure to Different Relative Humidity [2]**

| Composition | Time (hours) | Crushing strength (kg.cm$^{-2}$) of 20 × 20 mm pellet at relative humidity | | | | |
|---|---|---|---|---|---|---|
| | | 52% | 64% | 75% | 84% | 96% |
| Mg/KNO$_3$/phenolic resin | 0 | 564 | – | – | – | – |
| (38/51/11) | 24 | 552 | 472 | 401 | 387 | 361 |
| | 48 | 560 | 455 | 377 | 386 | 350 |
| | 72 | 550 | 441 | 369 | 376 | 325 |
| | 96 | 540 | 379 | 367 | 370 | 320 |
| Gunpowder (KNO$_3$/C/S) | 0 | 532 | – | – | – | – |
| | 24 | 520 | 364 | 348 | 284 | 229 |
| | 48 | 525 | 356 | 337 | 215 | 168 |
| | 72 | 525 | 345 | 302 | 112 | 136 |
| | 96 | 515 | 340 | 300 | 55 | 45 |

i. The height of the pellet (the normal length/diameter ratio should not exceed 1.5 and in extreme cases up to 2.0 in uniaxial pressing. This is because lower portion away from punch/drift remains less consolidated than upper portion)
j. The number of increments and its height
k. Clearance between pressing punch/drift and mould
l. Humidity (higher humidity reduces crushing strength)
m. Temperature (higher temperature increases compressibility due to softening of composition)

The effect of humidity on compaction may be seen in Table 24.1 where crushing strength reduces drastically due to humidity.

## 24.9 POROSITY AND COMPACTION COEFFICIENT

Consider a pyrotechnic composition containing $n$ ingredients, each weighing $m_i$ with densities $\rho_i$, ($i = 1, 2, 3....n$, respectively). Since volume = mass/density, the volume occupied by each ingredient would be $m_i/\rho_i$ the total volume thus occupied will be $\Sigma(m_i/\rho_i)$.

The theoretical maximum density $\rho_{th}$ thus attainable would be

$$\rho_{th} = \frac{\Sigma m_i}{\Sigma(m_i/\rho_i)} \tag{24.1}$$

$$\rho_{th} = \frac{1}{\Sigma(m_i/\rho_i)/\Sigma m_i} \tag{24.2}$$

However, $m_i/\Sigma m_i$ are ingredients mass fraction in the composition and may be denoted by $mf_i$, the above equation may be rewritten as

$$\rho_{th} = \frac{1}{\Sigma(mf_i/\rho_i)} \qquad (24.3)$$

Multiplying and dividing by 100, and since $mf_i \times 100 = P_i$ (percentage of ingredient)

$$\rho_{th} = \frac{100}{\Sigma(P_i/\rho_i)} \qquad (24.4)$$

Both the above equations are valid for any number of ingredients. However, the actual density $\rho_{actual}$ obtained during compaction is much less than $\rho_{th.}$

Let us now consider the porosity. A loose composition before compaction has a number of voids or air pockets or porosity. However, during compaction of composition, these voids are reduced substantially but not fully eliminated. These voids help in pyrotechnic combustion as hot combustion slags, gases and fluids are able to permeate heat to next layer of composition through these voids. Porosity is measured as

$$\text{Porosity } \% = \frac{\text{Volume of voids in compacted composition}}{\text{Volume of compacted composition}} \times 100 \qquad (24.5)$$

If $Vcomp$ = Practically attained volume of compacted composition
   $V_{th}$ = Theoretical volume of solid ingredients of composition
   Then, volume of voids in compacted composition = Practically attained volume of compacted composition - Theoretical volume of solid ingredients of composition

$$\text{Porosity } \% = \frac{(V_{comp} - V_{th})}{V_{comp}} \times 100 = (1 - Vth/V_{comp}) \times 100 \qquad (24.6)$$

Since volume is inversely proportional to density,

$$\text{Porosity } \% = (1 - \rho_{comp}/\rho_{th}) \times 100 \qquad (24.7)$$

Where $\rho_{comp}$ = density of compacted mass, $\rho_{th}$ = theoretical density of composition. The porosity in pellets values generally vary between 10% and 30%.

The actual compacted density "$\rho_{comp}$" is much less than theoretical density $\rho_{th}$ (as it does not consider voids in the composition). The ratio between these two densities is known as *compaction coefficient* ("$K$") and its value is always less than 1.

$$K = \rho_{comp}/\rho_{th} \qquad (24.8)$$

**TABLE 24.2**

**Effects of Compression on a Fine Fraction of Atomised Magnesium Powder of Consolidation Pressures to 10,000 p.s.i. [3]**

| Consolidation pressure (p.s.i) | Average permeability (Darcy) | Average porosity (% void) | Pressed density (g.cc$^{-1}$) |
|---|---|---|---|
| 0 | 0.008 | 33 | 1.169 |
| 2,000 | 0.007 | 30 | 1.218 |
| 3,000 | 0.006 | 29 | 1.236 |
| 4,000 | 0.004 | 27 | 1.276 |
| 5,000 | 0.003 | 25 | 1.312 |
| 6,500 | 0.003 | 21 | 1.359 |
| 8,000 | 0.002 | 18 | 1.399 |
| 10,000 | 0.001 | 15 | 1.464 |

Original average particle size by Fisher sub-sieve sizer: 28.3 microns (200/325 mesh), air permeability apparatus (Picatinny Arsenal), sample weight 5.22 g

Compaction coefficient is a measure of extent of compaction and its values will always be less than 1, generally varying from 0.7 to 0.9. A higher compaction coefficient indicates

a. Higher density
b. Lower porosity
c. Lower permeability (permeability is concerned with the relative distance between particles measured in terms of the degree of compactness of the consolidated composition)
d. Possibility of finer ingredients in the composition

Table 24.2 shows the effect of compression on density, porosity and permeability of atomised magnesium fuel.

Very high compaction resulting into almost negligible porosity/voids would lead to ignition and propagation failure since combustion products like hot fluid slags and gaseous products will be unable to move inside the composition for ignition of next layer of composition.

## 24.10   DENSITY VARIATION IN PELLET

The density of composition decreases away from the pressing punch/drift. Density is highest (low porosity) near the punch/drift and least (high porosity) at the other end. The rate of burning will, therefore, not be the same and will vary in the column of pressed pellet.

The density variation in uniaxial pressing of pellet is shown in Figure 24.9.

Density gradients result from differences in the compaction pressure inside the compact, due mainly to wall friction effects. Assuming the uniaxial pressure $P_{ax}$

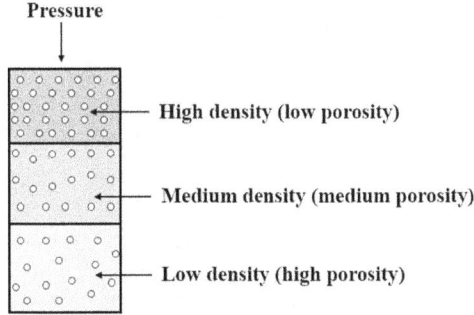

**FIGURE 24.9**   Density Variation in Uniaxial Pressing.

over the cylindrical cross-section of a diameter $d$, $P_{ax}$ can be calculated as a function of the distance $z$ from the upper punch/drift, according to equation [4]

$$P_{ax}(z) = P_{ax,o}. \ \exp(-4. \ \mu. \ k. \ z/d) \qquad (24.9)$$

$$or \ \ln P_{ax}(z) = \ \ln P_{ax,o} - (4 \ \mu \ k/d)z \qquad (24.10)$$

where $P_{ax, \ o}$ is the pressure at the upper punch/drift surface, $\mu$ is the friction coefficient which depends on the wall roughness and lubrication, and $k$ is the radial pressure coefficient (ratio of axial to radial pressure), which depends on the internal friction of the particles and is lower for irregular or unlubricated powders. The equation also shows a relationship between pressure and diameter of the pellet.

It is clear that the pressure on the pellet decreases as the distance increases. Hence density of pressed composition will vary in a pellet appreciably if the length ($z$) of the pellet is more. Hence, to have uniformity in density of composition, a better compaction method is to press the composition in a double acting press (Figure 24.10) where pressure from both top and bottom are applied equally and give more uniform compactness. The $L/D$ ratio may be increased in this case. It may be seen that the top floating punch/drift makes it easier for alignment and avoids scratch/friction during pressing. The bottom punch is kept longer for ejection of pressed pellet.

It may be noted that "*the achieved density of a pressed pellet is governed by diameter of the pellet, mechanical properties of the composition, frictional coefficient between mould and punch and pressing cycle like rate of pressing, rate of decompression and dwell time*".

## 24.11   CRACKS IN PELLET

Any crack in the pellet is detrimental to its performance. The cracks in the pellet during handling or transportation or shock of discharge may occur if the pressing load and/or dwell time is very low. The cracks may also occur if the resin-based composition is pressed late or too early where the resin has excessively matured or not matured to the desired level.

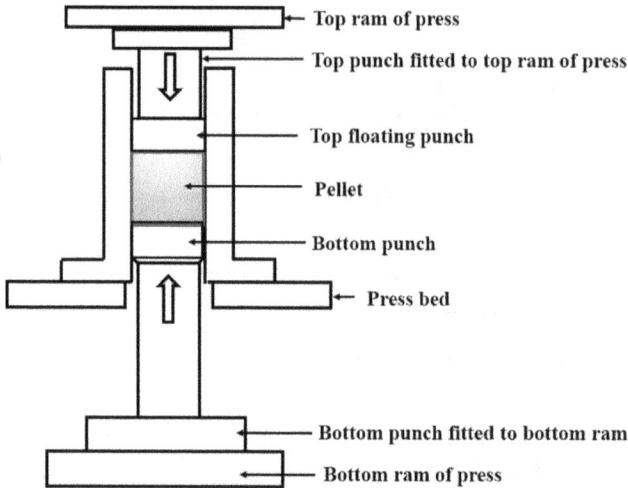

**FIGURE 24.10**   Pressing of Pellets in Double Acting Press.

As an example, in illuminating compositions, the igniting flame may penetrate through the crack in the pressed illuminating composition pellet and a chunk of the composition may fall out, thereby reducing the burn time in illuminating ammunition (Section 4.2.14). Similarly, the pressure of the propellant gases and the propellant flame acting on the tracer may penetrate through the tracer composition through these fine cracks and disintegrates the tracer. The tracer composition therefore will not burn layer by layer and give undesired burn time. Hence, tracers are pressed at very high pressure than most of the other pyrotechnic compositions. Base bleed igniters are also pressed at high pressure.

## 24.12   ASSEMBLY OF FILLED AMMUNITIONS

The manufacturer is required to make a standard operating procedure like standing instructions or general safety directives or work instructions for undertaking production of each ammunition. The assembly depends upon the type of ammunition. It is different for different ammunitions since each has different designs, different components, different compositions and different methods of assembly.

The simple general arrangement for ammunition manufacturing is shown in the Figure 24.11.

The ingredients are tested, sieved if required and mixed and the composition is dried (if required) and tested. The empty filling components are tested/proved (if required) and the pyrotechnic composition is filled/pressed in filling components. The filled components are tested/proved as required and assembled as complete ammunition using other components.

The assembled ammunition is inspected during assembly stage. The bulk lot is then painted and stencilled. Packages like steel boxes, wooden boxes and steel

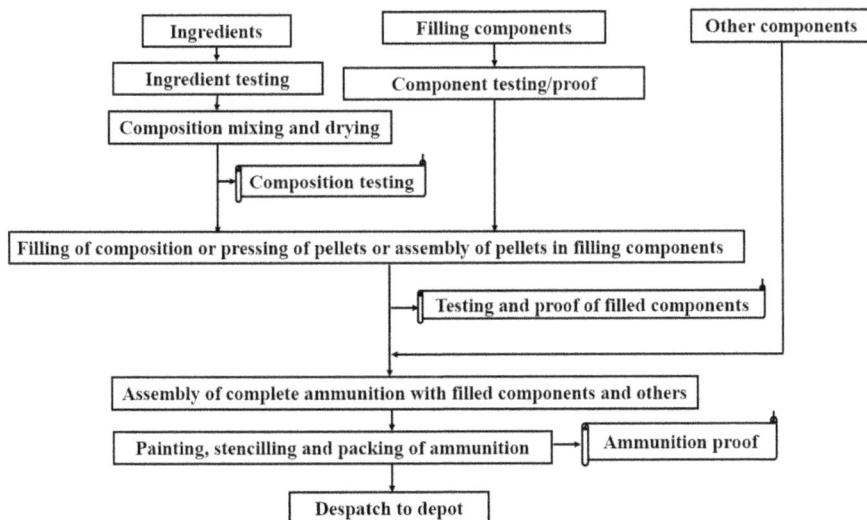

**FIGURE 24.11** Manufacture of Ammunition.

carriers are suitably painted and marked. It is ensured that the approved packing is snug tight, using specified fitments. These are stored in storage buildings.

Proof samples as per random sampling are picked from the bulk lot and packed in same boxes as bulk, and sent for proof. The proof results with sentencing of the lot once received are studied for improvement in future lot production, if required. Accepted lots are forwarded to depots with relevant acceptance documents.

The maturity in producing ammunitions with consistent quality over the years is attained by well-planned work instructions, supervision, training, a well laid out quality control plan and safety plan. Safety aspects in ammunition manufacturing requires robust safety arrangements. A pyrotechnician is required to maintain equal balance of the trio, namely *Safety, Quality and Reliability* of pyrotechnic ammunitions.

## REFERENCES

[1] Kapil R. Bare, Yogesh Kumar A. Biranwar, Hemant D. Salunke, Prajkta L. Ughade, and Dheeraj T. Baviskar, *"Compression Parameters involved in Powder Compression and Manufacturing of Tablet"*, International Journal of Pharmaceutical Sciences Review and Research, March–April 2011, Volume 7, Issue 2, Page 73–78.

[2] A.S. Redkar, V.A. Mujumdar, and S.N. Singh, *"Study on Magnesium Based Pyrotechnic Composition as a Priming Charge"*, Defence Science Journal, January 1996, Volume 46, Issue 1, DESIDOC.

[3] Doris E. Middlebrooks and Seymour M. Kaye, *"The Effects of Processing on Pyrotechnic Ingredients Part 1: Compressibility of Powdered Magnesium and Sodium Nitrate at Consolidated Pressures to 10,000 psi"*, Technical Report 3252, September 1965, Picatinny Arsenal Dover, New Jersey.

[4] Ralf Riedel and I-Wei Chen, *"Ceramics Science and Technology: Synthesis and Processing"*, Volume 3, First Edition, @Wiley-VCH Verlag GmbH & Co. KGaA, Reproduced with permission, 2012.

# 25 Sealing of Pyrotechnic Ammunitions

## 25.1 GENERAL

Ammunition consists of a large number of components which are assembled to make one compact unit. These assemblies involve various joints between components, some of them are:

a. Push fit/tight fit joints (example Impulse cartg. and decoy case joint)
b. Thread assembly joints (example shell and fuze joint)
c. Turn over joints (example paper/metallic cartg. case turn over with disc joint)
d. Crimp joints example (cartg. case and bullet joint)
e. Ringing joints (example cartg. case and rim fire cap joint)
f. Disc joint (example cap and disc joint)

Most of such joints need to be sealed to prevent the seepage of moisture as well as increase the strength of the joint between the two joined component assemblies. Generally, a fluid or paste is used as sealant followed by drying. A proper selection of sealant is essential to ensure proper functioning of the ammunition. Such sealants provide the following:

a. *Sealing of gaps* leading to protection of the ammunition by avoiding seepage of moisture leading to corrosion and thus enhance shelf life
b. *Bonds the parts of the component of the ammunition*
c. *Overcomes the stress during handling and transportation* of ammunition and firing stresses (set back force, set forward force, acceleration, deceleration, rotational force, etc.)
d. *Pack up gaps,* as required in ammunition

Choice and quantity of sealants for protection depends upon location of its application, type of joints and environmental factors, and needs proper curing time after cleaning any extraneous sealant.

The factors affecting proper sealing are:

a. Type of joints and gap between the joints (factors of design of ammunition)
b. Thermal cycle which may lead to expansion of components disproportional to each other
c. Shock, jolt, vibration and long storage
d. Stability and shelf life of sealants in extreme weather conditions

DOI: 10.1201/9781003093404-25

## 25.2 TYPES OF SEALANTS

There are a large number of sealants used in ammunition sealing. These are:

a. *Varnish and Lacquers*: shellac varnish, nitrocellulose varnish, bituminous varnish, kaolin varnish, etc.
b. *Adhesive and Cement*: Thermoplastic cement, thermosetting cement, celluloid cement, anchoring cement, ABS cement
c. *Greases:* Silicone grease, grease graphite, mineral jelly, grease ammunition, etc.
d. *Lutings*: Various types
e. *Miscellaneous*: Loctite, araldite, dobecot cement, rubber rosin cement, etc.
f. Rubber ring

Some fuzes use aluminium foil or wax as sealant for venting purpose as given in Section 4.3.4.

Some paper cartridge ammunitions are brushed or partially dipped in nitrocellulose dope using metallic tray with holes to hold the brass cum paper cartridges and then dried so that the paper cartridge is fully varnished, thus not allowing seepage of moisture through the paper cartridge (Figure 25.1). A number of holes (around 50) can be made in the tray to increase the productivity. Some paper ammunition like signal fuzee is brushed with shellac varnish.

However, any improper varnish on paper cartridges containing magnesium-based compositions may lead to formation of gaseous products during storage causing bulging of the cartridge case or cracking up of the cartridge case.

There are also some drawbacks of some sealants. A typical rubber ring is used in some cartridge cases to avoid ingress of moisture (Figure 4.5). Rubber rings/gaskets deteriorate due to ageing. Varnish and lacquers take more time to dry and hence not suitable for mass production in small arms ammunition manufacture. These also deteriorate with passage of time.

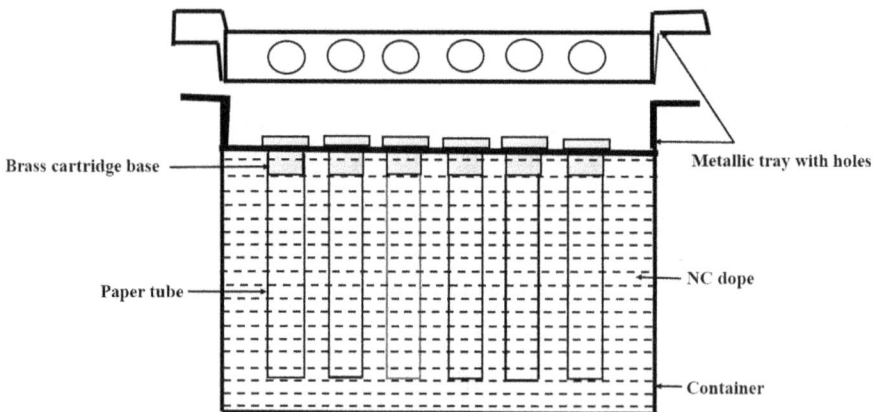

**FIGURE 25.1** *Varnishing of Paper Cartridge Ammunition.*

## 25.3 SEALANT SALIENT PROPERTIES

The sealant must possess the following properties:

a. The sealant must be stable and not be affected by very high and low temperatures and non-toxic and non-inflammable.
b. The sealant must have chemical compatibility with the metallic or paper or plastic components or chemical ingredients which it may have a direct or indirect contact during assembly or storage.
c. The sealant during maturation/drying should not produce any material/vapours to cause any irritating smell or reaction with composition or metallic components or cause corrosion.
d. The sealant should have no appreciable shrinkage due to solvent evaporation. The sealant should have good gap-filling properties and adequate consistency in its flow properties, as desired.
e. The sealant should have rapid room temperature cure.
f. The sealant should have adequate bonding strength.
g. The sealant must have adequate shelf life for long storage of ammunitions in extreme environment. The sealant should have excellent environmental resistance.
h. The sealant must not affect any electrical properties of the component.
i. The sealants rubber should have adequate hardness, tensile strength, etc.
j. The sealant must be able to withstand all the shock during production, transportation and proof (involving set back force, set forward force, spin force, etc.).
k. The sealant should be easy to manufacture and of less cost.

Some air force and naval ammunitions are heat sealed in polythene prior to packing in boxes while some are contained in metallic or plastic containers before packing in boxes. In addition to ammunition sealing, the package for ammunitions, especially naval ammunitions, are made hermetically sealed to avoid humidity effect. The packing fitments are ensured such that they do not lead to deterioration of ammunition. Section 27.8 may also be referred for measures to avoid ingress of moisture in composition.

The efficacy of sealing joints of few ammunitions are checked by leak test or vacuum test. Readers may refer [1] on sealants for details.

## REFERENCE

[1] R.W. Bryant, "*Handbook of Sealants for Ammunition*", Report No. 71, September 1971, Ministry of Defence, Explosives Research and Development Establishment, Ministry of Defence, Waltham Abbey, Essex.

# 26 Painting, Marking, Packing and Storage

## 26.1 GENERAL

Pyrotechnic ammunitions are required to be identified during proof, sentencing of proved lots, despatch of lots in packed boxes/pallets to depots, storage of packed boxes/ pallets at depots and during their use. In addition, ammunition lots are required to be linked to the corresponding inspection documents regarding their acceptance.

The process of ammunition identification broadly involves:

a. Type of ammunition
b. Class of high energy material used in ammunition
c. Mark like MK1, 2, etc. and modifications like Mod 1, 2, etc.
d. Lot numbers
e. Size, shape and weight
f. Presence of driving bands or tail units
g. Colour codes used to paint the ammunition like body colour paint, colour bands or symbols
h. Marking by lettering or stencilling or stamping on ammunition or components indicating the nomenclature, lot number, month and year of production
i. Sometimes marking of year of production is followed by month by letter A, B, C etc. for Jan, Feb, March etc.
j. Some high calibre ammunitions are marked with G for gun, H for howitzer, M for mortar and R for recoilless rifle after the calibre to indicate the firing equipment to be used.

Lot numbers are unique code numbers assigned to a particular lot of empty components, filled components and ammunitions produced at identical conditions and are therefore expected to give identical performance. (Similarly, compositions are assigned a unique batch number for each batch of composition).

Let us discuss the significance of an ammunition lot number. Lot number is required for traceability of the ammunition and its complete status of origin, performance and location. A unique number is assigned to each lot (consists of specified number of ammunition) during manufacture by a single production unit. These ammunitions are manufactured under uniform and stable production process based on specific drawing, specification and proved as per proof schedule of the ammunition. Lot is considered a homogeneous entity and expected to exhibit same performance. All the rounds in a particular lot shall have the same lot number. The next lot shall be allotted the next number. Same ammunition produced by another production agency will have its own serial lot number.

DOI: 10.1201/9781003093404-26

## 26.2 ROLE OF STAMPING OR MARKING OF COMPONENTS

The stamping or marking of the empty and filled components is done using monogram/abbreviations and numbers to ensure the following:

a. It gives the nomenclature of the component (e.g., Primer No.12 or Fuze 213Mk-5(M-3) with lot numbers.
b. It gives the monogram/abbreviation of the manufacturer.
c. In case of shells, heat treatment batches are also stamped/engraved.
d. It gives the month/year of manufacture of the component lot and also filled component lot and it is thus possible to know the shelf life of the filled components.
e. It is possible to take sample of the component lot, test and or prove for acceptance and assign an acceptance or rejection inspection report for the component lot.
f. It enables ease of storage and issue to filling and assembly line at the manufacturer's end.
g. It enables correct linking up of empty and filled component lots with their acceptance inspection reports and filled ammunition containing these filled components.
h. It is possible to link or trace the empty or filled component if any abnormality is noticed in performance of the ammunition.
i. In case of repair of an empty component lot or filled component lot, additional suffix like "R" or "RPD" is put to signify that it is a repaired lot. Generally, "R" is used for empty lots while "RPD" is used for filled lots.

The *empty hardware components* of ammunition like empty shell, cartridges, primers, fuzes, bombs, tail units, tracers, base bleed articles are all stamped/engraved/embossed so that these are not erased during handling and or processing. It is also easy to stamp or emboss them during lot formation in mechanical shop since the same are not filled with high energy materials. Some signal cartridges are given embossing, milling, knurling or notches on the base of cartridge for night identification (see Chapter 33). Top closing disc of some cartridges are embossed with a symbol to identify the special effects of the cartridge (see chapter 33). Some Illuminating ammunitions with parachute are, in addition to parachute symbol, are, marked with 'P' for parachute, preferably embossed on the nose cap, so that these may be identified in night.

*Filled components* like primers or fuzes may be stamped at the thick base of the primer or over the thick fuze body since the stresses involved are not capable of reaching to the filled composition. However, certain tracers are made of very fine metallic body and hence cannot be stamped. These are therefore stencilled.

Caps and squibs, being small and containing very sensitive composition, are neither stamped nor marked but kept in suitable containers/cardboard boxes with lot details, date and quantity.

## 26.3  ROLE OF PAINTING AND MARKING OF AMMUNITIONS

The painting of ammunition is done for protection against rust or corrosion as well as to signify the basic nature of contents of the ammunition for identification. These colours may vary in some countries. However, colour codes generally followed are as given below:

a. *Illuminating ammunitions with parachute* are painted white with a symbol of parachute in black and a red band on the body. Red band shows it is filled with explosives.

b. *Smoke ammunitions (including signalling colour smoke)* are painted brunswick green with red band with a white stencil marking.

c. *Inert ammunitions* are generally painted black with white stencil marking.

d. *Practice ammunitions* are generally painted blue.

e. *Signalling cartridge cases* (made of crimson red paper) are provided with colour band at the mouth like red, yellow, white or green as per the star colour. These signalling ammunitions also have night identification discs to indicate the colour of signal. A two star may have double colour band. Some signalling ammunitions with metallic cartridge case have anodised colour matching with the signal colour.

f. *Some ammunitions like flare trip wires* are painted green cum brown to camouflage its location.

g. *Infrared flare countermeasures* are generally made of aluminium body (occasionally anodised) with black stencil marking.

h. *Riot ammunitions* have the generic name of the active agent (CS or CN or DM) with or without its chemical name. If a colour of labelling is used, it shall conform to the convention that indicates agent identity: red for CN, blue for CS and green for DM.
   *CN: Red body with a white label and black marking*
   *CS: Blue body with a white label and black marking*
   *DM: Green body with all markings in black*

i. *Tracking and recovery ammunitions* are painted orange with black marking.

j. *Incendiary ammunitions* are painted light red with black stencil marking.

k. Some heavy calibre ammunitions like 155 mm are further marked with the *weight zone* for deciding on propellant charge for desired range as per the range table for the ammunition.

l. *Some small arms bullet tips* are painted for identification like blue for incendiary, silver for armoured piercing and red for tracer ammunition.

The nomenclature of ammunition is important for its recognition. This is done by stencil marking or screen printing of nomenclature on completed ammunition over the base paint, keeping in mind the simplicity, uniformity, ease of application and clarity of the same, without any ambiguity. Depending upon the size of the ammunition, the marking by stencil/screen printing sizes are varied. It helps in ascertaining following of ammunitions:

a. The calibre
b. The initials of manufacturer
c. The production month/year
d. The production lot number
e. Nomenclature of ammunition
f. Issue of complete lot to depots with quality control documents of the particular lot
g. Issue of complete lot to users from depots
h. Recall/segregation of lot if any defects are noticed in the lot
i. Recall of obsolete or shelf-life expired ammunition
j. Tracking/traceability of lot in depots
k. Tracking of ammunition till it is exhausted
l. Inventory management and logistics regarding quantity, location, condition, etc.

There is a requirement for *instructions to use the ammunition* in some cases. These are either printed on the ammunition itself (especially for distress signalling devices) or are printed as a leaflet and packed along with the ammunition for proper laying and use. These leaflets should be so placed inside the container or box that they do not get mutilated or crumbled during transport.

Some ammunitions, like flare trip wires, are provided with instructions to users using a metal tag marking. These instructions should be clear, legible and to the point without any ambiguity.

The mode of marking details varies from country to country. A good account of ammunition lot numbering can be seen in MIL-STD-1168C 11 March 2014 (Superseding MIL-STD-1168B) [1].

Complete identification of pyrotechnics is furnished by the nomenclature, including the model designation, the colour of the item, and lot number. Nomenclature of modified design use "A" or "Mod" or "Mark" or "Mk" after the original nomenclature.[2]

The U.S. Army Model Numbers [2] are used to identify differences between items with the same nomenclature. All army model numbers begin with the letter M. In M27, e.g., M means, "standard model." Types of model numbers are explained as follows:

XM67. The X means that the item is developmental or experimental.
XM67E1. The E1 added means that the experimental item has one experimental modification.
XM67E2. The item has its second experimental modification.
M67. When the X is dropped, it means the item is no longer experimental or developmental and has become a standard item.
M67E1. Standard items may sometimes have experimental modifications. E1 means it is a standard item with one experimental modification.
M67A1. This model is a standard item with one approved modification.
M67A1E1. This model has one approved modification and one experimental modification. When the experimental modification becomes standard, the new model number will be M67A2, meaning it has two approved modifications.

## 26.4   ROLE OF PACKAGES

The role of package is to ensure that the ammunition is kept intact (in its original form as packed by the manufacturer), when received by the depot or by the user without any detrimental external effects during transit and storage, over the years, so as to ensure that the ammunition gives the desired reliability, consistency and safety. Packages serve the following:

a. They hold the ammunition for ease of transportation.

b. They provide support so that ammunition may be transported easily.

c. They allow only fixed quantity to be packed in a box, thus making counting easy.

d. They allow the ammunition to be kept in a limited space by stacking packed boxes suitably in vertical columns.

e. They allow storing of the ammunition lot-wise so that lot-wise counting and issue is easy.

f. Details like nomenclature, lot number, quantity, consignee, consignor are marked on the packages. Stock numbers are also given in some countries.

g. These packages are designed and proved to ensure that the ammunition remains intact due to stress of transportation, vibrations, jolts, climatic temperature variations, dust, rain and humidity, etc. These packages should therefore be sturdy enough to withstand the stresses.

h. Packing system (separators, liners or felt pieces, etc.) ensure snug packing for minimal rattling due to shocks in transportation.

i. The package must be compatible with the ammunition/device as well as packaging material used in package.

j. It allows keeping special instructions (a leaflet) or equipment or complimentary materials to be stored together as these are required during use of the ammunition. Sometimes, data card providing details of ammunition and its components as well as instructions for loading and firing is kept in the packed boxes.

k. There are special packages which are leak-proof, especially for naval and air force ammunitions that prevent adverse effects of moisture on the ammunition.

l. It makes it easier to draw samples for review and/or proof to ensure that it has sufficient shelf life.

m. Identical ammunitions/devices are kept in similar boxes with similar markings. This helps in procurement, storage, handling and counting.

n. The package should just meet the adequate protection of the ammunition, with minimum cost. It should be of appropriate size and volume for human to lift for stacking or transportation.

o. Unlike civil packages where a few defectives are accepted during transport and storage, no such defects are permitted in ammunition packages and cent percent reliability is aimed.

## 26.5   TYPES OF PACKAGE

Packages may be of following types:

a. Wooden pallets
b. Wooden box
c. Steel box
d. Steel carriers
e. Fibre-reinforced polymer package

These packages must be resistant to corrosion, rusting in metal boxes and resistant to fungal infestation in wooden boxes. Some packages are designed to be hermetically sealed to avoid ingress of moisture. Packages are designed based on the size and quantum of ammunition. Thus, small calibre ammunitions will have different packages than those of high calibre. In addition, inner liners or packing fitments are also suitably chosen so that the ammunition is intact when packed as under:

a. Small filled components like caps are generally packed in cardboard boxes with felt sheets having holes to accommodate caps which provide safety during internal transport. Squibs and impulse cartridges are packed in cardboard boxes with suitable partition.
b. Sometimes the ammunition itself is packed in polythene tubes to avoid ingress of moisture and then packed in ammunition boxes, especially paper cartridge ammunitions and some naval ammunitions.
c. Some naval wooden packages have inner tin container with rubber gasket for hermetic sealing.
d. Some steel boxes are also provided with rubber lining for hermetic sealing.
e. Medium and large calibre ammunitions are packed in steel packages or wooden packages. Some of these are even packed with inner package of LDPE/HDPE or laminated paper liner followed by outer package, steel or wood.
f. Very large calibre ammunitions like 155 mm are packed in wooden pallets.

Packaging materials like felt, thick card board sheets and wooden fitments are used for snug packing. Wire iron galvanised with lead seals and steel tapes are also used for packing the boxes and crates. Steel boxes are preferred over wooden boxes as these boxes

a. May be recycled
b. Have clear and distinct marking through stencilling/screen printing
c. Have no possibility of insect/termite infestation, shrinkage or cracks during storage
d. Are more sturdy
e. Are not combustible (making wooden boxes non-combustible is costly)
f. More eco-friendly

Army make use of both metal and wooden packages but navy mostly uses metallic packages for the following specific reasons [3]:

a. Use of metal does not introduce additional combustibles into shipboard magazines.
b. If operationally necessary, metal packages can be discarded overboard without leaving a tell-tale trail of floating debris.

In general, all ammunitions which are quick-fired have fuzed ammunition due to time constraint. Navy mostly uses fuzed ammunition due to limited space on ship for fuzing operation. Army mortar ammunitions are mostly fuzed while artillery ammunitions and air force bombs are fuzed on as required basis. Hence, a separate fuze package is required for some ammunitions.

## 26.6  ROLE OF PACKAGE PAINTING AND MARKING

Packages are painted by brushing or spraying or dipping or flow coating. These are dried by air drying, stoving or infrared drying. Painting and marking of pyrotechnic ammunition packages are done to ensure the following:

a. It gives protection to the package against environment like corrosion of metallic boxes and moisture/insect infestation effect on wooden boxes.
b. It also provides an indication of the type of ammunition packed for identification.

The marking of the pyrotechnic ammunition packages is done to ensure the following:

a. It gives the nomenclature of the ammunition (like shell 105 mm BE smoke). It thus enables a user to know the type of ammunition, its functioning and special effects so that it may be deployed suitably.
b. It gives the ammunition lot number. It enables correct linking up of ammunition lots with their acceptance inspection report and ammunition receipt documents at depots.
c. It gives explosive category to enable depots to maintain the explosive category.
d. It enables depots for ease of sorting, storage and issue to field services.
e. It gives the names of consignor and the consignee.
f. Stock numbers are also given in some countries.
g. Part boxes containing less than the usual number of ammunitions in a package are painted differently or marked as PART BOX for ease of counting. This arises for many of the ammunition due to following reasons.

Ammunition lot size is not always fully divisible by the content in one package. For example, lot size of ammunition (say mortar bomb 51 mm illuminating ammunition) may be 2,000 numbers and one package (say Box B9A/L) contains 24 numbers. Hence, 83 boxes with serial number 1 to 83 will have 24 numbers each while the last box with serial number 84 will have only 8 numbers. This particular box is painted differently or marked as *PART BOX*.

Similarly, sometimes, even though the full lot need not have a part box, but due to reproof of the lot or repair of the lot, the lot size may become lower than usual, necessitating part boxes.

## 26.7   DEFECTS IN PAINTING, MARKING AND PACKING OF AMMUNITION AND PACKAGE

The likely defects of painting, stencilling and marking are colour shade, marking defects like missing letters, illegible or incorrect marking, incorrect or damaged label and peeling of paint. The most likely defects in steel boxes, containers and carriers may be rust, dents, cracks, broken hinges, handles and latches and damaged label. The defects in wooden boxes are cracks, gaps between planks, warping and insect infestation. The likely defects in packing are wrong quantity of ammunition in box, ammunition quantity and number of boxes not matching and mix-mixing up of lot number of ammunitions.

## 26.8   STORAGE OF AMMUNITION PACKAGES

Following are the precautions taken for storage of ammunition packages. Rules for handling of ammunition must be followed.

a. Only approved magazines with approved explosive limits shall be used for storage of ammunition.
b. Prohibited articles like mobiles, camera, cigarette, matches, inflammable liquids shall not be allowed.
c. Improper handling of ammunition may lead to safety hazards and failure of ammunition.
d. No ammunition should be dragged, dropped or allowed to get contaminated with sand, dust, oil, grease, etc.
e. No repair work or external force be used on ammunition. Ammunition should not be tampered during storage. No repair work to be carried out along with bulk storage of ammunition lots. Any cracked, corroded, deformed, dented, punctured or loose joints like fuze joints, projectile joints or other loose joint ammunition shall be segregated. Any broken box must be repaired (and tested for leakage proof, if stipulated for the box) in a separate area. The ammunition in such packed box must be examined properly.
f. No ammunition shall be exposed to flame or fire or smoking or spark or electrical lines in the near vicinity of ammunition.
g. Misfired/rejected ammunition should be segregated and it should not be tinkered with.

h. Precautions shall be taken to ensure that initiating devices like caps/primers are not activated due to accidental hit or electricity or spark.

i. Sufficient space should be left between ammunition stacks and the floor, ceiling and walls of the magazine to permit air circulation as under. The minimum space clearance to be observed is as follows.

    i. 0.6 m between walls and stack. If the floor area of store house is less than 40 sq. m, clearance may be reduced to 0.3 m.

    ii. Width of the gangways between the main stacks will be 1.2 m.

    iii. Clearance around the pillars should be 0.5 m.

    iv. Clearance between sub stacks should be 0.2 m.

    v. Clearance between ceiling and the top of the stack should be 0.6 m.

    vi. The stacking height of ammunition stacks should normally not exceed 3.5 m.

    vii. Ammunition stacks should be at least 1 m from doorways (to protect from sunlight, rain, etc. when doors are kept open).

j. Explosive/ammunition should be transported in closed type vehicle and not be transported in open vehicle exposing to sunlight or rain. These vehicles should be equipped with spark arrestor and suitable chemical fire extinguisher of approved type and capacity. During loading, the engine should be stopped.

k. Each ammunition package should be stacked as per lot number provided on the packages. This helps in picking the lots in serial order as well as to segregate any lot considered rejected.

l. Only non-sparking tools should be used in the depot.

m. Only trained personnel should handle the ammunition

n. The packages must be kept in such a way so as to retrieve any ammunition with lot details easily.

o. Ammunition should not be allowed to fall from height on hard ground.

p. The bulk storage house should not have any misfired or partially fired ammunition. It should also not be stored along with empty boxes.

q. The ammunition when used in field conditions should not be kept in open conditions, thereby risking its exposure to sun rays, rain, snow and mist. These may be kept under fire proof tarpaulins with sufficient space for free air movement. These should not be kept under tall trees or towers to avoid lightning, especially in rainy and cloudy weather.

Paper, cambric, plastic and rubber components in packages deteriorate during storage. These are required to be inspected periodically.

Readers may refer [3] for packaging and pack engineering. Reference [4] gives the review on the standardizing ammunition packaging report covering ammunition reliability, safety, history and rationale of development of joint conventional ammunition packaging policies and procedures.

# REFERENCES

[1] *"MIL-STD-1168C 11 March 2014 (Superseding MIL-STD-1168B 10 June 1998)"*, Department of Defense, Standard Practice, Ammunition Lot Numbering and Ammunition Data Card, Department of Defence, USA.

[2] *"Interpreting Ammunition Markings and Color Codes, Sub course MM2597, Edition 6"*, United States Army Combined Arms Support Command Fort Lee, Virginia.

[3] *"Packaging and Pack Engineering"*, Engineering Design Handbook, AMCP 706-121, March 1972, Headquarters, US Army Material Command, Washington DC.

[4] *"Standardizing Ammunition Packaging"*, Prepared by Joint Conventional Ammunition Program, Packaging and Preservation Task Group, 29 June 1977.

# 27 Classification, Shelf Life and Compatibility

## 27.1 GENERAL

All high energy materials are hazardous. The hazards are of three types, namely

a. Fire and explosive hazard
b. Health hazard
c. Environmental hazard

## 27.2 CLASSIFICATION

The purpose of explosive hazard classification codes helps in the following:

a. Understanding the nature of explosives and its hazard consequences by referring to the classification code
b. Reducing the explosive hazard by standardised system of storage and transportation of ammunitions and its components
c. Safety of human and environmental health

The compatible groups for storage and transport are based mostly on the following factors:

a. Chemical and physical properties
b. Design characteristics
c. Inner and outer packing configurations
d. Quantity-distance division
e. Net explosive weight
f. Rate of deterioration
g. Sensitivity to Initiation
h. Effects of deflagration, explosion or detonation

The overall classification, division and compatibility of dangerous goods is as given below.

a. *Class*:
   Class 1 – Explosives
   Class 2 – Gases, compressed
   Class 3 – Liquids, flammable
   Class 4 – Solids, flammable

DOI: 10.1201/9781003093404-27

Class 5 – Oxidising substances and organic peroxides
Class 6 – Poisonous (toxic and infectious)
Class 7 – Radioactive substances
Class 8 – Corrosives
Class 9 – Miscellaneous
  b. *Division:*
  Further Class 1 explosives are divided in following 6 divisions.
  Division 1.1: Substances and articles which have a mass explosion hazard
  Division 1.2: Substances and articles which have a projection hazard but not a mass explosion hazard
  Division 1.3: Substances and articles which have a fire hazard and either a minor blast hazard or a minor projection hazard or both, but not a mass explosion hazard
  Division 1.4: Substances and articles which present no significant hazard
  Division 1.5: Very insensitive substances which have a mass explosion hazard
  Division 1.6: Extremely insensitive articles which do not have a mass explosion hazard
  c. *Compatibility Groups:*
  The 13-compatibility groups of high energy materials are:
  A – Primary explosive substances (most sensitive)
  B – Detonators: articles containing a primary substance
  C – Propellant explosives (including smokeless powder)
  D – Secondary detonating explosives (including black powder, detonating cord)
  E – Explosive substances with a propelling charge
  F – Explosive substance with a propelling charge and initiator
  G – Pyrotechnic articles
  H – Explosive substance with white phosphorous
  J – Explosive substance with flammable liquid or gel
  K – Explosive substance with a toxic chemical agent
  L – Explosive substance or article presenting a special risk
  N – Articles containing only extremely insensitive detonating substances
  S – Safety explosives (one article will not initiate another and the effect is localised to the packing)

As seen from above, pyrotechnics are assigned "G" in compatibility group of high energy materials. The UN hazard classification code (HCC) for an explosive or type of ammunition consists of a combination of:

  a. The Hazard division
  b. The Compatibility groups

It is not possible to have separate explosive store house for each type of ammunition. Table 27.1 provides the general explosive compatibility groups for storage and transport with certain conditions, though variations may be observed in some countries. This makes it possible to gainfully utilise the space in storage houses.

**TABLE 27.1**

**General Explosive Compatibility Groups**

| Explosive group | Hazard classification codes | Total hazard classification codes | Compatible explosive group(s) for storage |
|---|---|---|---|
| A | 1.1A | 1 | A |
| B | 1.1B, 1.2B, 1.4B | 3 | B, S, (C *, D *, E *, F *, G *) |
| C | 1.1C, 1.2C, 1.3C, 1.4C | 4 | C, D, E, S, (B *. F *, G *, N *) |
| D | 1.1D, 1.2D, 1.4D, 1.5D | 4 | C, D, E, S, (B *. F *, G *, N *) |
| E | 1.1E, 1.2E, 1.4E | 3 | C, D, E, S, (B *. F *, G *, N *) |
| F | 1.1F, 1.2F, 1.3F, 1.4F | 4 | F, S, (B *, C *, D *, E *, G *) |
| G | 1.1G, 1.2G, 1.3G, 1.4G | 4 | G, S, (B *, C *, D *, E *, F *) |
| H | 1.2H, 1.3H | 2 | H, S |
| J | 1.1J, 1.2J, 1.3J | 3 | J, S |
| K | 1.2K, 1.3K | 2 | K * |
| L | 1.1L, 1.2L, 1.3L | 3 | L * |
| N | 1.6N | 1 | C *, D *, E *, N *, S * |
| S | 1.4S | 1 | B, C, D, E, F, G, H, J,, N *, S |
| Total 13 | Total 35 | Total 35 | – |

*Notes*

\* With certain conditions

Majority of pyrotechnic ammunitions belong to 1.2 G and 1.3 G. Some ammunition and their components are stored separately, such as ammunition of unknown origin/damaged condition or deteriorated ammunition, white phosphorous and filled detonators.

## 27.3  SHELF LIFE OF AMMUNITION

Ammunition is desired to serve its intended purpose for long years and stored in depots for eventual use. The ammunition and its explosives are expected to produce desired performance as per specification even after years of storage and that there should not be any deterioration in long storage.

The pyrotechnic compositions are all chemical ingredients. Just like medicines have an expiry date after a period, the pyrotechnic compositions also have a shelf life. Thus, all the pyrotechnic ammunitions have a shelf life during which they must be used. High energy materials have a *service life* during which they can be stored, handled and used. This may be subdivided as

a. *Chemical shelf life or safe (storage) life:* It is the period during which the explosive can safely be stored without any likely hazard to its surroundings.
b. *Functional life*: It is the period during which the explosive can be used safely and the functional requirements are fully met.

Both chemical shelf life and functional life are affected by

i. Deterioration of explosives due to ageing
ii. Incompatibility with other substances
iii. Manufacturing process causing ingress of moisture or volatile matter
iv. Design features of ammunition
v. Storage conditions
vi. Temperature cycles of storage

The stored ammunitions are generally reviewed during shelf life through periodic proof and after expiry of their shelf life for performance and if found satisfactory, shelf life is further increased for short periods. In case of unsatisfactory performance, the ammunition lot is taken off from operational use and considered for training purpose, if found suitable. If not, the same are disposed. In ammunition shelf life, functional aspects are considered important as they take into account all the factors of the ammunition system.

*Shelf life of the ammunition is the period up to which the ammunition is expected to perform as specified.* It provides a basis for arriving at the quantity to be stored in depots (vis a vis rate of use) and projecting future ammunition requirements. In other words, there shall not be any deterioration of the ammunition in functioning during shelf life. However, there appears no correlation between high energy materials degradation and functional aspect of the ammunition.

This is arrived based on various rigorous tests and climatic trials on the filled ammunition during development stage like jolt test, jumble test, vibration test, shock test, water immersion test, high and low temperature test, humidity test, etc. (see Chapter 23).

*Joint Service Guide JSG1012* provides two types of tests – one in unpacked condition and other in packed condition. There are two modes of *Intensified Simulating Alternating Tests (ISAT)* – ISAT (A) and ISAT (B) – and any one test is chosen by the designer, depending upon the type of ammunition and the likely stress the same is expected to be exposed. The shelf life is taken as 6 times or 9 times the period of test. Thus, if the ammunition is able to withstand x months in ISAT (A), the shelf life is assigned as 6x months. Similarly, if the ammunition is able to withstand y months in ISAT (B), the shelf life is assigned as 9y months. The brief process is as given below.

Specified numbers of samples are subjected to these tests. A few samples are taken at intervals and subjected to functional test as well as chemical stability test. The observation of any deterioration in chemicals and or performance are noted.

Therefore, if the performance becomes erratic after 11 months in ISAT (A), then the ammunition would be assigned a shelf life of 5 years (11 × 6 = 66 months).

Similarly, in case the chosen test is ISAT (B) and the performance is erratic after 10 months, then the ammunition would be assigned a shelf life of 7 years (10 × 9 = 90 months).

### 27.3.1 FACTORS AFFECTING SHELF LIFE OF AMMUNITION

The following factors affect the shelf life:

a. The formulation of the composition
b. Compatibility of ingredients amongst themselves and their hygroscopicity
c. Processing method for production
d. The moisture and volatile matter inside the composition
e. Compatibility of components with which the composition is in contact
f. Design features of the ammunition
g. The effectiveness of sealing of ammunition
h. The effectiveness of the sealing of the package
i. The storage condition and temperature cycles of storage
j. Number of years of storage, i.e., ageing
k. Corrosion of metallic components
l. Deterioration of rubber and plastic components
m. Deterioration of insulating ability of insulating materials

A better shelf life is achieved by certain measures incorporated during design, manufacture and packing. While making a composition for an ammunition system, one must review the following aspects, for desired performance of the ammunition.

a. Likelihood of chemical reaction between the ingredients during long storage
b. Effect on thermal stability of composition during long storage
c. Effect on sensitivity of composition during long storage
d. Likelihood of corrosion of metallic fuels and components used during long storage
e. Effect of corrosion products, if any, during long storage
f. Effect of left-over moisture and leftover solvents in the composition during long storage

The shelf life of pyrotechnic ammunitions are mostly 5 to 10 years. However, naval pyrotechnic ammunitions mostly have a shelf life of 3 years due to heavy presence of moisture in the sea atmosphere.

Apart from service life, there is an *installed life* for some pyrotechnic devices as the period during which the device is assembled in the system like impulse cartridges. For example, some air force impulse cartridges are given storage service shelf life of 10 years in hermetic sealing, 6 months after opening and 1 month as installed life on board aircraft.

The shelf life of a few ammunitions/devices are given in Table 27.2.

**TABLE 27.2**
**Shelf Life of Pyrotechnic Ammunitionsand Devices [1]**

| Ammunition/Device | Shelf life (Years) |
|---|---|
| Cartg. Signal 16 mm Red/Green/White | 5 |
| Cartg. Signal 1 inch Red/Green/Illuminating | 10 |
| Cartg. Signal 11/2-inch Green/Red/Yellow | 5 |
| Mortar Bomb 51 mm Illuminating | 10 |
| Mortar Bomb 51 mm Smoke 2A | 10 |
| Mortar Bomb 51 mm Red/Green | 10 |
| Mortar Bomb ML 2" Illuminating | 8 |
| Mortar Bomb 81 mm Illuminating 2A | 10 |
| Mortar Bomb 120 mm Illuminating | 10 |
| Shell 84 mm Illuminating | 7 |
| Shell 105 mm Illuminating | 10 |
| Shell 105 mm BE Smoke | 10 |
| Shell 105 mm Colour Smoke Red/Orange/Blue | 10 |
| Shell 155 mm Illuminating (MIRA) | 10 |
| Shell 155 mm Illuminating M1A1 | 10 |
| Generator Smoke Orange 3A | 5 |
| Generator Smoke No. 5 | 8 |
| Marker Smoke White A/S N3 | 5 |
| Grenade Hand Smoke Red/Green | 12 |
| CSES Red/Green/Yellow | 5 |
| Flare Ground Indicating Mk-1 Yellow | 10 |
| Signal Distress day/Night | 5 |
| Signal Flare (239-025) and Squib (239-31-028) | 1 |
| Bomb 3 Kg Practice | 7.5 |
| Rd. 14.5 mm Artillery Training Ammunition (w/o PD Fuze) | 7 |
| Flare Trip wire Mk-1 | 8 |
| Signal Fuzee | 5 |
| Signal Fog | 5 |
| Hand Flare Red Mk-II | 3 |
| Para Flare Red | 3 |
| Buoyant Smoke Orange | 3 |
| Training Smoke Generator | 3 |
| Thunder Flash MK 4 | 5 |
| Port Fire Friction | 10 |

The life extension and assessment programme (LEAP) to determine the effect of ageing on explosives and explosive devices (removed from ten tactical missiles from the army depot) has been given in [2]. A typical test plan for bridge wire explosive devices is given in Figure 27.1. The results of the investigation formed

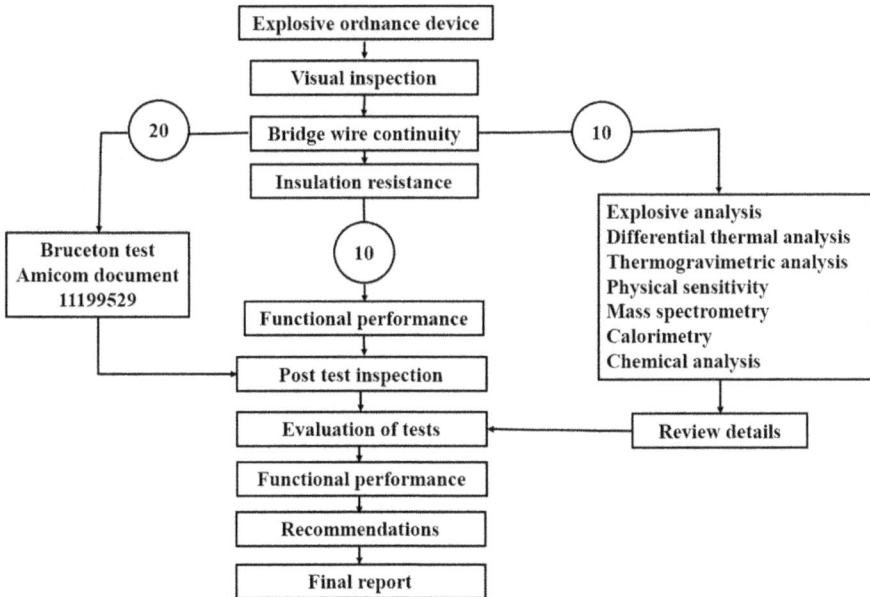

**FIGURE 27.1** Typical Test Plan for Life Extension and Assessment of Explosive Devices (Reprinted from [2]).

the basis for recommendation of service life, design improvements and quality control in production.

Such tests must be planned for aged ammunition and devices for improvement in design and shelf life of ammunitions and devices.

The major causes which affect the shelf life, during storage of pyrotechnic ammunitions are:

a. Compatibility
b. Moisture/humidity and degree of exposure
c. Temperature variation and ageing
d. Volatility of ingredients

## 27.4 COMPATIBILITY

A pyrotechnic composition in direct contact (or nearby the range of another material) may cause change in one or more properties of the composition and one or more properties of the contact material. Thus, desired effect of composition is not obtained. This phenomenon of affecting the physico-chemical properties of other materials is an important parameter during designing of the ammunition. A compatible composition means that neither it gets affected nor does it affect other materials used in the ammunition. Incompatibility results in reaction of one ingredient with other ingredient resulting in unwanted composition products which have detrimental effect on shelf life of ammunition.

Factors affecting compatibility of high energy materials [3] are:

a. Chemical reactivity in presence of moisture, volatile substances
b. Thermal stability of volatile substances like camphor, hexachloroethane
c. Effect of large charge size of composition (like main compositions evolving volatile substances) on small composition charges size (like caps, first fire charges, etc.)

Desirable properties for compatible materials are [3]:

a. Chemically non-reactive and stable
b. Low water content, non- hygroscopic
c. Neutral reaction
d. Low volatility and absence of volatile impurities
e. Absence of soluble components
f. No tendency to absorb soluble substances
g. No tendency to acquire electrostatic charge
h. Causes no friction or other mechanical hazard
i. Specified composition, adequately controlled

The pyrotechnic composition should be compatible not only with its own ingredients but also with other components, whether metallic, non-metallic, plastic, rubber or paper. The empty metallic component or container body containing the composition should not react with the composition during long storage.

The metallic components are therefore plated, if required, with materials which are not reactive with composition. In some cases, the composition is pressed in paper liners and assembled with the metallic containers. In this cases, it serves additionally in not allowing the heat dissipation through container body. A few incompatibilities are:

a. Sulphur (fuel) having low melting temperature and potassium chlorate (oxidiser) due to its low decomposition temperature are not compatible since these two ingredients make composition very sensitive. This is due to sulphur forming its oxide which in presence of moisture forms sulphurous acid ($H_2SO_3$ acid) and the later reacts with chlorate to form chloric acid ($HClO_3$) which may decompose spontaneously.
b. Sulphur and magnesium are incompatible.
c. Potassium chlorate is incompatible with ammonium nitrate due to formation of ammonium chlorate.
d. Potassium chlorate is incompatible with phosphorous.
e. Tetryl (composition exploding) is incompatible with sulphur, gunpowder and antimony sulphide. This is important for ammunition where tetryl is used as detonating material.
f. Initiatory compositions used in caps for initiating pyrotechnic ammunition also have some incompatibilities with use of certain metals as components.

i. Mercury fulminate is incompatible with aluminium, copper and zinc.
ii. Lead azide is not compatible with copper and copper alloys, zinc, steel and nickel. It forms hydrazoic acid ($HN_3$) in presence of moisture, as under [3].

$$Pb(N_3)_2(s) + H_2O(l) \rightleftharpoons Pb(OH)N_3(s) + HN_3(g) \tag{27.1}$$

Hydrazoic acid reacts with copper and its alloys.
Carbon dioxide in presence of moisture further degrades lead azide as under.

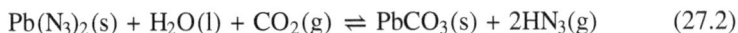

$$Pb(N_3)_2(s) + H_2O(l) + CO_2(g) \rightleftharpoons PbCO_3(s) + 2HN_3(g) \tag{27.2}$$

iii. Lead styphnate is incompatible with sulphur and sulphides.
iv. Tetrazene is incompatible with sulphur, chlorates, perchlorates and phosphorous and nitro compounds of high explosives.

Incompatible and hazardous combinations of several materials used in pyrotechnic compositions are given in reference [4] (Table 27.3)

---

**TABLE 27.3**
**Incompatible, Potentially Hazardous Combinations of Several Compounds Used in Pyrotechnic Composition [4]**

| Compound | Chlorates | Perchlorates | Aluminium | Magnesium | Zinc |
|----------|-----------|--------------|-----------|-----------|------|
| $ClO_3^-$ | NA[a] | –[b] | X [c] | X | X |
| $ClO_4^-$ | – | NA | ?[d] | ? | – |
| Aluminium | X | ? | -NA | – | – |
| Magnesium | X | ? | – | NA | – |
| Zinc | X | – | – | – | NA |
| Acids | X | – | – | X | |
| Water | – | – | ? | X | ? |
| $NH_4^+$ | X | – | – | X | |
| $Cu^{2+}$ | ? | – | ?[e] | X[e] | X[e] |
| S | X | X | – | X | X |
| $S^{2-}$ | X | X | – | – | – |

*Notes*
a  NA: not applicable,
b  little if any hazard
c  X: significantly hazardous combination,
d  ? Potentially hazardous combination depending on circumstances,
e  requires the presence of traces of $H_2O$

## 27.5 MOISTURE/HUMIDITY AND DEGREE OF EXPOSURE, TEMPERATURE VARIATION AND AGEING

High temperature, moisture and ageing alter the characteristics of pyrotechnic compositions, affecting activation energy, ignition temperature, strength of compaction, reactivity and ultimately performance. Moisture is the biggest problem for pyrotechnic ammunition manufacture. Improper sealing of ammunition or bad package may lead to ingress of moisture. It depends upon hygroscopicity of certain ingredients and is more if the hygroscopic ingredient is in fine powder form. Its effect is less when the composition is compacted in the form of star or candle. Effect of moisture leads to following:

a. In some composition, the left-over moisture (or volatile matter like acetone, toluene or xylene) may react with other materials.
b. In some compositions containing coated magnesium, a low degree of protection by the binder causes early attack of moisture on magnesium.
c. The moisture causes dissolution of some moisture prone constituents (like sodium nitrate or potassium nitrate) and thus causes changes in the density and shape of the pressed pellets.
d. It reacts with soluble oxidising salts and together corrodes the metal part of the component of the ammunition. These also reduce the content of fuel and oxidiser.
e. It forms lumps of oxidising material from powdered material.
f. It reacts with the metal powders forming oxide layer on the fuel surface, which plays a role as a barrier to heat propagation, reducing the contact area for combustion.
g. It loosens the pressed composition and thereby reduces the compaction strength.
h. It leads to uncontrolled, inconsistent and unreliable combustion, i.e., unpredictable burning.
i. It leads to another course of reaction and reaction products.
j. Excess moisture may lead to even non-combustion of the composition.
k. The moisture takes away heat during ignition to vaporise the moisture and thus increases the activation energy of the composition and time to ignition.
l. Paper cartridges made of cardboard are prone to moisture absorption.

Ageing has effect on physico-chemical-thermal properties of pyrotechnic composition. It causes

a. Variation in the activation energy
b. Variation in ignition temperature
c. Variation in burn rate
d. Variation in heat of reaction
e. Variation in purity of ingredient like oxide formation on fuels
f. Formation of unwanted products and micro-cracks in pellets
g. Reduction in the shelf life of ammunition

The following examples will explain the extent and type of deterioration of composition due to temperature variation, humidity and ageing.

A. *Ageing of MTV composition* [5]

Teflon/viton binder coating provides only some protection against ageing. Ageing causes conversion of magnesium to magnesium hydroxide

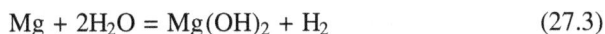

$$Mg + 2H_2O = Mg(OH)_2 + H_2 \qquad (27.3)$$

The volume increase is due to formation of magnesium hydroxide which has a different density than magnesium. During ageing process, magnesium hydroxide shells are *"peeled off"* from the magnesium particles, thereby increasing the magnesium surface area. These shells also push the teflon/Viton matrix aside, hence enlarging the porosity of the pellets. The mass of pellets increases with ageing. The mass increases by a factor of 1.91 on complete conversion. The volume increases by a factor of 3.68 on complete conversion (Table 27.4).

A significant increase in pressure is due to heating of magnesium hydroxide with formation of water vapour.

$$Mg(OH)_2 = MgO + H_2O \qquad (27.4)$$

The burning characteristics of aged pellets change with respect to unaged pellets and the formation of hydrogen increase the risk of accidents.

B. *Ageing of magnesium sodium nitrate laminac flare composition* [6]:
   The reaction proceeds through following steps.

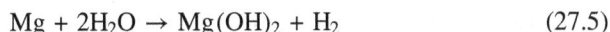

$$Mg + 2H_2O \rightarrow Mg(OH)_2 + H_2 \qquad (27.5)$$

$$4Mg + NaNO_3 + 7H_2O \rightarrow 4Mg(OH)_2 + NaOH + NH_4OH \qquad (27.6)$$

**TABLE 27.4**

**Increase in Mass and Volume of MTV Pellets During Ageing at 80°C and 80% RH [5]**

| Ageing time (hours) | Mass increase (%) | Volume increase (%) |
|---|---|---|
| 32 | 4.2 ± 0.1 | 1.038 ± 0.002 |
| 81 | 38.9 ± 0.5 | 1.552 ± 0.020 |
| 143 | 72.7 ± 0.5 | 2.272 ± 0.022 |
| 336 | 91.1 ± 0.2 | 3.68 ± 0.13 |

$$NH_4OH \rightarrow NH_3 + H_2O \tag{27.7}$$

Combined reaction is

$$5Mg + NaNO_3 + 8H_2O \rightarrow 5Mg(OH)_2 + NH_3 + H_2 + NaOH \tag{27.8}$$

The evolved gaseous products ammonia and hydrogen may cause rupture of sealings and split cases.

C. *Ageing of aluminium flare composition (35% atomised Al, 53% NaNO₃, 7% W, and 5% Laminac 4116 polyester resin binder)* [6].

The reaction proceeds [6] as under.

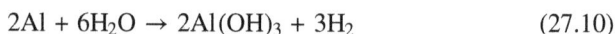

$$3NaNO_3 + 8Al + 18H_2O \rightarrow 8Al(OH)_3 + 3NH_3 + 3NaOH \tag{27.9}$$

$$2Al + 6H_2O \rightarrow 2Al(OH)_3 + 3H_2 \tag{27.10}$$

The unwanted gaseous products may cause rupture of sealings and split cases. Reaction (27.10) produces heat and is thus hazardous as it may cause spontaneous ignition.

D. *Ageing of composition SR 562 as flare* [7]:

SR 562 contains magnesium, sodium nitrate and calcium oxalate. On accelerated ageing for 30 days at 70% humidity at 70°C showed the following thermal cum kinetic and morphology of the aged composition.

Activation energy of the aged composition decreased to 189 kJmol$^{-1}$ from 239 kJmol$^{-1}$ for the fresh composition. Thermal decomposition temperature increased by 51°C after the ageing. X-ray diffraction showed the presence of magnesium hydroxide in the aged composition showing that some of the magnesium reacted with water vapours at high temperatures. Some micro-sized cracks were seen in the scanning electron microscopy micrographs of the aged composition.

E. *Ageing of composition SR 524* [8]:

Composition *Mg/NaNO₃*, used as a tracer at 70% humidity and 70°C temperature showed the following.

There is significant change in the thermal behaviour, kinetic parameters and the morphology of the aged composition. The kinetic parameters of the pyrotechnic mixture changed considerably after the ageing. Thermal analysis showed that the decomposition peak temperature of the aged composition lowered by 14°C. The activation energy of aged composition decreased nearly by 57% and the frequency factor also decreased noticeably. The ageing process made the composition easy to ignite by lowering the activation energy. However, the overall reactivity of the composition reduced due to decrease in the rate constant. The rate constant decreased due to a large decrease in the value of the frequency factor. XRD results suggest that the high humidity levels resulted in the production of magnesium hydroxide as the reaction product of the ageing process. The combined effect of high temperature and high humidity resulted in the development of micro-sized cracks in the bulk composition. The cracks made it relatively unsafe for storage and ignition as compared to the fresh composition.

F. *The influence of the fuel-to-oxidant ratio on the ageing behaviour of the magnesium–strontium nitrate pyrotechnic system in the presence of water vapor* [9]:

This has been investigated by isothermal microcalorimetry in conjunction with chemical and thermal analysis measurements. The following have been reported. The major products of ageing were identified as magnesium hydroxide and strontium nitrite. Measurement of the ratio of the amount of magnesium hydroxide formed to that of strontium nitrate reacted has shown that for compositions containing from 10% to 60% magnesium, a major component of the ageing process could be represented by the equation

$$3Mg + Sr(NO_3)_2 + 4H_2O \rightarrow 3Mg(OH)_2 + Sr(NO_2)_2 + H_2 \qquad (27.11)$$

DSC studies have confirmed that the formation of strontium nitrite during ageing has a significant influence on the high temperature pyrotechnic reaction between magnesium and strontium nitrate.

G. *Ageing causing variation in ignition temperature* [10]:

It has been reported that pyrotechnic composition containing magnesium and sodium nitrate (100 mg sample at 50°C and 65% humidity for 28 days) degrade during storage, causing changes the performance of the composition. It has been reported that

a. Ignition temperature decreased as the amount of magnesium increased.
b. Ageing increased the ignition temperature for each magnesium percent.

The overall decrease in ignition temperature due to ageing was about 100°C (Table 27.5).

On the other hand, the ignition temperature for 50% magnesium and 50% sodium nitrate composition showed that the ignition temperature increased rapidly over the first three days of ageing. After this it slowed to only a small increase. These results (Table 27.6) show that a significant part of the degradation reaction occurred during the first few days of ageing.

H. *Study on the aging mechanism of boron potassium nitrate (BKNO₃) for sustainable efficiency in pyrotechnic mechanical devices* [11]

Boron, potassium nitrate and laminac (23.7:70.7:5.6) were studied at 71°C and 50% RH for 8 and 16 weeks. It was observed that the heat of reaction and the reaction rate decreased by 18% and 67% over 16 weeks of ageing (Table 27.7), respectively.

**TABLE 27.5**

**Ignition Temperature for Magnesium and Sodium Nitrate [10]**

| Magnesium (%) | Mean ignition temperature °C | | |
|---|---|---|---|
| | Unaged | Aged | Difference |
| 20 | 640 ± 0 | 649 ± 2 | 9 |
| 30 | 622 ± 4 | 649 ± 2 | 27 |
| 35 | 614 ± 1 | 647 ± 0 | 33 |
| 40 | 602 ± 2 | 645 ± 1 | 43 |
| 45 | 594 ± 2 | 646 ± 0 | 52 |
| 50 | 592 ± 0 | 648 ± 1 | 56 |
| 60 | 570 ± 3 | 643 ± 1 | 73 |
| 70 | 572 ± 1 | 643 ± 1 | 71 |
| 80 | 541 ± 5 | 638 ± 4 | 97 |

**TABLE 27.6**

**Ignition Temperature for 50% Magnesium–50% Sodium Nitrate Composition [10]**

| Age (days) | Mean ignition temperature °C |
|---|---|
| 0 | 583 ± 7 |
| 1 | 607 ± 4 |
| 2 | 630 ± 3 |
| 3 | 641 ± 2 |
| 7 | 643 ± 3 |
| 10 | 644 ± 1 |
| 22 | 645 ± 1 |
| 48 | 647 ± 1 |

**TABLE 27.7**

**The Relative Heats, Peak Reaction Rates, and Thickness of Oxide Shells of B/ KNO$_3$ Under Accelerated Ageing [11]**

| Ageing (weeks) | Relative ΔH released | Peak reaction rate $(10^{-3}\text{-/s})$ | Thickness of oxide shell (nm) |
|---|---|---|---|
| 0 | 68.51 | 5.86 | 0 |
| 8 | 61.53 | 2.17 | 26 |
| 16 | 56.44 | 1.93 | 39 |

*Note:* Relative heat released (= reaction heat/melting heat of KNO$_3$) and reaction rates are converted from DSC signals using the AKTS program.

**TABLE 27.8**

**Deterioration of Mercury Fulminate [12]**

| Storage temperature °C | Time required to reduce purity to | | | | | |
| --- | --- | --- | --- | --- | --- | --- |
| | 95% | | | 92% | | |
| | Days | Months | Year | Days | Months | Year |
| 80 | 0.5 | – | – | 1 | – | – |
| 50 | – | 8 | – | – | 11 | – |
| 30–35 | – | – | 1.7 | – | – | 5.8 |
| 20 | – | – | 7 | – | – | 9 |
| 10 | – | – | 8 | – | – | 10 |

This is attributed to the oxide shells on the surface of boron particles. The formation of oxide shells could be confirmed using X-ray photoelectron spectroscopy and transmission electron microscopy–energy dispersive spectroscopy. In conclusion, surface oxide formation with the ageing of B/KNO$_3$ will decrease its propulsive efficiency; oxidation reduces the potential energy of the system and the resulting oxide decreases the reaction rate.

I. *Deterioration of mercury fulminate at high temperature* [12]:

Mercury fulminate is used in initiatory compositions. Deterioration of purity of mercury fulminate at varied temperature is shown in Table 27.8.

## 27.6  VOLATILE INGREDIENTS

Volatile ingredients lead to sublimation of ingredients like hexachloroethane and camphor. The problem gets aggravated if storage temperatures are consistently high. It also gets affected at high altitudes due to low atmospheric pressure, aiding in sublimation.

## 27.7  COMPATIBILITY TESTS

Readers may refer [13] for Standard NATO Agreement (STANAG) on chemical compatibility of ammunition components with explosives. This STANAG is limited to matters relating to chemical compatibility of munition components with explosives. It is not concerned with compatibility between ingredients in explosive compositions and the consequent stability of such compositions. A statement of applicability in each procedure provides guidance on the suitability of the procedure for the different classes of explosives. Readers may also refer [14] for compatibility of propellants, explosives and pyrotechnics with plastic and additives.

## 27.8    REMEDIAL MEASURES TO AVOID DETERIORATION OF PYROTECHNIC AMMUNITIONS

The important measure is to use compatible ingredients in the composition formulation with compatible components. The next measure is to control moisture or humidity since even slight moisture ingress shall lead to lower linear burn rate, lower burn temperature for all pyrotechnic compositions.

The following measures are taken to avoid moisture effect on ingredients, compositions and ammunitions:

a. Use of air-conditioned building at specified humidity for production. A humidity of 55%–65% has been found to be reasonable from safety point of view as well as from production point of view since humidity higher than 65% may lead to moisture effect on composition, while a humidity lower than 55% can lead to safety hazard in production.
b. Coating the fuel with binders to avoid exposure of its surface with moisture. The presence of organic binders in the composition will hinder the effect of moisture to the constituents of the composition. This is the reason that tracers and many illuminating compositions containing binders are better in stability than the photo mixture type binary composition which do not use binders.
c. Compacting of the composition (as it reduces the exposure of powdered ingredients) since ingredients with higher surface area shall have more moisture absorption tendency
d. Sealing of filled components and ammunition (see Chapter 25) by sealants. There are a good number of sealants. Nitrocellulose varnish is good sealant for paper cartridges. Particular care is required in case of use of shellac varnish due to presence of traces of water in alcohol used as shellac solvent.
e. A gasless pyrotechnic delay composition stabilisation process by conditioning at elevated temperature for a week has been proposed (Section 17.7.8).
f. Packing in hermetically sealed boxes is done for some naval ammunitions. Sometimes outer package is made of wood or metal but inside package is made of fibre or plastic or metal to avoid moisture effect.
g. Pyrotechnic ammunition boxes should be opened only when required so as to avoid effect of ambient conditions, especially moisture or humidity.

The corrosion in magnesium powder in atmosphere is more predominant since the porous and loose oxide coating over it is unable to protect it from further corrosion. However, in case of aluminium, aluminium oxides with thick coat of oxide on surface prevents it from further oxidation. Composition containing no metal powders and no hygroscopic salts are not appreciably affected by moisture.

It has been observed that certain cartridges bulge after prolonged exposure to high altitude and/or presence of humid conditions, especially paper cartridges containing magnesium or aluminium as one of the ingredients. The paper cartridge cases containing magnesium powder exhibiting bulge is attributed to reaction of magnesium or aluminium in the composition with water vapour (moisture) releasing hydrogen and causing cracking of the cartridge cases.

In view of humidity effect, certain air force ammunitions have been assigned certain limits of exposure when loaded in an aircraft (installed life) like number of sorties or number of hours of exposure to extreme cold temperatures (e.g., in the air at 10,000 m with temperatures around (-)50°C). The ammunition needs to be discarded after the above limits even if the ammunition appears visually satisfactory.

Readers may refer [15] for ageing studies and lifetime extension of materials and reference [16] on life cycle environmental assessment for smoke/obscurants.

## REFERENCES

[1] *"Product Brochure"*, Indian Ordnance Factories, Ordnance Factory Dehuroad, Pune-412113, Maharashtra, India.

[2] Edward M. Storma and Martin Marietta, *"Analytical and Functional Tests of Naturally aged Explosive Ordnance in the PERISHING LEAP Programme"*, 21-25 August 1972, Proceedings, Third International Pyrotechnics Seminar, Colorado, Springs, Colorado.

[3] N.J. Blay and I Dunstan, *"Compatibility Testing of Primary Explosives and Pyrotechnics"*, Technical Report Number 115, February 1973, Explosives Research and Development Establishment Waltham Abbey, Essex, England.

[4] Karina Rosa Tarantik, *"Investigation of New More Environmentally Benign, Smoke-reduced, Red- and Green-light Emitting Pyrotechnic Compositions Based on Nitrogen-rich Coloring Agents"*, Dissertation, 2010 Ludwig-Maximilians-University Munich.

[5] Chris Van Driel, Jeannette Leenders, and Jan Meulenbrugge, *"Ageing of MTV"*, 26th International Annual Conference of ICT, 4–7th July 1995, Karlsruhe Federal Republic of Germany.

[6] B. Jackson Jr., F. R. Taylor, R. Motto, and S. M. Kaye, *"Substitution of Aluminium for Magnesium as a Fuel in Flares"*, Technical Report 4704, January 1975, Picatinny Arsenal Dover, New Jersey.

[7] Zaheer-Ud-Din Babar and Abdul Qadeer Malik, *"Accelerated Ageing of SR-562 Pyrotechnic Composition and Investigation of its Thermo Kinetic Parameters"*, Fire and Materials, March 2016, Volume 41, Issue 2, Page 131–141, © 2016 John Wiley & Sons, Ltd. Reproduced with permission.

[8] Z. Babar And A.Q. Malik, *"Investigation of The Thermal Decomposition of Magnesium–Sodlum Nitrate Pyrotechnic Composition (SR-524) and the Effect of Accelerated Aging"*, Journal of Saudi Chemical Society, 2015, 10.1016/j.jscs.2015.06.005.

[9] I.M. Tuukkanen, S.D. Brown, E. L. Charsley, S.J. Goodall, P.G. Laye, J.J. Rooney, T.T. Griffiths, and H. Lemmetyinen, *"A Study of the Influence of the Fuel to Oxidant Ratio on the Ageing of Magnesium–Strontium Nitrate Pyrotechnic Compositions Using Isothermal Microcalorimetry and Thermal Analysis Techniques"*, Thermochimica Acta, February 2005, Volume 426, Issues 1–2, Pages 115–121, © 2004 Elsevier B.V. All rights reserved.

[10] Trevor T. Griffiths, Robert P. Claridge, Edward l. Charsley, Jim J. Rooney, Sarah J. Goodall, and Irmeli M. Tuukkanen, *"Investigation of the Thermal and Pyrotechnic Properties of Coloured Signal Compositions"*, Fraunhofer Institute of Chemical Technology, 2003.

[11] Lee J., Kim T., Ryu S.U. et al. *"Study on the Aging Mechanism of Boron Potassium Nitrate (B/KNO₃) for Sustainable Efficiency in Pyrotechnic Mechanical Devices"*. Sci Rep 8, 11745 (2018). https://www.ncbi.nlm.nih.gov/pmc/articles/PMC6078969/ (Creative Commons Attribution 4.0 International License).

[12] *"Military Explosives"*, Technical Manual No. 9-1300-214, Headquarters Department of the Army No. 9-1300-214 Washington, DC. 25 September 1990.

[13] *"Chemical Compatibility of Ammunition Components with Explosives (non-nuclear applications)"* STANAG 4147(Edition 2), 5 June 2001.

[14] Conference on *"Compatibility of Propellants, Explosives and Pyrotechnics with Plastic and Additives"*, 3–4 December 1974, Picatinny Arsenal, Dover, New Jersey.

[15] Leslie G. Mallinson, ed., *"Ageing Studies and Lifetime Extension of Materials"*, Kluwer Academic/Plenum Publishers, 2001, ©2001 Springer Science+ Business Media, New York.

[16] Roy L. Yon, Randall S. Wentsel, and John M. Bane, *"Programmatic Life Cycle Environmental Assessment for Smoke/Obscurants, Volume 2 - Red, White, and Plasticized White Phosphorus"*, Environmental Assessment ARCSL-EA43004, July 1983, US Army Armament, Munitions and Chemical Command, Abardeen Proving Ground, Maryland.

# 28 Classification and Nomenclature

## 28.1 GENERAL

There are a large number of pyrotechnic ammunitions and devices and it is essential to have classification and nomenclature of each type with lot number for ease of proper identification, production, storage and despatch from production units to concerned depots. It may be noted that lot number is not a part of the nomenclature of the ammunition or device. Also, in case of any failure observed during storage at depot, the full lot quantity may be retrieved from various depots for further action. The acceptance report of ammunition and devices as well as proof performance also requires classification and nomenclature of ammunition along with lot details. Barring a few pyrotechnic devices, all pyrotechnic ammunitions are identified based on the following.

a. Overall general classification
b. Distinct nomenclature
c. Mixed nomenclature

## 28.2 *OVERALL GENERAL* CLASSIFICATION

These may be grouped in the following three categories.

### 28.2.1 AS PER SHAPE

Ammunitions are classified based on shape as under.

a. Small arms (incendiary, tracer, spotter, etc.)
b. Cartridges (signalling, illuminating, distress, riot, smoke, photoflash, infrared flare etc.)
c. Grenades (smoke, illuminating, incendiary, signalling, riot, etc.)
d. Bombs (smoke, illuminating, riot, photoflash, incendiary, etc.)
e. Shells (smoke, illuminating, incendiary, riot, etc.)
f. Container pots (smoke)

### 28.2.2 AS PER SERVICE USE

This is a classification based on service use.

a. Service ammunition (used in combat)
b. Training/practice ammunition

DOI: 10.1201/9781003093404-28

    c. Drill or dummy ammunition (inert ammunition for training, handling and loading)
    d. Blank ammunition (ceremonial saluting purpose without projectile)

### 28.2.3  As Per Tactical Use/Type of Functioning

This classification is as per overall performance. A few examples are:

    a. Screening smoke ammunition
    b. Signalling colour smoke
    c. infrared smoke ammunition
    d. Illuminating ammunition
    e. Infrared illuminating ammunition
    f. Infrared flare decoy
    g. Incendiary ammunition
    h. Photoflash ammunition
    i. Distress signalling ammunition
    j. Signalling flare ammunition
    k. Riot devices/ammunition
    l. Training/practice ammunition
    m. Simulating ammunition
    n. Miscellaneous ammunition

Since the end use or end effect is most important, this classification is used predominantly while discussing any pyrotechnic ammunition. Therefore, if pyrotechnic ammunition produces illumination on functioning; it is classified as illuminating ammunition. If pyrotechnic ammunition produces screening smoke on functioning, it is classified as screening smoke ammunition.

## 28.3  DISTINCT AMMUNITION NOMENCLATURE

Most pyrotechnic ammunitions are generally recognised by their distinct nomenclature. They are categorised as per ammunition or device shape, bore dimensions and type of functioning.

For example, nomenclature *"Mortar bomb 81-mm illuminating ammunition"* signifies that the bomb is the shape, 81-mm is the bore diameter and illumination is the special effect.

Similarly, *"Shell 105-mm BE smoke ammunition"* where shell is the shape, 105 mm is the bore diameter of the shell, BE signifies base ejection and screening smoke is the special effect.

The nomenclature is changed if any modification is carried out in the overall ammunition system, thereby having a better performance than the earlier version. An example is modification of *"Bomb 81-mm illuminating Mk 1A"* to *"Bomb 81-mm illuminating Mk 2A"* where range has been increased substantially with a lot of modifications. Some ammunition includes fuze nomenclature as well. This nomenclature is used as standard nomenclature of most of the pyrotechnic ammunitions.

Modifications in ammunition design in some countries are shown by letter "M" for army and "Mk" for the navy followed by model number. Modifications in some countries are also shown by letter "A" or "Mod" for army and navy, respectively. Common pyrotechnic ammunitions for army and navy are shown as "AN" followed by model number. Examples are AN-M9A1 and AN-Mk5 Mod 3 See more details in section 26.3.

## 28.4  MIXED NOMENCLATURE

Many distress signalling, training, simulating, riot, infrared flare decoys and mis-cellaneous ammunitions do not follow a fixed pattern of nomenclature and have their own nomenclature containing a mix of some of the special effects, type, shape and size, etc.

Though nomenclature of some ammunitions are given in Chapters 30 to 39, a list of few are given for ready reference in Table 28.1.

**TABLE 28.1**

**Nomenclature of Some of Pyrotechnic Ammunitions**

| | |
|---|---|
| Illuminating Ammunitions | Mortar Bomb 51 mm Illuminating |
| | Mortar Bomb 81 mm Illuminating |
| | Mortar Bomb 82 mm Illuminating |
| | Mortar Bomb 120 mm Illuminating |
| | Mortar Bomb 107 mm Illuminating |
| | Bomb 120 mm: Illuminating, M91 with Fuze, Mechanical Time Superquick: M776 |
| | Bomb 120 mm: Illuminating M930 with Fuze, Mechanical Time Superquick: M776 |
| | Shell 84 mm Illuminating |
| | Shell 105 mm Illuminating |
| | Shell 155 mm Illuminating |
| Photoflash Ammunitions | AN-M46 Photoflash Bomb |
| | Bomb photoflash 100 Pound, M122 |
| | Bombs, Photoflash: 150 Pound, M120 and M120A1 |
| | Cartridge, Photoflash: M112A1, 1-, 2- and 4-Second Delay |
| Smoke Ammunitions | 36mm Grenade Smoke |
| | Mortar Bomb 51 mm Smoke (Hexachloroethane based) |
| | Mortar Bomb 60 mm Smoke (Plasticised White Phosphorous based) |
| | Mortar Bomb 60 mm Smoke (Hexachloroethane based) |
| | Mortar Bomb 60 mm Smoke (Red Phosphorous based) |
| | Mortar Bomb 81 mm Smoke (Plasticised White Phosphorous based) |
| | Mortar Bomb 81 mm Smoke (Hexachloroethane based) |
| | Mortar Bomb 81 mm Smoke (Red Phosphorous based) |
| | Mortar Bomb 81 mm Smoke (TTCl based) |

*(Continued)*

**TABLE 28.1 (Continued)**
**Nomenclature of Some of Pyrotechnic Ammunitions**

|  |  |
|---|---|
|  | Mortar bomb 82 mm Smoke (Plasticised White Phosphorous based) |
|  | Mortar Bomb 107 mm Smoke (Plasticised White Phosphorous based) |
|  | Mortar Bomb 120 mm Smoke (Plasticised White Phosphorous based) |
|  | Mortar Bomb 120 mm Smoke (Hexachloroethane based) |
|  | Mortar Bomb 120 mm Smoke (TTCl based) |
|  | Shell 105 mm BE Smoke (Hexachloroethane based) |
|  | Shell 105 mm BE Smoke (Red Phosphorous based) |
|  | Shell 155 mm smoke (Red phosphorous based) |
| Infrared Smoke Ammunitions | Smoke FFV 007 IR |
|  | 120 mm Mortar Infra-Red Smoke |
|  | 81 mm AT-Al smoke Grenade |
| Signalling Ammunitions | M18 Coloured Smoke Yellow Hand Grenade |
|  | Cartg. Signal 16 mm Red/Green/White |
|  | Cartg. Signal 1" Red/Green/Illuminating |
|  | Cartg Signal 38 mm Red/Green/Yellow |
|  | Mortar Bomb 51 mm Red /Green |
|  | Mortar Bomb 81 mm Red/Green |
|  | 105 mm Base Ejection Colour Smoke Red/Orange/Blue |
| Incendiary Ammunitions | Cartridge, Caliber 0.30, Armor Piercing Incendiary, M14 |
|  | Cartridge, Caliber 0.50, Incendiary M1 |
|  | Cartridge, Caliber 0.50, Ball, Armor Piercing Incendiary, M8 |
|  | Cartridge 0.50 Armour Piercing Incendiary M23 |
|  | Cartridge 23 mm Ghasha API |
|  | Round 30 mm BMP-II HEI |
|  | 60 mm Incendiary Mortar Bomb |
|  | 82 mm Incendiary Mortar Bomb |
|  | 120 mm incendiary Mortar Bomb |
|  | 105 mm Incendiary Shell |
|  | 152 mm Incendiary Shell |
|  | 155 mm Incendiary Shell |
|  | AN-M14 TH3 Incendiary Hand Grenade |
|  | NR 12 Incendiary Hand Grenade |
| Riot Ammunitions | Cartridge 40 mm Non–Lethal M 1006 |
|  | Cartridge 40 mm Crowd Dispersal M1029 |
|  | M 47 CS Riot Control Hand Grenade |
|  | ABC-M7A2 and M7A3 Riot-Control Hand Grenades |
|  | Bomb, 4.2-Inch: Tactical CS, M630 |
|  | 64mm M742 CS riot-control projectile |
|  | Cartridge, 105 mm: Tactical CS, M629 |
|  | 155mm Tactical CS XM631 |

**TABLE 28.1 (Continued)**
**Nomenclature of Some of Pyrotechnic Ammunitions**

|  |  |
|---|---|
|  | Rubber Bursting CS Hand Grenade |
| Distress Signalling Device | Signal distress day/night |
|  | Hand flare red |
|  | Para flare red |
|  | Buoyant smoke orange |
|  | Fuzee, warning, railroad: Red, M72 |
|  | Signal Fog |
|  | Marker Man Overboard |
|  | Signal Kits, Personnel, Distress Red M185 and Various Colours, M186 |
|  | Signal, Smoke and Illumination, Marine: AN-MK13, MOD 0 |
|  | Signal, Illumination, Marine: Two-Star, Red, AN-M75 |
| Training/Practice Ammunitions | Bomb 3 kg Practice |
|  | Training Smoke Generator |
|  | M69 Practice Hand Grenade |
|  | Cartridge, Photoflash: Practice, M124 |
|  | Cartridge, Photoflash: Practice, M121 |
|  | Cartridge, Caliber 0.5O, Ball, Plastic Practice, M858 |
|  | Cartridge, 14.5 MM, Trainer-Spotter, M181A1 |
|  | Cartridge, 14.5 MM, Trainer-Spotter, M182A1 |
|  | Cartridge, Photoflash: Practice, M121 |
|  | Practice Bomb Signal Cartridges |
| Simulating Ammunitions | Thunder Flash Mk-4 |
|  | Bicat Strips |
|  | Grenade Hand 90 Mk-3 |
|  | Simulator Gun Flash No.3 Mk1 |
|  | Simulators, Explosive Boobytrap: Flash, M117; Illuminating. M118; Whistling, Ml19 |
|  | Simulator Flash Artillery M110 |
|  | Simulator, Projectile, Air Burst: M74A1 and M74 |
|  | Simulator, Projectile Ground Burst M115A2 |
|  | Simulator, Hand Grenade: M116A1 |
|  | Detonation, Simulator, Explosive: M80 |
|  | Cartridge SA 7.62 mm Blank L.A. |
|  | Cartridge 105 mm blank M395 |
| Miscellaneous Ammunitions and Devices | Flare Trip Wire (various marks) |
|  | Flare Ground Indicating No. Mk-1 Yellow |
| Infrared Illuminating Ammunition | Bomb 60 mm Illuminating IR M767 |
|  | Shell 40 mm M992 Infrared Para (BA03) |
|  | Bomb 81 mm Cartridge Illumination, Infrared M816 |
|  | Shell 105 mm Infrared Illuminating M1064 |

*(Continued)*

**TABLE 28.1 (Continued)**
**Nomenclature of Some of Pyrotechnic Ammunitions**

|                |                                              |
| -------------- | -------------------------------------------- |
|                | Bomb 120 mm Cartridge Infrared M983          |
|                | Shell 155 mm Infrared Illuminating XM1066    |
| Infrared Decoys| Flare countermeasure 118 Mk 3 Type 1         |
|                | Flare countermeasure 218 Mk 3 Type 1         |
|                | Cartridge countermeasure 55mm Typhoon IR Decoy |
|                | Cartridge countermeasure 26 mm               |
|                | Cartridge countermeasure 50 mm               |

## 28.5 VARIATION IN CLASSIFICATION

It may be noted that there are several compositions, filled components and pyrotechnic ammunitions which are known by other names as well. Examples are

a. *Compositions:*
   i.   Priming compositions are elsewhere also referred as ignition composition, quick composition, ignition charge, first fire charge or starter mixture.
   ii.  Initiatory compositions are elsewhere also referred as priming mixtures.
   iii. Booster compositions are elsewhere also referred as intermediate composition or transitional composition or intermediate fire.

b. *Filled components with different classification:*
   i.   Percussion caps are elsewhere also referred as percussion primers.
   ii.  Electrical squibs are elsewhere also referred as electric igniter or fuse head.
   iii. Primers are elsewhere also referred as igniters. Primers in this book refers to filled components used to ignite propellants in conventional mortar and artillery ammunitions.
   iv.  Igniters in this book refers to filled component meant for ignition of solid propellants in missiles and rockets.
   v.   Flares and candles are used without much differentiation.

c. *Similar sound classification:* Some filled components and devices are having similar sound classification like fuse, fuze and fuzee, each a different entity.

d. *Pyrotechnic Ammunition and Device:* Sometimes pyrotechnic ammunitions in literature are referred to as pyrotechnic device or pyrotechnic system or pyrotechnic store. It appears logical to name it ammunition if used for military combat operations while as devices if used in non-combat operations like those of distress signalling devices used by navy, merchant navy or railways. and riot devices used by police.

# 29 Proof of Pyrotechnic Ammunitions

## 29.1 GENERAL

It is very difficult to design and manufacture a *zero-defect* ammunition which would never show any dismal performance. This is due to the fact that any unintentional minor change in the raw material and/or process or proof or storage conditions may result in non-conformance of the desired parameters of the ammunition. Hence, it is not desirable to simply manufacture the ammunition and send the same to users without further checking the performance. Such performance checking or tests is generally termed as proof. It ensures that the ammunition is functional and safe for the users. It may be mentioned that defects in ordnance testing/proof is mostly of mechanical nature but defects in pyrotechnic ammunition testing/proof are both mechanical and chemical in nature.

It is a destructive process. Proof of ammunition establishes the standard or quality of the ammunition. Proof is considered a necessity to ensure that the ammunition and its components shall perform in the desired manner. The samples are taken from the bulk quantity, known as *"Lot." Lot is a quantified bulk of ammunition having same size, shape, composition and functioning characteristics manufactured under identical conditions*. The necessity of proof may be understood as follows.

a. It provides assessment with respect to safety, reliability and effectiveness.
b. It shows the effectiveness of quality management of manufacturer for consistent production of ammunition.
c. The defects, if any, are noticed and corrective remedial measures are taken by the manufacturer of the lots. This remedial measure is implemented in subsequent future lots also. This avoids loss to the manufacturer .
d. It reveals that all mechanical and chemical components are functioning satisfactorily. The compositions used are neither too sensitive to cause premature or too insensitive to cause non-functioning.
e. It ensures that the ammunition is safe for transportation, loading and firing.
f. It reveals that the cumulative effects of various dimensions and material properties are not affecting ammunition performance.
g. The serviceability can be determined only through proof.
h. Shelf life can be extended or otherwise only through proof.
i. The performance of ammunition for consistency, can be determined only through proof in a proof range.
j. The ammunition is required to function satisfactorily at extreme ambient temperatures (both cold and hot temperatures) and such requirements are possible to

DOI: 10.1201/9781003093404-29

verify by proof firing the ammunition after subjecting the ammunition to such temperatures.

k. Proof of empty and filled components are required before they are accepted for assembly in the ammunition to avoid any subsequent failure of ammunition due to the empty component.
l. No unserviceable ammunition is issued to the user.

During such proof firing, pyrotechnic ammunitions are required to meet several requirements (Section 23.1) and must exhibit *safety, reliability and consistency* with environmental concern in performance. But occasional defects in performance are displayed by pyrotechnic ammunitions.

Automation in manufacturing technology, improved instrumentation for proof and testing, use of non-hygroscopic and non-toxic ingredients, improved safety in process have progressed in leaps and bounds over the past few decades, providing improved version of ammunition. There are generally two types of proof (apart from laboratory tests).

a. *Static Proof*: The filled components or ammunition are held static at static proof range and its functioning is observed. Examples of filled components are caps, primers, igniters, squibs, tracers, delays, illuminating candles, infrared flares. Examples of ammunitions are smoke pots, signal fuzee, thunder flash, fog signals, flare trip wire etc.
b. *Dynamic Proof*: The ammunition or its components are dynamically fired at dynamic proof range either through hand or from the firing equipment like hand-held projector, pistols, mortars, howitzers, guns and dispensers (infrared decoys) for functioning of ammunition as well as effectiveness of ammunition design to sustain stress from proof firing In dynamic proof testing, the ammunition is exposed to all the stresses of setback, set forward, spin and temperature and pressure etc. which otherwise is not possible in static mode. Also, some ammunitions are randomly (say one in ten lots) proof tested at simulated high and low temperatures by exposing the ammunition to such temperatures for stipulated time. This ensures that the ammunition shall perform at extreme temperatures. However, some ammunition, like signalling ammunition 16 mm red/green/white and 38 mm red/green/white, having small ranges may be fired in static proof range, if the range permits.

Ammunition performance are observed or measured, depending upon the type of ammunition, through various performance measuring equipment. The ammunition during design stage is also subjected to some of the environmental tests and mechanical handling tests/proof in static and dynamic proof ranges (Section 23.5 and 23.6). *The acceptance in proof of an ammunition implies acceptance of its dimensions, functioning, reliability, interchangeability and fitment with firing equipment.*

## 29.2   TEST/PROOF OF COMPOSITION AND COMPONENTS

There are several compositions filled in components in each ammunition. Each of these composition batches and filled component lots are required to be tested for quality of performance. Some of these tests are required to be done in laboratory and or static proof or dynamic proof range. Adequate care should be taken during these tests. The performance assessment of the composition, empty components sub-assemblies and filled components is accomplished as given below to avoid any failure of the ammunition due to the composition and components and also avoids loss of material and manpower.

a. *Composition Testing*:

The compositions are tested in a laboratory for specified composition ingredient percentage, moisture/volatile content, rise time of pressure, peak pressure, etc. as specified for compositions except such compositions which use binders which require quick utilisation of the composition. Most of the compositions are required to be filled/pressed in components and tested for quality of performance like burn time, luminosity, delay time, pellet crush load, smoke quality and density, sound, etc. at static proof range before bulk filling to avoid rejection of filled components.

b. *Empty Component Test/Proof*:

There are several small components used in the ammunition. Some of these components are required to be tested for performance acceptance. Specified number samples of these components are filled as empty filled (i.e., before bulk filling of the entire hardware lot with explosive composition) and tested in static or dynamic proof, as specified, so as to ensure that the empty components are satisfactory. These empty components include, amongst others

   i. Primer empty
   ii. Cartridge case empty
   iii. Fuze empty
   iv. Tail unit empty etc.

If the results are found satisfactory, the lot of empty hardware components is accepted for filling/assembly. However, some component test/proof may be waived based on consistent satisfactory proof performance or subjected to prove as one in ten lots.

Some hardware components like empty caps, empty detonators, empty tracers and empty delays are not subjected to empty hardware test.

c. *Filled Component Test/Proof*:

Empty components used are mostly taken for filling only after these pass tests/proof as empty components referred above. The filled components are a part of the ammunition system and to ensure that the ammunition performs satisfactorily, it is necessary that each filled component performs satisfactorily. These filled components are proved in static and or dynamic proof. Filled components requiring test/proof include, amongst others

     i. Filled caps (static)
    ii. Filled detonators (static)
   iii. Filled primer (static and dynamic)
   iv. Filled primary cartridge (static and dynamic)
    v. Filled tracer (static and dynamic)
   vi. Filled fuze (static and dynamic)
  vii. Filled delays (static)
 viii. Filled squibs (static)
   ix. Filled igniter(static)

For example, testing of filled delays, illuminating candles/infrared candles and tracers is given below.

i. The filled delay is tested in static mode. The delay is fixed in a device and initiated through squib. The functioning and delay time are recorded manually. In case of delay of lower values, the same is tested through an electronic timer device (see Chapter 5).

ii. Some of the tests are done in *simulated dynamic mode*. For example, the luminosity of the candle (or *infrared radiation of infrared flare pellet*) *and burning time* is done in tunnel where the equipment is having air exhaust which provide a current of air simulating the one encountered by a candle or flare descending in the air and hence give the burn time and luminosity/infrared radiation almost similar to the one expected in actual field tests (see Chapter 5).

iii. Similarly, tracers are proved in static mode with *simulation of spinning*. The tracers are made to spin (at approximately same rpm at which the projectile spins during firing) by placing them on spin equipment. The tracer is ignited by squib and the tracer burn time is observed while the tracer is under spin. Such values give almost correct value of tracer burning time in dynamic mode. It could be seen that as the spin increases, the tracer burning time gets reduced. If such facilities are not available, then the static burn time for tracer should be maintained much more than actual dynamic burn time (see Chapter 5).

If the results are found satisfactory, the lot of filled components are accepted for ammunition lot assembly.

## 29.3 DEPLOYMENT AND FUNCTIONING OF AMMUNITIONS

Before discussing proof of ammunition, let us understand the varieties of deployment and functioning of military pyrotechnic ammunitions Pyrotechnic ammunitions are fired through various means (hand, pistols, revolvers, mortars, guns, howitzers, aircrafts) at various locations (ground, hilly terrain, snow-capped mountains, air and over the sea as well as under the sea) and under various environmental conditions.

A few examples are as under.

a. *The ammunition which are fired statically on ground without any weapon*:
Examples are smoke generators producing smoke, signal fog producing
sound, flare ground indicating and flare trip wire producing flame and bicat
strips as simulating ammunitions

b. *The ammunition which are hand-held (without weapon system) during firing*:
Examples are signal fuzee (giving red flame), hand flare red (giving red
flame), Para flare red (ejecting flare with parachute), signal distress day and
night (giving red flame and orange smoke), etc.

c. *The ammunition which are fired through hand-held weapons or thrown by
hand*:
Examples are 16 mm signal cartridge, 1" signal cartridge and 38 mm signal
cartridge (all giving coloured flame) and fired through hand-held devices like
projector/revolver, etc. Example of hand-thrown ammunition is smoke grenades (coloured smoke and screening smoke).

d. *The ammunitions which are fired over water/sea*:
Examples are buoyant smoke orange (for distress signalling with orange
smoke), marker smoke white (for screening purpose), marker man overboard
(distress signalling giving orange smoke and light), etc.

e. *The ammunition which are fired under water/sea*:
Examples are cartridge smoke ejector signal (CSES) (ejects coloured flares and
coloured smoke) and squib and flare for torpedo (ejects twin red flares in air), etc.

f. *The ammunition which are fired through a mortar or gun or howitzer*:
Examples are illuminating mortar bomb, smoke mortar bombs, illuminating
shells, incendiary shells and smoke shells, etc.

g. *The ammunition flare pellets which are ejected from dispenser of the aircraft*:
Examples are cartridges signal flares, cartridge infrared flare decoys

h. *The ammunition which are dropped from aircraft*:
Examples are photoflash bombs, incendiary bombs and parachute flares etc.

i. *Ammunitions which are fired from ship*:
Examples are illuminating ammunitions, smoke ammunitions, etc.

The general deployments of pyrotechnic ammunitions with some examples of
ammunition are given in Figure 29.1.

These pyrotechnic ammunitions generally function as under.

a. *Bursting*: Examples are incendiary, photoflash, grenades, simulator projectile
air burst, simulator projectile ground burst and some riot devices.

b. *Emission*: Examples are smoke emission through pots like generator smoke
white, generator smoke orange, buoyant smoke orange and coloured light
emission from hand flares, signal fusee, flare trip wires, etc.

c. *Ejection cum emission*: Examples are ejection and light emission from stars of
16 mm red/green/white signals, screening smoke from canisters of 105 mm
BE smoke, colour smoke from canisters of 105 mm BE colour smoke, illumination from candles of 105 mm illuminating and 120 mm illuminating.

d. *Ejection from dispenser of aircraft*: Examples are infrared decoys emitting
infrared emission

e. *Penetration cum emission*: Examples are some fin-stabilised riot devices

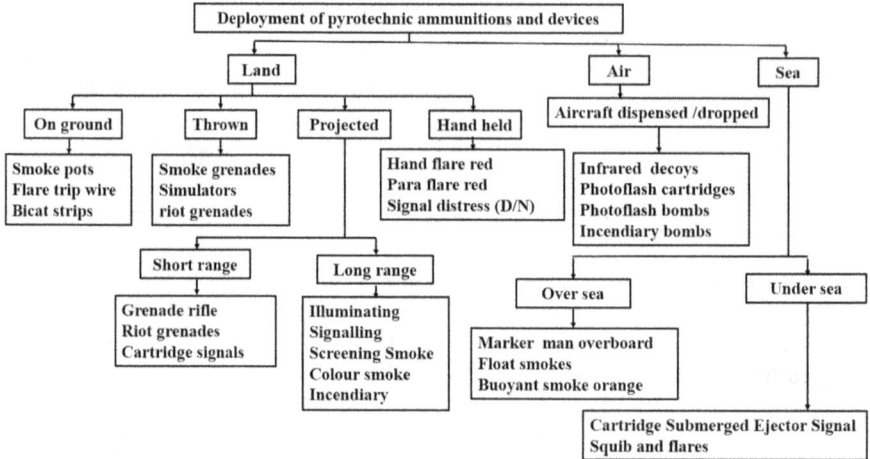

**FIGURE 29.1** Deployment of Pyrotechnic Ammunitions.

Ammunitions fired in ascending order of range are through small arms, mortar, howitzer or gun/cannon. The weights of the weapons increase in the order.

a. *Small Arms*: This includes pistols, revolvers, rifles, machine guns and sub-machine guns
b. *Mortar*: Low velocity, high angle, indirect fire. Use fin stabilised ammunition which are muzzle loaded. They are smooth bored and light. weight. These do not have recoil system like howitzers and gun/cannon. Also, these are the only muzzle loading weapons.
c. *Howitzer*: Medium velocity, medium angle up to 65 degree, direct or indirect fire. They are towed or self-propelled and heavy duty. Ammunitions are breach loaded and barrels have rifling to provide spin stability to projectile.
d. *Gun/Cannon*: High velocity, low angle generally up to 20 degrees, long range, direct fire. They are similar to howitzers with long rifled barrels.

## 29.4  PARAMETERS FOR BULK-FILLED LOT PROOF

The bulk-filled lot of the ammunition is required to assess its performance for acceptance. These are proved at static and or dynamic proof ranges, including tunnels (some small arms ammunitions) as per specification. The parameters for tests depend upon type of ammunition. These are tested for overall functioning capability of the ammunition. All these tests require suitable instruments at the proof ranges. These tests may include, amongst others

a. Velocity
b. Range
c. Accuracy
d. Burn time (illuminating candle, smoke candle, smoke pot, tracer, delay, infrared flare, signal flare etc.)

 e. smoke quality (colour concentration like dense or rare)

 f. Flare colour quality

 g. Height of burst

 h. Luminosity (star or candle)

 i. Infrared flare output

 j. Flame length

 k. Extreme temperature (high and low) functioning etc.

## 29.5   TYPES OF FILLED LOT PROOF

The filled lot proofs are of the following types though the nomenclature of each type of proof may vary among various countries.

 a. *Developmental Lot Proof*: The proof samples of developmental ammunition lot are proved during design-development stage. Developmental stage of ammunition requires a series of proofs and based on results, further modifications in the ammunition design is done. It ensures that the design is safe, reliable and consistent for pilot lot manufacture.

 b. *Pilot Lot Proof/Users Proof*: Once developmental trials are satisfactory; a short *pilot lot* is made. Higher proof samples are taken for proof. When pilot lot proof/users' proof is satisfactory, the design (drawing, specification, proof schedule etc.) are sealed for *regular out turn lots* and sealed particulars are provided to manufacturer by the designing agency.

 c. *Regular Out Turn Lot Proof*: The proof samples of regular ammunition lots are taken (as specified in proof schedule) from the premises of manufacturer for proof and sentence the lot based on proof performance as accepted, re-proof or rejected. The regular out turn proof is done in static and or dynamic proof range.

 d. *Periodic Proof*: Proof for ammunitions in depots (for ammunitions improperly stored due to rough handling, inadequate storage conditions at depots, corrosion/rust formation due to seepage of moisture, etc.) for assessment for further shelf life of the ammunition.

 e. *Check Proof*: The samples of accepted ammunition lot stored at depot and sent by depot for assessing the performance of outlived ammunition lots. In general, to economically utilise ammunitions, these samples are sent a year prior to shelf-life year and then after 5 years and subsequently every year depending upon the condition of the lot.

## 29.6   DEFECT CLASSIFICATION

The defects of the ammunition are noted during development stages. Some of these are anticipated. The designer, therefore, includes some defects which are likely to be observed during proof despite all precautions. These defects are classified in three groups, each having common factors. Thus, a group with risk hazard is termed as *critical defect*. Another with functional failures as *major defect* while one with not having any functional effect on functioning of ammunition as *minor defect*.

Proof schedule provides the specification, methods of proof, proof equipment and instruments required, lot size, proof sample size, acceptance quality level and defect classification level for acceptance of the empty and filled components as well as complete ammunition lot.

### 29.6.1  CRITICAL DEFECTS

A defect that is likely to result in hazardous or unsafe conditions for individuals using, maintaining or depending on the item, or a defect that is likely to cause the destruction of and/or serious damage to the weapon or launcher under normal training or combat conditions. The critical defects are

   a. *Premature*: When the ammunition functions before the desired place and time and which is likely to affect the firing crew and or the weapon system or civilian population.
   b. *Lodging/break up/bursting of shell in barrel*: This is likely to affect firing crew and weapon system

Any critical defect entails the rejection of the entire lot.

### 29.6.2  MAJOR DEFECTS

A defect other than critical that is likely to result in failure in tactical use or which precludes or reduces materially the usability of the item for its intended use. These are performance defects. Some typical major defects of various ammunitions are:

   a. Misfire
   b. Blind
   c. Star or candle burst in air
   d. Pierced cap or cap blown out
   e. Low burning time, low luminosity
   f. Quality of screening smoke not good for obscuration (Low dense smoke)
   g. Colour of smoke not dense
   h. Non ejection of pay load
   i. Infrared output and or burn time not within specified limits
   j. Delay time low or high
   k. Weak sound
   l. Failure of ammunition components like cartridge case cracking, tail unit crack, parachute failure
   m. Range and standard deviation in range etc.

### 29.6.3  MINOR DEFECTS

A defect other than critical or major that is not likely to result in failure during use or reduce the intended use of the item, but which should be corrected in future lots. These are defects which do not hamper performance of the ammunition but continued failure needs to be reviewed and suitable measures to be taken. Some typical minor defects are

    a. Rate of descent of parachute, partial opening of parachute, variation in height of deployment of parachute, burning on ground after satisfactory performance in air for some illuminating ammunitions

    b. Initial delay in smoke emission in smoke ammunition

    c. Cartridge hard extraction, i.e., cartridge slightly bulge after firing

    d. Function of cap after two hits only

    e. Marking defects

    f. Ammunition box gaps, stenciling defects, rust, broken hinges etc.

The performance of each type of pyrotechnic ammunition is different since it depends upon the design of the ammunition and its components. Accordingly, separate acceptance criteria are made for each type of ammunition. Empty and filled components and ammunition lots may be sentenced after conduct of proof, based on proof schedule of ammunition, as under.

    a. *Serviceable*: Accepted for next stage of assembly (for empty component and filled component proof) or accepted bulk ammunition lot for issue to depots.

    b. *Reproof*: Proof to be repeated with same sample size as first proof.

    c. *Rejected*: Reject the empty component lot or filled component lot or ammunition lot.

However, a different method of defect classification has been adopted [1] where defect codes are assigned to identify and clarify the serviceability of ammunition. Defect codes are composed of six alpha-numeric characters: First character is the percent defective indicator; the second and third characters identify the type of assembly; fourth character indicates the classification of defect; and the fifth and sixth characters are the defect narrative. Readers may refer to [1] for details.

## 29.7  LOT SIZE AND SAMPLING PLAN

The ammunition quantity is made into lots depending upon the type of ammunition and the confidence of the manufacturer in meeting the desired specification of the ammunition. This is reflected in specification and proof schedule of the ammunition. Generally small arms ammunition lot sizes are big while heavy calibre ammunition lot sizes are smaller.

    During the design and development stage, the lot size is made small. When the developmental design is finalised for bulk production or when bulk production of new ammunition is made based on *TOT* (*Transfer of Technology*), the lot is termed as *"Pilot Lot"* and subjected to *users' proof*. The subsequent lots after clearance of *"pilot lot"* are termed as *Regular Outturn Lots*. The proof sample size for the pilot and bulk lots may vary; the former has bigger sample size.

    Ammunition proof is a destructive test designed to assess the desired/expected performance of components, sub-assemblies or complete ammunition. Since this is a destructive test, this is performed on sampling basis for which sampling plan (single or double sampling plan) has been evolved.

  a. The lot size for *single sample plan* is taken as the *lot size + proof sample*.
  b. The lot size for d*ouble sample plan* is taken as the *lot size + first proof sample + second proof sample*.

Samples are drawn as per sampling plan. The lot is spread out and samples are taken at random. Each sample of the lot is marked or assigned a number as 1P, 2P, 3P, etc. to nP where n stands for proof sample size. *Random selection of proof samples means that any sample in the lot shall have a chance of being selected, no matter the location and content of the material in the sample. It is independent of the samples already taken or yet to be sampled.*

In a single sampling plan, there is no provision of reproof. Lots are to be accepted based on performance of single sample. Most of the ammunition proof are based on double sampling plan.

Sampling plan is described by sample size (nos.) and acceptance (nos.) and rejection (nos.). When the rejection (nos.) are equal or less than the acceptance (nos.) specified, the lot is accepted. If the rejection (nos.) are equal or more than the rejected (nos.) specified, the lot is sentenced rejected. If the rejected (nos.) are more than accepted (nos.) but less than rejected (nos.) specified, the lot is sentenced reproof.

A typical double sampling plan and acceptance criteria is given in Table 29.1.

Let us consider major defects of an ammunition lot with lot size of 1,026. If no major defects are observed in the first proof sample (qty. 13), the lot is sentenced serviceable (since acceptance criteria is shown as (0) zero). However, if 2 or more major defects are observed, the lot is sentenced rejected. However, if 1 major defect is observed in the first proof sample (qty. 13), the lot is sentenced as reproof. The same numbers of samples (qty. 13) are again proved. However, if there is no further major defect, the lot is sentenced serviceable or accepted. But if 1 or more major defect is observed, taking the cumulative defects to 2 (or more) the lot is sentenced rejected since out of 26 samples, only one major defect is acceptable as per the plan.

---

**TABLE 29.1**

**Double Sampling Plan and Acceptance Criteria**

Lot size (1,000 + 26 proof)

| Sampling plan | | | Major defects (AQL2.5%) | | Minor defects (AQL 6.5%) | |
|---|---|---|---|---|---|---|
| Sample | Sample size | Cumulative sample size | Acc | Rej. | Acc | Rej. |
| First | 13 | 13 | 0 | 2 | 1 | 3 |
| Second | 13 | 26 | 1 | 2 | 3 | 4 |

If the lot passes in first proof, the quantity issued against the ammunition lot would be 1,000 + 13, i.e., 1,013 numbers. However, if the lot passes in second proof, the lot quantity issued against the ammunition lot would be 1,000 numbers only.

Proof reports shall be prepared in prescribed format and sentenced as accepted, reproof, or rejected. In case of any problem or accident, it is sentenced as *suspended*. Many ammunitions follow *skip proof* where no critical or major defects are noticed in continuous large number of lots. In such cases, sample from one lot is taken amongst 10 lots and the 10 lots are sentenced based on performance of this one lot.

## 29.8  SOME MAJOR DEFECTS IN PROOF OF AMMUNITIONS

Every ammunition has a specified proof schedule. Each ammunition lot must conform to the proof schedule for its acceptance. Performance requirement of each type of ammunition is different and it is not within the scope of this book to include all. Hence as an example, only all-likely defects in proof of four typical type of ammunitions viz. generator smoke, signalling cartridge, 105mm BE screening smoke and 81mm illuminating bomb are given.

a. *Generator Smoke: Likely defects*: Non-ignition of composition, smoke emission time low, poor quality of smoke, flaming of composition.

b. *Signalling Cartridge Ammunitions*: Likely defects: Misfire, pierced cap, gas escape around cap periphery, cap blown out, non-pick up of ignition by star, low  burning time of star in air, burning of star on ground, hard extraction of cartridge case

c. *105 mm BE Screening Smoke Ammunition*: Likely defects: Smoke emission time low, quality of smoke not dense, non-ejection of canisters, non-pick up of ignition by canisters

d. *81 mm Illuminating Ammunition*: Likely defects: Non-ejection of candle with parachute, candle performance defects (non -pick up of ignition by candle, low burn time of candle, breaking of chunk of candle composition while burning in the air thereby giving low burn time), parachute defects (non-opening of parachute or fast descent of candle causing ground burst due to partial opening of parachute or breaking of some rigging lines of parachute or parachute cloth cut/torn during ejection or burning of rigging lines of para-chute or detachment of candle from parachute or entanglement of para with tail unit or with its own rigging lines), fuze defects (blind, ground burst, delay time), range and standard deviation

It may be stated that the ammunition manufacturing organisation with its expertise, skill, experience and robust quality management ensures that the ammunition lots conform to the specified proof schedule. However, occasional failures may crop up.

Each proof range is equipped with various firing euipments and performance measuring instruments. Static proof requires a short range. while dynamic proof requires long range and wide area as it proves the ammunition in dynamic mode and the range varies as per the ammunition. For example, with available ammunition in the ammunition industry, 51-mm ammunition would require minimum range of +2 km; an 81-mm would require a minimum range of +6 km while a 155 mm would require a minimum range of +32 km. However, the actual requirement of range is much more than the above values to allow for any wild round for safety reasons.

Readers may refer [2] regarding surveillance and in-service proof and [3] for inspection aspects of artillery ammunition design.

## REFERENCES

[1] *"Joint Conventional Ammunition Policies and Procedures"*, Joint Ordnance Commanders Group, Department of Defense, USA, October 2017.

[2] *"International Ammunition Technical Guideline, Surveillance and In-service Proof"*, IATG 07.20, Second edition, 2 January 2015.

[3] *"Inspection Aspects of Artillery Ammunition Design"*, Engineering Design Handbook, Ammunition Series 5, AMCP 706-248, Headquarters, United States Army Material Command, Washington DC, March 1966.

# 30 Illuminating Ammunitions

## 30.1 GENERAL

The role of illuminating ammunition is to illuminate the battlefield at night for target detection, recognition and identification by way of observation of troop movement and tanks or equipment. Searchlights have drawback that their position is revealed and it is not possible to have illumination if there are obstacles in front of the searchlight like tall trees or hill top. Illuminating mortar bombs and shells when fired in air illuminate the ground from above the ground and hence do not reveal the position of the gunner and assist air forces and ground forces.

All illuminating ammunitions can be divided into following categories:

  a. Grenade illuminating ammunitions
  b. Mortar bomb illuminating ammunitions
  c. Shell illuminating ammunitions
  d. Cartridge illuminating ammunitions

Composition in most of the cartridges is pressed as star while composition is pressed as candle in bombs and shells. All ammunitions functioning at high altitude need a suitable parachute attached to candle to reduce the rate of fall of candle, thereby increasing the burn time in air and consequently increase the illumination time on ground. Thus both star and candle are used as illuminants. Table 30.1 gives the differences between a star and a candle.

The burn time and illumination area of some typical illuminating ammunitions are given in Table 30.2.

The following are desired for illuminating ammunition:

  a. The illumination should be moderately high for better visibility of the target.
  b. The illumination should be near the target surface for better visibility.
  c. The burn time should be so adjusted that burning should stop before the parachute/star touches the ground.
  d. For illuminating ammunitions with parachute, the parachute should be able to open and hang in air with candle load and without pendulum effect and descend at specified rate.

## 30.2 GRENADE ILLUMINATING AMMUNITIONS [2]

Illuminating hand grenades are used primarily for illumination but can also be used for signalling and incendiary purposes against flammable targets. A typical example

DOI: 10.1201/9781003093404-30

## TABLE 30.1
## Differences between a Star and Candle

| Star | Candle |
|---|---|
| Used in small calibre ammunition | Used in medium and large calibre ammunition |
| Small size and small mass of composition | Large size and large mass of composition |
| Shorter range | Higher range |
| Designed to function at lower height | Designed to function at higher height |
| Burn from all sides | Burns linearly layer by layer in the candle |
| Lower luminosity | Higher luminosity |
| Lower burn time | Higher burn time |
| Descends in air as free fall burning star | Descends in air slowly due to parachute |

## TABLE 30.2
## Burn Time and Illumination Area of Typical Illuminating Ammunitions [1]

| Ammunition | Diameter of usable range of illumination (metre) | Burn time (seconds) |
|---|---|---|
| White Star Parachute | 450 | 36 |
| Illuminating Grenade | 200 | 25 |
| Trip Flare | 300 | 55 |
| 40 mm White Star Parachute | 150 | 15 |
| 60 mm Mortar | 800 | 25 |
| 81 mm Mortar | 1,100 | 60 |
| 120 mm Mortar | 1,500 | 50 to 60 |
| 105 mm Howitzer | 1,000 | 60 |
| 155 mm Howitzer | 2,000 | 120 |
| Air Force Drop Flare | 1,500 | 160 |
| Naval Gun Fire 5 Inch | 350 to 550 | 45 to 52 |

is *Grenade, Hand: Illuminating, Mk1* (Figure 30.1). The body is made in two pieces. The illuminating charge is pressed into the lower half of the body and covered with a layer of first fire composition. This, in turn, is covered with an igniter charge. The fuze is an integral part of the grenade. The body contains a primer and quick-match bushing. Assembled to the body of the fuze are a striker, striker spring, safety lever and safety pin with pull ring. The split end of the safety pin has an angular spread. The safety clips are not required with illuminating hand grenades.

Removal of the safety pin permits release of the safety lever. When the safety lever is released, it is forced away from the grenade body by a striker acting under

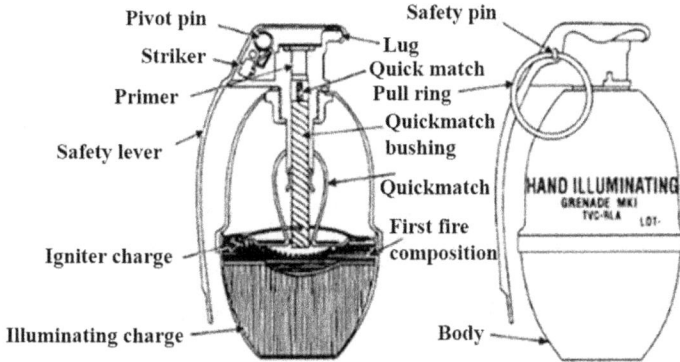

**FIGURE 30.1** Grenade, Hand: Illuminating, Mk1 (Reprinted from [2]).

the force of a striker spring. The striker rotates on its axis and strikes the percussion primer. The primer initiates the quick-match, which burns for seven seconds, and then ignites the igniter charge. The igniter charge ignites the first-fire composition which, in turn, ignites the illuminating charge. Gas pressure produced by burning of the illuminating composition causes the upper half of the grenade body to separate from the lower half. This exposes the burning illuminating charge. The grenade will burn for 25 seconds with approximately 55,000 candlepower and will illuminate an area of 200 m (656 feet) in diameter.

## 30.3 MORTAR ILLUMINATING BOMB

All mortar bombs are fin-stabilised. The illuminating ammunition is loaded into the mortar, a smooth bore muzzle loading weapon. On loading the low range mortar bombs like 51-mm illuminating mortar bombs, the ammunition has four delays of 3.5 sec., 6.0 sec., 9.5 sec. and 11.5 sec. for four ranges 300–900 m. The bomb slides down the barrel and primary cartridge at its base is fired when it hits the firing pin and ejects the bomb. After set delay, the candle with parachute is ejected to provide illumination for 30 seconds with 250,000 candela.

The details of the typical cartridge, 60 mm illuminating, M721 (actually an il-luminating mortar bomb) Figure 30.2. [3] is as follows.

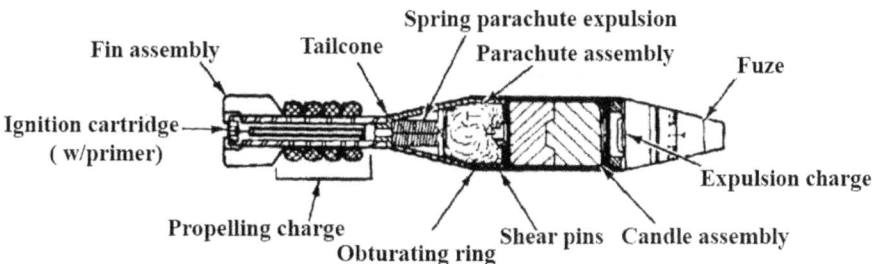

**FIGURE 30.2** Typical Cartridge, 60 mm Illuminating, M721 (Reprinted from [3]).

The cartridge has a mechanical time super quick fuze with an expulsion charge, a candle/parachute assembly of a four-increment propelling charge and an ignition cartridge. The round provides 400,000 average candlepower illumination for about 40 seconds. Loaded fin-end first into the mortar barrel, the cartridge slides down the barrel and strikes the firing pin. The ignition cartridge functions and ignites the propelling charge. Combustion gases from the ignition cartridge and propelling charges propel the cartridge out of the barrel. At a pre-set time the fuze functions in flight. The expulsion charge ignites and ejects the candle assembly. A spring ejects the parachute from the tail cone. The parachute opens, slowing the descent of the burning candle which illuminates the target.

In case of bomb 81 mm illuminating, the time fuze (mechanical set time/electronically programmed time) functions at set delay time, causing the burster bag with gunpowder to function and provide flame to the candle and also create high pressure inside the bomb. This causes breaking of shearing pins, thereby separating the lower half of the bomb body (consisting of lower body and the tail unit) and the upper body with fuze. The half cup with parachute and the candle is ejected out due to spring force. The parachute gets deployed and the candle with parachute descends with specified rate, providing illumination over the ground with 900,000 candela for 30 seconds. The following Figure 30.3 shows deployment of candle for bomb 81 mm illuminating ammunition.

In case of very long-range typical mortar bombs like 120 mm mortar bomb, the fuze (mechanical set time/electronically programmed time) function after specified set time and gives flame to ignite the delay and also opens up the assembled bomb body in air, thus ejecting the auxiliary parachute with the canister in the air. The auxiliary parachute opens up in the air holding the candle in air. After delay

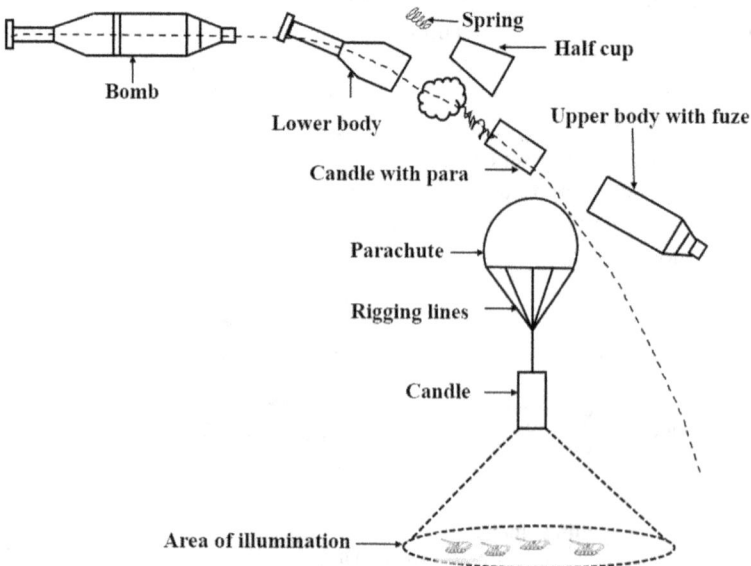

**FIGURE 30.3** Functioning of Bomb 81 mm Illuminating.

functioning, burster charge functions and main parachute gets deployed suspending illuminating candle and illuminates the area for specified time.

*Identification*: Bomb bodies for lower ranges (which do not contain fuzes) have nose caps embossed with "P" (for night identification) indicating it is with parachute, to differentiate it with smoke ammunition of similar type. The bombs are painted white with a black symbol of parachute for ease of identification. A red ring over bomb indicates that the bombs are explosive filled, i.e., live bombs. Examples are as under:

*Bomb 2" Illuminating Ammunition*
*Bomb 51mm Illuminating Ammunition*
*Bomb 52mm Illuminating Ammunition*
*Bomb 60 mm: Illuminating, M83A3, M83A2 and M83A1 [3]*
*Bomb 60 mm: Illuminating, M721 [3]*
*Bomb 81 mm: Illuminating, M301A2 and M301A1 [3]*
*Bomb 81 mm: Illuminating, M301A3 [3]*
*Bomb 81 mm: Illuminating, M853A1 [3]*
*Bomb 105mm Illuminating Ammunition*
*Bomb 107 mm Illuminating Ammunition*
*Bomb 4.2-Inch: Illuminating, M335A1 and M335 [3]*
*Bomb 120 mm: Illuminating, M91 with Fuze, Mechanical Time Superquick: M776 [3]*
*Bomb 120 mm: Illuminating M930 with Fuze, Mechanical Time Superquick: M776 [3]*
*Bomb 122 mm Illuminating Ammunition*
*Bomb 130 mm Illuminating Ammunition*
*Bomb 152 mm Illuminating Ammunition*

A typical bomb 81 mm illuminating M301A3 with fuze is shown in Figure 30.4.

## 30.4 SHELL ILLUMINATING AMMUNITIONS

All the high calibre shells are spin-stabilised ammunitions. The driving band of the shell engages with the rifling, which imparts spin as well as prevents gas leak. The shells are fired in howitzer or gun by setting the fuze time and using specified

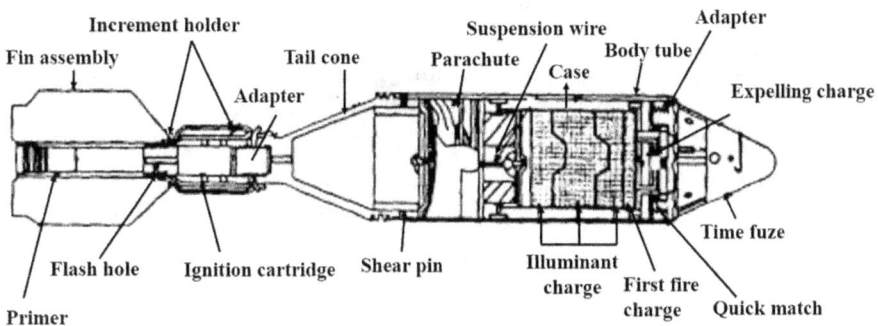

**FIGURE 30.4** Bomb 81 mm Illuminating M301A3 (Reprinted from [3]).

propellant charge mass and angle of fire to achieve the range. Shell illuminating ammunitions are assembled with fuze and contain the candle with parachute or candle with parachute inside the canister with auxiliary parachute/drogue parachute/ brake parachute. These are also known as carrier shells and carry candle or candle in canister with parachute. All artillery illuminating shells generally have following main components.

a. *Projectile*: This contains the main pyrotechnic composition in the form of pressed candle or candle in canister with parachute. Candle gets ignited on functioning of the fuze or through delay element.

b. *Candle/Canister*: Candle with parachute or canister (with candles and main parachute) with drogue/brake/auxiliary parachute are used in illuminating shells.

c. *Base bleed igniter*: Some heavy calibre ammunitions have base bleed charge and base bleed propellants to enhance the range of the ammunition, by reducing the effect of base drag on projectile during dynamic firing of projectile (Section 22.6.2).

d. *Main parachute and drogue/brake/auxiliary parachute:* The canister in heavy calibre ammunition initially hangs out with drogue/break/auxiliary parachute and after delay component functioning, ejects the candle with main parachute.

A typical 105 mm illuminating shell is shown in Figure 30.5. It has an average luminosity of 450,000 candlepower with burn time of 60 seconds.

The anti-spin flaps are provided in typical 84 mm illuminating shell so that on ejection, the anti-spin flaps open up and reduce the spin. The swivel assembly ensures that the parachute opening is not affected by spin of the candle. It provides luminosity of 650,000 candela for 30 ±5 seconds.

The 155 mm illuminating ammunition has very high velocity and spin. These illuminating ammunitions use an additional *auxiliary parachute also known as drogue parachute or brake parachute* which helps in holding the candle in a

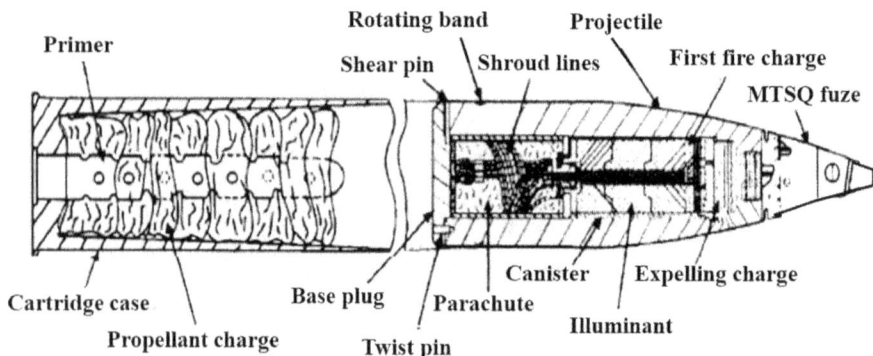

**FIGURE 30.5**   Cartridge 105 mm Illuminating M 314, M314A2, M314A2B1 (Reprinted from [3]).

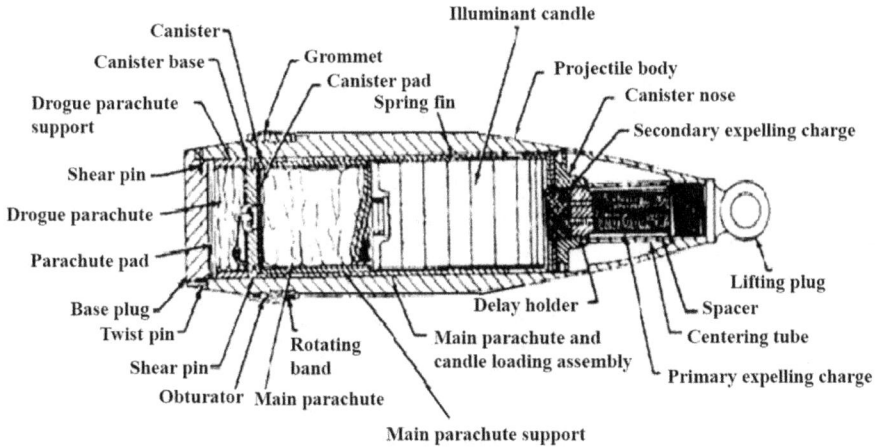

**FIGURE 30.6**  Projectile, 155 mm, Illuminating, M485 Series (Reprinted from [3]).

canister for some small period of time before the main parachute is opened fully through delay element. These parachutes are of smaller size than the main parachute and their performance is over once the main parachute is deployed. A typical 155 mm illuminating ammunition is shown in Figure 30.6.

The projectile [3] is a hollow steel shell containing an illuminant canister, a canister expelling charge in the nose, and a drogue parachute in the base. The illuminant canister contains the main parachute and lines, the illuminant candle assembly, a secondary expelling charge and a delay element holder. The outer shell of the canister is fitted with four longitudinal fins. The fins extend under spin forces when the canister is ejected from the projectile. The base of the projectile is closed with a press-fitted steel plug retained by shear and twist pins. A gilding metal rotating band and a plastic obturating band encircle the projectile near the base and are protected by a grommet during shipment and handling. The projectile uses an MT type fuze. The fuze cavity is fitted with a lifting ring plug for shipment and handling.

*Functioning* [3]: When the weapon is fired, the rotating band engages the barrel rifling to impart spin to the projectile for stability in flight. The obturator band expands to prevent leakage of gas pressure past the projectile. The burning propellant charge produces rapidly expanding gases to propel the projectile through the barrel with the velocity required to reach the desired point of function. When the fuze functions, the primary expelling charge ignites forcing the drogue parachute and canister assembly against the base plate, rupturing the base pins and expelling the canister and parachute. The drogue parachute then deploys, and the canister fins extend. These actions combine to decelerate the canister and stop rotation, the expelling charge also ignites the delay element in the canister nose, the delay element ignites the secondary expelling charge within the canister after 8 seconds when velocity has been safely reduced. The secondary expelling charge ignites the candle illuminant, and expels the main parachute and candle-loading assembly. With the main parachute and candle open, the illuminant candle descends at 15fps and 120 seconds producing 1,000,000 candlepower.

These ammunitions may be made partially as incendiary by allowing the fuze to function on the ground by setting the fuze suitably. However, failures due to soft ground target or marshy land may hamper such incendiary performance.

*Identification*: The shells are painted white with a black symbol of parachute for ease of identification. A red ring over shells indicates that the shells are filled with explosives, i.e., live. Large heavy calibre ammunition like those of 155 mm are also weight zone marked so that desired propellant charges (as per *Range Tables*) may be used to obtain desired range.

Examples are:

*Shell 84 mm Illuminating Ammunition,*
*Cartridge, 105 mm: Illuminating, M314, M314A2, M314A2B1* [3]
*Cartridge, 105 mm: Illuminating, M314A3* [3]
*Projectile, 155 mm: Illuminating, M118 Series* [3]
*Projectile, 155 mm: Illuminating, M485* [3]
*Shell 122 mm HOW Illuminating*
*Shell 155 mm Illuminating MIRA (18 km)*
*Shell 155mm Illuminating MIRA ER (24 km)*

The characteristics of some illuminating ammunitions are given in Table 30.3.

The burning time with luminosity and range of a few ammunitions [5] are given in Table 30.4.

It could be seen that varying ranges are obtained due to propelling charge and fuze setting time while burn time and luminosity is due to main illuminating composition type, mass of composition, diameter of candle and pressing load.

## 30.5  CARTRIDGE AND AIRCRAFT ILLUMINATING AMMUNITIONS

Cartridge illuminating ammunition consists of a paper or metallic cartridge. A cap is assembled at the base and some propelling composition like gunpowder is filled and a star pellet is assembled. The cartridge is turned over using mill board or metallic or plastic discs. When fired through hand-held weapon, the cap functions, providing ignition to the propelling composition and ejects the burning star. The star goes up in the air and falls through gravity, illuminating the area but extinguishes before touching the ground. Example is *Cartridge 1" Illuminating*.

Aircraft flares are released from aircraft and used for battlefield illumination, target-marking, bombardment, observation and reconnaissance. All flares have very high luminosity and are attached with a parachute for smooth descend. These are generally of two types [6].

i. Flares which are designed to fall a pre-determined distance below the launching aircraft before functioning. Examples are Aircraft Parachute Flare Mk 5 and Mods, Aircraft Parachute Flare Mk 6 and Mods, Aircraft Parachute Flare Mk 10 and Mods, and Flares, Aircraft, Parachute, M26 and AN-M26.

**TABLE 30.3**

**Characteristics of Artillery and Mortar Illuminating Ammunitions [4]**

| Ammunition | Candlepower | Burn time (s) | Rate of descent (f.s⁻¹) | Diameter of area illuminated (m) | Height of burst (m) | Range (m) | Continuous illumination (rds. per minute) |
|---|---|---|---|---|---|---|---|
| Mortars* | | | | | | | |
| 81 mm M301A2 | 5,00,000 | 12 | 12 | 1,100 | 400 | 3,300 | 2 |
| 4.2-inch M335A2 | 850,000 | 12 | 12 | 1,500 | 400 | 5,500 | 2 |
| Artillery* | | | | | | | |
| 105 mm How M314 series | 600,000 | 35 | 35 | 1,000 | 750 | 8,500 | 2 |
| 155 mm How M118 series | 500,000 | 35 | 35 | 1,000 | 750 | 11,600 | 2 |
| 1555 mm How M485 series | 1,000,000 | 15 | 15 | 2,000 | 600 | 14,000 | 1 |

*Notes*

* Reference TM 9-1300-203

**TABLE 30.4**

**Burning Time, Luminosity and Range of Illuminating Ammunitions (Compiled) [5]**

| Ammunition | Burn time (Seconds) | Minimum luminosity (lakh candela) | Range (metre) |
|---|---|---|---|
| Bomb 2" Illg. | 25–30 | 2.0 | – |
| Bomb 51 mm Illg. | 30 | 2.5 | 300–900 (as per delay setting) |
| Bomb 81 mm Illg | 30 | 9.0 | 4,800 |
| Shell 84 mm Illg | 30 ±5 | 6.5 | 300 -2,100 |
| Shell 105 mm Illg. | 25 | 7.0 | 170,00 |
| Bomb 120 mm Illg | 40 | 10.0 | 6,000 |
| Shell 155 mm Illg. (MIRA) | 47 | 21 | 1,80,00 |
| Shell 155 mm Illg. (Naschem) | 90 | 7.5 | 240,00–320,00 (with Base Bleed Unit) |

ii. Designed to function as soon as they are clear of the launching aircraft. Examples are Aircraft Parachute Flares Mk 4 and Mods, Aircraft Parachute Flares Mk 8 and Mods, and Aircraft Parachute Flares Mk 11 and Mods.

Figure 30.7 shows aircraft parachute functioning. of both types

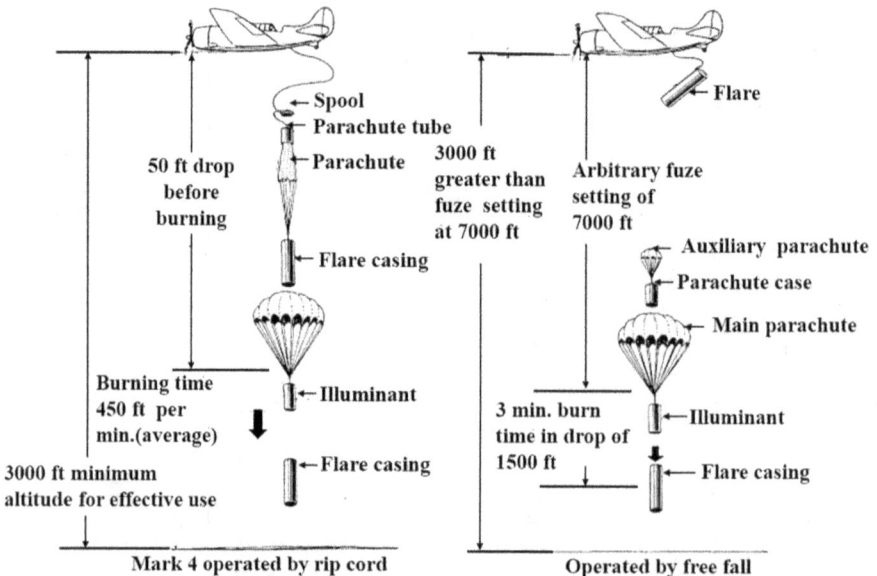

**FIGURE 30.7**   Aircraft Parachute Flare Functioning (Reprinted from [6]).

**TABLE 30.5**

**Aircraft Pyrotechnic Flare Performance (compiled) [6]**

| Pyrotechnic Item | Illumination candlepower (cp) | Burn time (minutes) | Delay before ignition (seconds) | Approximate weight (lbs) |
|---|---|---|---|---|
| **Aircraft Parachute Flares** | | | | |
| MK4 and Mods | 300,000 | 3 | None | 18 |
| MK 5 and Mods | 600,000 | 3 | Variable | 18 |
| MK 11 and Mod 0 | 1,000,000 | 3 | Variable | 30 |
| MK 8 Mod 0 and 1 | 500,000 | 3 | 90 | 18 |
| MK 8 Mod 2 | 500,000 | 3 | 120 | 18 |
| MK 10 Mod 0 | 750,000 | 4.5 | Variable | 30 |
| MK 6 and Mods | 1,000,000 | 3 | Selective | 30 |
| MK 26 and AN-M26 | 800,000 | 3 | Variable | 53 |
| High Altitudes Parachute Flare Mk 20 Mod 0 | 85,000 | 1 | 5 | 4 |
| Aircraft Parachute Flare 1.5 Minute (Electrically operated) | 110,000 | 1.5 | 1.5 | 4.25 |
| Aircraft Parachute Flare 3 Minute (Electrically operated) | 200,000 | 3 | 1.5 | 22 |
| **Float Flares (Aircraft)** | | | | |
| MK 17Mod 0 | 1,000,000 | 4.5 | 60 | 88 |
| MK 17 Mod1 | 1,000,000 | 4.5 | 330 | 88 |

Some other aircraft parachute flares and float flares (aircraft) performance are given in Table 30.5 [6].

Figure 30.8 shows typical float flare (aircraft) and aircraft parachute flare.

A typical float flare aircraft is as under.

*Float Flare (Aircraft) Mk 17 and Mods* [6]: The flare floats on the surface of the water during the period of illumination. The light produced by the Float Flare Mk 17 and Mods is yellowish white and has an intensity of approximately 1,000,000 candlepower for a period of about 4.5 minutes. A safety fuse causes a delay between launching and ignition of 60 seconds for the Float Flare Mk 17 Mod 0, and 330 seconds for the Float Flare Mk 17 Mod 1.

Two typical examples of functioning of aircraft parachute flares are given here as under.

a. *Flare, Aircraft: Parachute, M9A1* (Figure 30.9) [7]: Provides illumination for aerial reconnaissance. The firing pin of the pistol strikes the primer, the propelling charge is ignited. The charge ignites the quick match and projects

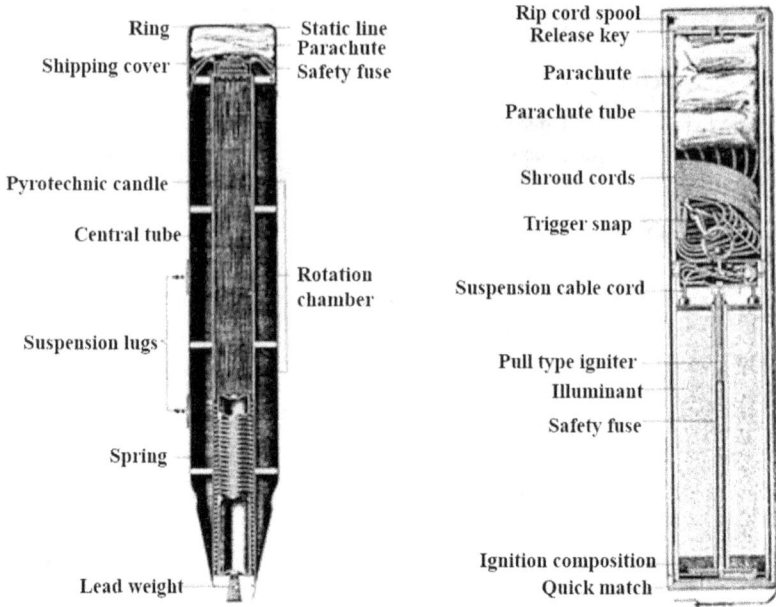

**FIGURE 30.8**  Float Flare (Aircraft) Mod 17 and Mods and Aircraft Parachute Flare Mk 8 and Mods (Reprinted from [6]).

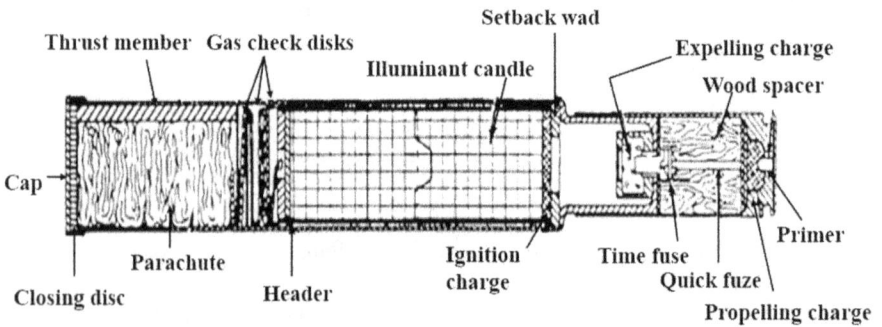

**FIGURE 30.9**  Flare, Aircraft: Parachute, M9A1 (Reprinted from [7]).

the flare case assembly containing the expelling charge, illuminate candle, and parachute some 80 feet from the aircraft. At this time, the delay fuse ignites the expelling charge, and the candle, parachute assembly is blown out through the end of the tube simultaneously: the expelling charge ignites the illuminant. The parachute opens upon leaving the case to lower the flare slowly. Delay 2.5 seconds, burning time 60–80 seconds candlepower 60,000.

  b. *Flare, Aircraft: Parachute, Mk45 Mod O and Mod O w/adapter [7]:* Restricted to launch from aircraft operating below 70 knots indicated air speed. These provide delay of 2.0 seconds min for MK45 Mod O, and 3.0

**FIGURE 30.10** A Range of Illuminating Ammunitions (Reprinted from [5]).

seconds min. for MK45 Mod O w/adapter and burning time 210 seconds nom., and candlepowder 2 million nom.).

A range of illuminating ammunition like cartridge 1", mortar bomb 51 mm, mortar bomb 81 mm, shell 84 mm, shell 105 mm, mortar bomb 120 mm and shell 1,55 mm illuminating ammunitions are shown in Figure 30.10.

Readers may refer [8] for improved illuminating flares. Details on infrared illuminating ammunitions are given in Chapter 39.

## REFERENCES

[1] https://www.globalsecurity.org/military/library/policy/army/fm/3-21-94/appb.htm
[2] "Army Ammunition Data Sheets for Grenades", Technical Manual TM 43-0001-29, Headquarters, Department of the Army, Washington DC, 30 June 1994.
[3] "Technical Manual, Army Ammunition Data Sheets, Artillery Ammunition, Guns, howitzers, Mortars, Recoilless Rifles, Grenade Launchers and Artillery Fuzes" (Federal Supply Class 1310, 1315, 1320, 1390) TM 43-0001-28, Headquarters, Department of The Army, Washington DC, 27 October 2003.
[4] "Battlefield Illumination", Field manual FM 20-60, Headquarters, Department of the Army, Washington, DC, 13 January 1970.
[5] "Product Brochure", Indian Ordnance Factories, Ordnance Factory Dehuroad, Pune, Maharashtra, India.
[6] "Aircraft Pyrotechnics and Accessories", OP 998(second revision), 29 May 1947, Navy Department, Bureau of Ordnance, Washington 25 DC.
[7] "Army Ammunition Data Sheets for Military Pyrotechnics", (Federal Supply Class 1370), Technical Manual No. 43-001-37, Headquarters, Department of the Army, Washington DC, 6 January 1994.
[8] Bernard E. Douda, "Improved Illuminating Flare", RDTR 121, July 1968, US Naval Ammunition Depot, Crane, Indiana.

# 31 Photoflash Ammunitions

## 31.1 GENERAL

These ammunitions are used in the battlefield by combustion of photoflash composition for a few milliseconds sufficiently above the ground surface, providing enough light for photographic purpose. These are generally of two types.

a. *Bomb*
b. *Cartridge*

This makes it possible to avoid very low flight by surveillance aircrafts (which is fraught with danger at low heights for possible detection and enemy fire) as they can now deploy photoflash cartridges or photoflash bombs from a higher altitude, allowing the flash cartridge or bomb to function with bright flash light after a set delay incorporated in the photoflash cartridge or bomb, at lower heights, allowing capturing the aerial photograph of the terrain. Generally, a cartridge is used at an altitude of maximum 30,000 feet while a bomb is used from 30,000 feet to 60,000 feet.

## 31.2 PHOTOFLASH BOMBS

These are mostly made of thin metal bodies with tail fins at the end to stabilise its flight and a nose fuze in the front section Some photoflash bombs are:

i. *AN-M46 Photoflash Bomb* [1].
Description. Length-451/2 inches; diameter = 8 inches; weight = 51.9 pounds. Contents = 25 pounds flashlight powder; maximum candlepower = 1,000,000,000.
The photoflash bombs shown in Figure 31.1 are filled with photoflash compositions (which are sensitive against friction and shock) and are assembled with a mechanical time flare fuze M111 which allows functioning of the bomb at 3,000 feet when released from aircraft at any altitude above 20,000 feet. On release from the plane, the fuze arms in several seconds, and then functions at the predetermined time according to the altitude of the plane. When the fuze functions, the 70-grain black powder booster in the fuze sets off the 140-grain adapter charge in the bomb. The pressure set up by the adapter charge ruptures the gilding metal disc and ignites the flashlight powder charge for aerial photography.

ii. *Bombs, Photoflash: 150-Pound, M120 and M120A1* [2]
The 150-pound photoflash bombs, M120 and M120A1, are cylindrical in shape with a short ogival nose and a box-type tail fin. The bomb is filled with photoflash powder, contains a booster charge for bursting the bomb and uses a mechanical-time

DOI: 10.1201/9781003093404-31

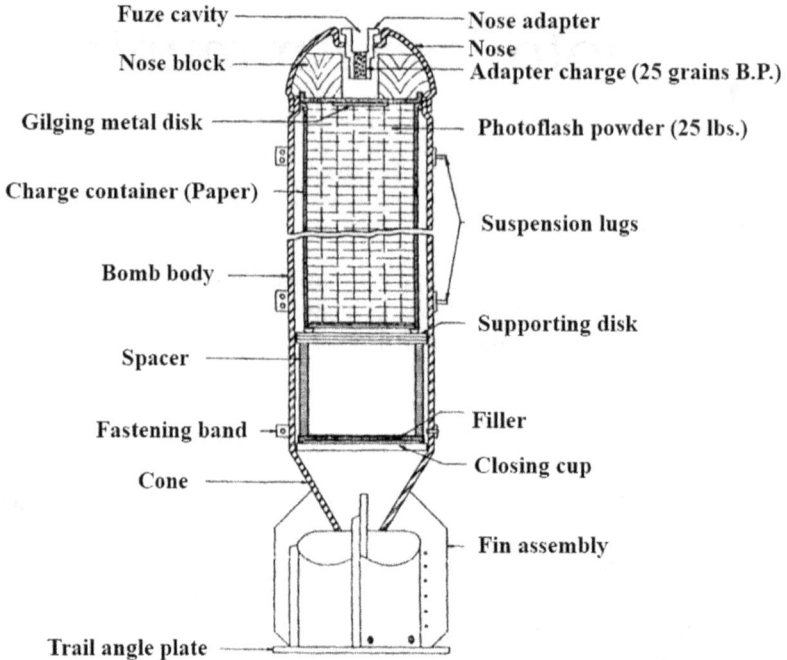

Fuze cavity — Nose adapter
Nose block — Nose
Adapter charge (25 grains B.P.)
Gilging metal disk — Photoflash powder (25 lbs.)
Charge container (Paper) —
Suspension lugs
Bomb body —
Supporting disk
Spacer —
Fastening band — Filler
Cone — Closing cup
Fin assembly
Trail angle plate —

**FIGURE 31.1**   Bomb Photoflash, AN-M46 (Reprinted from [1]).

nose fuze M907 series. The M120 and M120A1 photoflash bombs are identical in external configuration. They differ in weight of filler: the M120, 69 pounds; the M120A1, 82 pounds. The M120 photoflash bomb develops a peak intensity of 3.4 billion candlepower with an average light output of 72 million candlepower seconds. The M120A1 photoflash bomb develops a peak intensity of 4.1 billion candlepower with an average light output of 92 million candlepower-seconds.

## 31.3  PHOTOFLASH CARTRIDGES

The photoflash cartridge contains a charge case containing photoflash composition and a fuze. The charge case with fuze is ejected by expelling charge initiated by a cap. The fuze functions after set delay in air, imparting flash over the terrain for aerial photography.

Photoflash cartridges may be fired mechanically or through electrically operated squib. A typical mechanically fired photoflash cartridge [3] is shown in Figure 31.2. It consists of a cartridge case assembled with primer cap and expelling charge assembled with charge case consisting of fuze surrounded by photoflash composition. On firing the primer cap, the expelling charge gets ignited and ejects the charge case. The fuze functions with some delay, providing intense flash over the target area.

A typical squib-operated photoflash cartridge [3] is shown in Figure 31.3. It consists of a charge case with delay unit filled with high explosive. The charge case is assembled to squib assembly containing expelling charge. On firing, the squib

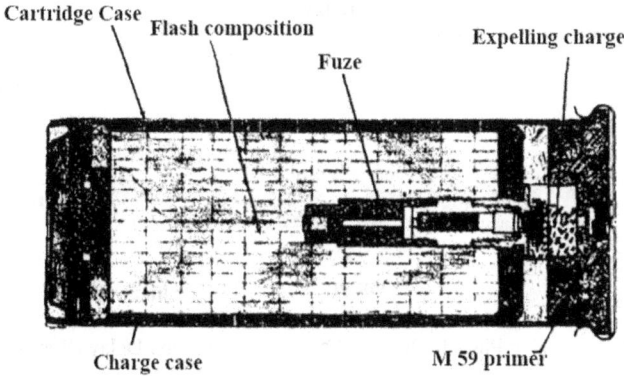

FIGURE 31.2   Typical Photoflash Cartridge (Reprinted from [3]).

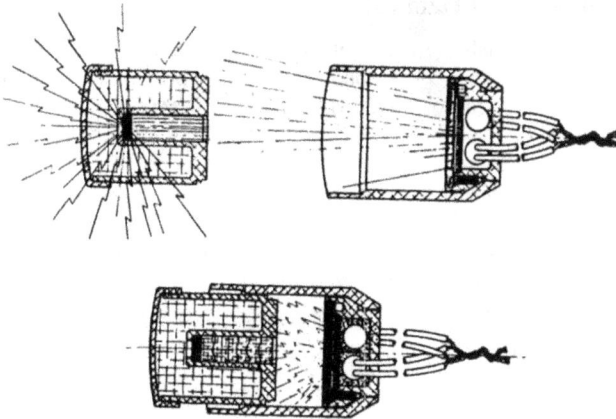

FIGURE 31.3   Typical Squib-Operated Photoflash Cartridge (Reprinted from [3]).

flash ignites the expelling charge which ejects the charge case. The charge case provides intense flash on functioning of the delay unit. It is used to obtain information on trajectory of a missile by firing photoflash cartridge at sufficient distance from the missile.

Some typical photoflash cartridge ammunitions are as under.

   i. *Cartridge, Photoflash: M112A1, 1-, 2- and 4-Second Delay* (Figure 31.4) [4] The cartridge is used to provide illumination for night aerial photographic reconnaissance. The delays are 1, 2, 4 seconds. Burn time 0.4 seconds, peak candlepower 100,000,000.
  ii. *Cartridge, Photoflash: M123A1, 2-, 4- and 6-Second Delay* [4]. The cartridge is used to provide illumination for night aerial photographic reconnaissance. The delays are 2, 4 and 6 seconds. Burn time 0.4 seconds, peak candlepower 400,000,000.

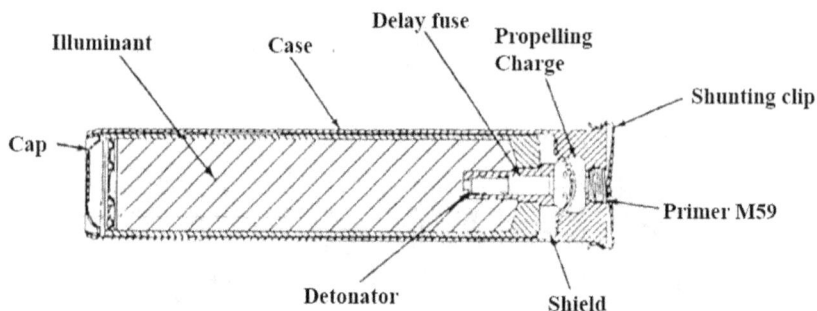

**FIGURE 31.4**  Cartridge Photoflash 112A1 1-, 2- and 4-second Delay (Reprinted from [4]).

**TABLE 31.1**

**Photoflash Ammunitions Data [5]**

| Item | Bomb photoflash M120 | Bomb photoflash M120A1 | Cartridge photoflash M112A1 | Cartridge photoflash M123A1 |
|---|---|---|---|---|
| Flash Comp. wt. | 69 lb | 82 lb | 7 Oz | 1.75 lb |
| Peak candlepower intensity | 3.4 billion | 4.25 billion | 110 million | 260 million |
| Time to peak (sec.) | 0.005 | 0.005 | 0.0015 to 0.003 | 0.004 |
| Burn time (sec) | 0.170 | 0.180 | 0.040 | 0.040 |
| Effect (cp.sec$^{-1}$) | 72 million (1) | 92 million (1) | 1.4 million | 5.5 million |
| Fuze | M146A1 | M146A1 | – | – |
| Delay (sec) | 0.2 | 0.2 | 1–4 (2) | 2–6 (2) |
| Length (inches) | 39.66 | 39.66 | 7.73 | 8.45 |
| Dia. (inches) | 8 | 8 | 1.57 | 2.88 |
| Weight (lb) | 152 | 165 | 1.47 | 4.3 |

*Note:* (1) Within first 0.04 seconds (2) ±10%

The data on some photoflash ammunitions are at Table 31.1.

Readers may refer [6] on development in aerial reconnaissance. Aerial photography was earlier used for main military intelligence of enemy installations. The use of these cameras, flash bombs and cartridges have now lost much of its relevance as better systems like night-time optics and satellite imagery are available.

## REFERENCES

[1] "*Ammunition Inspection Guide*", War Department Technical Manual TM 9-1904, Washington 25 DC, 2 March 1944.

[2] *"Bombs and Bomb Components"*, TM 9-1325-200, Department of the Army Technical Manual, Departments of The Army, The Navy and The Air Force, Washington, DC, 29 April 1966.

[3] *"Theory and Application"*, AMCP 706-185, Engineering Design Handbook Military Pyrotechnic series, Part ONE, Headquarters, US Army Material Command, Washington DC, April 1967.

[4] *"Army Ammunition Data Sheets for Military Pyrotechnics"*, *(Federal Supply Class 1370)*, *Technical Manual No. 43-001-37*, Headquarters, Department of the Army, Washington DC, 6 January 1994.

[5] *"Pyrotechnic, Screening, and Dye Marking Devices"*, NAVORD OP 2213, First Revision Change 19, Department of the Navy, Ordnance Systems Command, Washington, 1 January 1969.

[6] George W. Goddard, *"New Developments for Aerial Reconnaissance*\* Photographic Laboratory, Engineering Division Hq., Air Materiel Command, Wright-Patterson Air Force Base, Dayton, Ohio.

# 32 Screening Smoke Ammunitions

## 32.1 GENERAL

The smoke ammunition is used for screening and thermal attenuation. White/grey screening smoke ammunitions are produced from the following types of ammunition:

a. Grenade smoke (combustion of smoke composition)
b. Mortar smoke bombs (combustion or bursting of smoke composition)
c. Artillery smoke shells and cartridges (ejection of ignited smoke pellet filled canisters or ejection of smoke pellets with burning of pellets)
d. Generator smoke or container/ pot smoke ammunitions (combustion products allowed to pass through a venturi for cooling and screening effect)

There are certain ammunitions containing red phosphorous-based compositions which produce smoke as well as incendiary effects. Smoke compositions may be filled by

a. Direct filling into the hardware (like PWP composition)
b. Pelleting of composition at high tonnage

The latter is preferred for the ammunitions that have to withstand strong firing stresses during firing and in flight.

## 32.2 GRENADE SMOKE

A typical AN-M83 white smoke hand grenade [1] is shown in Figure 32.1 which contains 11 ounces of terephthalic acid and gives white smoke for 25 to 70 seconds. It has a forest green body with light green marking, a light blue band and white top.

The AN-M8 HC white smoke [1] contains 19 ounces of HC smoke composition type C and it can be thrown up to a distance of 30 m. It provides dense smoke for 105 to 150 seconds. It has a light green body with black marking and white top (Figure 32.2).

## 32.3 MORTAR SMOKE BOMBS

The mortar smoke bombs contain smoke composition. The range is achieved through primary cartridge, augmenting cartridge, delay and angle of fire.

The small calibre mortar ammunition like 51 mm, 52 mm and 60 mm do not contain any fuze since the range is less and composition pellet may be ignited easily by the delay mechanism.

DOI: 10.1201/9781003093404-32

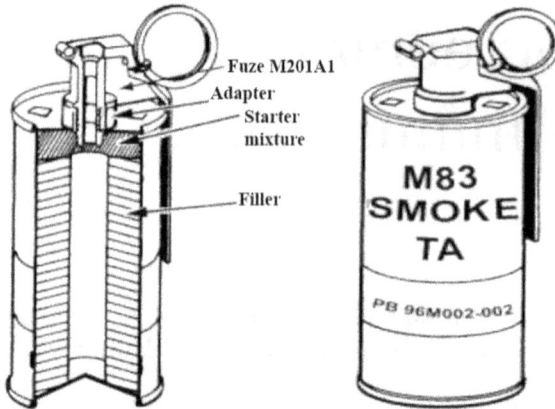

**FIGURE 32.1**   AN-M83TA White Smoke Hand Grenade (Reprinted from [1]).

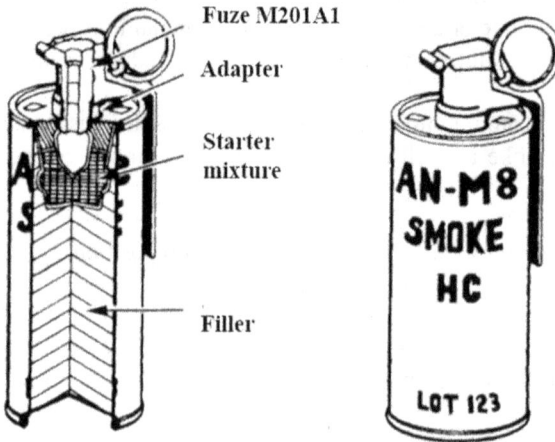

**FIGURE 32.2**   AN-M8 HC White Smoke Grenade (Reprinted from [1]).

In a typical case of composition filled with PWP composition, a central cavity is made for assembly of high explosive pellets. The central explosive pellets enhance the blasting power of fuze and causes the composition to burst on the ground, and generate smoke cloud. A typical 120 mm smoke (W) XM929 with Fuze Point Detonating M745 is shown in Figure 32.3.

120 mm (W) M929 with Fuze Point Detonating M745 functions [2] as follows. When the cartridge (actually a mortar bomb) is dropped down the mortar tube, the firing pin at the bottom of the tube initiates the percussion primer and charge in the ignition cartridge. The charge in the ignition cartridge flashes through the holes in the shaft of the fin assembly and ignites the propelling charge. The gases from the burning propellant expand and propel the cartridge out of the mortar tube. The fuze functions on point detonating and the booster ignites the burster charge in the centre

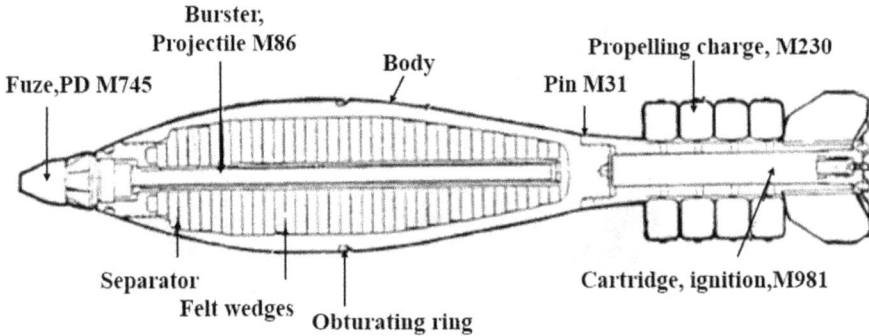

**FIGURE 32.3** Bomb 120 mm smoke (W) M929 with Fuze PD M745 (Reprinted from [2]).

part of the projectile body. The burster charge fragments the projectile body and disperses 144 felt wedges impregnated with WP, which burns immediately on contact with air. It is intended for use as an incendiary device and to produce a smoke screen.

Examples of few bombs producing screening smoke ammunition are:

*Bomb 51 mm Smoke Ammunition*
*Bomb 52 mm Smoke Ammunition*
*Bomb 60 mm Smoke Ammunition*
*Bomb 60 mm Smoke (WP), M722* [2]
*Bomb 60 mm Smoke, WP, M302* [2]
*Bomb 60 mm Smoke, WP, M302A1 (M302E1) andM302A2* [2]
*Bomb 81 mm Smoke, WP, M57A1 and M57* [2]
*Bomb 81 mm Smoke, WP, M370* [2]
*Bomb 81 mm Smoke, WP, M375A2 and M375A1* [2]
*Bomb 81 mm Smoke, WP, M375A3* [2]
*Bomb 81 mm Smoke, RP, M819* [2]
*Bomb 81 mm Smoke, WP, M375* [2]
*Bomb 82 mm Smoke Ammunition*
*Bomb 4.2 inch: Smoke, WP, M328A1 and M328* [2]
*Bomb 4.2 inch, Smoke PWP or WP, M2A1 or M2* [2]
*Bomb 120 mm Smoke (WP), M68 with Fuze, PD: M935* [2]
*Bomb 120 mm Smoke (W) XM929 with Fuze, Point-Detonating: M745* [2]
*Bomb 122 mm Smoke Ammunition*
*Bomb 130 mm Smoke Ammunition,*
*Bomb 152 mm Smoke Ammunition*

A typical 51 mm smoke ammunition [3] is given in Figure 32.4. It provides white smoke for 120 seconds.

## 32.4  AIRCRAFT SMOKE BOMBS [4]

These are:

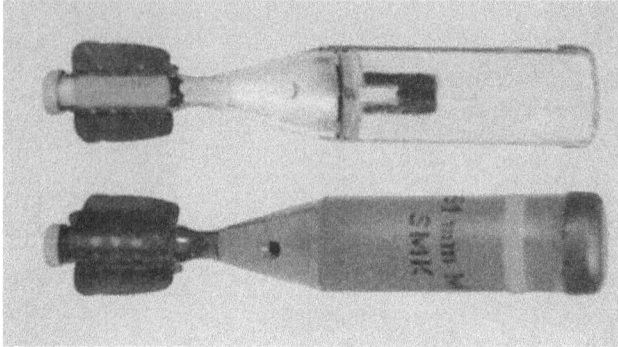

**FIGURE 32.4**  51 mm Smoke Ammunition (Reprinted from [3]).

a. *Aircraft Floating Smoke Bombs [4]*

Three variants of aircraft floating smoke bombs are

  i. 100-lb Aircraft Smoke Bomb Mark 3 Mod 0 [4]
 ii. 50-lb Aircraft Smoke Bombs Mark 1 Mod 1 [4]
iii. 50-lb Aircraft Smoke Bombs Mark 1 Mod 2 [4]

The aircraft floating smoke bombs for naval use are filled with HC smoke mix (Type A, composed of hexachloroethane, zinc dust, ammonium perchlorate, ammonium chloride and magnesium carbonate), which upon burning produces a dense white smoke. The 100-lb Smoke Bomb Mark 3 Mod 0 produces smoke for 6 to 10 minutes, the 50-lb Smoke Bomb Mark 1 Mod 1 produces smoke for 3 to 5 minutes, and the 50-lb Smoke Bomb Mark 1 Mod 2 produces smoke for 6 to 7.5 minutes. All floating smoke bombs are safe for take-offs and landings anywhere, including carriers. These smoke bombs are floating bombs designed for dropping from aircraft for the purpose of creating a smoke screen, primarily on the surface of a body of water. By means of the fuze adapter Mark 1 Mod 0 these bombs will also function on land impact.

b. *100-lb Quick-Opening Cluster E44 of 14 1O-lb. HC Smoke Bombs M77 [4]*
   The l00-lb Quick-Opening Cluster E44 (designated as the M25 until January 1945), consists of 14 numbers 10-lb HC Smoke Bombs designed to provide frontal amphibious operations or troop movement to blanket large areas of enemy positions with smoke. They are for use over land only. Tests have shown that each cluster opens a few feet below the releasing plane and when dropped from 500 feet altitude disperses its bombs over a somewhat elliptical area around 50 yards across. The bombs require two to four minutes after hitting the ground to build an effective screen, and will produce smoke for eight to twenty minutes.

c. *100-lb Smoke Bomb M47A2 and AN-M47A3* [4]

The l00-lb Smoke Bomb M47A2 and ANM47A3 consists of a 100-lb Bomb body M47A2 filled with either 100 pounds of WP (white phosphorus) or 72 pounds of PWP (plasticised white phosphorus – a mixture of white phosphorus and synthetic rubber in xylene). This bomb is equipped with a tetryl or black powder buster, and an instantaneous functioning; air-arming nose Fuze AN-Ml26Al.An effective white smoke screen is produced for 1 to 2 minutes (WP) and 3 to 5 minutes (PWP). The only difference between the M47A2 and the AN-M47A3 is in the length of the tail fins. The tail fins of the AN-M47A3 are three inches longer for increased flight stability.

## 32.5   ARTILLERY SMOKE SHELLS AND CARTRIDGES

Smoke canisters (Figure 32.5) are assembled in the carrier shell. The smoke composition is generally pressed as bare pellets with central hole and then assembled in the canisters. Primed cambric cloth is used in the sleeves. Priming composition is also filled in the canisters. The burning method of smoke composition in canister has been shown in Section 4.2.16 and Figure 4.7.

A typical shell 105 mm BE smoke is shown in Figure 32.5 where grease material is used inside the shell for ease of sliding of the canisters during ejection.

Fuze is assembled at the time of firing by removing transit plug. On firing, the fuze functions in the air after desired fuze set time. This causes burster bag containing gunpowder to pick up flash resulting into a very high flash and pressure inside the bomb. This causes separation of the base plate and ignition of the four canisters, which are ejected out, generating smoke. The pellets in canisters burn radially and produce large volume of gaseous products.

The canisters fall in an area over the target. The canisters display a streak of white grey smoke as they fall through the air to ground and generate white grey smoke on the ground (Figure 32.6).

A typical Cartridge 90 mm Smoke WP M313 and M313C [2] is shown in Figure 32.7. This cartridge is used in 90 mm guns for spotting and screening purposes and has a limited incendiary effect.

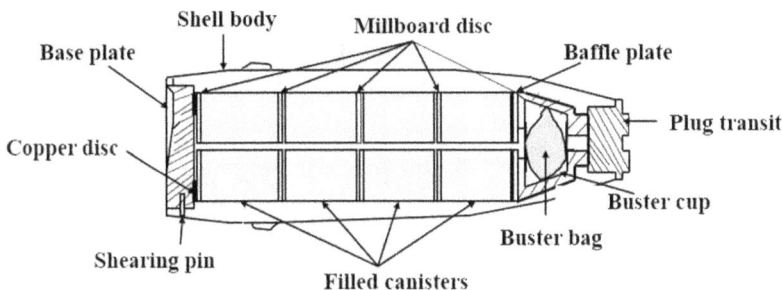

**FIGURE 32.5**   Shell 105 mm BE Smoke.

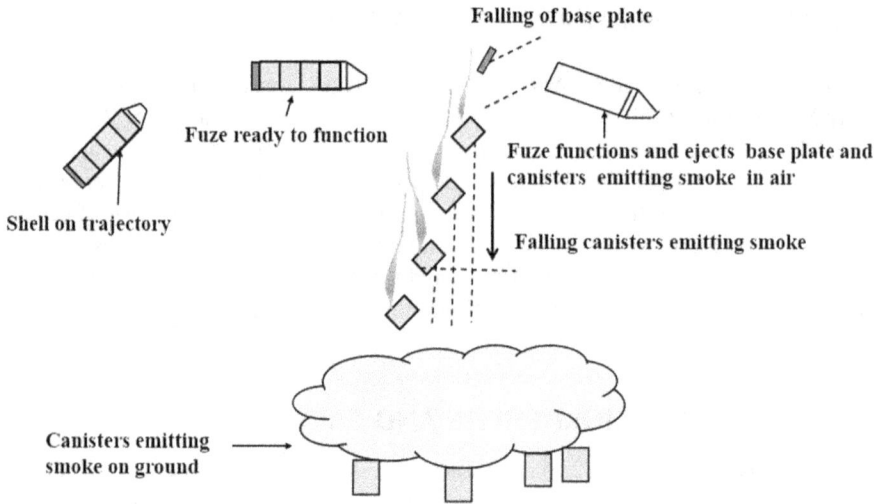

**FIGURE 32.6**   Ejection of Filled Canisters from 105 mm BE Smoke Shell.

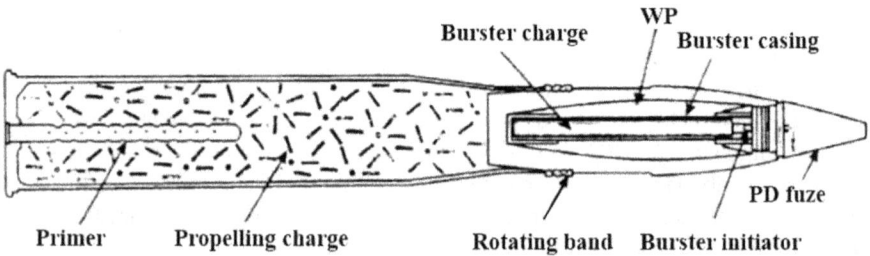

**FIGURE 32.7**   Cartridge 90 mm Smoke WP M313 and M313C (Reprinted from [2]).

M313C has a lower propellant charge weight and thus low wear and tear of weapon system. The burster is Tetrytol (mixture of tetryl and TNT) and the fuze is PD M48A3, M57; MTSQ, M 501 series. The muzzle velocities are 821 and 730 m per second, respectively.

Typical 155 mm HC smoke ammunition [2] is shown in Figure 32.8. This projectile is fired from 155 mm howitzers and is used for screening, spotting and signalling. Fuzes used are MT, M565; MTSQ, M577; ET, M762. An effective smoke cloud is produced within 30 seconds, and maximum smoke emission occurs in about one minute.

The 155 mm shell (Naschem) ammunition contains a large number of red phosphorous pellets assembled in the shell (Figure 32.9). It expels burning smoke pellets in the air which descends slowly on the ground like burning star and lead to smoke screen as well as incendiary effect on the ground.

For firing high calibre ammunition, a variety of fuzes are used. These fuzes may be of *mechanical time fuze, electronic time fuze, point detonating fuze or proximity fuze.*

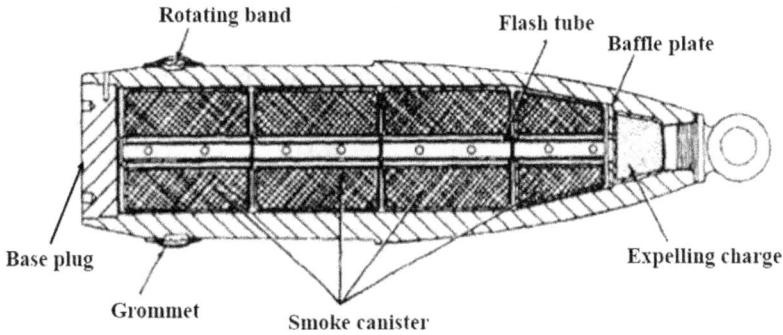

**FIGURE 32.8** Projectile 155 mm Smoke HC 116A1 (Reprinted from [2]).

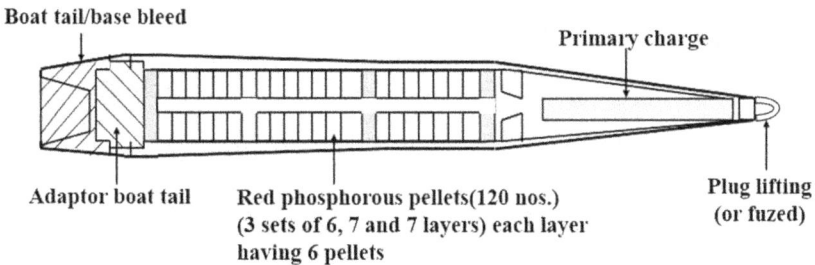

**FIGURE 32.9** Shell 155 mm Red Phosphorous.

Mechanical and electronic time fuzes may be set at desired time so that after firing from the mortar or gun, they function after desired set timing. The point detonating fuzes function on impact at the target while the proximity fuzes have sensitive electronic devices which may be set to function above the target at pre-determined height for ensuring better effect.

Some further examples of white screening smoke artillery ammunition are:

*Cartridge, 57 mm Smoke, WP, M308A1 and M308* [2]
*Cartridge,76 mm Smoke, WP, M361A1or M361* [2]
*Cartridge, 75 mm Smoke, WP, M311A1 and M311* [2]
*Shell 84 mm Smoke Ammunition*
*Cartridge, 105 mm Smoke, WP-T, M416* [2]
*Cartridge, 105 mm Smoke, WP, M60 Series* [2]
*Cartridge, 105 mm Smoke H. C., BE, M84 Series* [2]
*Projectile, 155 mm Smoke, WP, M110 and M110E1* [2]
*Projectile, 155 mm Smoke WP, M110A1 (M110E2) M110A2 (M110E3)* [2]
*Projectile, 155 mm Smoke, BE, M116 and M116B1, HC* [2]
*Projectile, 155 mm Smoke, HC, M116A1* [2]
*Projectile, 155 mm Smoke, WP, M825 and M825A1* [2]
*Shell 160 mm Smoke Ammunition*
*Smoke FFV 007 (18 km)*
*Smoke FFV 007 ER (24 km)*

**TABLE 32.1**

**Field Artillery and Mortar Smoke Capabilities and Effects [5]**

| Ammunition | Type of round | Nomenclature | Fuze | Time to build effective smoke (seconds) | Burn time (seconds) | Average obscuration length per round (metres) wind direction | |
|---|---|---|---|---|---|---|---|
| | | | | | | Cross | Head or tail |
| 155 mm | WP | M110A2 | M557 | 30 | 60–90 | 100 | 50 |
| | Smoke | M116B1 | M501A1 | 60–90 | 240 | 350 | 75 |
| | Improved Smoke | M825 | M577 | 30 | 420 | 100 | 95 |
| 105 mm | WP | M60A1 | M557 | 30 | 30–90 | 75 | 50 |
| | Smoke | M84B1 | M501A1 | 60–90 | 180 | 250 | 50 |
| 107 mm | WP | M328A1 | – | 30 | 60 | 150 | 40 |
| 81 mm | WP | M375A2 | – | 30 | 60 | 100 | 40 |

*Note:* 107 mm mortar WP projectile produces better smoke than 105 mm howitzer WP projectile.

The field artillery and mortar smoke capabilities of some ammunition are given in Table 32.1.

## 32.6  GENERATOR SMOKE OR CONTAINER /POT SMOKE AMMUNITIONS

These smoke-generating containers/pots (Figure 32.10) emit smoke on ground. The smoke composition is pressed in the paper container having a central recess to

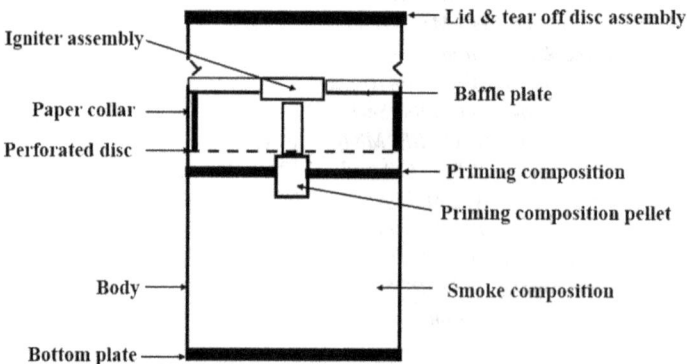

**FIGURE 32.10**  Generator Smoke or Container / Pot Smoke.

accommodate a priming composition pellet and then paper container with pressed pellets are assembled in the container/pot. Some blank space over the pellet is maintained. The container is seamed. For firing, the tear off band is taken off and the composition is ignited by port fire friction. The composition burns inside the container and emits white screening smoke for specified time.

The perforated disc and baffle allow the gases to mix and escape. The blank space over the composition surface helps in cooling of the combustion products. In addition, the exhaust hole also results in adiabatic cooling due to pressure drop of the exiting gases than the gases which are inside the container and thus avoid flaming. A central core hole in the composition also allows the exposure of more surface area as well as allows the tar material to be accommodated.

The screening smoke emission time of smoke of a few typical smoke ammunitions [3] are given in Table 32.2.

**TABLE 32.2**
**Typical Screening Smoke Emission Time of Some Ammunitions [3]**

| Ammunition | Smoke emission time (seconds) | Colour |
|---|---|---|
| Mortar Bomb 51 mm Smoke | 120 | White |
| Generator Smoke no. 5 | 1,080 | White |
| Marker Smoke White A/S N3 | 960 | White |
| 105 mm BE Smoke | 45 | White |

## 32.7  INFRARED SCREENING SMOKE AMMUNITIONS

Examples of infrared smoke ammunition are as under.

a. *81 mm Anti-Thermal-Anti-Laser Smoke Grenade Mk1* [6]:

HEMRL, Pune (India) has designed and developed bursting and burning type of 81 mm Anti-Thermal-Anti laser smoke grenade Mk-I which bursts in mid-air/ground at a distance of 50–70 m from fire tank and produce a dense white smoke screen. This smoke screen is capable to obscure the visual and all bands of infrared regions (0.4 to 14 microns) and also confuse laser range finders.

The sectional view of a grenade with various components is shown in Figure 32.11. The grenade is electrically ignited by 24 V DC supply from tank, which initiates primer, which in turn initiates propellant, which builds up sufficient pressure behind the grenade for its ejection from SGD at a velocity of 20–35 m/s. The delay tube is initiated by propellant gases. After a pre-set delay of approximately 3.8 seconds, during which the grenade travels a distance of approximately 60 m, bursting composition is initiated. The smoke grenades burst open and smoke screen is created from the burning compositions.

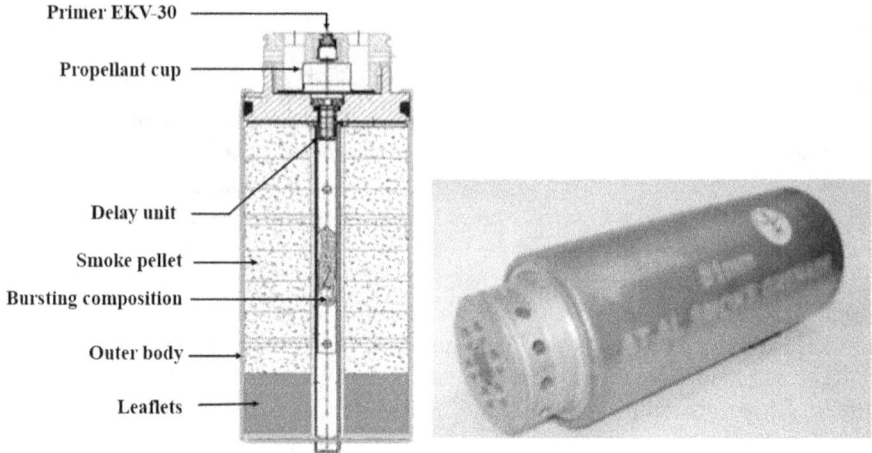

Primer EKV-30

Propellant cup

Delay unit

Smoke pellet

Bursting composition

Outer body

Leaflets

**FIGURE 32.11**   81 mm AT-Al Smoke Grenade Mk1 (Reprinted from [6]).

The grenade consists of three different red phosphorus-based smoke composi-tions, for instantaneous smoke screen the leaflet technology is developed for the first time in smoke technology. Two types of red phosphorus-based pellets, namely fast burning and slow burning are developed for achieving the smoke screen for quick and longer duration, respectively. When grenade is electrically actuated from the grenade discharger, the explosive train starts. The sequence starts with electrical primer, the propellant charge, delay charge, burster charge and finally pellets and leaflets. Propellant charge propels the grenade to a distance of 60 ±10 m in about 3.8 ± 0.3 seconds and explodes in mid-air, approximately 8 m above the ground. The close variation in height of burst is achieved by developing an improved ac-curacy pyrotechnic delay element. A white dense smoke screen of approximately 15 m width × 10 m height is formed with single grenade which is capable of obscuring the night vision equipment in the range of 0.4 to 14 microns for duration of not less than 20 seconds. The size and effectiveness of the smoke screen depends on the number of grenades fired, relative humidity of the air, ambient wind speed and direction. The launching of grenades should be in the line of the target. On bursting in mid-air, the burning leaflets float in air producing instantaneous and well-spread smoke screen. It merges with emerging smoke of pellets burning on ground. The smoke screen blinds effectively the TI sight and LRF of the enemy AFVs. To facilitate easy bursting, the outer metallic container is performed with four equi-distance longitudinal grooves on the inner wall. The smoke screen is formed due to the burning of red phosphorus which produces $P_2O_5$ and subsequently converts into ortho-phosphoric acid by reaction with atmospheric moisture. Therefore, the change in atmospheric humidity affects the grenade performance.

b. 120 mm Infrared Smoke Round: Contains red phosphorous in smoke pots with time fuze and provides visual and infrared screening for several minutes. It has muzzle velocity 128 to 391 m.sec$^{-1}$ and corresponding range of 0.3 to 8.3 km.

c. 66 mm and 76 mm smoke screening IR grenade are launched from main battle tanks (MBT's) or armoured fighting vehicles (AFV's) as smoke vehicle protection system (SVPS) for visual obscuration and infrared attenuation of the tank.

## REFERENCES

[1] *"Grenades and Pyrotechnic Signals"*, Field Manual no. 3-23.30, Headquarters, Department of the Army, Washington, DC, 7 June 2005.

[2] *"Technical Manual, Army Ammunition Data Sheets, Artillery Ammunition, Guns, howitzers, Mortars, Recoilless Rifles, Grenade Launchers and Artillery Fuzes"*, (Federal Supply Class 1310, 1315, 1320, 1390) TM 43-0001-28, Headquarters, Department of The Army, Washington DC, 27 October 2003.

[3] *"Product Brochure"*, Indian Ordnance Factories, Ordnance Factory Dehuroad, Pune, India.

[4] *"Aircraft Smoke Bombs"*, OP 1050 Second Revision, Navy Department Bureau of Ordnance, Washington 25 DC, 28 November 1945.

[5] *https://www.globalsecurity.org/military/library/policy/army/fm/6-30/f630_7.htm*

[6] *"Technology Focus"*, Volume 24, Issue 4, July-August 2016 (ISSN No 0971-4413), Defence Scientific Information and Documentation Centre (DESIDOC), DRDO, Ministry of Defence, New Delhi, India.

# 33 Signalling Ammunitions

## 33.1  GENERAL

Signalling ammunitions are used for signalling from land, sea and air for the purpose of communication. The mode of signalling is through following means.

a. Light of varied colours, intensities and duration
b. Smoke of varied colours and densities
c. A combination of light and smoke
d. Fluorescent dyes on water surface

Signal ammunition may be divided into following categories:

a. Grenade sigalling
b. Mortar signalling bombs
c. Artillery signalling shells
d. Miscellaneous signalling ammunitions/devices

These are

i. Fired in air
ii. Fired under sea
iii. Fired through aircraft
iv. Fired on ground

Various types of compositions producing coloured smoke and coloured light are used in signal ammunitions.

## 33.2  GRENADE SIGNALLING

Grenades have their own means of initiation and hence do not require any external means of initiation. These grenades are filled with signalling colour smoke compositions and assembled with a striker assembly with a delay element. On throwing the grenade after removal of the safety pin, the grenade emits a dense signalling colour smoke for specified time. There are also some that are thrown using a rifle grenade launcher.
    Typical examples are:

*Grenade Hand Smoke (Red)*
*Grenade Hand Smoke (Green)*
*Grenade Hand Smoke (Yellow)*

DOI: 10.1201/9781003093404-33

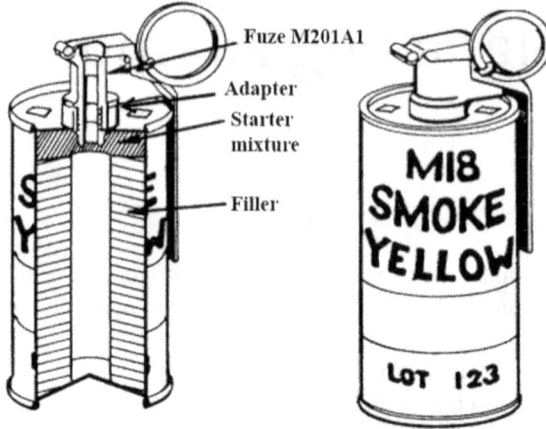

**FIGURE 33.1**  M18 Coloured Smoke Yellow Hand Grenade (Reprinted from [1]).

A typical hand grenade smoke yellow [1] is shown in Figure 33.1 which can be thrown approximately 35 m and emits coloured smoke for 50 to 90 seconds.

Another typical grenade hand smoke red [2] is shown in Figure 33.2 below which is useful in any terrain including snow-bound areas. It functions with a delay of 2 to 3 seconds and gives red smoke for a minimum of 25 seconds. Another version is the grenade hand smoke green.

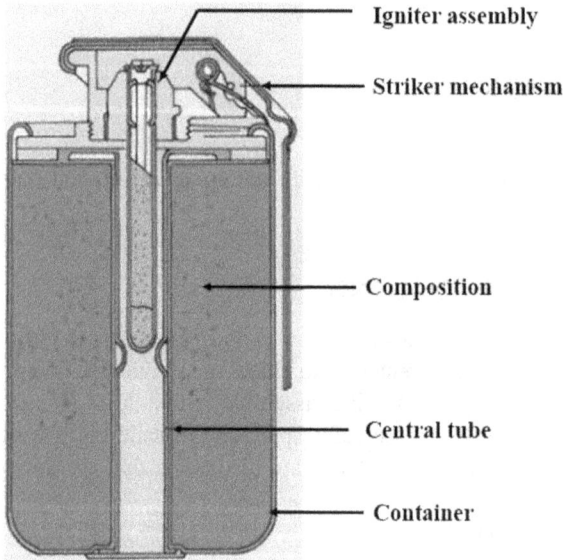

**FIGURE 33.2**  Grenade Hand Smoke Red (Reprinted from [2]).

## 33.3   MORTAR SIGNALLING BOMB

Mortar bombs like 51 mm and 81 mm are assembled with signalling composition filled candles. The candle is ignited by delay component or fuze. It functions similar to illuminating ammunition except that it provides coloured light signals like red or green. Examples are

*Mortar Bomb 51 mm Red [2]*
*Mortar Bomb 51 mm Green[2]*
*Mortar Bomb 81 mm Red*
*Mortar Bomb 81 mm Green*

A typical 51 mm mortar bomb red and green [2] is at Figure 33.3 which produce red flare (20 ±3 seconds) or green flare (25 ±3 seconds) attached to parachute and used for communication or locating position. It has four delay settings of 3.5 seconds, 6 seconds, 9.5 seconds and 11.5 seconds to achieve range of 300 to 900 m.

**FIGURE 33.3**   Mortar Bomb 51 mm Red Green (Reprinted from [2]).

## 33.4   ARTILLERY SIGNALLING SHELLS

Artillery signalling ammunitions like shell 105 mm colour red/orange/blue are assembled with signalling colour smoke filled canister (similar to screening smoke ammunition) with a time mechanical fuze. The fuze may be mechanically set up to 80 seconds and hence smoke canisters may be ejected up to 17 km range. A buster bag containing gunpowder is kept below the fuze to build pressure and eject the candle. The ejected candles fall from air to ground with a streak of signalling colour smoke and emit signalling colour smoke on the ground for specified time. Examples are:

*105 mm Base Ejection Colour Smoke Orange[2]*
*105 mm Base Ejection Colour Smoke Blue [2]*
*105 mm Base Ejection Colour Smoke Red [2]*

**FIGURE 33.4**   Shell 105 mm BE Colour Smoke Red (Reprinted from [2]).

A typical shell 105 mm BE Colour Smoke red [2] is given in Figure 33.4.

## 33.5   MISCELLANEOUS SIGNALLING AMMUNITIONS

These signalling ammunitions do not fall under above categories and each has a different design. These include all coloured flares (stars), smoke and dyes. We may consider them based on firing.

### 33.5.1   Ammunition Fired in Air

Mostly typical coloured flare signal ammunition cartridges are made of aluminium or crimson red ammunition paper with brass base. The base of cartridge is assembled with a cap and anvil and filled with gunpowder as propelling charge. The pressed composition (with priming and booster composition) known as star is inserted in the cartridge case (ensuring that its priming side is facing the gunpowder) and closed with millboard cylinder and metallic disc or paper disc and turned over.

The star exhibits desired red, green, yellow and white colour light signals on firing. A typical schematic diagram for cartridge signal is shown in Figure 33.5.

*Function*: These are fired through hand-held weapon. The cartridges are fired through a mini hand-held projector or pistol. The cap gives flash to the propelling gunpowder which ignites the star. The pressure built up inside the cartridge case ejects

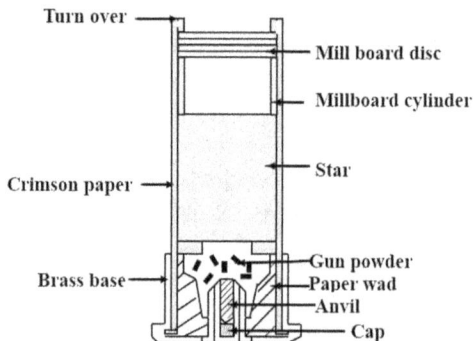

**FIGURE 33.5**   Cartridge Signal.

the burning star pellet to go up in the air and give bright coloured light for specified duration. The star burn time is so adjusted that the pellet burns in the air only and does not burn on the ground. This ensures that there is no fire on the ground due to burning of signalling star. This is very important as it avoids burning of grass or bushes on ground or burning on ship floor and also avoids identification of location of fire.

*Identification*: The metallic cartridges are passivated with dye to impart red, green white and yellow colour to the cartridge case for day identification and are assembled with suitable metallic index disc with markings for night identification. This helps to know the colour of the signal cartridge by simple finger feel/touch during night operations. There are many methods to have identification as given below.

i. For identification of colour, embossed single dot, doble dot and triple dot for red, green and white are used on closing cap. In another case, embossing shapes are varied like embossed dot, embossed V or embossed line for red, green and white on closing cap.
ii. Identification symbols with alphabets are also used, with colour of bands on the closing caps as under.

R = Red star, Y = Yellow star, G = Green star, RR = Double Red Star, YY = Double Yellow Star, GG = Double Green Star, RG = Double Red-Green Star, GY = Double Green-Yellow Star and RP = Red Star with Parachute.

iii. Some 16 mm signalling ammunitions have closing discs embossed with symbols +, Δ or W to indicate that the star colour is red, green or white.

Figure 33.6 depicts some typical methods of identification on closing disc of signal cartridges.

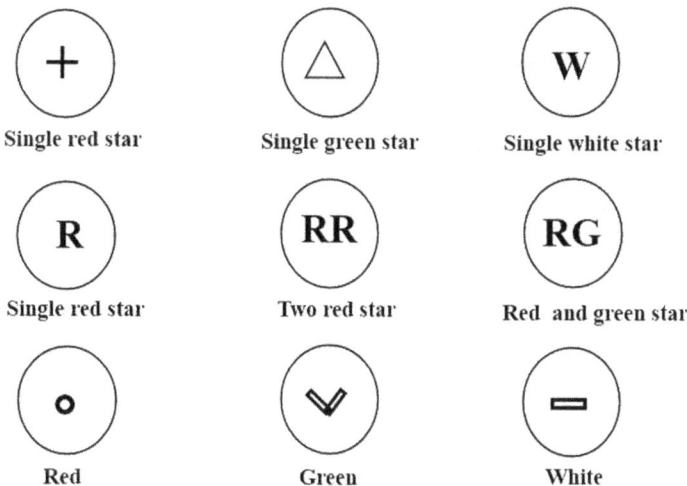

FIGURE 33.6    Methods of Identification of Signal Cartridges.

iv. The crimson red paper cartridges of 1" and 38 mm signalling ammunitions are given a stencil mark of red, green and white star on cartridge case to indicate colour of the signal. Also, the base of the brass cartridge is kept fully serrated or half serrated or plain for night identification for red, green and illuminating signalling ammunitions.

Some signalling ammunitions are:

a. *Signals, Illumination, Ground: Green Star, Parachute, M19A2, AN-M19A2B2* [3]: For night signalling, burns with delay of 5.5 seconds for 20–30 seconds with illumination 5,000 candela.
b. *Signals, Illumination, Ground: Clusters, Green Star, M125A1; Red Star, M158; White Star, M159* [3]: For day or night signalling, burns with delay of 5 seconds for 6–10 seconds. The candlepower is 9,000, 30,000 and 30,000 for M125A1, M158 and M159, respectively.
c. *Signals, Illumination, Ground: Parachutes, Red Star, M126A1; White Star, M127A1; Green Star, M195* [3]: For day and night signalling and night illumination, burns with delay of 5 seconds for 50, 25 and 50 seconds and the candlepower is 10,000, 125,000 and 5,000 for M126A1, M127A1 and M195, respectively.
d. *Signals, Illumination, Ground: Red, M187; White, M188; Green, M189; Amber, M190* [3]: For surface to air or air to surface signalling, burns for 5 seconds and the candlepower is 3,200.
e. *Signal, Smoke, Ground: Red, M62; Yellow, M64; Green, M65; Violet, M66* [3]: For daylight signalling, burns with a delay of 5.5 seconds for 4–8 seconds with red (M62), green (M64), yellow (M65) and violet (M66) smoke.
f. *Signals, Smoke, Ground: Parachute, Green, M128A1; Red, M129A1; Yellow, M194* [3]: For day and night signalling, burns with delay of 5 seconds with burning time 6–18 seconds (M128A1 and M129A1) and 9–18 seconds (M194) of coloured smoke.
g. Following signal cartridges are fired from 13 mm hand projector where the ammunition is screwed and fired. It is used by army, air force and navy.
   i. *Cartg. Signal 16 mm Red/Green/White Ammunition*: Burn time 3.5 seconds [2]
   ii. *Cartg. Signal 1" Red/Green/Illuminating Ammunition*: Burn time 6.5 seconds [2]
   iii. *Cartg Signal 38mm Red/Green/Yellow Ammunition*: Burn time 6–8 seconds [2]

Some typical cartridge signal 16 mm and cartridge signal 1" are given in Figure 33.7.

## 33.5.2 Ammunition Fired Under Sea

These are fired under the sea. Summary of performance of some submarine launched pyrotechnics are given in Table 33.1.

**FIGURE 33.7**   Cartridge Signal 16 mm and Cartridge Signal 1" (Reprinted from [2]).

**TABLE 33.1**
**Submarine Launched Pyrotechnics Summary [4]**

| Item | Display | Burn time (seconds) | Delay (seconds) | Length (inch) | Weight (pounds) |
|---|---|---|---|---|---|
| Signal Smoke, Marine (Submarine), Mk2 (Signal, Float, Submarine) | Red, Black, Green or Yellow smoke on surface | 15 | 30–46 | 18.5 | 3.0 |
| Signal Illum. Marine, Mk3 (subm. Emerg. Ident. Signal) | Red, Yellow or Green parachute susp. star,300 ft. altitude | Red and Yellow:25 Green :35 | 54–70 | 18.5 | 3.15 |
| Marker Location Marine, Mk26 (Marker, Location, Submarine) | Yellow flame and white smoke on surface | 180–300 | 130–165 | 18.5 | 3.3 |
| Signal Illum. Marine Mks 41, 45 and 46 | Red, Yellow or Green parachute susp. star,350 ft altitude | 25 | 100–140 | 24 | 4.36 |
| Marker Location Subm. Mks 21, 22, 23 and 24 | Red, Yellow, Green or Black smoke on surface | 14–45 | None | 18.5 | 3.4 |
| Signal Smoke and Illum. Marine Mks 66, 67 and 68 | Red, Green or Yellow smoke and parachute susp. star, 400 ft altitude | Smoke :15 Star :23 | None | 36 | 6.8 |
| Marker Location Marine Mk 28 | Yellow-Green fluorescent slick on surface | Slick persists for 90 minutes | None | 36 | 7.5 |

Two typical examples are as under.

a. *Cartridge Submerged Ejector Signal Red/Green/Yellow* [2]: The submerged submarines are provided with cartridge submerged ejector signal. These are fired when the submarine intends to move up to the surface of the sea or proceed further so that nearby ships may move away from the path or to pass on precoded message.

As CSES is launched from submarine, it moves up under sea water, and during this process, the impeller unscrews and falls off. On the sea surface, the side flaps are straightened, allowing the steel ball to roll out, thus allowing striker to hit the primer.

When fired, emit dense signalling colour smoke for 20 seconds minimum as well as eject a signal star (with parachute) to around 300-m height with burn time 30 seconds minimum. Thus, smoke is visible during day time and flare signal is visible during night time (Figure 33.8).

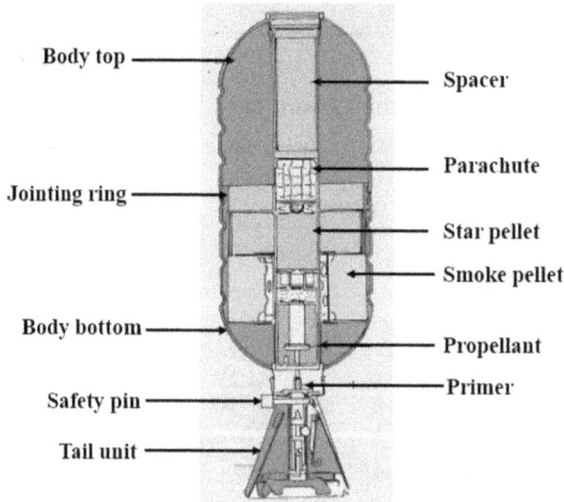

**FIGURE 33.8** Cartridge Submerged Ejector Signal (CSES) (Reprinted from [2]).

b. *Cartridge Squib and Flare* [2]: The training torpedoes are fitted with cartridge squib and flare and are used for locating the practice torpedoes (Figure 33.9).

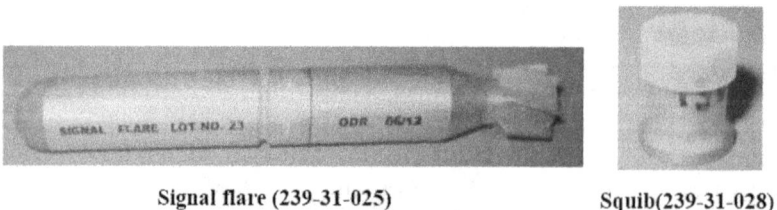

Signal flare (239-31-025)                          Squib(239-31-028)

**FIGURE 33.9** Cartridge Signal Flare and Squib for Torpedo (Reprinted from [2]).

When fired (through a squib through electric D.C. voltage 27 ±2 volts) successively under sea from a depth of 10–15 m from a moving torpedo, it ejects the projectile to around 20 m above sea height and then ejects two stars giving red signal flare in the air for 5 seconds indicating the location and movement of the torpedo.

### 33.5.3 Ammunition Fired through Aircraft

Some paper as well as metallic cartridges are fired through aircraft for signal purposes. These eject burning star for the specified time. Some examples are

    a. *Signals, Illumination, Aircraft: Double Star, AN-M37A2- AN-M42A2* (Figure 33.10) [4]:

These devices are fired by the pistol firing pin striking the primer. The primer ignites the propelling charge (2.5 grams black powder) which projects the stars and ignites the quick match. The quick match ignites the stars. These signals are fired from Pyrotechnic Pistol, AN-M8 either from aircraft or from the surface for day or night identification for signalling. Each signal projects two stars – burning time 7 seconds – of the same or different colours to an altitude of approximately 250 feet above the point of launching. The 25,000-candlepower stars are visible from 2 to 3 miles in daylight and 5 miles at night in clear weather. These have two ¼" colour band for identification. These signals have no delay feature. These double star signals are used in following model number

    *AN-M37A2 red/red; AN-M38A2 yellow/yellow; AN-M39A2 green/green; AN-M40A2 red/yellow; AN-M41A2 red/green and AN-M42A2 green/yellow*

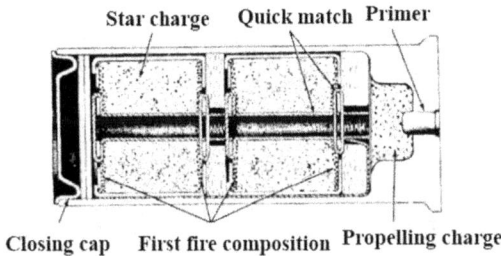

**FIGURE 33.10** Aircraft Illumination Signal, AN-M37A2 to AN-M42A2 (Reprinted from [4]).

    b. *Signals, Illumination, Aircraft: Single Star, AN-M43A2 Red, AN- M44A2 Yellow and AN-M45A2* Green[4]:

These devices are similar to above. These have one 1/2" single colour band for identification.

    c. *Signals, Illumination, Aircraft: AN-M53A2- AN-M58A2 Series, Tracer Double Star,* [4]:

These signals project a tracer in addition to two stars, the colour of the display is indicated by a ¼-inch band near the closing wad for the tracer and two ½-inch bands near the centre of the signal body for the two stars, and these signals have a 4-second delay and a shorter burning time. The various models are

*AN-M53A2 tracer yellow, stars red/yellow: AN-M54A2 tracer green, stars red/ red; AN-M56A2 tracer red, stars green/green; AN-M57A2 tracer red, stars red/red; AN-M58A2 tracer red, stars green/red.*

Some air- and surface-launched dye marking devices are:

d. *Air- and Surface-Launched Dye Marking Devices* [4]: These devices produce coloured chemical displays in or on water. Some of these dye devices do, however, contain explosive charges which serve to burst the outer container, thus spreading the dye particles on the water. Others are broken open simply by impact. All are expressly intended for making visible reference points for any daytime situation that calls for a relatively stable and long-lasting mark. The summary of a few devices is given in Table 33.2.

A typical marine location marker, AN-Mk1 Mod1 containing fluorescein dye is shown in Figure 33.11.

**TABLE 33.2**
**Dye Marking Devices Summary [4]**

| Item | Launching method | Display | Display time (minute) | Length (inch) | Diameter (inch) | Weight |
|---|---|---|---|---|---|---|
| Marker Location Marine AN-Mk1 Mods | Pyro Pistol AN-M8 | Green Dye (Fluorescein) | 30 | 3.85 | 1.68 | 4 oz. |
| Marker Location Marine Mk1 Mod 2 | Hand Launched from surface or aircraft | Green Dye (Fluorescein) | 45 | 11.87 | 3.5 | 3.7 lb. |
| Marker Location Marine Mk1 Mod 3 | Hand dropped from surface or aircraft | Lemon yellow (stearated chrome) | 45 | 11.87 | 3.5 | 3.7. lb. |
| Marker Location Marine Mk8 Mod 0 | Hand dropped from surface craft | Lemon yellow (stearated chrome) | 60 | 7.25 | 3.5(max) | 31oz. |
| Marker Location Marine AN-M59 (obsolete) | Hand launched from aircraft | Green Dye (Fluorescein) | 120 | 10.5 | 3.0 | 2.9 lb. |

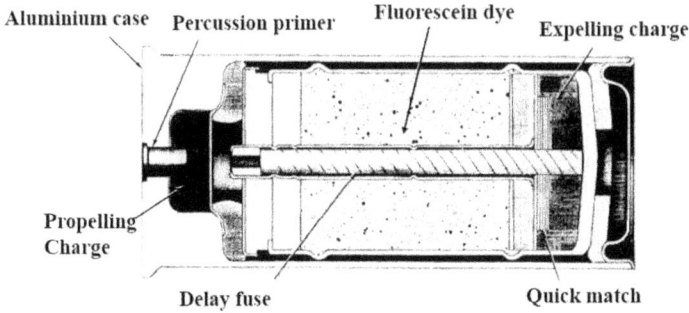

FIGURE 33.11   Marine Location Marker AN-Mk1 Mod 1(Reprinted from [4]).

### 33.5.4   AMMUNITION FIRED ON GROUND

These are signalling ammunitions which are not fired through any weapon but are laid on the ground and made to function on the ground only. These are mainly signalling for specific purpose like emergency landing of aircraft, warning signal for infiltrating troops or for location.

Examples are

a. *Flare ground indicating Mk1 Yellow* (Figure 33.12) [2]: It is used for signalling ground emergency landing of aircraft. The intense bright yellow light signals for minimum 170 seconds and are visible from a long distance. The device consists of a metallic cylinder filled with illuminating composition and a nail arrangement to allow it to be fixed on the ground. The device is arranged on both sides of the makeshift air strip for aircraft to land in the night. It can also be used for distress signalling.

b. *Flare, Surface: Airport, M76* [3] : It illuminates' runways or landing areas for aircraft operating at night. The primer may be initiated either by an electric current to the firing squib or manually by a quick pull on the lanyard to operate the firing pin. The primer ignites a first-fire charge at the top of the

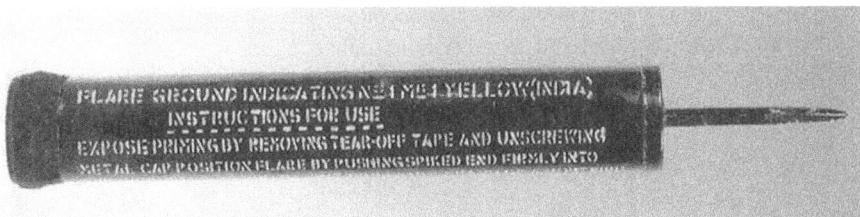

FIGURE 33.12   Flare Ground Indicating MK-1 Yellow (Reprinted from [2]).

illuminant candle to ignite the illuminant composition. The flare provides a minimum of 500,000 candlepower for 5 to 7 minutes. Visibility of the flare is up to 30 miles on a clear night.

c. *Flare Trip Wire Mk-1* (Figure 33.13) [2]: It is laid on the fence using fine wires. The ammunition functions with intense red light for 30 seconds minimum when any intruder breaks the wire of the fence.

**FIGURE 33.13**   Flare Trip Wire (Reprinted from [2]).

d. *Generator Smoke Orange 3A* [2]: It emits orange smoke for signalling in snow-bound areas. The filling and performance are similar to generator smoke white except that signalling colour smoke composition is used which gives deep orange smoke for 360 seconds. The orange colour suits well against white background of snow-bound areas.

The burning time and colour of a few typical signalling ammunitions [2] are given in Table 33.3

**TABLE 33.3**
**Burning Time and Colour of Signalling Ammunitions and Devices** [2]

| Ammunition | Burning time (minimum) (seconds) | Colour |
|---|---|---|
| Cartg. sig. 16 mm | 3.5 | Red/Green/White |
| Cartg. sig. 1" | 6.5 | Red/Green/White |
| Cartg. signal 1½" | 6.0–8.0 | Red/Green/Yellow |
| Flare Ground Indicating Yellow | 170 | Yellow |
| Flare Trip Wire MkI | 30 | Red |
| Cartridge Submerged Ejector | 30 | Red/Green/Yellow (flare) |
| Signal | 20 | Red/Green/Yellow (smoke) |
| Mortar Bomb 51mm Red | 20±3 | Red |
| Mortar Bomb 51mm Green | 25±3 | Green |
| Shell105mm BE Colour Smoke | 45 | Red/Orange/Blue |
| Generator Smoke Orange 3A | 360 | Deep orange |
| Grenade Hand Smoke Colour | 25 | Red/Green |

## REFERENCES

[1] *"Grenades and Pyrotechnic Signals"*, Field Manual No. 23-30, Headquarters Department of the Army Washington, DC, 1 September 2000.
[2] *"Product Brochure"*, Indian Ordnance Factories, Ordnance Factory Dehuroad, Pune, India.
[3] *"Army Ammunition Data Sheets for Military Pyrotechnics"*, Federal Supply Class 1370,Technical Manual No. 43-001-37,Headquarters,Department of the Army, Washington DC, 6 January 1994.
[4] *"Pyrotechnic, Screening, and Dye Marking Devices"*, NAVORD OP 2213, First Revision, Change 19, 1 January, 1969, Department of the Navy, Ordnance Systems Command, Crane, Indiana.

# 34 Incendiary Ammunitions

## 34.1 GENERAL

These are widely used to destroy buildings, installations, oil fields, ammunition depots, documents and equipment including causing jungle fire. All incendiary ammunitions can be divided into following types.

a. Small Arms Incendiary Ammunitions
b. Mortar Incendiary Bombs
c. Artillery Incendiary Shells
d. Grenade Incendiary Ammunitions
e. Gel incendiary ammunitions

## 34.2 SMALL ARMS INCENDIARY AMMUNITIONS

These cartridges contain propellants. It is assembled with a cap at its base. A bullet is press fitted into the cartridge mouth. Two typical incendiary cartridge bullets are shown in Figure 34.1.

   i. *Cartridge Calibre, 0.50 Incendiary, M1 [1]:*

It creates a pressure of 54,000 p.s.i. and moves out with a velocity of 2,950 fps at 78 m from muzzle. Upon impact with a hardened or armoured target, the incendiary composition bursts into flame and ignites any inflammable material. It is identified with a blue tip.

   ii. *Cartridge, Calibre, 0.50, Ball, Armor Piercing Incendiary, M8 [1]:*

It creates a pressure of 59,000 p.s.i. and moves out with a velocity of 2,910 fps at 78 m from muzzle. The cartridge combines the function of the M2 Armour piercing bullet and the incendiary bullet. It is used against flammable targets and light-armoured or unarmoured targets, concrete shelters and similar bullet resisting targets. The cartridge is identified by an aluminium bullet tip.

Further examples are:

*Cartridge, Caliber 0.30, Armor Piercing Incendiary, M14*
*Cartridge, Caliber 0.50, Incendiary M1*
*Cartridge, Caliber 0.50, Ball, Armor Piercing Incendiary, M8*
*Cartridge 0.50 Armour Piercing Incendiary M23*
*Cartridge, Caliber 0.50, Ball, Armor Piercing Incendiary – Tracer, M20.*

DOI: 10.1201/9781003093404-34

Cartridge, caliber .50, incendiary,M1

Cartridge, caliber .50, ball, armour piercing incendiary,M8

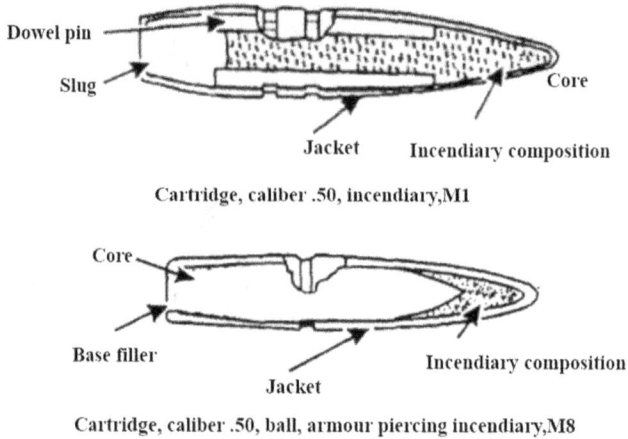

**FIGURE 34.1**   Two Typical Incendiary Cartridges (Reprinted from [1]).

*Cartridge S.A. 12.7 mm API and APIT (Armour Piercing Incendiary and Armoured Piercing Incendiary Tracered)*
*Cartridge S.A. 14.5 mm API and APIT (Armour Piercing Incendiary and Armoured Piercing Incendiary Tracered)*

## 34.3   MEDIUM CALIBRE INCENDIARY AMMUNITIONS

Examples of medium calibre incendiary ammunitons are
*Cartridge, 20 mm, Armor Piercing Incendiary Tracer, M52E1*
*Cartridge, 20 mm, Armor Piercing Incendiary, M53*
*Cartridge, 20 mm, Incendiary, M96*
*Cartridge, 20 mm, Armor Piercing Incendiary Tracer, M601*
*Cartridge 23 mm Schilka APIT Shell*
*Cartridge 23 mm Ghasha API*
*Round 30 mm BMP-II HEI*

## 34.4   MORTAR INCENDIARY BOMBS

These bombs are filled with the incendiary composition and cause fire. Examples are

*60 mm Incendiary Mortar Bomb*
*82 mm Incendiary Mortar Bomb*
*120 mm Incendiary Mortar Bomb*

A typical 120 mm incendiary bomb [2] (Figure 34.2) is filled with thermite canisters in a WP matrix. When these rounds burst, they produce some fragmentation and dense smoke from the WP. The thermite canisters scatter and burn intensely, causing fires in exposed ammunition and fuel.

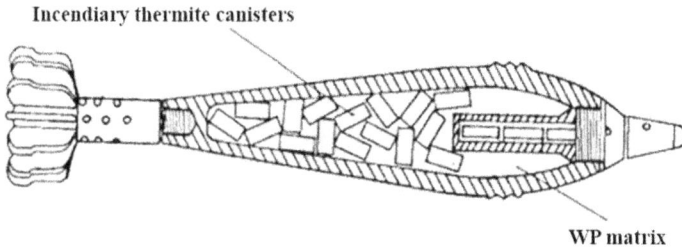

FIGURE 34.2    120 mm Incendiary Mortar Bomb (Reprinted from [2]).

## 34.5    ARTILLERY INCENDIARY SHELLS

The shells are filled with incendiary composition or incendiary pellets, which fall over an area and cause fire. The 155 mm red phosphorous-based composition pellets provide both fire and smoke. Examples are

*105 mm Incendiary Shell*
*152 mm Incendiary Shell*
*155 mm Incendiary Shell*

## 34.6    GRENADE INCENDIARY AMMUNITIONS

The grenades are filled with incendiary compositions which cause fire. A typical incendiary hand grenade [3] is shown in Figure 34.3. It can also damage, immobilize, or destroy vehicles, weapons systems, shelters, or munitions. This grenade does not have a safety clip. The filler has 26.5 ounces of thermate (TH3) mixture. The average soldier can throw this grenade to about 25 m. A portion of thermate mixture is converted to molten iron, which burns at 4,000°F. The mixture

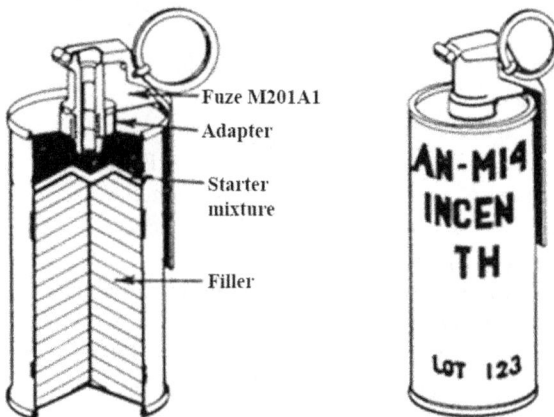

FIGURE 34.3    AN-M14 TH3 Incendiary Hand Grenade (Reprinted from [3]).

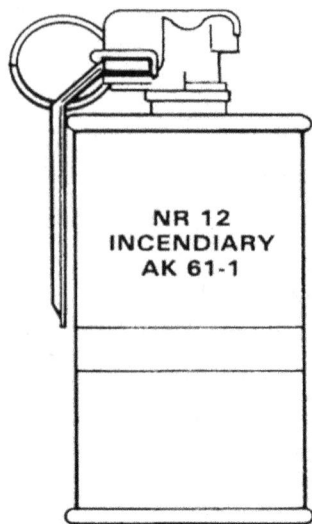

**FIGURE 34.4** NR 12 Incendiary Hand Grenades (Reprinted from [4]).

fuses together the metallic parts of any object that it contacts. Thermate is an improved version of thermite; the incendiary agent used in hand grenades during World War II. The thermate filler can burn through a 1/2-inch homogeneous steel plate. It produces its own oxygen and burns under water.

The NR12 Incendiary Hand Grenade [4] (Figure 34.4) is made of tinned steel and filled with thermite composition. It has a striker release delay fuze of 1 to 3 seconds and range 40 m and burns for 40 seconds with peak intensity 2,200°C.

## 34.7  GEL INCENDIARY AMMUNITIONS

A typical miscellaneous incendiary ammunition is container bomb with incendiary gelled fuel composition. The gelled fuel (petroleum based) is ignited with a percussion fuze and an igniter. It is dropped from the aircraft, the percussion fuze functions on ground impact, giving flash to igniter. The igniter burns for minimum 15 seconds, thus allowing the gelled fuel to catch fire and spread over a large area, causing fire and damage.

HEMRL Pune, India [5] has developed Mine Anti-Personnel Inflammable (MAPI) with the use of instant gel and gasoline. MAPI can be effectively used to prevent the movement of enemy personnel across water surfaces like canal, ditch-cum-bunds (DCBs) and similar water obstacles by generating a high temperature flame that and floats on water.

The mine deployed under water burst open on initiation and gelled petrol floats on water surface. The floating gelled petrol catches fire from pieces of burning sodium metal and the flame characteristics are circular diameter of flame 10–12 m, height of flame 2–3 m, burning time 2–3 minutes and flame temperature 1,100–1,200°C. The assembled mines may be deployed in DCBs or canals, in a row and fired as a single mine or in a group. *Six* such mines deployed in a row usually provide a frontage of 100 m. The burning and floating gelled fuel may also cling to any surface or object and thus effectively transmit fire to an object present within the flame zone.

Readers may see reference [6] on incendiaries and flame throwers in detail.

## REFERENCES

[1] *"Army Ammunition Data Sheets, Small Caliber Ammunition"*, FSC 1305, 29 April, 1994, Headquarters, Department of the Army, Washington, DC.

[2] *https://www.globalsecurity.org/military/library/policy/army/fm/7-90/Ch7.htm*

[3] *"Grenades and Pyrotechnic Signals"*, Field Manual No. 3-23-30, Headquarters Department of the Army Washington, DC, 1 September 2000.

[4] *"Grenades and Pyrotechnic Signals"*, Field Manual no. 3-23.30, Headquarters, Department of the Army, Washington, DC, 7 June 2005.

[5] Agrawal, J.P., *"High Energy Materials: Propellants, Explosives and Pyrotechnics"*; Page 379, 2010 Copyright Wiley-VCH GmbH & Co. Weinheim, Reproduced with permission.

[6] *"Fire Warfares: Incendiaries and Flame Throwers"*, Summary Technical report of Division 11 NDRC, Volume 3, Office of Scientific Research and Development, Washington DC, 1946.

# 35 Riot Control Devices/ Ammunitions

## 35.1 GENERAL

These devices/ammunitions have riot control composition or plastic pellets/rubber balls or flash and bang composition for riot control. These are considered less lethal device/ammunition as they do not kill a person. However, the user has to be sufficiently trained and minimum distance maintained should be approximately 5–10 m, otherwise it may also lead to serious injuries. Use of riot devices/ammunitions is prohibited for use in armed conflict by chemical weapon convention (see Section 14.1). Riot devices/ammunition may be divided into following general types:

   a. Grenades
   b. Cartridges and shells
   c. Heavy calibre ammunitions
   d. Other devices

The tear smoke composition may be used as single component or in two components or three components (which bursts in two or three parts); the latter two give more coverage area. These may also have two parts, one tear smoke unit and other flash and bang unit or three parts with two parts tear smoke unit and one-part flash and bang which functions over the crowd causing temporarily stunning of the crowd.

These are mostly of bore size 37 mm, 38 mm, 40 mm and 60 mm though even higher calibres are available. The aerosol emission time is generally from 20 seconds to 60 seconds with delay varying from 1 second to 5 seconds. These are sometimes used with indelible ink which remains for 48 hours to identify the trouble creators. These may also have hardened plastic pellets or rubber balls for direct impinging on the rioters.

## 35.2 GRENADES

A few examples are as under.

   a. *Rubber Bursting CS Hand Grenade* [1]: It is a cylindrical hand grenade (Figure 35.1) made of rubber and packed with 23 CS pellets. On release of striker, there is a 2 to 2.4 second delay and CS pellet burn time is 12 seconds. It can be thrown to 25 to 35 m with effective radius of 15 m. Since it bursts, it cannot be thrown back.

DOI: 10.1201/9781003093404-35

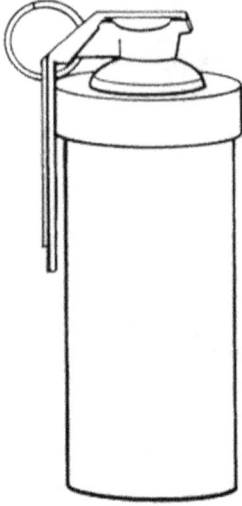

**FIGURE 35.1**   Rubber Bursting CS Hand Grenade (Reprinted from [1]).

b. *M25A1 and ABC-M25A2 Riot-Control Hand Grenades* [1]: The ABC riot-control hand grenade is a bursting munition with an integral fuze (Figure 35.2). The M25A2 grenade is an improved version of the M25A1 grenade. The two types of grenades differ primarily in body construction. They are used to deliver all three types of riot control agents presently used in hand grenades. The body of this grenade is compressed fibre or plastic sphere. All fillers are mixed with silica aerosol for increased dissemination efficiency. The average soldier can throw the grenade to about 50 m. The M25 series of riot-control hand grenades have a radius burst (visible cloud grenade) of about 5 m, but fragments of the grenade are occasionally projected up to 25 m.

c. *Tear Smoke Grenade (Plastic)* [2]: Tear smoke grenade (Figure 35.3) generates a large volume of tear smoke and designed for crowd dispersal. It is

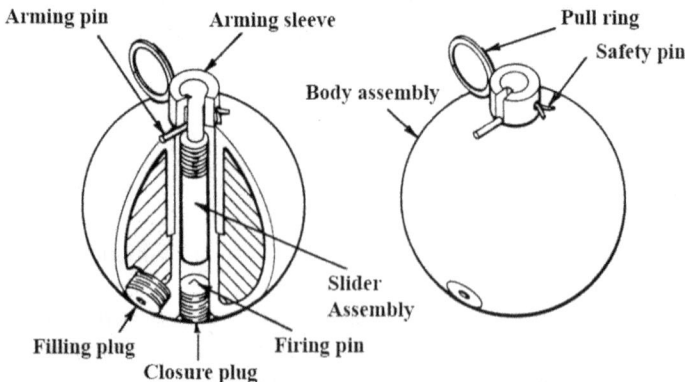

**FIGURE 35.2**   The ABC-M25A1 Riot-Control Hand Grenade (Reprinted from [1]).

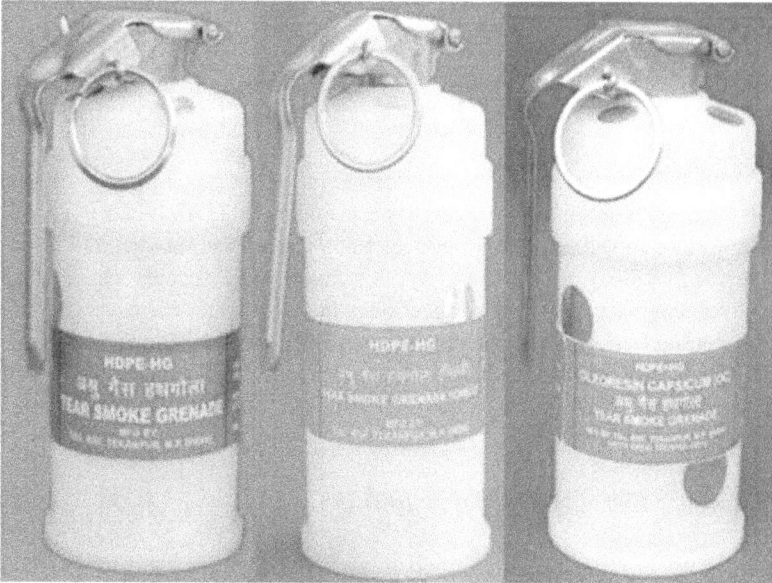

**FIGURE 35.3** Tear Smoke Grenade (Plastic) (Reprinted from [2]).

**FIGURE 35.4** Stun Grenade (Plastic) (Reprinted from [2]).

thrown by hand/using launcher cup by 12 bore PAG and 38 mm gas gun. The plastic body starts melting on ignition and does not allow miscreants to throw it back on law enforcing agencies.

d. *Stun Grenade (Plastic)* [2]: A non-fragmentation grenade (Figure 35.4) explodes with a resounding bang and blinding flash causing temporary stunning of miscreants. It is thrown by hand.

## 35.3 CARRIDGES AND SHELLS

These are of following types.

a. *Cartridge 40 mm Crowd Dispersal M1029* [3]:

These contain multiple plastic pellets or soft rubber balls or foam batons or wooden batons, etc. A typical *cartridge 40 mm crowd dispersal M1029* [3] is shown in Figure 35.5 below.

It incapacitates a person without penetrating in the body. It has aluminium cartridge case with percussion primer and a propellant and 48 rubber balls of 0.08 oz. It has a muzzle velocity of 375 ±25 feet per second and a range of 100 m with effective range 15–30 m. Some old versions have steel sphere coated with thin rubber layer weighing 14.0 g or the new ones made of metal cylinders coated with rubber weighing 15.4 g. These devices/ammunitions are available in various shapes and sizes. These are of lower density, large size and fired with lower velocity to ensure that only a severe impact is provided by the round, without penetrating the human body.

b. *Cartridge 40 mm Non-Lethal M 1006* [3]:

It (Figure 35.6} also incapacitates a person without penetrating the body or causing fatal injury. It has a projectile with foam rubber nose and high-density plastic body. It is fired by M 203 40 mm grenade launcher. It gets a spin due to projectile having rotating bands and has a velocity of 265 ± 20 feet per second.

c. *Tear Smoke Shell Soft Nose Plastic (Long Range/Short Range)*[2]:

It is a plastic version of tear smoke shell and has soft sponge fitted at the head of the shell body, which provides cushioning effect on impact, thereby minimising chances of serious injury when hits directly. The plastic bodies start melting immediately on landing, with emission of smoke making it difficult to throw it back. It is fired through

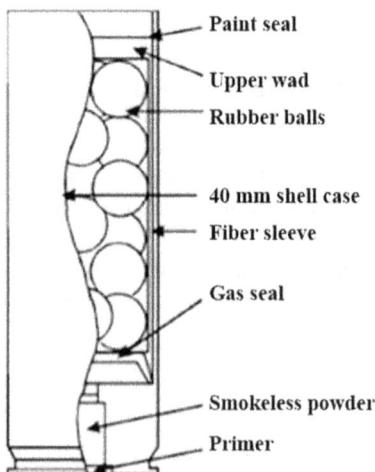

Paint seal
Upper wad
Rubber balls
40 mm shell case
Fiber sleeve
Gas seal
Smokeless powder
Primer

FIGURE 35.5   Cartridge 40 mm Crowd Dispersal M1029 (Reprinted from [3]).

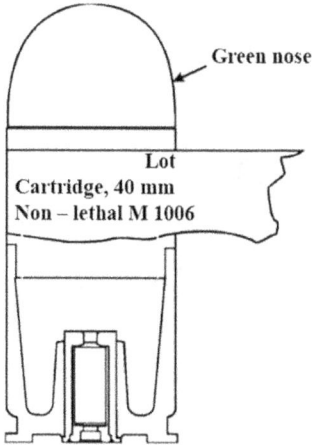

**Green nose**

Lot
Cartridge, 40 mm
Non – lethal M 1006

**FIGURE 35.6**  Cartridge 40 mm Non-Lethal M 1006 (Reprinted from [3]).

38 mm Gas Gun & Multi Shell Launcher. This ammunition (Figure 35.7) is very effective for maintaining law and order.

   d. *Wood Piercing Shell* [2]:

The wood piercing shell (Figure 35.8) used against barricaded miscreants in flushing out operations. The shell is fin stabilised and can penetrate timber of the

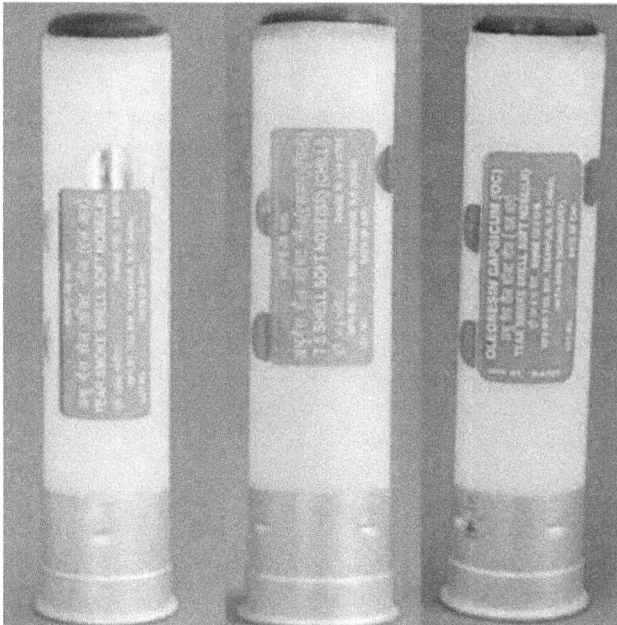

**FIGURE 35.7**  Tear Smoke Shell Soft Nose Plastic (Long Range/Short Range) (Reprinted from [2]).

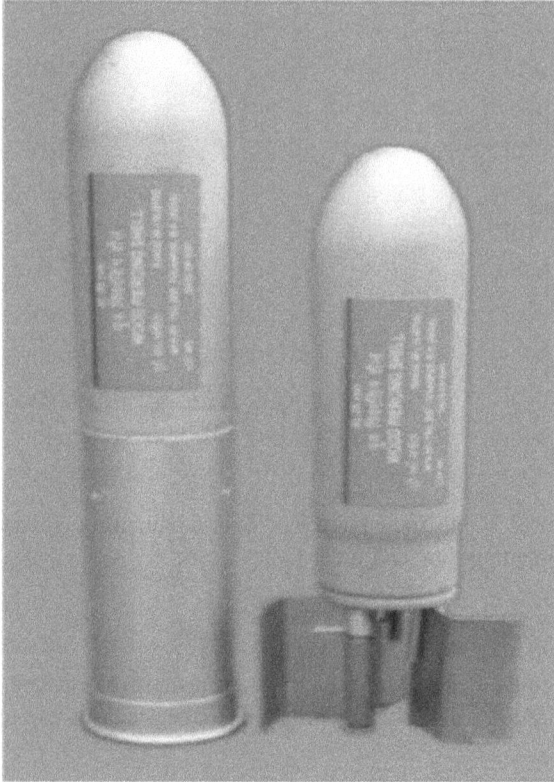

**FIGURE 35.8**   Wood Piercing Shell (Reprinted from [2]).

window or door of one inch thickness from a range of 50 to 70 m. After penetration the shell produces tear smoke which makes miscreants uneasy and forces them to come out. During hostage situation it is highly effective to limit collateral damage. It is fired through 38 mm gas gun.

   e. *Tear Smoke Shell (Metal)* [2]:

Tear smoke shell (Figure 35.9) has been designed to generate tear smoke at the target end from a long distance for crowd dispersal. The normal version of this shell is fired from 38 mm gas gun and multi shell launcher whereas its electrical version is fired with multi launcher (Agnivarsha).

   f. *Stun Shell (Plastic)* [2]:

Stun shell (Figure 35.10) explodes in air with a loud bang and blinding flash causing temporary stunning of miscreants. The normal version of this shell is fired from 38 mm Gas Gun and Multi Shell Launcher whereas its electrical version is developed for firing with Multi Barrel Launcher (Agnivarsha).

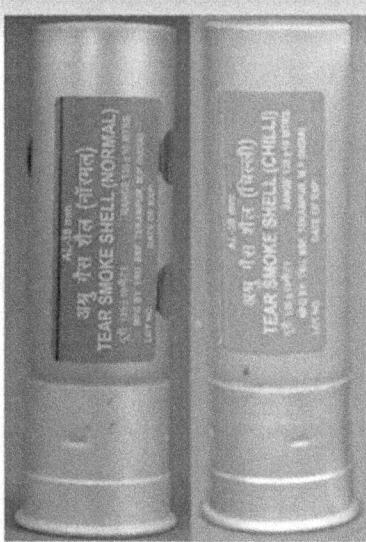

**FIGURE 35.9** Tear Smoke Shell (Metal) (Reprinted from [2]).

**FIGURE 35.10** Stun Shell (Plastic) (Reprinted from [2]).

Figure 35.11 shows some assorted riot devices.

The technical parameter of some riot control devices [2] are given in Table 35.1.

## 35.5 HEAVY CALIBRE AMMUNITIONS

Some heavy calibre ammunitions are as under but their use is restricted due to Article II(9)(d) of the 1993 Chemical Weapons Convention (see Section 14.1).

**FIGURE 35.11**  Assorted Riot Devices (Reprinted from [2]).

**TABLE 35.1**
**Technical Parameters of Riot Control Devices [2]**

| Parameters | 1 | 2 | 3 | 4 | 5 | 6 |
|---|---|---|---|---|---|---|
| Delay time (seconds) | 3 ± 1 (LR) 2 ± 1 (SR) | 1.25 ± 0.75 | 3 ± 1 | 4 ± 1 | 1.1 ± 0.3 | 2.10 ± 0.65 |
| Area covered (meters) | 20–30 | 20–30 | 20–30 | – | 40–50 | – |
| Chemical used | CS/PAVA/OC | CS | CS/PAVA | Mg and sodium nitrate | CS/PAVA/OC | Mg and sodium nitrate |
| Smoke emission time (sec.) | CS/PAVA 20 ± 5 OC55 ± 5 | 20 ± 5 | 20 ± 5 | – | CS/PAVA 25 ± 5 OC 60 ± 5 | – |
| Shell/Grenade body | Plastic | Aluminium | Aluminium | Plastic | Plastic | Plastic |
| Range (meters) | 135 ± 10 (LR) 50 ± 10 (SR) | 60 ± 10 | 135 ± 10 | 100 ± 10 | Grenade launcher cup 100 ± 10 Hand thrown 35 ± 10 | 35 ± 10 |

1 = Tear Smoke Shell Soft Nose Plastic (Long Range/Short Range), 2 = Wood Piercing Shell, 3 = Tear Smoke Shell (Metal), 4 = Stun Shell (Plastic), 5 = Tear Smoke Grenade Plastic, 6 = Stun Grenade

a. *Cartridge, 105 mm: Tactical CS, M629* uses fuze MTSQ M548 or MT M565 [3]. The projectile functioning is dependent upon the fuze used and may function on above ground either at a predetermined height based upon time of flight, or function in proximity with the target area and ejects four canisters which burn for 60 seconds (Figure 35.12).

b. Projectile *155 mm Tactical CS XM631* [3]: It uses fuze MTSQ M548 which on functioning ejects canisters filled with with CS-based composition and burns for 90 seconds, causing persons affected with burning of eyes, coughing and difficulty in breathing (Figure 35.13)

c. *Cartridge 4.2-Inch: Tactical CS, M630* [3]: It has an ignition cartridge M2A2 and propelling charge M36A1 and time fuze MT, M565; MTSQ, M548 with four canisters filled with irritating composition. The composition burns for minimum 60 seconds.

d. 64 mm M742 CS riot-control projectile [3] and 64 mm M743 kinetic-energy riot-control projectile [3]: These are fired with launcher M234. The maximum

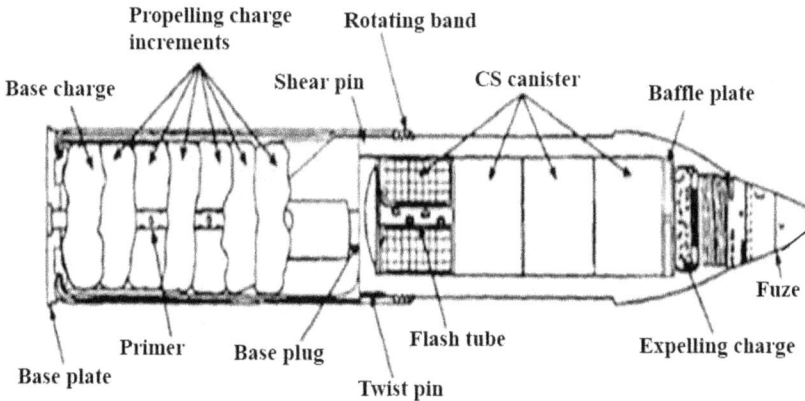

**FIGURE 35.12** Cartridge 105 mm Tactical CS M629 (Reprinted from [3]).

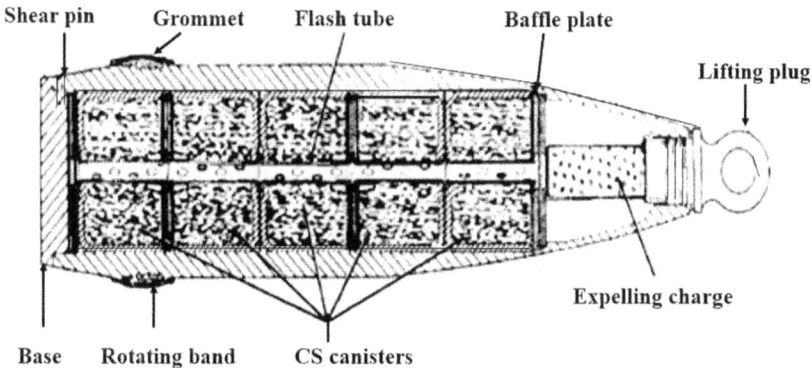

**FIGURE 35.13** Projectile 155 mm Tactical CS XM631 (Reprinted from [3]).

range is 100 m (on group of individuals 60 m and on individual 40 m) and 4–8 rounds can be fired per minute.

## 35.6   OTHER DEVICES

a. *Floating Device*: The container provides tear smoke on the surface of water for under water miscreants.
b. *Motor Vehicle Device*: The riot device/ammunition is fitted on vehicle and can be initiated from the dash board of the vehicle, providing tear smoke and may be moved over a large area.
c. *Spray Bottle*: Oil of capsicum as aerosol is used for small crowd.

A good report on various riot control management tools may be found in [4]. Readers may refer [5] for chemical ammunitions guide, which provides various riot ammunition details (including other ammunitions) with figure and colour code of ammunitions.

## REFERENCES

[1] *"Grenades and Pyrotechnic Signals"*, Field Manual No. 3-23.30, Headquarters Department of the Army, Washington, DC, 7 June 2005.
[2] "Brochure", BSF Takenpur Smoke Unit, Gwalior, Madhya Pradesh, India.
[3] *"Technical Manual, Army Ammunition Data Sheets, Artillery Ammunition, Guns, howitzers, Mortars, Recoilless Rifles, Grenade Launchers and Artillery Fuzes"*, (Federal Supply Class 1310, 1315, 1320, 1390), TM 43-0001-28, Headquarters, Department of The Army, Washington DC, 27 October 2003.
[4] *"Study Report on Development and Testing of Effective Non-lethal Weapons/ Technologies and Tactics for Countering Public Agitation with Minimum Force"*, Bureau of Police Research & Development, India.
[5] *"Chemical Munitions EOD/TE Study Guide"*, July 1976, US Army Missile and Munitions Center and School, Munitions Department Redstone Arsenal, Alabama.

# 36 Training /Practice Ammunitions

## 36.1 GENERAL

Training ammunition is required by user's (army, air force, navy, paramilitary forces, police, etc.) for:

a. Understanding the ammunition visually by shape, type, weight, packing box and its weight and marking details (like manufacturer's initial, lot number, nomenclature, colour code, etc.)
b. Understanding the method of fuzing, loading the training/practice ammunition and firing the training/practice ammunition before using the main ammunition
c. Observing some of the performance of the ammunition like smoke, light, heat, sound, etc. to a limited extent
d. Target practice
e. Understanding safety standing instructions for personnel and equipment while using the ammunition
f. Financial savings as training/practice ammunitions are mostly less costly

Generally training/practice ammunition (also known as target practice ammunition) contain a small amount of explosive material.

## 36.2 TYPES OF TRAINING/PRACTICE AMMUNITIONS

Almost all ammunitions have training/practice ammunition. Training/practice ammunitions are in the form of

a. Grenades
b. Cartridges
c. Infrared flares
d. Smoke pots
e. Bombs
f. Dummy and cut dummy model

Some examples of training/practice ammunitions are as under.

### 36.2.1 GRENADES

*A typical M69 Practice Hand Grenade*) [1] is shown in Figure 36.1. This grenade provides realistic training and familiarises the soldiers with the functioning and

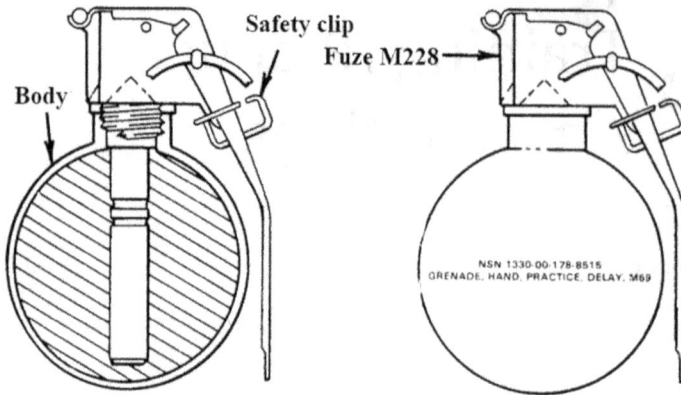

**FIGURE 36.1** M69 Practice Hand Grenade (Reprinted from [1]).

characteristics of the fragmentation hand grenade. The average soldier can throw the M69 hand grenade to about 40 m. After a delay of 4 to 5 seconds, the M69 emits a small puff of white smoke and makes a loud popping noise. The grenade body can be used repeatedly by replacing the fuze assembly.

### 36.2.2 CARTRIDGES

There are several training/practice cartridges as under.

a. *Cartridge, Photoflash: Practice, M124* [2]: For training in the use of M1123A1 photoflash cartridges. Actuates through an electric primer and contains black powder propelling charge. The remaining space is occupied by an inert charge.

b. *Cartridge, Photoflash: Practice, M121* [2]: For training in the use of M112A1 photoflash cartridges. Actuates through an electric primer and contains black powder propelling charge. The remaining space is occupied by an inert charge.

c. 14.5 *mm Artillery Training Ammunition* [3]:

Cartridge 14.5 training artillery ammunition has three versions.

i. *Cartridge, 14.5 mm, Trainer-Spotter, M181A1 (with 3±15% second delay)* [3]

ii. *Cartridge, 14.5 mm, Trainer-Spotter, M182A1 (with 6±15% second delay)* [3]

Both are similar except for the delay time. The cartridge has a chemical-filled (smoke composition) projectile. It is identified by the markings on the cartridge case and a yellow projectile tip. The cartridge has no fuze and it uses a time delay element to provide an air burst. When the cartridge is fired, the heat of the propellant gases ignites the delay element. The delay element burns 3 or 6 seconds and, in turn, ignites the smoke composition. The heat and pressure produced by the

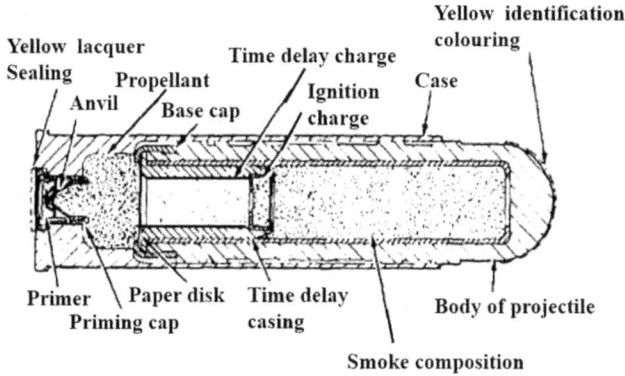

**FIGURE 36.2** Cartridge, 14.5 mm, Trainer-Spotter, M182A1 (Reprinted from [3]).

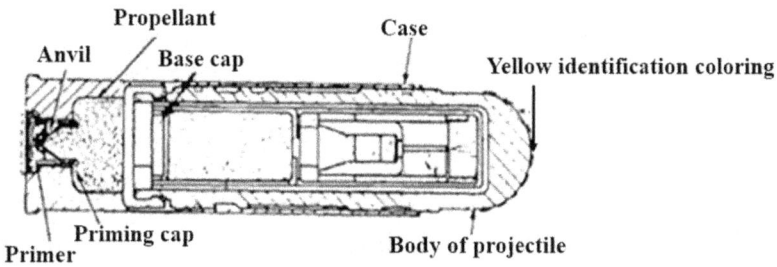

**FIGURE 36.3** Cartridge, 14.5 MM, Trainer-Spotter, M183A1 (Reprinted from [3]).

burning filler rupture the plastic closing disc at the projectile base and a visible puff of smoke is emitted. Figure 36.2 shows cartridge, 14.5 mm, trainer-spotter, M182A1.

iii *Cartridge, 14.5 MM, Trainer-Spotter, M183A1* [3]:

The cartridge consists of a cartridge case, a chemical filled (smoke composition) projectile and a fuze with a small explosive charge to provide a visible and audible verification at point of impact. The cartridge is identified by the markings on the cartridge case and a yellow projectile tip. The projectile impacts with the target. The point detonating fuze functions and an explosion charge within the fuze ignites the smoke composition and ejects it out the base of the projectile. Figure 36.3 shows cartridge, 14.5 mm, trainer-spotter, M183A1.

The 14.5 artillery training ammunition (Figure 36.4) characteristics are given in Table 36.1 [4].

### 36.2.3 Infrared Flare

Some training infrared flare cartridges are as under:

**FIGURE 36.4** Round 14.5 mm ATA 6 Seconds Delay (Reprinted from [4]).

**TABLE 36.1**

**14.5 mm Artillery Training Ammunition Characteristics [4]**

| Propellant charge | Nominal mass of propellant (g) | Velocity (metres/sec) | Range (metres) | Chamber pressure |
|---|---|---|---|---|
| Charge I | 0.2 | 115 | 894 | Pressure not exceeding |
| Charge II | 0.3 | 125 | 994 | 1,300 ±2 bar at 294°K |
| Charge III | 0.33 | 135 | 1,146 | |

a. *Flare, Ballistic Aerial Target; Infrared Tracking Mk33, Mod 0* [2]: This Flare is strapped to a 2.75-inch rocket motor to increase the infrared signature. Two flares are used and up to five rocket motors. The rocket motors and flares are assembled to the ballistic aerial target and launched to provide an aerial target during air defence gunnery training. The flare is clamped to a 2.75-inch rocket with the aluminium taped end aft of rocket motor. The rocket motor flame burns through the aluminium tape to initiate the flare composition. The flare

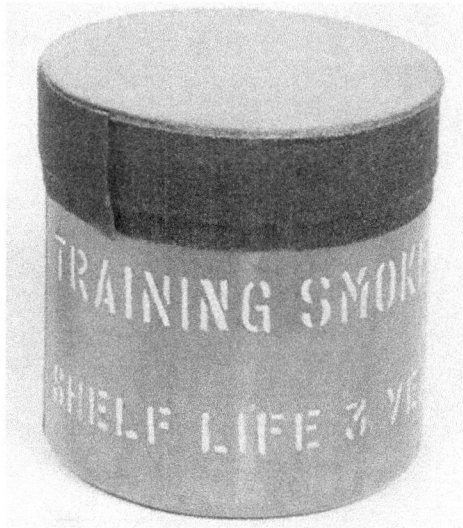

**FIGURE 36.5**   Training Smoke Generator (Reprinted from [4]).

provides the infrared signature needed for heat-seeking munitions. The flare also provides a source of light for day and night operations.

b. *TAU-15/B Infrared Target Flare* [5]: This is an air force expendable aerial target flare designed to provide an airborne infrared target for utilisation in aircrew training and to checkout systems employing infrared seeking missiles. The TAU-15/B flare evolved into the Navy Mk 28 Mod 0 tracking flare.

c. *M-50 Tow Target Flare for Army* [5]: This Army flare provides a target for both night and day practice firing of antiaircraft guns. It is 22.8 inches in length by 2.62 inches in diameter and weighs 7.13 pounds.

### 36.2.4   SMOKE POTS

An example is *Training Smoke Generator* [4], which is shown in Figure 36.5. It is use by the navy and provides non-toxic white smoke for 300 seconds for training the personnel for fire-fighting on board ship.

### 36.2.5   BOMBS

A typical *Bomb 3 kg Practice Ammunition* [4] is shown in Figure 36.6. It is used for target practice by air force which is fired through an aircraft and gives 6 seconds dense orange smoke on hitting the target.

The smoke emission time and colours of smoke of a few training ammunitions [4] are given in Table 36.2.

A typical Cartridge, 81 millimeter: Training, M445 (T32E1) is used for training loading and firing of the 81 mm mortar. It consists of an empty projectile bomb body, a training fuze, and a fin assembly designed to hold an ignition cartridge and a

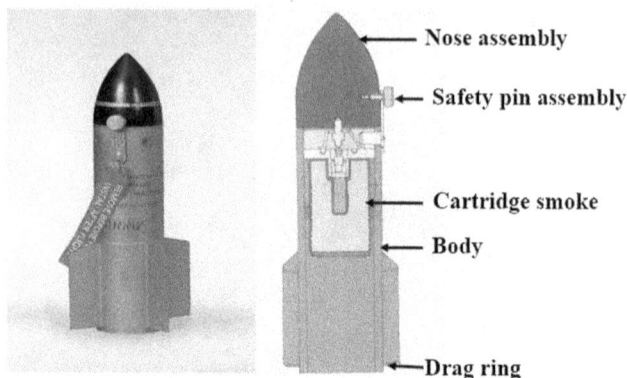

**FIGURE 36.6**   Bomb 3 kg Practice Ammunition (Reprinted from [4]).

**TABLE 36.2**
**Smoke Emission Time and Smoke Colour of Training Ammunitions [4]**

| Ammunition | Smoke emission time (seconds) | Colour | User |
|---|---|---|---|
| Training Smoke Generator | 300 | White | Navy |
| Bomb 3 kg Practice | 6 | Orange | Air force |
| Rd. 14.5 mm Artillery Training Ammunition 3 Seconds Delay | 2.55 to 3.45 | White | Army |
| Rd. 14.5 mm Artillery Training Aammunition 6 Seconds Delay | 5.55 to 6.45 | White | Army |

smoke cartridge [6]. The empty bomb body round is fired only at Charge 0, the gases from the ignition cartridge expel the projectile from the mortar tube and propel it to the target. The smoke projectile detonates on impact providing a spotting charge. The ignition and smoke cartridge are replaceable, and the round is designed for reuse.

## 36.3   DUMMY AND CUT DUMMY MODEL

Dummy(drill) ammunitions do not contain any explosive and are inert filled. They have same size, shape and weight as of the service ammunition. These are used for training in handling, loading, unloading and other drill purposes. These can be used several times. These are mostly used for display during training in lecture rooms. Sometimes, cut model of dummy (cut at 45° angle) are also used which allows to see the internal features of the components and compositions in the ammunition to a limited extent.

# REFERENCES

[1] "Grenades and Pyrotechnic Signals", Field Manual No. 3-23-30, Headquarters Department of the Army Washington, DC, 1 September 2000.

[2] "Army Ammunition Data Sheets for Military Pyrotechnics", (Federal Supply Class 1370), Technical Manual No. 43-0001-37, Headquarters, Department of the Army, Washington DC, 6 January 1994.

[3] "Army Ammunition Data Sheets Small Caliber Ammunition", Federal Supply Class 1305, Technical Manual 43-0001-27, Headquarters, Department of the Army, Washington, DC, 29 April 1994.

[4] "Product Brochure", Indian Ordnance Factories, Ordnance Factory Dehuroad, Pune-412113, Maharashtra, India.

[5] Bernard E. Douda, "Genesis of Infrared Decoy Flares", Report 26 January, 2009, Naval Surface Warfare Center, Crane Division Crane, Indiana USA.

[6] "Technical Manual Army Ammunition Data Sheets Artillery Ammunition Guns, Howitzers, Mortars, Recoilless Rifles, Grenade Launchers, and Artillery Fuzes" (FSC 1310, 1315, 1320, 1390) TM 43-0001-28, Headquarters, Department of the Army, Washington DC, April 1994.

# 37 Distress Signalling Devices

## 37.1 GENERAL

This chapter provides role and types of distress signalling devices used in air by airmen, used in sea by marine navy or merchant navy, used on land by ground personnel or troops. It also provides SOLAS (safety of life at sea) regulations for naval distress signals.

Distress signal devices provide visual or audible signals in the form of either red flare or orange smoke or sound. Some devices provide both signals of red flare and orange smoke. The intensities of red flare and orange smoke densities varies in in various devices. These are used for distress signalling or marker purpose.

The basic aim of these devices is to convey distress message through red flare, orange smoke and sound. Essentially, it is a visual/audible means of sending SOS signals. The aim is to

   a. Intimate that someone is in distress and needs help
   b. Indicate the location of distress for help

Orange smoke is chosen since it gives better contrast with grey/blue air and blue sea water while red flares for its better visibility due to its longer wavelength. The signals must be capable of conveying distress for day time or night time, though some are capable of distress signalling in both day time and night time. Burning time for smoke, flare or flame and the intensity of the signal is kept sufficiently high so as to ensure that the distress signal is visible for rescue by providing its location.

Basic requirements of a flare and or smoke distress signalling device is

   a. Strong visibility for longer time even at a far distance or unfavourable environmental conditions
   b. The instructions and illustrations for firing should be simple and easy to operate
   c. Reliable ignition even in unfavourable conditions by having a robust design and that no burning material should fall on ground during firing
   d. Should not function accidentally
   e. Should have proper sealing of joints with extra precautions
   f. Shelf life/expiry should preferably be marked

The drawbacks are:

   a. Audibility and visibility distance limitation
   b. Rescuers must be within audible or visible distance

DOI: 10.1201/9781003093404-37

c. Short shelf life of distress signalling devices
d. Hazards of using high energy material causing personnel injury and or damage to the boat, raft or ship

A special feature is that most of the distress signals have an internal means of
ignition. Therefore, there is no need for separate firing equipment for stimulus.
However, a few require projector.

These display flares are hand held or ejected stars hanging through parachute, or
exhibiting smoke and light or sound. These are used on land (ground personnel or
troops), air (airmen) and water (marine) as well as by aircrafts.

These are highly useful for boats, rafts and ships in the event of lost direction,
breakages, engine failures, fire, leakages or injuries to personnel.

## 37.2  DIFFERENCES BETWEEN DISTRESS SIGNALLING DEVICES AND SIGNALLING AMMUNITION

Table 37.1 shows the differences between distress signalling and signalling
ammunitions.

## 37.3  TYPES OF DISTRESS SIGNALLING DEVICES

There are a large number of distresses signalling devices. A few examples are given
below

a. *Air (used by Airmen):*
   i. *Signal Kits, Personnel, Distress Red M185 and Various Colours, M186*
   [1]: Distress signalling device is used by downed airmen or others. Each

**TABLE 37.1**
**Differences between Distress Signalling Devices and Signalling Ammunitions**

| Distress Signalling Devices | Signalling Ammunitions |
|---|---|
| Devices are used for distress for messaging help/rescue operation. | Ammunitions are used as coded language for messaging in combat operation. |
| Devices are labelled with printed instructions for method of firing. | These are not provided with printed instructions for method of firing. |
| Burning time for orange smoke or red flare or flame is kept sufficiently high to ensure that it is seen by others with certainty. | Burning time for smoke or red flare even for short time is sufficient to pass on the message (though some may have higher duration of functioning). |
| Most have internal means of stimulus for firing and ignition and hence not dependent on any other firing device. They are ready to fire at any time. However, a few require firing equipment. | They need external means of stimulus like pistol or projector or mortar or howitzer or gun for firing. |
| SOLAS regulations govern distress signalling devices meant for navy and merchant navy. | No such regulations for signalling ammunitions. |

of these kits consists of one hand-held projector, seven ground illumination signals and an instruction sheet. M185 is supplied with seven red ground M187; kit M186 is equipped with three red ground illumination signals M187, two green ground illumination signals M188, and two white ground illumination signals M189. Burning time 5 seconds, candlepower 3200.

ii. *Signal Kits, Personnel Distress: Foliage Penetrating, Red, M260* [1]: A distress signalling device used by downed airmen or others exposed to emergency escape and evasion situations. The kit comes equipped with a hand-fired projector and a bandolier assembly which contains a plastic-moulded bandoleer holding seven red signals. The burning time is 10 seconds, candlepower 10,000.

b. *Water (used by Marine):*
   i. *Signal, Smoke and Illumination, Marine: AN-MK13, MOD 0* (Figure 37.1) [1]: For day or night use by aircraft crewmen downed at sea. It is initiated by friction wire. The smoke composition produces smoke for 18 seconds for day time use and the night time illuminant candle will burn with a red flame for 18 to 20 seconds with a minimum of 3,000 candlepower. Use of either end does not impair the future usefulness of the other.
   ii. *Signal, Illumination, Marine: Two-Star, Red, AN-M75* [1]: Hand-held distress signal used by aircrew personnel forced down over water. The signal may be seen up to 3 miles on a clear day or 15 miles on a clear night. Delay 2 to 4 seconds for first star, 4 to 8 seconds for second star. Burning time is 4 to 6 seconds for each star, candlepower 8,000 (each star).
   iii. *Signal Distress Day/Night* (Figure 37.2) [2]: It is used by the navy during day and night operations. It is one side on initiation gives dense orange smoke for 15 to 23 seconds and other side on initiation a vivid red flare for 17 to 25 seconds.
   iv. *Buoyant Smoke Orange* (Figure 37.3) [2]: Used in ships, life raft and life boats. Once ignited and dense smoke starts within 4-5 seconds, it is thrown in sea, it emits dense orange signalling colour smoke for minimum 3 minutes.

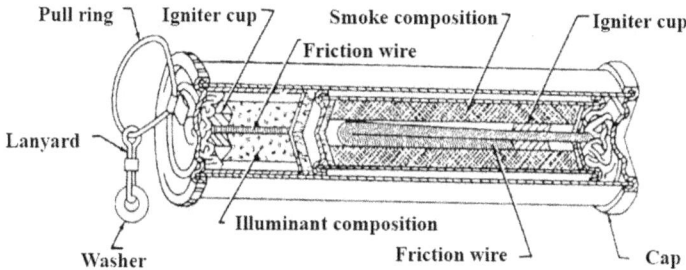

**FIGURE 37.1** Signal, Smoke and Illumination, Marine: AN-MK13, MOD 0 (Reprinted from [1]).

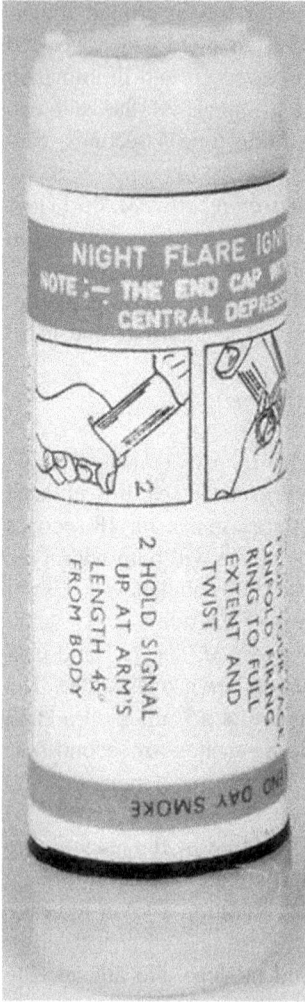

**FIGURE 37.2** Signal Distress Day/Night (Reprinted from [2]).

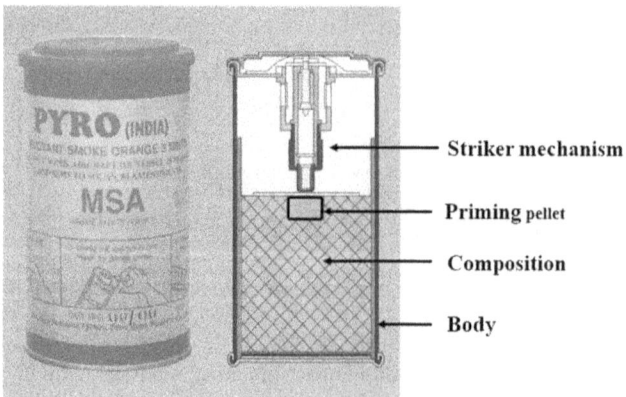

**FIGURE 37.3** Buoyant Smoke Orange (Reprinted from [2]).

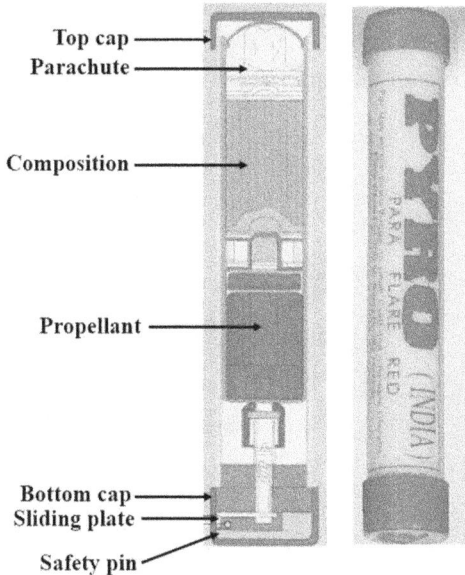

FIGURE 37.4    Para Flare Red (Reprinted from [2]).

The emission of smoke continues even after it is immersed in water (see SOLAS regulations).

v.   *Para Flare Red* (Figure 37.4) [2]: Used in ships, life raft and life boats. On releasing the striker mechanism, propellant ignition takes place, leading to ejection of the burning vivid red star burning for minimum 40 seconds, suspended from parachute with minimum luminous intensity 30,000 candela, height of deployment 300 m when fired vertically and 200 m when fired at 45° angles with speed of desent not more than 5 meter per second (see SOLAS regulations).

vi.   *Hand Flare Red* (Figure 37.5) [2]: Used in ships, life raft and life boats. It is a hand-held device and can be used during day as well as night, produces vivid red signal for minimum 60 seconds with luminous intensity 15,000 candela (see SOLAS regulations).

vii.   *Marker Man Overboard* (Figure 37.6) [2]: It is a lifesaving aid used for marking the position of a person fallen overboard and also aid search by vessels during day and night. It provides dense orange signalling colour smoke for minimum 15 minutes and glow two battery operated bulbs (12 volts, 6 watts) for minimum 90 minutes. These are visible from 2 nautical miles in clear weather. (see SOLAS regulations).

c.   *Land (Used by Ground Personnel or Troops):*
   i.   *Fusee, Warning, Railroad: Red, M72* (Figure 37.7) [1]: Used for railways: Provides vivid red light for minimum 10 minutes and does not get quenched even in heavy rain.

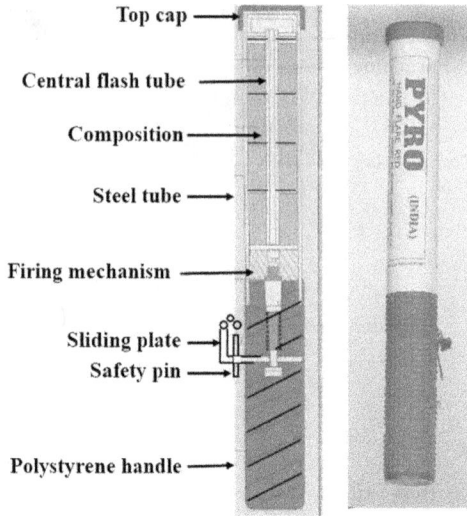

**FIGURE 37.5**   Hand Flare Red (Reprinted from [2]).

**FIGURE 37.6**   Marker Man Overboard (Reprinted from [2]).

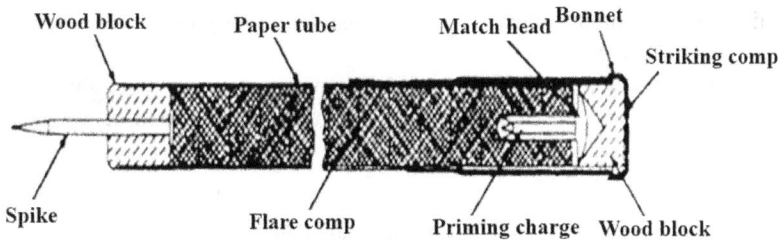

**FIGURE 37.7**   Fuzee, Warning, Railroad, Red, M72 (Reprinted from [1]).

**TABLE 37.2**
**Burning/Smoke Emission Time of Distress Signalling Devices Compiled [2]**

| Devices | Burning/smoke emission time (seconds) | Special effects |
| --- | --- | --- |
| Signal Distress Day/Night | 20 ± 2 | Red Flare |
|  | 20 ± 2 | Orange Smoke |
| Signal Fuzee | 600 | Red Flare |
| Hand Flare Red | 60 | Red Flare |
| Para Flare Red | 40 | Red Flare |
| Buoyant Smoke Orange | 180 | Orange Smoke |
| Signal Fog | – | Loud Sound |

ii. *Signal, Illumination, Ground: Red Star, Parachute, M131* [1]: The signal is a hand-held device containing a single red star illuminant candle, a parachute to suspend the candle and a small rocket propulsion motor. For use on ground for distress signalling to aircraft. Delay 5 seconds, burning time 30 seconds, candlepower 10,000, height of parachute deployment 1,500 feet, visible distance 35 miles.

The burning time or smoke emission time and colour of a few distress signalling devices [2] are given in Table 37.2.

## 37.4   SAFETY OF LIFE AT SEA (SOLAS) REGULATIONS

SOLAS (Safety of Life at Sea) has provided certain regulations for some marine pyrotechnic signal devices, for navy and merchant navy. Some main regulations are as under.

a. *Buoyant Smoke Orange:*
   i. It must be contained in water-resistant buoyant container with clear diagram for operating instruction.
   ii. It should not ignite explosively if used as per the described operating instruction.

   iii.  Smoke emitted must be of high visible colour with a uniform rate and minimum period of 3 minutes.

   iv.  It must emit only smoke and not flame when floating in calm water.

   v.  If immersed in water, it must emit smoke for a period of minimum 10 seconds under 100 mm of water.

b. *Para Flare Red:*

   i.  It must be contained in a water-tight casing.

   ii.  The minimum vertical height of rocket, when fired from the operating point, must be 300 m.

   iii.  A parachute must be activated when it reaches top or near to the top of its trajectory level.

   iv.  It must burn with red flame for a period not less then 40 seconds with minimum luminous intensity of 30,000 candelas.

   v.  The rate of decent after opening of the parachute must be minimum 5 m/seconds.

   vi.  Arrangement must be such that it must not damage or burn the parachute when the flare is activated.

   vii.  A brief description with figure of use of flare shall be printed on casing.

c. *Hand Flare Red:*

   i.  It must be contained in a water tight casing.

   ii.  It must have self-ignition system.

   iii.  It should not cause discomfort to the person using it.

   iv.  It must continue to burn after having immersed for a period of 10 seconds under 100 mm of water.

   v.  It must illuminate with a bright red colour for not less than 60 seconds with average luminous intensity not less than 15,000 candelas.

   vi.  A simple and brief operating instruction diagram must be there on its casing.

As per SOLAS, Buoyant Smoke Orange, Para Flare Red and Hand Flare Red devices are required to be kept in the ratio of 2:4:6 numbers in all life boats, rafts, etc.

d. *Marker Man Overboard:*

   i.  At least one man overboard marker to be mounted on each bridgewing.

   ii.  Once released the marker will float in sea attached to the lifebuoy and produce dense orange smoke for at least 15 minutes and two lights will burn for at least 2 hours as a day and night signal.

   iii.  Marker man overboard should be able to be operable from a mounting height of at least 30 m.

Readers may see reference [3] on suggested design criteria of distress signalling devices.

An alternative to signalling pyrotechnics is electronic flares, also referred as *Electronic Visual Distress Signals (EVDS)* using LED technique (yet to be approved by SOLAS) which require change of battery periodically but can be used several times and lasts for several hours and does not give flame or smoke.

# REFERENCES

[1] *"Army Ammunition Data Sheets for Military Pyrotechnics"*, (Federal Supply Class 1370), Technical Manual No. 43-001-37, Headquarters, Department of the Army, Washington DC ,6 January 1994.

[2] *"Product Brochure"*, Indian Ordnance Factories, Ordnance Factory Dehuroad, Pune, India.

[3] G. M. Simpson, *"Pyrotechnic Devices for Use by Water-Sport Enthusiasts Amateur Sailors and Small Craft in Coastal Waters"*, Proceedings of the sixth international pyrotechnics seminar, July 1978.

# 38 Simulating Ammunitions

## 38.1 GENERAL

Simulating ammunitions are widely used as alternatives to deceive enemy and train troops. These are cost-effective as well. These ammunitions are used to

a. Produce sounds and flash and sometimes smoke of a battle field to deceive the enemy
b. Training exercises for battle field experience/familiarisation of unseasoned troops with battle conditions without exposing them to main ammunition

These mostly use body material as kraft paper, plastic, chipboard, polyethylene etc. and thus has less fragmentation hazard and are less costly.

## 38.2 TYPES OF SIMULATING AMMUNITIONS

These are of three types

a. Burst type: These are of *"air burst"* type which burst with delay in air with sound and exhibit light, puff of smoke and sound, simulating the artillery ammunition burst
or *"ground burst"* type which function on the ground and provide flash or big sound of burst.
b. Whistling types: These produce whistling sound of ammunition firing
c. Blank ammunition: These are without projectile and produce big sound for simulation and ceremonial firings giving flash, smoke and loud sound.

Examples are as follows:

i. *Simulators, Explosive Boobytrap: Flash, M117; Illuminating. M118; Whistling, Ml19* (Figure 38.1) [1]: As safe boobytraps during maneuvers and in troop training to teach the installation, detection and boobytraps, and to install caution in troops exposed to traps set by an enemy. M117 functions instantaneously (filled with loose flash composition) with explosion and flash while M118 (filled with pressed illuminating composition) functions for 28 seconds giving illumination and M119 (filled with whistle composition) functions with whistle of 2.5-5 seconds.
ii. *Simulator, Flash, Artillery: M110* (Figure 38.2) [1]: Used for effect battle conditions in artillery maneuvers and as a decoy in forward combat areas. Its flash closely resembles those of the 90 mm Gun M2 series and the 155 mm

DOI: 10.1201/9781003093404-38

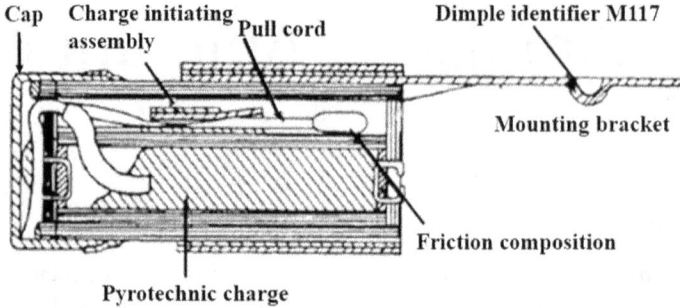

**FIGURE 38.1**  Simulator, Explosive Boobytrap; Flash, M117; Illuminating, M118; Whistle, M119 (Reprinted from [1]).

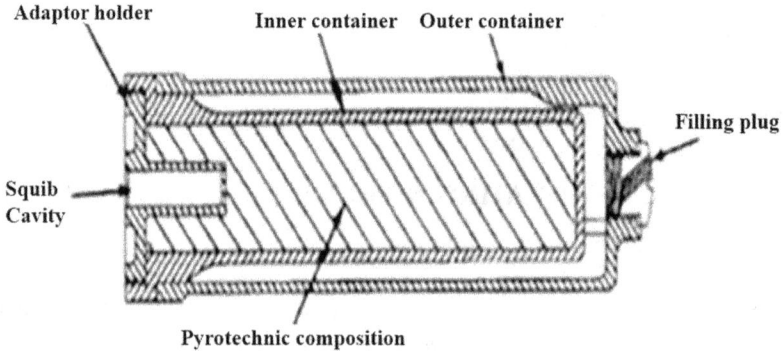

**FIGURE 38.2**  Simulator Flash Artillery M110 (Reprinted from [1]).

Howitzer M1 series (particularly with Charge 5). The electric squib ignites the pyrotechnic charge. The simulator flash lasts slightly longer than that of the actual weapon. Its report is loud, but not comparable to the actual weapon.

iii. *Simulator, Flash, Artillery: M21* [1]: Simulates the acoustic (bang) and optional (flash and smoke) signature of tank main gun. Firing is from the Simulator Tank Gunfire. The simulator is mounted on the tank's main gun tube and each cartridge is activated electrically upon depressing the main gun trigger. The cartridges function in place to produce the audible and visual simulation of the tank gun.

iv. *Simulator, Launching, Anti-Tank Guided Missile and Rocket, M22* [1]: It provides a credible simulation of the weapon signature including report, flash and smoke. Provides flash for 400 milliseconds duration at 4,000°C and bang and white smoke.

v. *Simulator, Projectile, Air Burst: M27A1B1* [1]: Simulate the airburst of an artillery projectile for high-burst ranging, practice. Launching is from an M76 grenade launcher attached to an M14 rifle firing a 7.62 mm grenade

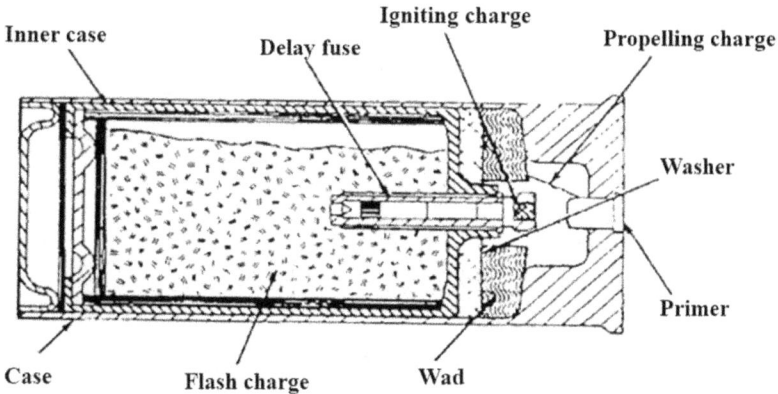

**FIGURE 38.3**  Simulator Projectile Air Burst M74A1 (Reprinted from [1]).

cartridge M64. It has a delay of 5 seconds and burns instantaneously giving flash, a puff of gray smoke and a loud report.

vi. *Simulator, Projectile, Air Burst: M74A1 and M74* (Figure 38.3) [1]: Simulates artillery fire. Aimed at a 45-degree elevation, the height of burst is about 100 feet and burns instantaneously with a delay of 2–3 seconds with candlepower 600,000. The difference between the two ammunitions is that M74A1 has a better reliable fuze.

vii. *Simulator, Projectile, Air Burst: Charge, Smoke Puff, White* [1]: Visually simulates the bursting of an artillery projectile near the ground. Burns instantaneously.

viii. *Simulator, Projectile, Ground Burst: M115A2* (Figure 38.4) [1]: Simulates battle noises and effects (shells in flight and ground explosions) during troop maneuvers on land. Simulator consists of a cylindrical paper tube containing a photoflash charge and a whistle assembly. Safety fuze

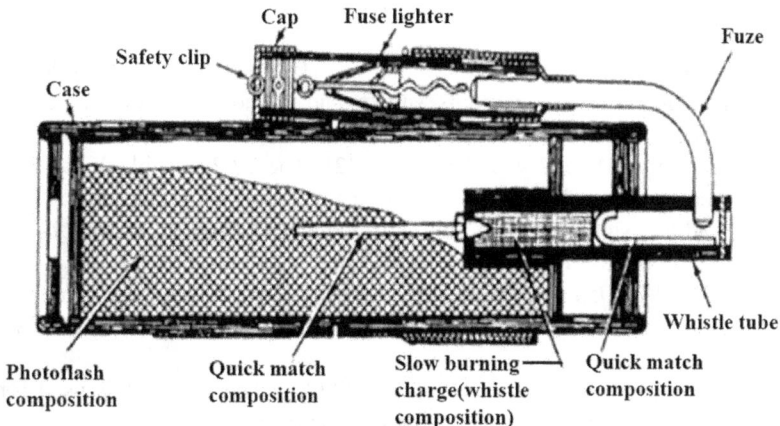

**FIGURE 38.4**  Simulator, Projectile Ground Burst M115A2 (Reprinted from [1]).

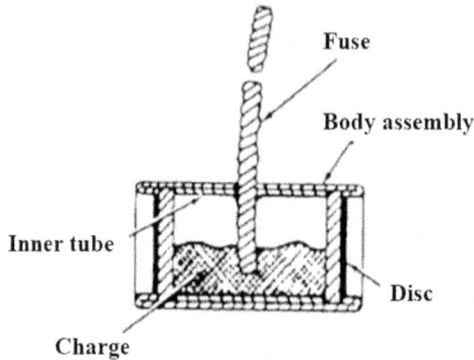

**FIGURE 38.5**   Detonation, Simulator, Explosive: M80 (Reprinted from [1]).

provides a delay of 6–10 seconds after ignition for whistle composition which provides whistle for 2–4 seconds, a delay of 8–14 seconds for photoflash composition which bursts giving flash and loud report.

ix. *Simulator, Hand Grenade: M116A1* [1]: A hand-thrown ammunition which simulates battle field noises and effects during troop manoeuvres on land. It contains photoflash composition and functions with a delay of 5–10 seconds.

x. *Detonation, Simulator, Explosive: M80* (Figure 38.5) [1]: This simulates rifle or artillery fire, hand grenades, boobytraps or land mines. This simulator is a paper cylinder containing 3 grams of flash composition. Each end of the cylinder is closed by a disk, crimp sealed in place. A length of fuze extends from the side of the cylinder. The simulator is fired by lighting the fuze with a match or similar source of flame. The burning time of the fuze provides a 4 to 7 second delay, and directly ignites the flash charge at the completion of the delay. When used for simulating boobytraps or land mines, a firing device such as a pull type, pull-release type or pressure type may be substituted for fuze.

xi. *Simulator, Atomic Explosion: M142* [1]: Simulates a ground-detonated nuclear explosion, contains smoke and sound composition, burn time 1.8–2.4 seconds, cloud of fireball, 10 feet minimum diameter.

xii. *Cartridge SA 7.62 mm Blank L.A.* [2]: The 7.62 mm blank cartridge is used for gun salutes. It is also used for gunnery training purpose, simulated fire, signalling, etc. This cartridge is fired from a Light Machine Gun (LMG) or rifle for training, signalling gun salutes, etc.

xiii. *Cartridge 105 mm Blank M395* [3]: These cartridges are used for saluting and simulated firing. Cartridge 105 mm Blank M395 is shown in Figure 38.6.

xiv. *Cartridge, Caliber 0.30, Blank, M1909* [4]: The cartridge is used for simulated firing in training or for saluting purposes.

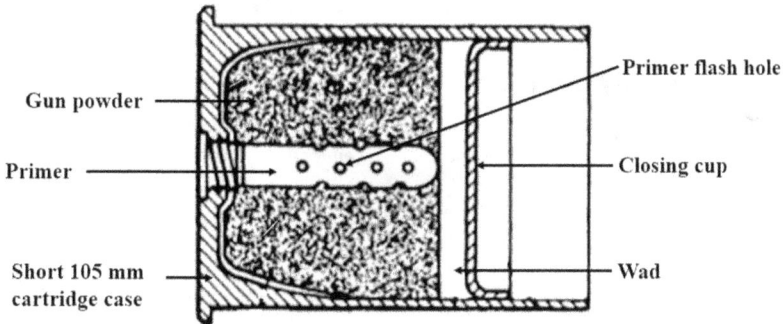

**FIGURE 38.6** Cartridge 105 mm Blank M395 (Reprinted from [4]).

xv. *Cartridge, Caliber 0.45, Blank, M9* [4]: The cartridge is used for simulated firing in training manoeuvres and for saluting purposes.
xvi. *Cartridge, Caliber 0.50, Blank, M1* [4]: The cartridge is used to simulate firing in training exercises.
xvii. *Cartridge, Caliber 0.50, Blank, M1A1* [4]: The cartridge is used to simulate firing in training exercises.
xviii. *Cartridge, 75 mm Blank, M337A2 (M337A1E1) M337A1, M337* [3]: These cartridges are used for saluting and simulated firing.
xix. *Cartridge 76 mm Blank M335A2* [3]: This cartridge is used for salutes and simulated fire in 76 mm guns.
xx. *Cartridge 90 mm Blank M394* [3]: This blank cartridge is provided for saluting purposes and simulated firing in 90 mm guns.

Table 38.1 provides the summary of some of the simulators.

**TABLE 38.1**
**Simulators Summary Compiled [1]**

| Ammunition | Method of actuation | Special effect | Delay (seconds) | Burn time |
|---|---|---|---|---|
| Simulator Booby Trap | Tripwire | | | |
| M117 flash comp | | Explosion | 0 | instantaneous |
| M118 illuminating comp. | | Flame | 0 | 28 seconds min. |
| M119 whistle comp. | | Whistle | 0 | 2.5–5.0 sec |
| (Kraft paper) | | | | |
| Simulator Flash Artillery M110 | Fired from tube | Flash | 0 | Instantaneous |
| flash comp. | but actuated by | Sound | | |
| (plastic) | blasting | | | |
| | machine or | | | |
| | battery | | | |

*(Continued)*

## TABLE 38.1 (Continued)
## Simulators Summary Compiled [1]

| Ammunition | Method of actuation | Special effect | Delay (seconds) | Burn time |
|---|---|---|---|---|
| Simulator Flash Artillery M21<br>Flash comp.<br>(Polyethylene) | Electrical | Flash | 0 | Instantaneous |
| Simulator Hand Grenade M116A1<br>Photoflash comp.<br>(Kraft paper) | Manual pull cord | Flash | 5–10 | Instantaneous |
| Simulator, Launching Anti-Tank,<br>  Guided Missile and Rocket M 22<br>Smoke composition<br>Bang composition<br>(Case glass filled ABS) | Electrical/<br>Mechanical | Flash 400<br>  ms, 4000°F<br>smoke white<br>bang 165 db<br>@1foot | 0 | Instantaneous |
| Simulator Projectile Air Burst<br>  M27A1B1<br>Smoke comp.<br>(Plastic) | M 76 grenade<br>launcher | Flash, puff<br>  of gray<br>  smoke and<br>  sound | 5 | Instantaneous |
| Simulator Projectile Air Burst<br>  M 74A1<br>Flash comp.<br>(Aluminium) | Pyrotechnic<br>pistol AN-M8 | Flash and<br>  loud noise | 2–3 | Instantaneous<br>600,000<br>candle power |
| Simulator Projectile, Air Burst:<br>  Charge, Smoke Puff, white<br>Black powder comp.<br>(chipboard) | Percussion<br>primer in smoke<br>puff discharger | Bursting | 0 | Instantaneous |
| Simulator Projectile<br>Ground Burst M115A2<br>Photoflash comp.<br>Whistle comp.<br>(Kraft paper) | Hand pull cord | Whistle<br>Burst | 6–10<br>8–14 | whistle 2–4 sec.<br>instantaneous |
| Detonator, Simulator Explosive,<br>  M 80<br>Flash comp.<br>(Kraft paper) | Ignition of fuze<br>by match | Flash, smoke<br>  and sound | 4–7 | Instantaneous |

*Notes*
 (Body material in bracket)

# REFERENCES

[1] *"Army Ammunition Data Sheets for Military Pyrotechnics"*, Federal Supply Class 1370, Technical Manual No. 43-0001-37, Headquarters, Department of the Army, Washington DC, 6 January 1944.

[2] *"Product Brochure"*, Ammunition Factory Khadki, Pune, Maharashtra, India.

[3] *"Technical Manual, Army Ammunition Data Sheets, Artillery Ammunition, Guns, howitzers, Mortars, Recoilless Rifles, Grenade Launchers and Artillery Fuzes"*, Federal Supply Class 1310, 1315, 1320, 1390 TM 43-0001-28, Headquarters, Department of The Army, Washington DC, 27 October 2003.

[4] *"Army Ammunition Data Sheets, Small Caliber Ammunition"*, FSC 1305, TM 43-000 1-27, Headquarters, Department of The Army, Washington DC, 29 April 1994.

# 39 Infrared Flare Ammunitions and Devices

## 39.1  GENERAL

The infrared ammunitions are used for:

a. Tracking the enemy movement in night by infrared illumination
b. Decoys for infrared seeking missiles
c. Infrared illumination for thermal imaging
d. Training for firing of infrared seeking missiles

## 39.2  TYPES OF INFRARED FLARE AMMUNITIONS

Infrared ammunition can be divided into following types:

a. Mortar Infrared illuminating bombs with parachute
b. Artillery infrared illuminating shells with parachute
c. Hand held infrared flares with parachute
d. Infrared decoy flares for defensive purpose
e. Infrared tracking flares

### 39.2.1  MORTAR INFRARED BOMBS WITH PARACHUTE

Mortar infrared bombs are similar to normal illuminating bombs but filled with infrared composition. On firing, the primary cartridge functions giving flame to augmenting charges and the combined pressure of these two propels the bomb body through the mortar. After desired delay time or distance from the target, the fuze (which is either a mechanical time fuze or electronic fuze) functions and gives flash to the expelling charge and causes the bomb body to open up by shearing the pins, thereby allowing the candle (pressed pyrotechnic composition in candle) with the parachute to eject out and hang in the air. The candle pyrotechnic composition radiates infrared radiation at the target area and can be seen through night vision devices. A typical 60 mm mortar infrared bomb [1] with parachute is shown in Figure 39.1.

A typical 81 mm mortar infrared bomb with parachute [1] is shown in Figure 39.2

DOI: 10.1201/9781003093404-39

Cartridge ignition M702                          Base plate      First fire composition

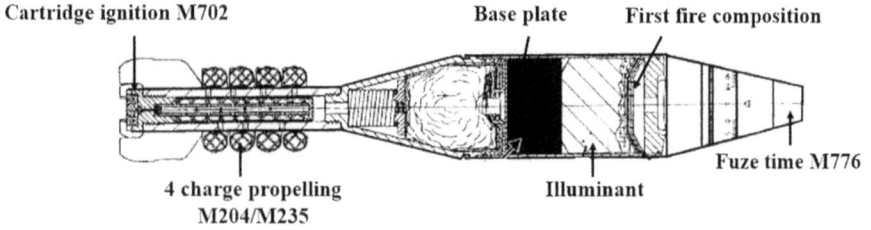

4 charge propelling                          Illuminant
M204/M235

Fuze time M776

**FIGURE 39.1**   Bomb 60 mm Illuminating IR M767 (Reprinted from [1]).

Fin, mortar, cartridge;                    Projectile, 81 mm: infrared
81 mm, M 29                                illuminating XM 816 loading assembly
                   Cartridge, ignition, M752A1
                                                              Fuze, MTSQ, M772

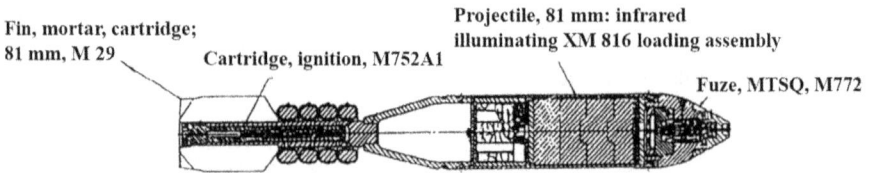

**FIGURE 39.2**   Cartridge, 81 mm: Illuminating, Infrared (IR), Ms16 W/Fuze, Mechanical Time Superquick, M772 (Reprinted from [1]).

## 39.2.2   ARTILLERY INFRARED ILLUMINATING SHELLS WITH PARACHUTE

Artillery infrared illuminating shells are akin to normal illuminating artillery shells but filled with infrared composition. The functioning is similar to illuminating ammunition except that it radiates infrared radiation over the target area and can be seen by night vision devices. The visible light produced are near infrared region. The performance of some infrared illuminating shells and bombs [2] is given Table 39.1[1]

Some typical infrared illuminating ammunitions [2] are at Figure 39.3.

## 39.2.3   HAND-HELD INFRARED FLARES WITH PARACHUTE

Examples of concealment index (see Section 18.10) of hand-held infrared flares are [3]

  i. 26.5 mm parachute flare (Rheinmetall) burn time 15 seconds, radiant intensity 20 w.sr$^{-1}$ (NIR) and 250 cd (VIS), concealment index $\chi$ of 54.
  ii. 40 mm hand-held parachute rocket (Rheinmetall) burns 28 seconds with NIR emission of 25 w.sr$^{-1}$, 250 cd and concealment index ($\chi$) of 68.

## 39.2.4   INFRARED FLARE DECOYS FOR DEFENSIVE PURPOSE

Target (fighter aircrafts, helicopters and war ships)-seeking technique of enemy missiles is through either infrared tracking or through radar tracking. Infrared flare decoys protect the aircraft, helicopter and war ship from the infrared -seeking missile threat.

**TABLE 39.1**

**Performance of Infrared Illuminating Ammunitions (Compiled) [2]**

| Ammunition | Burn time | Illumination (cd) | Infrared output (watt.sr$^{-1}$) | Minimum range (m) | Maximum range (m) |
|---|---|---|---|---|---|
| Shell 40 mm M992 Infrared Para (BA03) | 40 | – | – | – | – |
| Bomb 60 mm Illumination Infrared M767 | 40 | <450 | 30 | 300 | 3,175 |
| Bomb 81 mm Cartridge Illumination, Infrared M816 | 60 | <500 | 50 | 1,025 | 4,925 |
| Shell 105 mm Infrared Illuminating M1064 | 40 | <450 | 30 | 4,100 | 9,100 |
| Bomb 120 mm Cartridge Infrared M983 | 50 | <550 | 75 | 375 | 6,675 |
| Shell 155 mm Infrared Illuminating XM1066 | 120 | <650 | 75 | 2,800 | 17,500 |

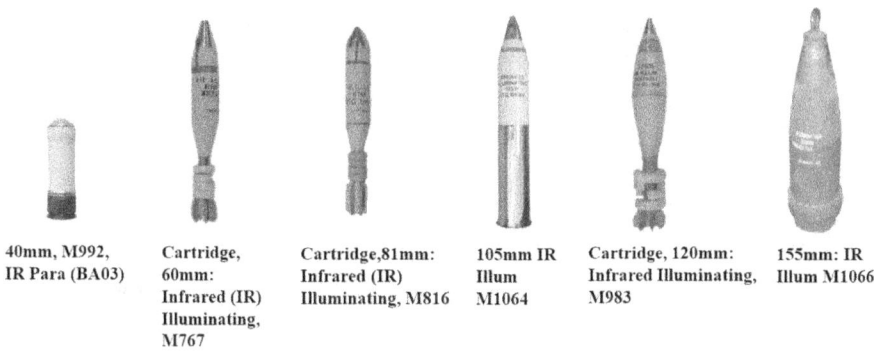

40mm, M992, IR Para (BA03)    Cartridge, 60mm: Infrared (IR) Illuminating, M767    Cartridge,81mm: Infrared (IR) Illuminating, M816    105mm IR Illum M1064    Cartridge, 120mm: Infrared Illuminating, M983    155mm: IR Illum M1066

**FIGURE 39.3** Some Typical Infrared Illuminating Ammunitions (Reprinted from [2]).

To avoid being hit by a heat seeking missile, pyrotechnic compositions have been developed, which produce almost similar infrared radiations as the aircraft or helicopter or war ship. These are made in the form of pellets and dispensed from the aircraft or helicopter as and when it is under attack from the infrared seeking missile.

Decoy flares when ejected and burn for sufficient duration, they produce intense heat a little more than the aircraft exhaust engine. This causes confusion and distraction to missile and induces the heat seeking missile to get locked to the infrared flare signature instead of the aircraft infrared signature. This gives enough time to pilot to move away from its path. The missile explodes on burning pellets (decoy flare) at a farther distance from the aircraft and thus avoids damage to aircraft. Thus, the main function of infrared flares is to attract the missile through its own infrared radiations rather than the missile getting attracted to the aircraft radiations.

### 39.2.4.1 Mechanism of Infrared Flare Decoys for Defensive Purpose
A missile can hit the aircraft only if

a. The aircraft is detected through its infrared radiation
b. The enemy missile is capable of tracking the aircraft flight trail
c. The missile is guided suitably for hitting the aircraft

Infrared decoys ensure that the above (b) and (c) are avoided by inducing the missile to its own infrared radiation and thus avoid the danger to the aircraft. Let us now consider an enemy aircraft and a heat seeking missile targeting the aircraft.

The infrared missile (generally with field of view of 1° to 2°) is able to lock on to enemy aircraft by the infrared radiation from the aircraft (exhaust gases, hot metal parts, etc.) and follows the aircraft and destroys it. The same is shown in the Figure 39.4.

The infrared decoy ammunition when ejected from the aircraft provides the infrared radiation similar to the aircraft and induces the missile to get locked to the flare instead of the aircraft and gets destroyed. The same is shown in the Figure 39.5.

Thus, infrared flare composition produces infrared radiation similar to the infrared radiation emitted by aircraft and act as decoy for the infrared missiles.

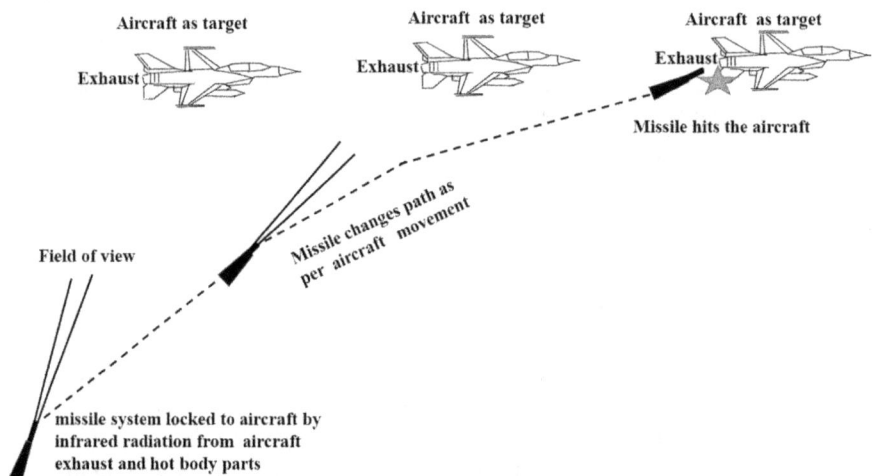

**FIGURE 39.4**  Infrared seeking Missile Attack on Aircraft.

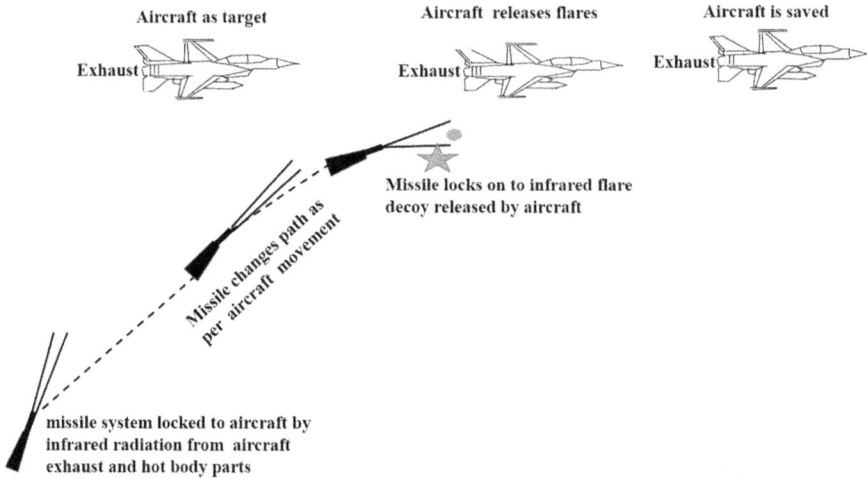

**FIGURE 39.5**  Protection of Aircraft Using Flare Decoy.

Let us understand the process of deploying infrared flares to deceive the heat-seeking infrared missile. When the infrared signature radiation from the aircraft is captured by a missile seeker, it locks on to the target aircraft. Once the presence of a "live" attacking IR missile is noticed, flares are released by the aircraft in an attempt to deceive the missile; some systems are automatic, while others require manual jettisoning of the flares. Flares are released from the dispenser. These dispensers can be programmed by the pilot to dispense flares in short intervals, one at a time, long intervals or in clusters through electronic panel (Figure 39.6).

Just after releasing the flares, the aircraft is taken out of the field of view of the heat-seeking missile by taking a sharp turn and reducing the engine power and thus causing a reduction in infrared flare signature of the aircraft engine. This causes the heat-seeking missile to follow the infrared flare instead of the aircraft due to change in temperature and new infrared signatures. Thus, infrared flares become the new false target for the heat-seeking missile and missile tracking command causes the missile tracking logic to lock with the infrared flares and unlock the aircraft. As the aircraft moves away quickly and the burning flares also shift its position, the missile completely gets deviated from the path of aircraft and gets destroyed on its new target infrared flares and thus aircraft gets saved.

### 39.2.4.2  Important Features of Deployment of Infrared Decoys

The three basic elements of aircraft defense are [4]

  a. Suppression of the aircraft's signature to reduce missile acquisition range
  b. Warning of a missile launch and cueing of an appropriate countermeasure
  c. Activation of a countermeasure, of which there are two types: decoys and jammers.

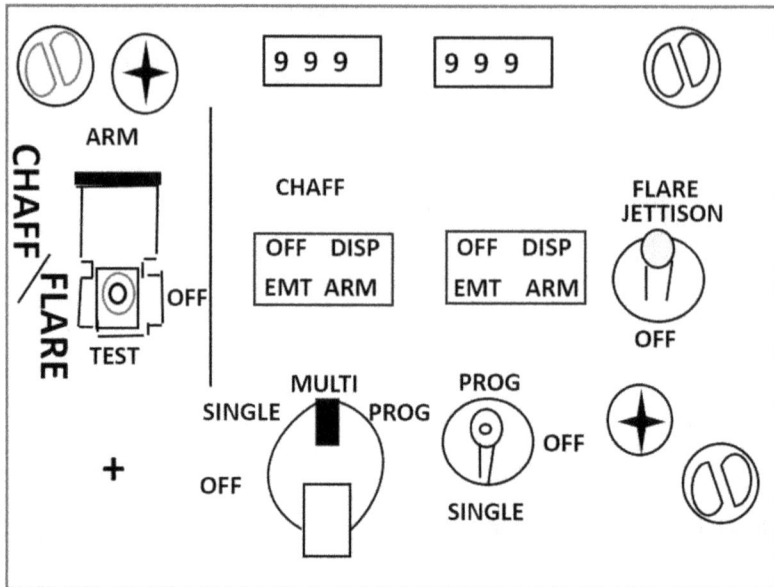

**FIGURE 39.6**   A Rough Sketch of Electronic Panel of a Fighter Aircraft.

Readers may also refer [4] on aircraft infrared principles, signatures, threats and countermeasures. An infrared decoy on activation undergoes following three phases.

a. Ignition (ejection and ignition of flare)
b. Deployment in air (giving infrared radiation of required intensity and duration)
c. Decoying enemy infrared missile (engagement with infrared enemy missile)

The important feature of an infrared flare are as follows.

a. *Timing of Deployment*: The flares are required to be released at the opportune time. If deployed early, the flare would move far away from the aircraft and thus out of field of view of missile seeker which is not desirable. On the other hand, a late deployment shall result in engagement of the missile with the flare near the aircraft and fragments of the missile may damage the aircraft.
b. *Flare Ejection Velocity*: The flares are ejected due to functioning of the impulse cartridge containing the propellant inside impulse cartridge. A low ejection velocity shall cause flare burn near the aircraft and this may pose hazard to aircraft when missile engages with flare and detonates near the aircraft. A very high ejection velocity shall cause flare to burn far away but it may be out of field of view (FOV) of the missile seeker and hence not useful. A velocity of approximately 20 m sec$^{-1}$ has been found to be very suitable.
c. *Rise Time*: It is defined as the time taken by a flare to reach its peak intensity,

or a defined fraction of that intensity, measured in milliseconds. It must be very fast to ensure that the flare is capable of luring the enemy missile at the earliest as the main target instead of the aircraft. Due to low atmospheric pressure at high altitudes, the flare takes more time to reach its peak intensity and thus increase the rise time. This is detrimental to flare becoming the target in short time. Suitable compositions are required to be developed to meet the above requirements.

d. *Burn-Time:* It is the overall time that the flare produces specified output, usually of the order of 3–6 seconds. A very low flare burn time would cause missile to re-capture the aircraft signals and in such cases further infrared flares have to be deployed. A sufficiently long burn time would ensure positive locking with the enemy missile. This requires loading of more composition but space restriction in aircraft does not allow the same. Hence maintaining flare burn time is very essential. A flare shall burn for more time with lower intensity at higher altitude. A longer burn time shall enhance the probability of inducing the infrared missile by the flare. A number of flares may be dispensed to overcome this problem but it would reduce the number of decoys available with the aircraft.

e. *Peak Intensity:* It is the maximum power output of the flare over time, measured in kilowatts/steradian in a given waveband. The intensity of the flare must be larger than the aircraft to ensure it as a main target than the aircraft. It must meet the intensity ratios of alpha band (2–3 µm) and beta band(3–5 µm) of the aircraft where α- band and β-bands exhibits radiation intensity of hot components of aircraft and exhaust, respectively.

Key decoy performance requirements [5] are shown in Table 39.2.
Following are the drawbacks of infrared decoys [6]:

a. Burns at the necessary intensity for only a short period of time.
b. Gravity quickly separates the flares from the dispensing aircraft removing them from the missile seeker's field of view-thus limiting or reducing their effectiveness.
c. IR flares can also be identified by some missiles and rejected because they tend to initially provide more intense IR emissions than the aircraft.
d. Furthermore, some missiles can also identify IR emitting flares by their IR spectrum.

### 39.2.4.3  Infrared Decoy Flares Assembly

Pressed pellets or extruded pellets used in infrared decoys.

(i) *Pressed Pellets:* Infrared decoy flares contain pressed pellets and are grooved (cutting and milling) for increased surface area and grooves are linked with flash pick up hole for better pick up of ignition. These grooves are primed preferably under nitrogen atmosphere. The pellets are in many cases wrapped with aluminium foil to avoid friction during assembly as well as during ejection on firing. The pellets are assembled in a metallic body, may be cylindrical or circular, along with a closing cap and an optional pyrotechnic sequencer mechanism with delay. The top

**TABLE 39.2**

**Key Decoy Performance Requirements [5]**

| Performance characteristics requirement | Typical objective | Design issues |
|---|---|---|
| Peak intensity | Exceed target platform signature to attract target | Primary driver for decoy weight, volume and cost |
| Initiation rise time | Achieve effective intensity quickly enough to capture seeker before leaving field of view | Usually requires special ignition materials. Excessive rapid rates may trigger counter-countermeasures |
| Spectral characteristics | Differences between decoy and target may be exploited by target | Dependent on threat characteristics. Often difficult to verify. May have severe effect on choice of fuel and radiation mechanism |
| Function (burn time) | Maintain credible signature until target is no longer in threat field of view | Also, strong impact on weight, volume and cost |
| Ejection velocity | Generate sufficient separation rate from target so that threat is decoyed beyond lethal radius | Launcher must be able to withstand recoil |
| Aerodynamic properties | Maintain credible trajectory to avoid rejection by threat | Some threat-imposed requirements have substantial impact on complexity and decoy size |

is closed with a circular disc or rectangular disc, with felt, as the case may be, and crimped. The infrared cartridge is then assembled with impulse cartridge at the base when required.

The pyrotechnic sequencer mechanism (PSM) or pyrotechnic safety mechanism (PSM) contains a slider (with spring) and a delay pellet and piston. This pyrotechnic sequence mechanism blocks the flash hole of the infrared pellet. On firing of the impulse cartridge, the delay pellet receives the flash and at the same time the pellet is pushed outward from the metallic or plastic casing. Just outside the casing, the slider (with spring) slides away, allowing the delay pellet in alignment with the flash hole for direct contact with the infrared pellet. After specified delay, the flash from delay pellet ignites the infrared pellet, thus ensuring that the infrared pellet burns away from the aircraft. A typical rectangular infrared flare is shown in Figure 39.7.

Another conventional cylindrical infrared flare decoy is shown at Figure 39.8.

(ii) *Extruded Pellets:* MTV (magnesium, teflon and Viton) extruded cake is grooved by sawing and milling to form grooves in pellet and dipped in fast burning MTV slurry composition in acetone and dried. Aluminium wrapper is used to wrap the pellet which avoids friction during ejection. Infrared flare burning

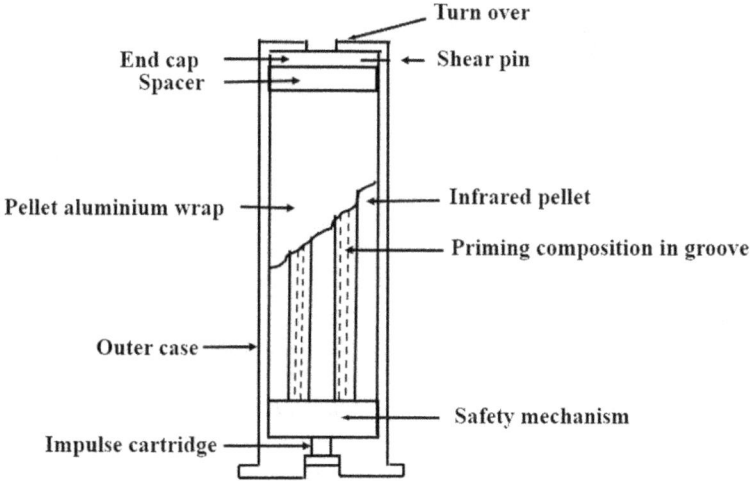

**FIGURE 39.7** Typical Rectangular Infrared Flare Ammunition.

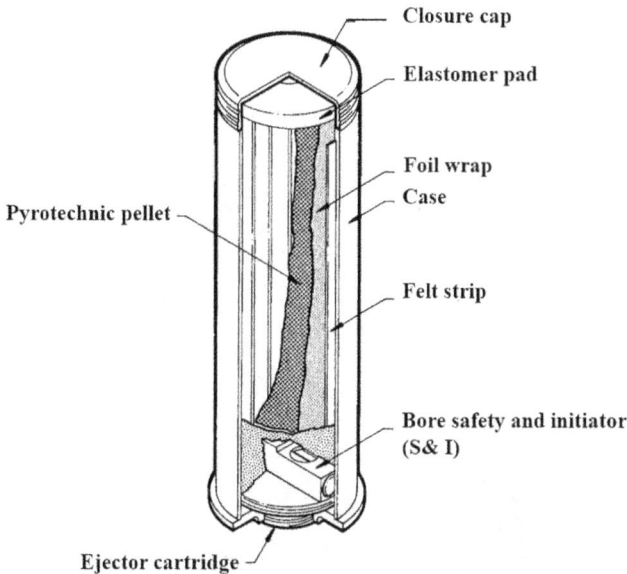

**FIGURE 39.8** Components of a Conventional Flare (Reprinted from [5]).

characteristics are checked at this stage. Pyrotechnic sequencer mechanism with ignition pellet is assembled to pellet. Pellet with PSM is then assembled to cartridge case.

X-ray and ejection test are carried out at this stage. Impulse cartridges are assembled or issued separately as per user's requirement.

There are a wide range of calibres and shapes of infrared flares. This is due to variety of shapes and sizes of aircraft, helicopters, etc. and each having its own infrared signature. Due to volume storage restrictions on board platforms, many aircrafts use square infrared flare cartridges. However, cylindrical cartridges are also used. The American aircrafts mostly use square shaped infrared flares, apart from some cylindrical shaped ones. The cylindrical ones are used mainly on-board French aircrafts and Russian aircrafts. Typical MTV based cylindrical infrared flares are of 1", 1.5", 2" and 2.5" while rectangular infrared flares are 1" × 1" × 8", 2" × 1" × 8" and 2" × 2.5" × 8". The air force impulse cartridges are generally given storage service shelf life of 10 year in hermetic sealing, 6 months after opening and 1 month as installed life on board aircraft.

Figure 39.9 gives infrared flares in action [4] in a typical fighter aircraft.

Figure 39.10 shows Lockheed C-130 Hercules aircraft in action [7].

Warships exhibit higher intensity of infrared radiation. Hence warships use heavy calibre decoys like 130 mm decoy variants for anti-ship missiles. However, some reduction in infrared signature of a warship is possible by cooling the exhaust gases, hull and other hot parts with cold water. It is necessary that the decoy flare is launched in the direction of the infrared seeker missile.

DOD public domain release.

**FIGURE 39.9**   Flare Salvo From C-17. (A C-17 Globemaster III Aircraft Releases Flares Over the Atlantic Ocean during a Local Exercise Over Charleston, South Carolina, 6 May 2006.) (Reprinted from [4].

**FIGURE 39.10**   Lockheed C-130 Hercules Aircraft Jettisoning Decoy Flares (Reprinted from [7]).

#### 39.2.4.4   Infrared Flare Dispensing Towed Decoy [6]

A towed IR decoy flies the same profile as the aircraft it is protecting, such that the decoy remains in the IR-guided missile's field of view unlike current aircraft deployed flares which quickly fall away from the aircraft. This decoy also exhibits the same IR spectral characteristics such that the attacking missile cannot discriminate between the decoy and the aircraft to be protected on the basis of these characteristics. Furthermore, this decoy is able to vary its radiant intensity so as to provide an irresistible distraction to the incoming missile. Finally, the decoy is long-lived so that it provides protection against a possible missile attack over an appropriate period of time. This allows the towed IR decoy to be used pre-emptively (i.e., without need of warning of missile attack) at the option of the aircrew whenever they are likely to be immediately vulnerable to IR missile attack. The block diagram of the major components is at Figure 39.11.

### 39.3   DIFFERENCES BETWEEN INFRARED DECOYS AND CHAFF DECOYS

The infrared as well as chaff decoys are deployed by aircrafts to save the aircraft from enemy infrared guided missiles as well as from radar guided missiles. The differences between infrared flare decoys and chaff decoys are at Table 39.3.

Readers may refer [7] for genesis of infrared decoy flares- the early years from 1950 into the 1970s and [8] for electronic warfare fundamentals.

**Host aircraft** — **Forward section** — **Aft section**

Host aircraft | Launch controller | Power supply | Towline | Forward section | Modem | Power Conditioning circuitry | Motor control circuitry | Motor | AFT Section | Payload dispensing mechanism | Screw | Piston | Payload of Pyrophoric material

Decoy

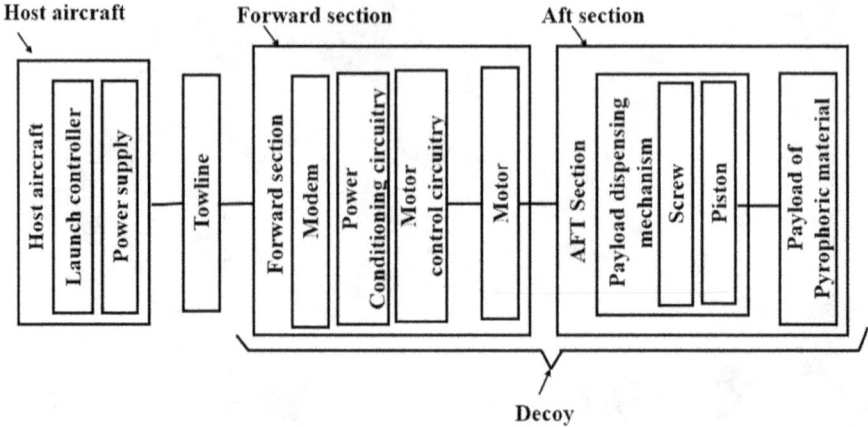

**FIGURE 39.11** The Block Diagram of Major Components of Flare Dispensing Towed Decoy (Reprinted from [6]).

## TABLE 39.3
## Differences between Infrared Decoys and Chaff Decoys

| Infrared flare Decoys | Chaff Decoys |
| --- | --- |
| Infrared flare is IR countermeasure | Chaff is a radar countermeasure |
| Protection from attack on aircraft by infrared guided missiles which are fitted with heat seeker and follows infrared signature of the aircraft | Protection from attack on aircraft by radar guided missiles which are fitted with a radar receiver and follows the RF (radio frequency) signature of the aircraft |
| Based on infrared radiation signature of aircraft | Based on RF signature of aircraft |
| Uses pyrotechnic composition | Uses aluminum coated glass dipoles dia. 23-28 micron |
| Special effect is by pyrotechnic combustion which produces infrared radiation | Special effect is by chaff dipoles |
| Combustion gives infrared radiation (invisible) and the target appears as a region of hot and cold areas | Cloud of chaff (invisible) radiates back radar frequency and appears as cluster of radio frequency emitters |
| Confuses and induces infrared guided missile towards infrared flare decoy | Confuses and induces radar guided missile towards cloud of chaff dipoles |

## REFERENCES

[1] *Technical Manual, Army Ammunition Data Sheets, Artillery Ammunition, Guns, howitzers, Mortars, Recoilless Rifles, Grenade Launchers and Artillery Fuzes,* Federal Supply Class 1310, 1315, 1320, 1390, TM 43-0001-28, Headquarters, Department of The Army, Washington DC, 27 October 2003.

[2] *"PEO Ammunition Systems Portfolio Book 2012–2013"*, Ammunition Program Executive Officer, Picatinny, Arsenal NJ.

[3] Susane Scheutzow, *"Investigations of Near and Mid Infrared Pyrotechnics Detonation Velocities of New Secondary Explosives"*, Ph.D. Thesis, 2012 Ludwig-Maximilian University, Munich, Germany.

[4] Jack R. White, *"Aircraft Infrared Principles, Signatures, Threats, and Countermeasures"*, NAWCWD TP 8773, Naval Air Warfare Center Weapons Division, Point Mugu, CA, September 2012.

[5] Neal Brune, *"Expendable Decoys"*, Chapter 4, Volume 7, Countermeasure Systems, The Infrared and Electro Optical Systems Handbook, Editor David H. Pollock, Executive Editor, Joseph S Accetta and David L. Schumaker, Copublished by Infrared Analysis Information Center, Environmental Research Institute of Michigan, Ann Arbor, Michigan, USA and SPIE Optical Engineering Press, Bellingham, Washington, USA. ©1993 The society of Photo Optical Instrumentation Engineers.

[6] L. Ray Sweeny, *"Electronically Configurable towed Decoy for Dispensing Infrared Emitting Flares"*, Patent Number US6,055,909A, United States, May 2 2000.

[7] Bernard E. Douda, *"Genesis of Infrared Decoy Flares-The Early Years from 1950 into the 1970s"*, Naval Surface Warfare Centre, Crane Division, Crane, Indiana, USA.

[8] *"Electronic Warfare Fundamentals"*, November 2000. https://falcon.blu3wolf.com/Docs/Electronic-Warfare-Fundamentals.pdf

Access the Support Material:
https://www.routledge.com/9780367554118
Chapter 40 Chaff Decoys
Chapter 41 Rejection Analysis of Pyrotechnic Ammunitions and Devices
Chapter 42 Safety in Pyrotechnic Ammunition Manufacture Nanoparticles
(supplementary to chapter 4)

# Index

For Product Safety Concerns and Information please contact our EU
representative GPSR@taylorandfrancis.com
Taylor & Francis Verlag GmbH, Kaufingerstraße 24, 80331 München, Germany

www.ingramcontent.com/pod-product-compliance
Lightning Source LLC
Chambersburg PA
CBHW060417220326
41598CB00021BA/2203

* 9 7 8 0 3 6 7 5 5 4 1 2 5 *